# Modern Origins

# Vertebrate Paleobiology and Paleoanthropology Series

Edited by

**Eric Delson**
Vertebrate Paleontology, American Museum of Natural History,
New York, NY 10024, USA
delson@amnh.org

**Eric J. Sargis**
Anthropology, Yale University
New Haven, CT 06520, USA
eric.sargis@yale.edu

Focal topics for volumes in the series will include systematic paleontology of all vertebrates (from agnathans to humans), phylogeny reconstruction, functional morphology, Paleolithic archaeology, taphonomy, geochronology, historical biogeography, and biostratigraphy. Other fields (e.g., paleoclimatology, paleoecology, ancient DNA, total organismal community structure) may be considered if the volume theme emphasizes paleobiology (or archaeology). Fields such as modeling of physical processes, genetic methodology, nonvertebrates or neontology are out of our scope.

Volumes in the series may either be monographic treatments (including unpublished but fully revised dissertations) or edited collections, especially those focusing on problem-oriented issues, with multidisciplinary coverage where possible.

For other titles published in this series, go to
www.springer.com/series/6978

# Modern Origins

## A North African Perspective

Edited by

## Jean-Jacques Hublin

*Department of Human Evolution, Max Planck Institute for Evolutionary Anthropology, Deutscher Platz 6, 04103 Leipzig, Germany*

## Shannon P. McPherron

*Department of Human Evolution, Max Planck Institute for Evolutionary Anthropology, Deutscher Platz 6, 04103 Leipzig, Germany*

 Springer

*Editors*
Jean-Jacques Hublin
Department of Human Evolution
Max Planck Institute for Evolutionary
  Anthropology
Deutscher Platz 6
04103 Leipzig
Germany

Shannon P. McPherron
Department of Human Evolution
Max Planck Institute for Evolutionary
  Anthropology
Deutscher Platz 6
04103 Leipzig
Germany

ISSN  1877-9077        e-ISSN 1877-9085
ISBN  978-94-007-2928-5     e-ISBN 978-94-007-2929-2
DOI 10.1007/978-94-007-2929-2
Springer Dordrecht Heidelberg New York London

Library of Congress Control Number: 2012931410

*Cover illustration:* The Aterian point is an original image from Abdeljalil Bouzouggar. The Dar es-Soltan skull is a surface model, created in the Department of Human Evolution at the Max Planck Institute for Evolutionary Anthropology. Cover image created by Marike Schreiber and Alyson Reid.

Printed on acid-free paper

Springer is part of Springer Science+Business Media (www.springer.com)

*A volume in the*

# Max Planck Institute
# Subseries in Human Evolution

Coordinated by

**Jean-Jacques Hublin**

*Max Planck Institute for Evolutionary Anthropology, Department of Human Evolution,
Leipzig, Germany*

# Preface

## Modern Human Origins in North Africa

In September 2007, we hosted a conference at the Max Planck Institute for Evolutionary Anthropology (Leipzig, Germany) on modern human origins from a North African perspective. In doing so, we brought together scholars working on climate, chronology, archaeology, and physical anthropology. The goal was to have an integrated view of current research in a geographical area and time period of importance for questions concerning the origins of modern humans, both biologically and culturally, and their subsequent spread from Africa to the rest of the world.

Currently within Africa the rich sub-Saharan archaeological record is the best documented for this time period, particularly in South Africa. However, one of the weaknesses of the South African record is the relatively few fossil hominins these deposits have yielded. In contrast, the North African record, especially that of the western Maghreb, is quite rich. The North African record, however, as many of the chapters in this volume make clear, is flawed by early and poorly published excavations that produced datasets either riddled with too many questions of provenience or too incomplete and biased to answer the questions we have today. Fortunately, the situation has started to change, particularly in the last decade, and there are now a large number of ongoing projects at many of the classic North African sites. Work is especially active in the Moroccan cave sites, including Jebel Irhoud, Rhafas, Taforalt (Grotte des Pigeons), Dar es-Soltan I, El Mnarsa, El Harhoura, Contrebandiers, Ifri n'Ammar, and Mugharet el 'Aliya. In Libya, Haua Fteah is now being re-excavated, and in Tunisia a program is underway to re-investigate the known sites and survey for new sites. In Algeria, new studies are being conducted at Bir-el-Ater, the eponymous site of the Aterian. These works have already forced a complete revision of the Aterian chronology, resulted in the discovery of significant new fossil material from well-documented archaeological contexts, and, more gradually, produced a better understanding of the lithic technologies and subsistence practices of these early modern humans.

## Climate

Our view on the ancient peopling of Africa is biased by the current development of desertic areas, and of the Sahara in particular. The emphasis often put on the high genetic diversity of sub-Saharan populations should be put into perspective. Today the Sahara is virtually empty of human occupation but this was not always the case during the Pleistocene. As demonstrated in the contribution by Smith (Chap. 3), the sedimentological and geochemical evidence suggest that it is only after 70 ka that humid events became less intense, although they were still significant. Marine sediment cores off the coast of North Africa (NW African margin and Alboran Sea) also provide indications of these climatic fluctuations during the late Middle and Late Pleistocene. The climatic pattern driven by monsoonal variations seems to be modulated after about 60–70 ka by the influence from high-latitude processes due to the lower amplitude of the precessional cycles. According to Moreno (Chap. 1), the data indicate an increase in Saharan wind intensity during the Dansgaard/Oeschger stadial periods and the Heinrich events. However, most of the Mousterian or Aterian sites mentioned or analyzed in this volume predate this period of increasing aridity, and their geographical distribution extends into areas that are today covered by desert. During Marine Isotope Stage (MIS) 5 in particular, large bodies of water sometimes covering tens of thousands of square km developed in the Sahara, allowing an intense human occupation and population exchange between the Maghreb and the rest of Africa. This period is crucial for the evolution of the modern populations involved in the last Out-of-Africa event and possibly for the emergence of modern behavior in Africa. According to Geraads (Chap. 4), although the first half of the Middle Pleistocene of northwestern Africa is characterized by open landscapes, some more humid and/or forested areas, likely inhabited by early *Homo sapiens*, are documented during the second half of this period. It is during the Late Pleistocene that an increase in aridity reduced the faunal diversity. However, it is still difficult to reject the permanent human occupancy of large parts of the area, at least at some elevation. As underlined by Larrasoaña (Chap. 2), North Africa holds a key location in the gateway to Eurasia. Although in northeastern Africa the "green Sahara" periods broadly correlate with Acheulian and Mousterian archaeological sites, Aterian sites are linked to spring deposits and mountain areas during a prolonged arid period. This author suggests that environmental fluctuations controlled the exchanges with the Near East and that the burst of arid conditions after 70 ka might have driven some of the late Middle Stone Age populations of northeastern Africa "out of the desert" into Eurasia. In Chap. 11, Hawkins also links this shift from wetter to drier conditions in northeastern Africa to a shift from MSA to Aterian. In her view, the Aterian is an adaptation to drier conditions requiring greater mobility and better projectile technologies to increase the effectiveness of hunting.

## Biology

A very rich fossil record documents human evolution throughout the Middle and Late Pleistocene in North Africa. The main questions surrounding the interpretation of this material relate to the first evidence of a modern morphology and to the continuity between late Middle Pleistocene populations and recent humans in the area. In Chap. 15, Bräuer supports a model of gradual emergence of *Homo sapiens* with different grades. When comparing the evolutionary processes in northern, eastern, and southern Africa, it is difficult to assess the distinct roles of the different regions in the emergence of modern populations. It is therefore suggested that interregional migration and/or gene flow during periods of a "green Sahara" might have led to complex patterns. Supporting previous studies on the Jebel Irhoud material, Harvati and Hublin (Chap. 12) confirm that the facial morphology observed on these late Middle Pleistocene hominins is close to that of early modern humans from the Near East. It also relates to later populations in the area associated with the Aterian assemblages at Dar es-Soltan II. In contrast, the later Ibero-Maurusian humans, post-dating the peak of the last glacial maximum, seem to be morphologically closer to European Upper Paleolithic populations. The culmination of aridity represented by MIS 2 likely resulted in the final isolation of the Maghreb from the rest of Africa, with the development of more population exchanges around the Mediterranean Basin via the Levant. The Aterian hominins are of utmost importance as they likely immediately predated the last Out-of-Africa exodus at the origin of the modern peopling of Eurasia. Although it is difficult to argue for any direct connection between the hominins who inhabited the western Maghreb and the first modern Europeans, they might give us a good picture of the populations who lived between the Nile Valley and the Atlantic Coast between 100 and 50 ka. Hublin and collaborators (Chap. 13) used the dental evidence to show that although very robust and still displaying some primitive features, these populations can be related to other early modern humans in Africa and the Levant. Importantly, this material sheds new light on the interpretation of the very first modern humans known in western Eurasia. The Nazlet Khater 2 (Egypt) skeleton is dated to MIS 3 and represents a population much closer to the gate of Eurasia. Crevecoeur's analysis of this specimen (Chap. 14) reinforces the notion that the populations of this time period still retained several archaic features, notably in the face and mandible.

## Archaeology

One of the themes arising from the conference and apparent in the chapters included here is Paleolithic nomenclature. It is unclear what we call or how we define some of the archaeological units that form the basis of the record for this time period. So, for instance, while there are a few researchers with strong opinions, there seems to be a large amount of uncertainty about what to call non-Aterian assemblages that post-date the Acheulian and pre-date the Iberomaurusian. For some these assemblages are Mousterian, for some they are Middle Paleolithic, and for others they are Middle Stone Age. The problem is that each of these terms carries with it certain implications. To call these assemblages Mousterian, or Moroccan Mousterian as suggested during the conference, links them to Europe and recalls the time when fossil hominins from North Africa were considered to be Neandertals. This nomenclature also reinforces the notion of the Sahara as a barrier that divides Africa with, in this case, a sub-Saharan Middle Stone Age versus a North African Mousterian. However, as discussed, it is clear that the Sahara was not a barrier during times relevant to the discussion.

Working with materials from the Western Desert of Egypt, Hawkins (Chap. 11) defines all Levallois based industries, including the Aterian, as Middle Stone Age. In Chap. 9, Garcea addresses the terminology problem directly and argues strongly for keeping the North African materials integrated into the African nomenclature, thus favoring a Middle Stone Age attribution. Garcea makes the point that the continued use of European terms mainly stems from European research traditions in North Africa and even considers that it could be derogatory to continue using European terms to describe African assemblages. Essential to her argument too is the association of biological types with stone tool industries. Thus, Middle Paleolithic assemblages, including the Mousterian, were made by Neandertals, and Middle Stone Age assemblages, including all of those in North Africa, were made by modern humans. In contrast, while primarily concerned with the definition of the Aterian, Richter and collaborators (Chap. 5) call these industries Middle Paleolithic.

The question of how exactly the Aterian should be defined (Bouzouggar and Barton, Chap. 7) and the related question of whether there are interstratified examples of Aterian and [other] MSA assemblages (Aouadi-Abdeljaouad and Belhouchet, Chap. 10; Richter et al., Chap. 5) remain problematic. Most cite Tixier's (1967) definition of the Aterian as a Mousterian industry with Levallois technology, including Levallois blades, a high frequency of faceted platforms, numerous side-scrapers, relatively abundant points, and a high frequency of end scrapers often made on blades. The key item for the Aterian, however, is tanged pieces. Tixier's definition included the tenet that a noticeable proportion of the pieces, sometimes as much as 25%, included proximal tangs, often bifacially prepared. Finally, the Aterian is also characterized by some bifacial foliates.

Bouzouggar and Barton (Chap. 7) recall that earlier definitions of the Aterian also noted the existence of small discoidal and Levallois cores. They further discuss their presence in the Aterian of Contrebandiers, El 'Aliya, El Mnasra and El Harhoura. Small flake production is now well-documented from a variety of Lower and Middle Paleolithic contexts based on a number of technologies including small Levallois and single surface cores, truncated-faceted techniques, Kombewa, and simple flaked-flake techniques. In some cases, small flake production is a response to raw material constraints (either size or availability), while in others small flakes were desired end-products. Bouzouggar and Barton also wonder whether the small cores themselves could have been used as tools. The techniques of small flake production, the use of small flakes, and whether these exist throughout the Aterian are areas of research that deserve more attention.

It is clear that the presence of a tanged piece makes an assemblage Aterian. What is less clear is what to do with industries that post-date the Aterian, based on either absolute dates or interstratifications, and that lack typical Aterian elements. In this volume, for instance, two sites are reported where MSA assemblages overlie Aterian assemblages. First, at the Moroccan site of Ifri n'Ammar (Richter et al., Chap. 5), a deep sequence contains interstratified layers of Aterian and MSA/MP industries with the only difference between the two being the presence of tanged pieces and the percentage of notched pieces. Second, the recently excavated open-air Tunisian site of Aïn El-Guettar in the Meknassy Basin contains a well-stratified Aterian level below a Middle Paleolithic horizon separated by 1.4 m of sterile deposit. Aouadi-Abdeljaouad and Belhouchet (Chap. 10) conclude that the Mousterian and the Aterian are contemporaneous cultural groups, while Richter and collaborators prefer to group both into a Middle Paleolithic that contains at least two variants, one with tanged pieces and the other without, without chronological significance.

Given how important tanged pieces are in defining Aterian assemblages, we need a better understanding of the prevalence of tanged pieces in Aterian assemblages in general, as well as of the functional role of the tang. With regard to the former, Bouzouggar and Barton (Chap. 7) show the percentage of tanged pieces varying between as low as 1.4% and as high as approximately 30% at Dar es-Soltan I, with the caveat that the latter is almost certainly elevated due to collection biases. These numbers, however, are relative to other tools in the assemblage. To better understand their prevalence, data are also needed on assemblage size, artifact densities, raw materials, and more difficult measures like site function or context (e.g., open-air versus shelter, proximity to quality raw materials, etc.). These kinds of data should appear in the coming years as the assemblages from new excavations are published.

With regard to the functional role of tanged pieces in Aterian assemblages, here again, relatively little is known. The presumption has been that tools are tanged to accommodate hafting and that tanged pointed pieces are evidence of stone-tipped spears used for hunting. The latter have been used to pull the Aterian into the African MSA with an emphasis on regional point traditions and, in part, to support the idea that more efficient hunting characterizes an emerging modern human behavioral package. There are certainly impressively symmetrical, well-made, tanged points that one could easily imagine as armatures, but this has yet to be convincingly demonstrated. In addition, use-wear studies to date provide alternative interpretations, and the underlying variability in tanged pieces paints a more complex picture. Garcea reviews this topic explicitly in Chap. 9. She notes that tangs are present on a wide variety of pieces, including those that are retouched and unretouched and those that are pointed and non-pointed. This aspect of tanged pieces is clear whenever one looks at a recently excavated or collected assemblage that does not suffer from collection bias. Garcea also argues that the stem portion of the tanged pieces in her studied assemblage was too short to have provided an effective haft for a projectile and that the tangs themselves show use-wear patterns indicative of scraping activities and not of hafting. In Chap. 7, Bouzouggar and Barton review data which show that tanged pieces were rarely used as projectiles and instead find evidence for working hard and soft materials and for cutting activities even when the piece has a pointed morphology. Garcea is careful not to exclude the possibility that some tanged pieces were in fact hafted, and evidence that tanged pieces were used for cutting or scraping does not exclude the possibility that they were hafted. Still, it seems that until evidence can be presented in support of Aterian points as armatures and even in support of hafting, we should remain cautious in our interpretation of the behavioral significance of the tang technique. Here again, we should expect more data soon on factors such as the technology of blank production for tanged pieces, blank selection for tanged pieces, retouch on tanged pieces, and the relative frequencies of techniques such as basal thinning in the assemblage.

At the same time, however, while the Aterian's distinctive tanged pieces can be used to recognize a techno-complex that extends over much of what is today arid North Africa, it is also clear that there is considerable regional and likely chronological variability within the Aterian such that we should not expect to be able to generalize findings from one set of sites to the whole of the Aterian.

## Chronology

Raynal and Occhietti (Chap. 6) and Richter et al. (Chap. 5) tackle chronological issues. Since the application of ESR, TL, and OSL methods to North Africa, it has become clear that the Aterian is older than previously supposed based on radiocarbon dates and chronostratigraphy. The question now is just how old the Aterian is. Along the Atlantic littoral zone south of Rabat a number of Aterian deposits have now been dated and shown to extend to the last interglacial MIS 5e. This finding is supported by Raynal and Occhietti with the additional technique of amino acid analysis. To find older Aterian one has to look elsewhere, as the basal deposits in these cave sites were formed by the high beach stands of MIS 5e. Thus, it is perhaps not surprising that still older Aterian has recently been announced from the site of Ifri n'Ammar (Richter et al., Chap. 5). The oldest Aterian at this site has been dated to $145 \pm 9$ ka based on thermoluminescence dating of 9 heated lithics. It remains to be seen how much older both the Aterian and the preceding MSA/MP will go or whether additional, equally old Aterian sites will be found. On the other hand there are now relatively few sites, especially in the Maghreb, with late dates, and there may be a gap between the final Aterian and the Ibero-Maurusian in these locations (Garcea, Chap. 9).

## Subsistence

One potentially important avenue to more fully explore is diet, especially considering the location of several of the better known Aterian sites that are currently being investigated along the coast south of Rabat where marine resource consumption is a possibility. As Steele notes in Chap. 8, debates over the effectiveness of MSA hunting strategies, and more recently the role of marine resources in the diet have figured prominently in discussions concerning the origins of modern human behavior. These debates have resulted in an emphasis on zooarchaeological studies. For North Africa, however, zooarchaeological studies have lagged behind, but this is changing. As in South Africa, there are numerous North African cave sites, especially in northwestern Africa, with good faunal preservation, being studied with a growing concern for identifying dietary changes. Until recently, most reports have emphasized the paleoenvironmental or chronological value of fauna and listed only the presence or absence of species. Despite numerous coastal sites, evidence for Aterian consumption of marine resources is still slim. In Morocco, coastal sites such as Dar es-Soltan I, El Harhoura I, and Murgharet el 'Aliya show the presence of several species of fish and of marine molluscs, and at Murgharet el 'Aliya, monk seal is also reported from the Aterian. This is the only marine mammal reported from a Late Pleistocene Moroccan assemblage (Steele, Chap. 8). However, the abundance and taphonomy of these occurrences are not yet well documented. Marine diets are easily detectable using stable isotope analysis, but to date there are no stable isotope studies on North African hominins. With current methods the problem will be poor collagen preservation in these relatively low latitude sites with high mean annual temperatures, and it remains to be seen whether this can be overcome. In the meantime, we should soon expect a series of more zooarchaeologically-oriented reports coming from the new excavations, and it will be especially interesting to get an evaluation of marine resources in the diet from the coastal sites south of Rabat currently under excavation.

## Conclusion

The Middle to Late Pleistocene record has become the focus of intense research in large part driven by questions and debates surrounding the origins of modern human behavior and anatomy. Despite a long history of research in the area, North Africa has, until now, not featured prominently in these debates, with the focus instead being on East and particularly South Africa. The assessment of the North African evidence has suffered from two biases. One is related to the current geographical situation in which the Maghreb is separated from the rest of Africa by a major natural barrier representing the largest desertic surface of the planet. Although today the Maghreb is primarily connected to the Mediterranean Basin, this situation is relatively recent at the scale of the Pleistocene. It is only during the Late Pleistocene that the aridity in the region dramatically increased, although relatively "green" episodes are documented during MIS 5 and 3. It is therefore important to realize that exchanges of fauna and human populations were possible among North, East, and Central Africa until the eve of the last Out-of-Africa event, as documented by the large number of archaeological sites dating from this period in various parts of the Sahara. A second important bias results from the history of research conducted in the area. More than anywhere else in Africa, the studies in the Maghreb were conducted by European archaeologists who exported European models and developed the notions that "Neandertaloid" populations produced the "Middle Paleolithic" industries in the area. This is partly because this local "Mousterian" was by default considered to be roughly contemporary to the European Mousterian assemblages, and the Aterian was assumed to represent a sort of "transitional" assemblage chronologically centered on the European Middle/Upper Paleolithic boundary. One of the main advances in African archaeological studies has certainly been the reassessment of the chronology of these industries and of the antiquity of their makers. The North African paleoanthropological record is gradually being reconnected to the rest of the African record.

We suspect that another conference may be required a few years from now to help begin to incorporate and integrate this new work. In the meantime, however, what we have tried to do here is to take a kind of snapshot of where we are.

## Reference

Tixier, J. (1967). Procédés d'analyse et questions de terminologie concernant l'étude des ensembles industriels du Paléolithique récent et de l'Epipaléolithique dans l'Afrique du Nord-Ouest. In W.W. Bishop & J.D. Clark (Eds.), *Background to evolution in Africa* (pp. 771–820). Chicago: University of Chicago Press.

# Acknowledgments

We thank each of the participants represented in this volume not only for their contributions present here but also for making the actual conference so interesting and productive. Additionally, some of the conference presenters are not represented in this volume but their contributions at the conference were greatly appreciated. We also thank discussants Sally McBrearty, Harold Dibble, and Chris Stringer for their contributions to the conference. A very special thanks go to Allison Cleveland, Silke Streiber, Diana Carstens, and Michelle Hänel for their efforts with the organization and logistics of the conference, to Regina Querner, Marike Schreiber and Sylvio Tüpke for their help with the figures and images, and to all of the many anonymous reviewers that gave their time to improve the chapters in this volume. Carolyn Rowney and Alyson Reid's work editing the volume has been truly invaluable. We thank the Max Planck Society for supporting the conference. We thank Eric Delson, Eric Sargis and Ross MacPhee for supporting this publication through their co-edited series, and Eric Delson for his thorough and detailed reading and editing of the manuscripts and for his substantial organizational and logistical support along the way to making this a finished volume. Finally, we thank Judith Terpos, Tamara Welschot and the typesetting team at Springer for their help in making this volume possible.

Leipzig, December 2010
Jean-Jacques Hublin
Shannon P. McPherron

# Contents

# Contributors

**Nabiha Aouadi-Abdeljaouad**
Institut National du Patrimoine, Musée National de Raqqada, 3100 Kairouan, Tunisia
aouadi73@yahoo.fr

**Shara Bailey**
Department of Anthropology, New York University, Rufus D. Smith Hall, 25 Waverly Place, New York, NY 10003, USA; Department of Human Evolution, Max Planck Institute for Evolutionary Anthropology, Deutscher Platz 6, 04103 Leipzig, Germany
sbailey@nyu.edu

**Nick (R.N.E.) Barton**
Institute of Archaeology, University of Oxford, 36 Beaumont Street,
Oxford, OX1 2PG, UK
nick.barton@arch.ox.ac.uk

**Lotfi Belhouchet**
Institut National du Patrimoine, 4 Place du Château, 1008 Tunis, Tunisia
lotfi_belhouchet@yahoo.fr

**Abdeljalil Bouzouggar**
Institut National des Sciences de l'Archéologie et du Patrimoine, Rabat-Instituts, Madinat Al Irfane Angle rues 5 et 7, Hay Riad, Rabat, Morocco; Department of Human Evolution, Max Planck Institute for Evolutionary Anthropology, Deutscher Platz 6, 04103 Leipzig, Germany
bouzouggar@eva.mpg.de

**Günter Bräuer**
Abteilung für Humanbiologie, Universität Hamburg, Allende-Platz 2, 20146 Hamburg, Germany
guenter.braeuer@uni-hamburg.de

**Isabelle Crevecoeur**
Laboratoire d'Anthropologie des Populations Passées et Présentes, UMR 5199 PACEA/ Université Bordeaux 1, Bâtiment B8, Avenue des Facultés, 33405 Talence Cedex, France; Laboratory of Anthropology and Prehistory, Royal Belgian Institute of Natural Science (RBINS), 29 rue Vautier, 1000 Brussels, Belgium
i.crevecoeur@pacea.u-bordeaux1.fr

**Elena A. A. Garcea**
Dipartimento di Lettere e Filosofia, Università di Cassino, Via Zamosch 43, 03043 Cassino, FR, Italy
egarcea@fastwebnet.it

**Denis Geraads**
UPR 2147—Centre National de la Recherche Scientifique, 44 rue de l'Amiral Mouchez, 75014 Paris, France
denis.geraads@evolhum.cnrs.fr

**Katerina Harvati**
Senckenberg Center for Human Evolution and Paleoecology, Institut für Ur- und Frühgeschichte und Archäologie des Mittelalters, Eberhard Karls Universität Tübingen, Rümelinstrasse 23, 72070 Tübingen, Germany
katerina.harvati@ifu.uni-tuebingen.de

**Alicia L. Hawkins**
Laurentian University, 935 Ramsey Lake Road, Sudbury, ON P3E 2C6, Canada
ahawkins@laurentian.ca

**Jean-Jacques Hublin**
Department of Human Evolution, Max Planck Institute for Evolutionary Anthropology, Deutscher Platz 6, 04103 Leipzig, Germany
hublin@eva.mpg.de

**Juan Cruz Larrasoaña**
Instituto Geológico y Minero de España, Unidad de Zaragoza, C/Manuel Lasala 44 9B, 50006 Zaragoza, Spain
jc.larra@igme.es

**Shannon P. McPherron**
Department of Human Evolution, Max Planck Institute for Evolutionary Anthropology, Deutscher Platz 6, 04103 Leipzig, Germany
mcpherron@eva.mpg.de

**Ana Moreno**
Pyrenean Institute of Ecology—CSIC, Apdo. 202, 50080 Zaragoza, Spain
amoreno@ipe.csic.es

**Johannes Moser**
Kommission für Archäologie Außereuropäischer Kulturen des Deutschen Archäologischen Instituts, Dürenstrasse 35-37, 53173 Bonn, Germany
moser@kaak.dainst.de

**Mustapha Nami**
Centre d'Inventaire et de Documentation du Patrimoine (CIDP), Ministère de la Culture, Rabat, Morocco
m.nami@caramail.com

**Serge Occhietti**
Université du Québec à Montréal, GEOTOP, C.P. 8888, Succ. Centre-Ville, Montréal, QC H3C 3P8, Canada; Université de Lorraine, CERPA, 23 Boulevard Albert 1er, BP 3397, 54015 Nancy Cedex, France
serge.occhietti@gmail.com

**Anthony Olejniczak**
Department of Human Evolution, Max Planck Institute for Evolutionary Anthropology, Deutscher Platz 6, 04103 Leipzig, Germany
anthony.olejniczak@gmail.com

**Jean-Paul Raynal**
Centre National de la Recherche Scientifique, Université Bordeaux 1, UMR 5199 PACEA, PPP, Bâtiment B18, Avenue des Facultés, 33405 Talence Cedex, France; Department of Human Evolution, Max Planck Institute for Evolutionary Anthropology, Deutscher Platz 6, 04103 Leipzig, Germany
jpraynal@wanadoo.fr

**Daniel Richter**
Department of Human Evolution, Max Planck Institute for Evolutionary Anthropology, Deutscher Platz 6, 04103 Leipzig, Germany
drichter@eva.mpg.de

**Fatima Z. Sbihi-Alaoui**
Institut National des Sciences de l'Archéologie et du Patrimoine (INSAP), Rabat, Morocco
fsbihialaoui@yahoo.fr

**Jennifer R. Smith**
Department of Earth and Planetary Sciences, Washington University, 1 Brookings Drive, Campus Box 1169, St. Louis, MO 63130, USA
jensmith@wustl.edu

**Tanya Smith**
Department of Human Evolutionary Biology, Harvard University, Peabody Museum, 11 Divinity Ave, Cambridge, MA 02138, USA; Department of Human Evolution, Max Planck Institute for Evolutionary Anthropology, Deutscher Platz 6, 04103 Leipzig, Germany
tsmith@fas.harvard.edu

**Teresa E. Steele**
Department of Anthropology, University of California-Davis, One Shields Ave, Davis, CA 95616-8522, USA; Department of Human Evolution, Max Planck Institute for Evolutionary Anthropology, Deutscher Platz 6, 04103 Leipzig, Germany
testeele@ucdavis.edu

**Christine Verna**
Department of Human Evolution, Max Planck Institute for Evolutionary Anthropology, Deutscher Platz 6, 04103 Leipzig, Germany; UPR 2147 CNRS, Paris, France
christine.verna@evolhum.cnrs.fr

**Mehdi Zouak**
Direction Régionale de la Culture (Tanger Tétouan), Inspection Régionale des Monuments Historiques et Sites, 2 Rue Ben Hsain, 93000 Tétouan, Morocco
m_zouak@yahoo.fr

# Part I
# Paleoenvironment and Chronology

# Chapter 1
# A Multiproxy Paleoclimate Reconstruction over the Last 250 kyr from Marine Sediments: The Northwest African Margin and the Western Mediterranean Sea

A. Moreno

**Abstract** Marine sediments are one of the best archives of past climate change because they are essentially continuous in character and their age can be determined relatively easily. The study of grain-size and geochemical composition of marine sediments off the shore of the Northwest (NW) African margin and in the western Mediterranean during the past years has shed some light on the forcing mechanisms for dust input and wind strength in the North Canary Basin, and the climatic teleconnections between high and low latitudes during the glacial–interglacial transitions. On an orbital scale, the records show that changes in insolation associated with precession and eccentricity cycles are the major drivers in controlling dust input to the North Canary Basin. A particular response is observed during glacial terminations when maxima in productivity and grain-size are detected as a likely consequence of subtropical anticyclonic circulation intensification due to higher insolation but lower sea surface temperature (SST) over the North Atlantic. Then, the strengthened trade winds forced upwelling and had the ability to carry coarser particles at terminations. On a millennial timescale, Saharan dust variations appear to be strongly related to periods of strengthened atmospheric circulation in high northern latitudes, that is Dansgaard-Oeschger (D/O) cold stadials. Therefore, the study of marine sediments off the coast of NW Africa and in the Alboran Sea provides evidence of changes in climate and landscape distribution on land during the last 250 kyr that can be useful for understanding the origin of human populations and the patterns of their dispersal out of Africa ca. 50 ka.

**Keywords** Marine sediments • Multiproxy approach • Northwest African margin • Paleoclimate reconstruction • Western Mediterranean

A. Moreno (✉)
Pyrenean Institute of Ecology—CSIC, Apdo. 202,
50080 Zaragoza, Spain
e-mail: amoreno@ipe.csic.es

## Paleoclimate Reconstruction from Marine Sediments

Understanding the main patterns that controlled climate variability in the northern African region over the last several hundreds of thousands of years is essential to place investigations about the evolution and dispersal of modern humans in a broad climatic context. Such a study requires paleoclimate archives with enough resolution and well-constrained chronologies to reveal climate variations on land that potentially influenced human evolution. Unfortunately, the lack of well-dated and continuous terrestrial records in that area prevents direct inferences of past climate changes. Some paleolake records (e.g., Schuster et al. 2005) and spring-deposited carbonate rocks (e.g., Smith et al. 2007) are available in the North African region and, although they are not continuous in time, their study is providing directly datable stratigraphic context for some archaeological investigations. Marine records from the NW African margin and the Mediterranean (see also Larrasoaña 2012) have contributed to the understanding of climate variability from that region complementing those on-land studies, since they are essentially continuous in character and their age can be determined relatively easily.

A number of parameters recorded in the marine sediments provide information about the climate and landscape distribution on land. For example, the pollen content accumulated in marine sediments off the coast of Africa is an indicator of changes in the vegetation cover of the neighboring lands (Hooghiemstra et al. 1992; Dupont et al. 2000). Other very useful indicators are those that can provide information about the source of the sediments. Thus, grain-size and mineralogy or geochemical composition of the terrigenous particles have proven to be dependent on the source area (Bergametti et al. 1989a; Grousset et al. 1998) and the type of transport (fluvial vs. eolian) (Matthewson et al. 1995; Stuut et al. 2005). In areas such as the NW African margin or the Mediterranean, under the influence of dust from the nearby Sahara Desert, discriminations in the time intervals when eolian or fluvial fractions in the sediments were dominant will indicate the alternation between wetter and drier periods (Moreno et al. 2001).

Paleoproductivity variability (i.e., the organic primary production in the ocean during the past) can be inferred from the associations of marine organisms accumulated in the sediments (diatoms, foraminifers, radiolarian, etc.) or by indirect proxies, such as some geochemical elements that are enriched under high-productivity areas (e.g., Barium, organic carbon) (Fisher and Wefer 1999). Along the NW African coast, oceanic productivity is linked to trade winds and the Canary Current systems through the outgrowth of coastal upwelling and upwelling filaments (Nykjaer and Van Camp 1994). Upwelling is an oceanographic phenomenon that involves the wind-driven movement of dense, cooler, and usually nutrient-rich water towards the ocean surface, replacing the warmer, usually nutrient-depleted surface water. Thus, in the case of NW African coastal upwelling, the location and pressure gradient of the Azores' high-pressure center are the main forces behind upwelling intensity and the resulting productivity pattern in this upwelling region. From the study of paleoproductivity variations in the sediments of the North Canary Basin, changes in trade winds and Canary Current intensity should be detectable. Finally, comparison with other studies from high latitudes and the analyses of the obtained proxy data in terms of cyclicities (Moreno et al. 2005) allow us to infer the order of investigated atmospheric and oceanic processes and explore the climate teleconnections among high and low latitudes.

In the NW African margin, due to the oceanographic setting and conditions on land, sedimentation rates recorded on most of the marine cores are generally low. In those cases, the studied cores allow a description of climate changes at the scale of the solar insolation variation (e.g., 10–100 kyr). However, higher sedimentation rates are attained closer to the coast, due to the influence of river inputs, and this allows us to resolve rapid climate variability (Kuhlmann et al. 2004b; Tjallingii et al. 2008). Thus, both orbitally-driven changes and rapid climate variability can be detected in the same area, which provides us with the opportunity to carry out the analysis of climate change at different time scales. In this chapter, I review the main data published recently on the NW African margin, although my primary focus is on two cores from the North Canary Basin (GeoB 5559-2 and GeoB4216-1) and one core from the Alboran Sea, western Mediterranean (MD95-2043) (Fig. 1.1), whose main data have been previously published elsewhere. GeoB 5559-2 and GeoB4216-1 cores were obtained from a region where a strong interaction between the atmospheric and ocean circulation systems is currently occurring: trade winds drive seasonal coastal upwelling

while dust storm outbreaks from the Sahara Desert are the major source of terrigenous sediments. However, this present-day situation was certainly different in the geological past, when river inputs were also very significant during some time periods (Holz et al. 2007). The results from these two cores allow us to describe climate variability over the last 250 kyr, in association with main changes in the amount of insolation that is received at that latitude (30°N) from the sun (solar insolation) and how the reconstructed climate oscillations were driven by the variability and cycles of the orbital parameters (eccentricity, with a cycle of 100,000 years, obliquity (41,000), and precession (19,000 and 23,000 years). The three main orbital parameters, also called Milankovitch parameters, are the response to the gravitational pull of other planets and represent variation in the Earth's orbit and rotation patterns.

To explore the abrupt climate changes that occurred over the last 50 kyr in detail, a marine core from the Alboran Sea (western Mediterranean) with enough resolution and well-constrained chronology was selected to be included in this review. The investigation of dust supply to the western Mediterranean on a centennial scale has allowed us to focus on the climate variability of this region during Oxygen Isotope Stage (OIS) 3 (Moreno et al. 2002a) and provide climate data that may help to explain modern human migrations. OIS 3 is a very peculiar time period because it was characterized by high climate variability observed as abrupt temperature shifts (called Dansgaard-Oeschger (D/O) cycles) in ice cores (Dansgaard et al. 1993), marine sediments (Bond et al. 1997) and terrestrial sequences (loess; Rousseau et al. 2007), lakes (Brown et al. 2007), and speleothems (Wang et al. 2008) from all over the world (Voelker 2002). These D/O cycles occur for as little as 10–20 years, and constitute a jump of 7–10°C in mean annual atmospheric temperatures over both Northwest Europe and the Greenland ice sheet. Nowadays, the most accepted hypothesis for D/O variability relates to instabilities in global thermohaline circulation (THC) (Clement and Peterson 2008), a phenomenon referred to as oceanic conveyor belt circulation by Wallace Broecker in the 1980s (Broecker et al. 1985). THC oscillations have major global climatic implications and are significant modulators of continental environmental conditions. The THC hypothesis for abrupt climate change at a D/O scale relies on modeling experiments that have demonstrated that different THC modes are possible, both at present and during the last glacial period (Ganopolski and Rahmstorf 2001). The three main modes are: (1) a warm mode with strong convection in the Nordic Seas, (2) a cold mode with weaker convection occurring south of the Greenland–Scotland sill, and (3) a "collapsed" mode with virtually no deep-water formation in the North Atlantic and severe cooling. Transitions between these modes appear to be most clearly associated

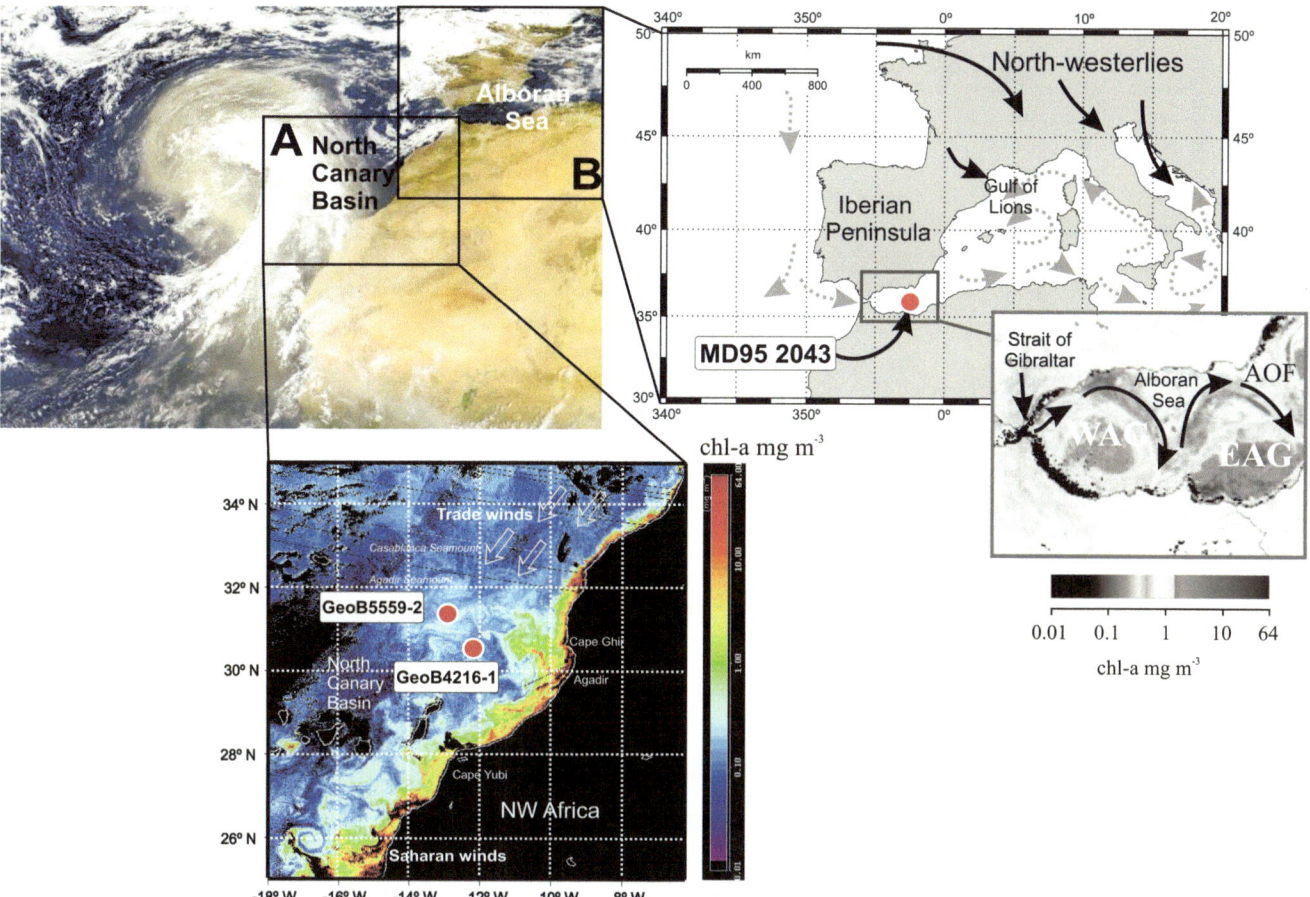

**Fig. 1.1** Location of the studied cores in the North Canary Basin and the Alboran Sea. The satellite image shows a massive Saharan dust plume that affected the NW African margin and the western Mediterranean. The photograph was taken February 26, 2000, by the SeaWiFS Project and the Distributed Active Archive Center, Goddard Space Flight Center, Greenbelt, MD, USA. **a** Surface pigment concentration (mg m$^{-3}$) in the North Canary Basin as observed by SeaWiFS during an extraordinary Cape Ghir filament event on the 19th March 1998 (processed by Davenport). **b** Location of IMAGES core MD95-2043 in the Alboran Sea and present-day dominant oceanographic circulation represented by dashed arrows. The enclosed satellite picture represents the surface pigment concentration (mg m$^{-3}$) as observed by SeaWiFS on May 1999 (Fabrés et al. 2002). Both anticyclonic gyres (Western Alboran Gyre and Eastern Alboran Gyre) and the position of the Almeria-Orán Front are also indicated

with D/O cycles. The record of D/O cycles in low latitudes in the NW African margin, clearly discussed in recent studies (e.g., Jullien et al. 2007; Mulitza et al. 2008; Tjallingii et al. 2008; Itambi et al. 2009), points to an important role for atmospheric circulation and, likely, for dust transport and its effects on climate, to transfer rapid climate variability on a hemispheric scale.

## Present-Day Oceanographic and Atmospheric Setting

The study region roughly lies in the eastern boundary current system, where the Azores Current transports the cold waters from the North Atlantic southwards. One branch of this current enters the Mediterranean Sea through the Gibraltar Strait, while the other branch goes south, where it forms the Canary Current. The Canary Basin lies in the recirculation regime linking the Gulf Stream with the North Equatorial Current via the Azores and Canary Currents (Fig. 1.1a). At this latitude, coastal upwelling is driven by the interaction between the trade winds and the Canary Current. Although the upwelling zone is restricted to the coast, satellite images have shown that large upwelling filaments develop at special coastal positions, such as Cape Ghir, Cape Blanc, and Cape Yubi (Davenport et al. 2002). In the Alboran Sea, the surface water mass that enters from the Atlantic Ocean through the Gibraltar Strait is called Modified Atlantic Water (MAW) (Millot 1999); this term describes a quasi-permanent anticyclonic gyre, the Western Alboran Gyre (WAG), in the west, and a more

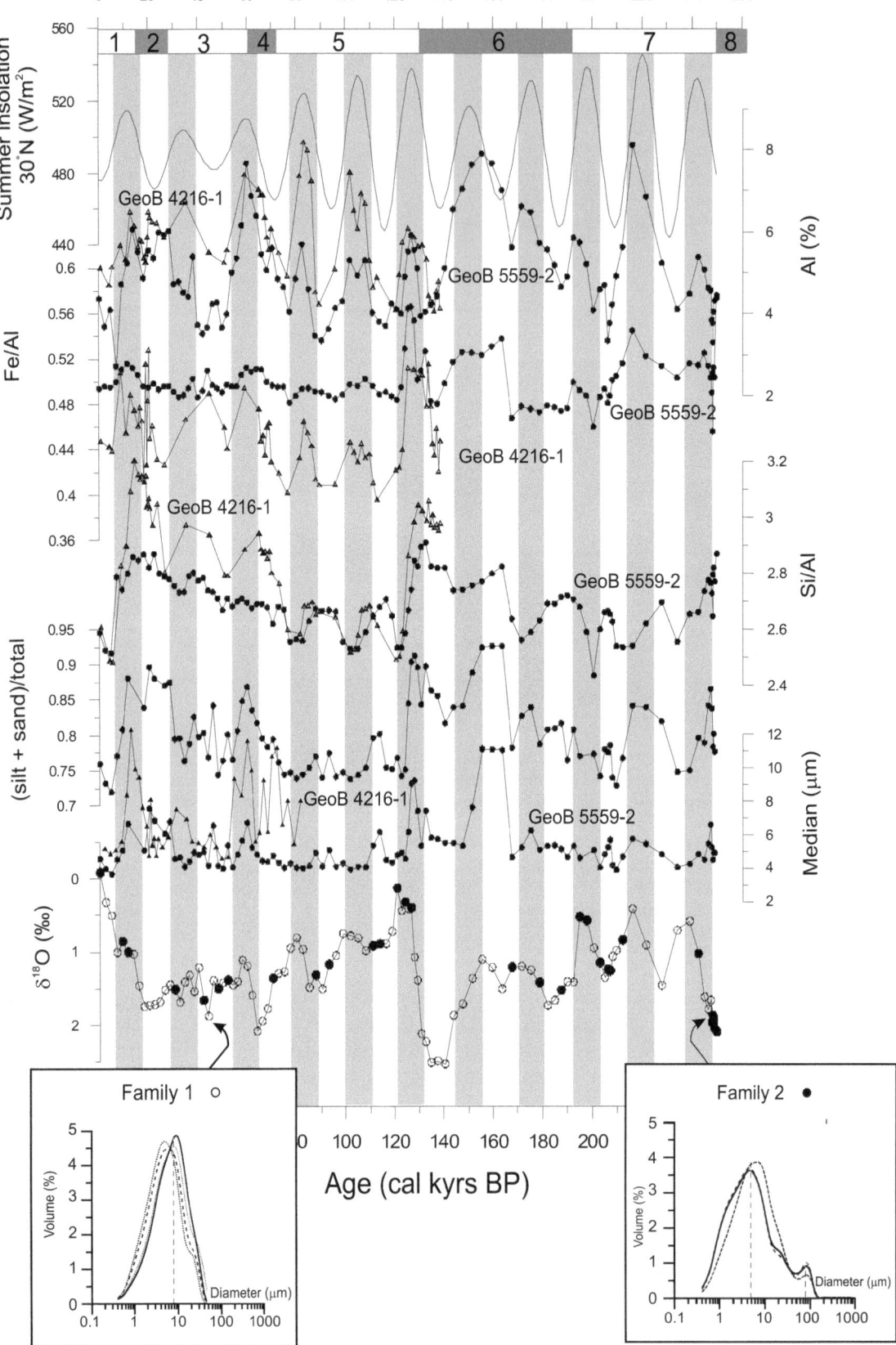

◄**Fig. 1.2** Downcore profiles of grain-size and geochemical proxies from GeoB 5559-2 and GeoB 4216-1 cores. From *bottom* to *top*: Oxygen isotope variability (GeoB 5559-2) where samples are classified in terms of grain-size distributions (*white dots*: Family 1-unimodal; *black dots*: Family 2-bimodal); Median of the terrigenous fraction (both cores); (Silt+Sand)/Total as a grain size indicator of coarser sediments and Si/Al and Fe/Al ratios and percentage of Aluminium (both cores).

Note that when data from both cores are represented, *black dots* are from GeoB 5559-2 and *black triangles* are from GeoB 4216-1. Geochemical study of Core GeoB 4216-1 has been carried out in the first 6 m. To compare with these results, variation of summer insolation at the latitude of the studied cores is also plotted. Intervals with maximum summer insolation (= minimum precession index) are shaded

variable circuit, the Eastern Alboran Gyre (EAG), in the east, (Fig. 1.1b). Two main factors influence surface productivity in this area: (1) the speed of the inflowing jet of cold and less-saline surface waters from the Atlantic Ocean that interacts with saltier and warmer waters from the Mediterranean (Perkins et al. 1990), and (2) westerly winds (so-called Poniente) associated with the progression of atmospheric low-pressure centers into the Mediterranean region (Parrilla and Kinder 1987). Both processes favor the development of anticyclonic gyres that promote upwelling and primary productivity. In addition, the availability of nutrients via fluvial input further supports high productivity in the Alboran Sea, as is clearly evidenced by the close correlation between high chlorophyll concentrations and fluvial discharge from southern Iberian rivers (Fabrés et al. 2002).

In the study area, summertime climates are dry and hot due to the influence of the atmospheric subtropical high-pressure belt. In contrast, the subtropical high is displaced southward during winter, which allows mid-latitude storms to enter the region from the open Atlantic and bring increased amounts of rainfall to the Mediterranean. Therefore, the westerlies are located over the Mediterranean area in the winter and displaced towards northern Europe in the summer. On a decadal scale, much of the present day climate variability in this region has been linked to a natural mode of atmospheric pressure variation, the North Atlantic Oscillation (NAO; Hurrell 1995). NAO activity is defined as the difference between normalized winter sea-level atmospheric pressure between the Azoric high pressure and Icelandic low pressure cells. Although this oscillation is more pronounced during winter, it is present throughout the year. Winters with high NAO indices are characterized by a deepening of the Icelandic low associated with a stronger Azores anticyclone. The related changes in the large-scale meteorological situation lead to a northward shift of the North Atlantic westerlies, which provide much of the atmospheric moisture to northern Africa and Europe. In the case of a high NAO index, this shift yields drier conditions over southern Europe, the Mediterranean Sea, and northern Africa, together with changes in the storm tracks in these regions (Hurrell 1995). During low-NAO years, northwesterly winds are weaker and are guided to mid-latitudes, thus bringing higher precipitation to the Mediterranean and large areas of North Africa.

The interplay between Saharan air masses and the Azoric high-pressure cell constitutes another meteorological pattern that defines the climate of this region. Both the North Canary Basin and the Mediterranean Sea are located in the path of long-range atmospheric dust transport off the coast of northwestern Africa (Prospero 1996). Dust input to the North Canary Basin is controlled by the northeast trade winds and the Saharan Air Layer (SAL) wind systems. At present, major dust outbreaks that carry dust towards the Atlantic Ocean and Europe are linked with the northern branch of the SAL wind system (Torres-Padrón et al. 2002). These Saharan winds transport particles from the rim of the South Sahara and Sahel regions. The highest occurrence of dust outbreaks occur in winter and summer, related to two dominant meteorological scenarios. In winter, dust events are favored to occur when the Azores High is weakened and the Intertropical Convergence Zone (ITCZ) is southwardly located. Bergametti et al. (1989b) observed that dust transport from the Sahelian regions occurs when the incursion of a polar depression cuts the Azores High into two anticyclonic cells, an oceanic and a continental one. In contrast, dust outbreaks appear in summer when the high pressure is centered on the Azores, in combination with a low pressure cell over northern Africa, which favors dust transport from a northern source (Torres-Padrón et al. 2002). Until now, two origins for dust have been accepted in the literature: the Sahelian origin and the northeastern Morocco origin (Chiapello et al. 1997). Evaluation of back trajectories and isobaric meteorological maps shows that Saharan air masses dominate the Mediterranean region whenever the Azores High is displaced westward and the North African High is strengthened and centered over Algeria (Rodriguez et al. 2001). The development of summertime thermal lows over the Iberian Peninsula apparently stimulates this meteorological setting through intense heating of the land surface. In addition, an apparent relationship was found among the NAO and the Saharan dust load transport, indicating larger dust load transport to the Atlantic when NAO is at its positive phase. It is demonstrated that the long-term increase of the desert dust load corresponds to the upward trend of the NAO since the early 1970s. Both interannual variability and decadal increases in African dust export, as observed in Meteosat images and mineral aerosol in Barbados, are thus apparently controlled by climatic conditions expressed by the NAO index (Moulin et al. 1997).

## Some Methods to Study Marine Sediment Cores

The NW African margin and the western Mediterranean Sea constitute a large area where many paleoclimate studies focused on the investigation of marine sediments have been carried out. In the North Canary Basin, located between 34° and 28° north latitude in the NW African margin, the recovery of nearly one hundred sediment cores in the framework of the Canary Islands, Azores, and Gibraltar Observations (CANIGO) European Project promoted high-quality paleoclimatic research. Two of these gravity cores, GeoB 5559-2 and GeoB 4216-1, were studied in order to investigate the forcing mechanisms for dust input and wind strength in the North Canary Basin, and the climatic teleconnections between high and low latitudes during the glacial–interglacial transitions (Moreno et al. 2001, 2002a, b, 2004, 2005; Moreno and Canals 2004) (Fig. 1.1a). The two cores are presented here with greater detail but many other cores from the same area were studied by other researchers (e.g., Bozzano et al. 2002; Freudenthal et al. 2002; Kuhlmann 2003; Kuhlmann et al. 2004a, b; Holz et al. 2007) and further south, off the coast of Cape Blanc, the number of records increases exponentially (e.g., de Menocal 1995; Matthewson et al. 1995; Martinez et al. 1999; de Menocal et al. 2000a, b). In addition, and with the aim of exploring the climate variations recorded during OIS 3 in greater detail, the main results from the MD95-2043 core obtained in the western Alboran Sea during 1995 IMAGES-I Calypso coring campaign aboard R/V Marion Dufresne (Fig. 1.1b) are reviewed in this chapter (for more detail, read previous papers in which the data set has been published; Cacho et al. 1999, 2000, 2006; Plaza 2001; Moreno et al. 2002a, 2004, 2005; Sánchez-Goñi et al. 2002; Fletcher and Sánchez Goñi 2008).

The chronological framework for the two North Canary Basin cores was obtained by correlating the measured $\delta^{18}O$ values with the SPECMAP $\delta^{18}O$ chronology (Martinson et al. 1987) spanning the last 250 kyr, from OIS 1 to 8. The age model for the core MD95-2043 was developed by Cacho et al. (1999) and is derived from graphically correlating the down-core sea surface temperature (SST) record with the climatic cycles displayed in the Greenland GISP2 ice core $\delta^{18}O$ record (Meese et al. 1997). According to this age model, the records presented in this study span the time interval from 28,000 to 48,000 cal yr BP with a mean sedimentation rate of 27 cm/1,000 years. The main indicators measured on the three cores and discussed in this chapter are *grain-size distributions* of both the total fraction and the carbonate-organic matter free fraction, and

*elemental composition* of the sediments (e.g., Al, K, Ca, Si, Mn, Ti, Fe, Mg Ba, Sr, Cu, Co, Ni, V). To aid in the interpretation of the grain-size records, we modelled end-member grain-size distributions using numerical–statistical algorithms developed by Weltje (1997). Grain-size end-members represent a series of fixed sediment grain-size compositions that can be regarded as discrete subpopulations within the data set from all analyses (Prins and Weltje 1999). In addition, spectral analyses of the entire set of proxies were carried out by the *Analyseries* package (Paillard et al. 1996) to obtain the main cyclicities recorded by each proxy. Thus, we are able to objectively separate the orbital parameters associated with each climatic process represented by the analyzed proxies.

## Supply of Saharan Dust to the NW African Margin: The Role of Orbital Parameters

Grain-size distributions of samples from North Canary Basin cores point to two different patterns. The first pattern is characterized by a unimodal distribution (Family 1 in Fig. 1.2), with the mode centred at 8 μm and only 5% of particles coarser than 63 μm. The second pattern (30% of the samples) is characterized by a bimodal distribution with about 10% sand content (Family 2 in Fig. 1.2). The first pattern more commonly appears during full glacial conditions while the second one is the most common at glacial terminations (Fig. 1.2). In addition to this classification, in Fig. 1.2 we show other grain-size parameters, such as the median of the grain-size distributions and a ratio that represents the coarser fraction, that is: (sand + silt)/total ratio (Moreno et al. 2001). In general, a clear precession control (23,000 year cycle) is observed (see comparison with insolation curve, shaded bars in Fig. 1.2), but the glacial–interglacial signal is more evident (100,000 year cycle), pointing to the presence of coarser terrigenous particles during the last three glacial terminations. To combine with the grain-size results, three geochemical markers are plotted in Fig. 1.2: (1) the Si/Al ratio as an indicator of wind strength because terrigenous quartz is more abundant in the coarse dust fraction (Guieu and Thomas 1996; Martinez et al. 1999), (2) the Al percentage, as a proxy of dust input in the NW African margin because it is incorporated into fine-grained, wind-borne clays (Matthewson et al. 1995), and (3) the Fe/Al ratio as a proxy for eolian particles coming from the Sahel area (Bergametti et al. 1989a; Balsam et al. 1995). The two cores shown here are located further to the coast and one of them (GeoB 5559-2) was recovered from a seamount, thus minimizing the amount of river-borne particles.

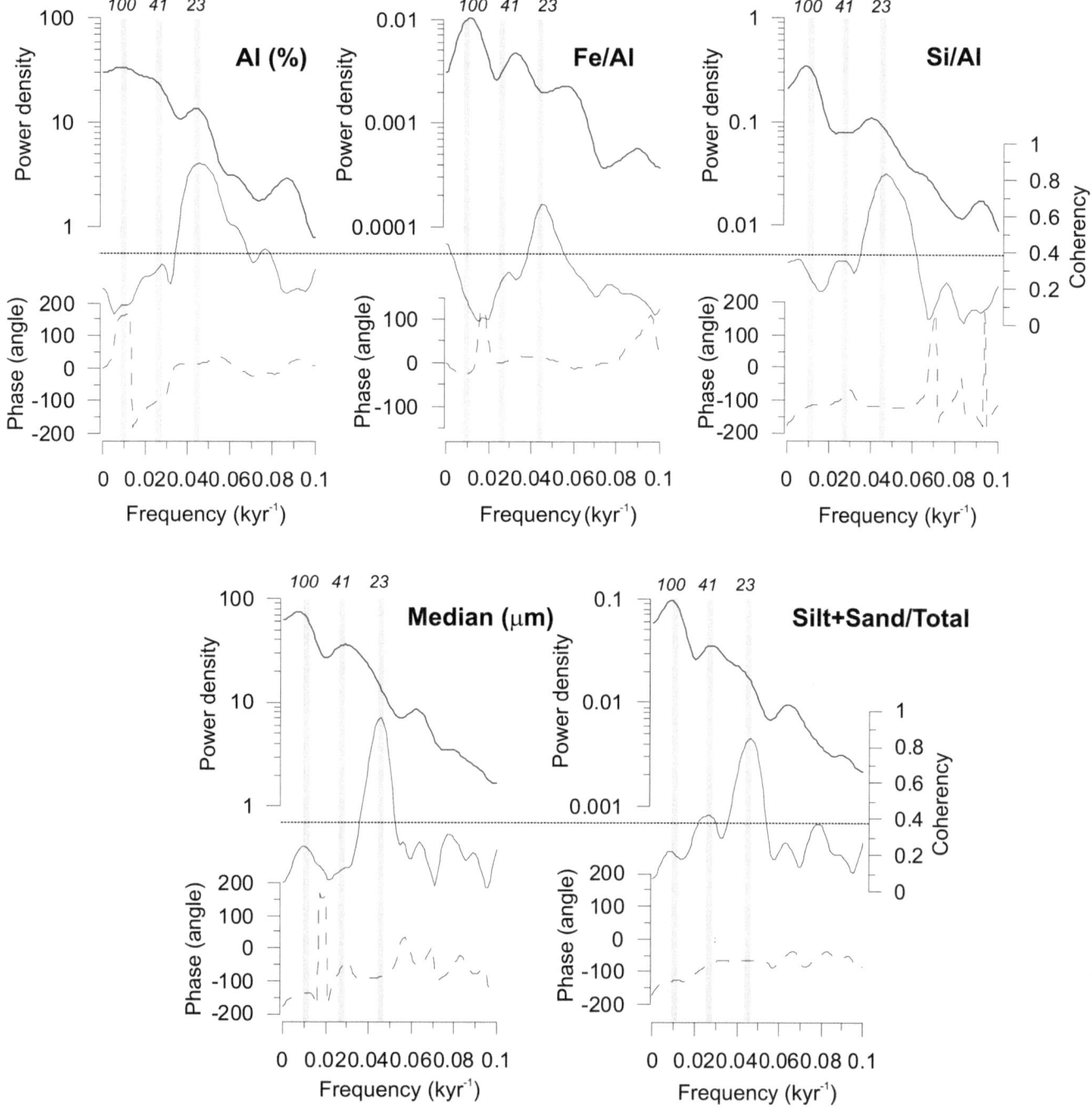

**Fig. 1.3** Variance spectra of the core profiles shown in Fig. 1.2, expressed as the logarithm of spectral power density versus frequency in cycles/kyr using the Blackman–Tuckey method (Paillard et al. 1996), in *blue*. The three main orbital periods of eccentricity (100 kyr), obliquity (41 kyr) and precession (23 kyr) are marked as vertical *gray* bands. The coherency plot (continuous *black line*) indicates what frequency components are shared between the proxies and the insolation curve (summer insolation at 30° north latitude, following (Laskar et al. 2004) solution). An 80% confidence level is set. The phase plot (*black dashed line*) indicates the phase angle between the proxies and the insolation curve at any frequency. Note, however, that only is significant for the frequency components shared between the proxies and the insolation curves (see Table 1.1)

Although both precession and eccentricity parameters seem to control the variations of both dust input and wind strength, precession is the main force behind dust input variations (Al% and Fe/Al ratio), and eccentricity (i.e., glacial–interglacial variability) dominates in the proxies of wind strength (grain-size parameters and Si/Al ratio) (see also Moreno et al. 2001). The study of cyclicities by means of spectral analyses is shown in Fig. 1.3, where a

**Table 1.1** Correlation and phase angle of the proxy records of dust in GeoB 5559-2 with respect to summer insolation at 30°N (Laskar et al., 2004) by means of Blackman and Tukey spectrum using a Bartlett window (bandwidth = 0.02; non-zero coherence > 0.383754)

| Period (kyr) | Al (%) | Fe/Al | Si/Al | Median | (Si+Sa)/T |
|---|---|---|---|---|---|
| *Coherence* | | | | | |
| 100 | 0.10 | 0.13 | 0.28 | 0.20 | 0.21 |
| 41 | 0.27 | 0.17 | 0.35 | 0.07 | 0.41 |
| 23 | 0.87 | 0.62 | 0.77 | 0.72 | 0.79 |
| *Phase* | | | | | |
| 100 | – | – | – | – | – |
| 41 | – | – | – | – | – |
| 23 | 13.01 | 10.69 | −120.03 | −90.76 | −66.96 |

Phase angle is shown for precession, the only parameter where coherency with insolation is higher than 0.6 at the 80% interval of confidence

plot showing the observed cycles from every record and the coherence and phase with summer insolation at 30° north latitude (Laskar et al. 2004) is indicated. These results clearly confirm (1) the influence of precession parameter in the amount of dust transported to the North Canary Basin for the last 250 kyr, and (2) the increase of grain-size at glacial terminations. Surprisingly, the phase relationship established between dust proxies and insolation variability points to enhanced dust supply during precession minima, i.e., during maxima insolation intervals (Table 1.1). This apparently contradicts the fact that, during a precession-minimum-summer, the monsoon regime is enhanced, bringing rainfall to the Sahel region (McIntyre et al. 1989). Nevertheless, it is evident that an increase in moisture also implies a change in the weathering patterns, and, in regions with hyper-arid conditions (100 mm annual rainfall), such as the Sahara Desert, some moisture is necessary to break down large minerals into clays of a size suitable for long distance transport (Rea 1994).

The effect of precession has thus been interpreted as the influence of seasonality on dust generation and its posterior transport to the North Canary Basin (Moreno et al. 2001). As can be presently observed, high seasonality favors dust production and export to the atmosphere. Thus, during hot, wet summers, monsoonal rains in the Sahel region provide the humidity that is needed for the generation of dust particles of a suitable size to be wind-transported. During cold, arid winters, the meteorological scenario facilitates the transport of dust particles by the Saharan winds (Torres-Padrón et al. 2002). Precession controls the seasonality in the source area and thus the generation and transport of dust particles, as observed by the coherent increase in Al (%) and Fe/Al ratio in Fig. 1.2 with insolation maxima (higher seasonality). Another hypothesis to explain this phase relationship takes into account the fluvial

origin of the particles, thus suggesting that the increases in Al (%) and Fe/Al ratio point to an enhanced fluvial discharge as a response to increased monsoonal rains during minimum in precession (Kuhlmann et al. 2004b).

One explanation for the observed maxima in grain-size of the terrigenous fraction and in the Si/Al ratio observed at glacial–interglacial transitions is the increase in sea-level that characterizes every glacial termination. At those intervals, when sea level is rising, it can mobilize shelf sediments that would be transported offshore, thus explaining the presence of coarser sediments. However, this hypothesis does not take into account the increase in paleoproductivity off the coast of NW Africa recorded during terminations that was attributed to a different phenomenon (Moreno et al. 2002b).

## Wind Strength and Productivity at Glacial Terminations

The North Canary Basin core sites GeoB 4216-1 and GeoB 5559-2 are located under the influence of the Cape Ghir upwelling filament, in a productivity gradient from the coast to the open ocean. Recent studies in the North Canary Basin have shown that the present-day productivity signal of Cape Ghir upwelling filament is transferred through the water column and preserved in surface sediments (Nave et al. 2001; Meggers et al. 2002). Therefore, the hemipelagic setting of Cape Ghir has a large potential to record climatically-induced productivity changes. Some selected paleoproductivity proxies are plotted in Fig. 1.4 to reflect variations in the upwelling system and the extent of the upwelling filament (Moreno et al. 2002b). Thus, both TOC (%) and $Ba_{excess}$ concentration increase at the three last glacial terminations, simultaneously with maxima in grain-size of the terrigenous fraction and in the Si/Al ratio (Fig. 1.2). The lack of coherence between TOC (%) and $Ba_{excess}$ in core GeoB 5559-2 is probably due to two factors that influence the TOC signal: (1) the proximity to coastal upwelling and the upwelling filament (closer in GeoB 4216-1), and (2) the sedimentation rate (lower in GeoB 5559-2). These two factors lead to less-organic carbon in the GeoB 5559-2 core due to lower productivity and/or enhanced oxidation produced by deeper oxygen penetration in low sedimentation settings (Rühlemann et al. 1999; Tyson 2001; Moreno et al. 2002b).

The observed variations at terminations (enhanced productivity and increase in grain-size) can be interpreted as being related to changes in the energy of the transporting wind. In addition, the presence of large particles with clear evidence of eolian transport (see SEM images in Fig. 1.4) supports the hypothesis. Terminations are unique intervals

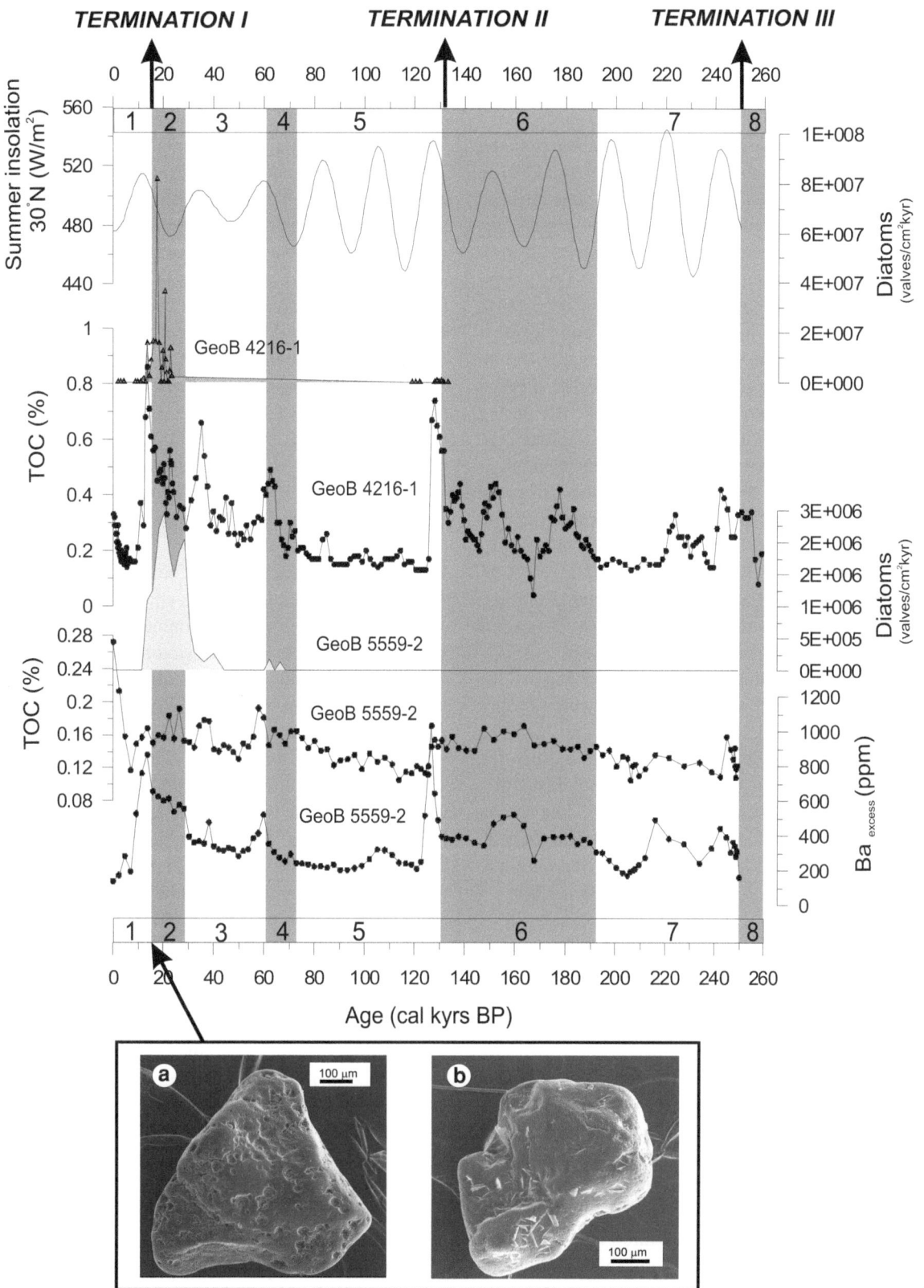

**Fig. 1.4** Downcore profiles of paleoproductivity proxies from GeoB 5559-2 and GeoB 4216-1 cores. From *bottom* to *top*: Ba$_{excess}$ (ppm), Total Organic Carbon—TOC (%) and Diatom accumulation rates (DAR) for core GeoB 5559-2; TOC (%) and DAR (note the different scale) for core GeoB 4216-1. To compare with these results, variation of summer insolation at the latitude of the studied cores (30° north) is also plotted. Glacial stages are shaded and the three glacial terminations are indicated. Scanning electron microscopy (SEM) microphotographs of eolian particles from sediments located at 28 cm depth (Termination I) in core GeoB 5559-2. Note their rounded edges and the signals of eolian impacts. **a** Dolomite, **b** Quartz

of climate change in which the climate switches from a glacial to an interglacial mode. In these intervals, maxima in boreal summer insolation, rapid ice-sheet melting, and fast rises in sea-level occur. These characteristics deserve attention when trying to explain the wind strength and productivity peaks found in GeoB 5559-2 and GeoB 4216-1. Researchers from the same setting but studying pollen contents (Marret and Turon 1994; Lèzine and Denèfle 1997) have interpreted an increase in trade winds during the last glacial–interglacial transitions. Using general circulation models, Overpeck et al. (1989) proposed that the lowering of the North Atlantic SST by glacial melt-water releases during deglaciation strengthened the North Atlantic high-pressure system, thus favoring the enhancement of trade wind velocities. This ocean-wind system connection can be explained by taking into account the higher thermal difference between land and sea that was reached during terminations. The temperature contrast may modulate the Azores' high-pressure intensity leading to an enhancement of the trade wind system. The hypothesis of a coupled tropical/ high latitude North Atlantic climate system operating during the last deglaciation is also supported by various tropical records (Hughen et al. 1996; de Menocal et al. 2000b). Therefore, high latitude low SST anomalies at terminations can enhance trade winds and thereby explain productivity events observed in the areas located under their influence.

Finally, this study postulates that the lowering of the North Atlantic SST by melt-water discharges, which in turn strengthened the Azores' high-pressure center and increased trade wind velocities, can be the mechanism to explain the enhancement of the coastal upwelling and associated filaments at terminations (Moreno et al. 2002b). Other nearby paleoproductivity records (Thomson et al. 2000; Kasten et al. 2001) also show an increase at terminations, thus pointing to a similar mechanism to the one inferred for NW Africa.

## Role of Low Latitudes in the Generation and Transfer of Abrupt Climate Change

The finding of increments of wind strength during glacial terminations is relevant to the investigation of the interactions among dust input and climate because it helps to determine whether the higher dust supplied at the end of glacial periods was caused by a climate change or whether this higher input of eolian dust was one of the potential triggering mechanisms for the glacial–interglacial switches because of interference by dust in the solar radiation that the Earth receives. The evidence presented here, together with several recent paleoclimate studies carried out at tropical latitudes (e.g., Dunbar 2003; Brown et al. 2007; Weldeab et al. 2007), demonstrates the participation of the tropics in global climate changes. The key is to find out whether low

latitudes act as a participant or a driving force in these abrupt climate changes. To delve further into this question, higher resolution studies of dust input and wind strength from tropical areas are needed to detect the temporal leads and lags between high and low latitudes.

Due to its location, high sedimentation rate (27 cm $kyr^{-1}$), and accurate age model (Cacho et al. 1999), the IMAGES core MD95-2043 from the Alboran Sea (western Mediterranean) provides an exceptional opportunity to explore the variations in dust supply from the Saharan region as an indicator of more arid climates at a centennial-scale resolution during the last glacial cycle (Moreno et al. 2002a). In addition to the well-known rapid climate change events, the D/O cycles, the interest in this time interval lies in the observation of important changes in insolation patterns at about 70–80 ka. As is plotted in Fig. 1.2, from 70 kyr on, insolation variability is less controlled by precessional forcing and, likely, high latitude processes will have a stronger imprint in tropical latitudes at that time (e.g., Brown et al. 2007). In addition, the beginning of OIS 4 (70 ka) is also significant in the history of modern humans from northern Africa because the first Aterian human remains were found from this time period, while there was a reduction in Egypt's population (Larrasoaña 2012). Unravelling the environmental changes that early humans had to face remains extraordinarily important to our understanding of the origin of human populations and the patterns of their dispersal out of Africa ca. 50 ka.

In Fig. 1.5 (modified from Fig. 1.2 in Moreno et al. 2005), some selected proxies from core MD95-2043 are represented and millennial variability linked to D/O cycles appears evident in the entire data set. Thus, we represent proxies related to Saharan wind strength (Si/Si+K and eolian end-member; Moreno et al. 2002a); indicators of aridity in the Iberian Peninsula (Aluminum and steppic plants percentages; Sánchez-Goñi et al. 2002); deep-water ventilation ($\delta^{13}C$ measured in benthic foraminifers and the ratio among two biomarkers; Cacho et al. 2000); paleo-productivity (Barium and concentration in alkenones; Moreno et al. 2004); and SST variability (SST $Uk'_{37}$ and the percentages of the cold water foraminifera *Neogloboquadrina pachyderma*; Cacho et al. 1999). All the indicators outlined above that were obtained from the IMAGES core MD95-4043 were later compared to two Greenland GISP2 core proxies: (1) the $\delta^{18}O$, related to atmospheric temperature and (2) the Polar Circulation Index, a proxy for the atmospheric circulation.

We propose that the observed proxy pattern, clearly following the D/O oscillations first recognized in Greenland ice cores, is best explained by the variability in wind systems and precipitation patterns over the Mediterranean region, which was driven by rapid switches between two modes of atmosphere circulation over the North Atlantic region. Therefore, our results can be interpreted according

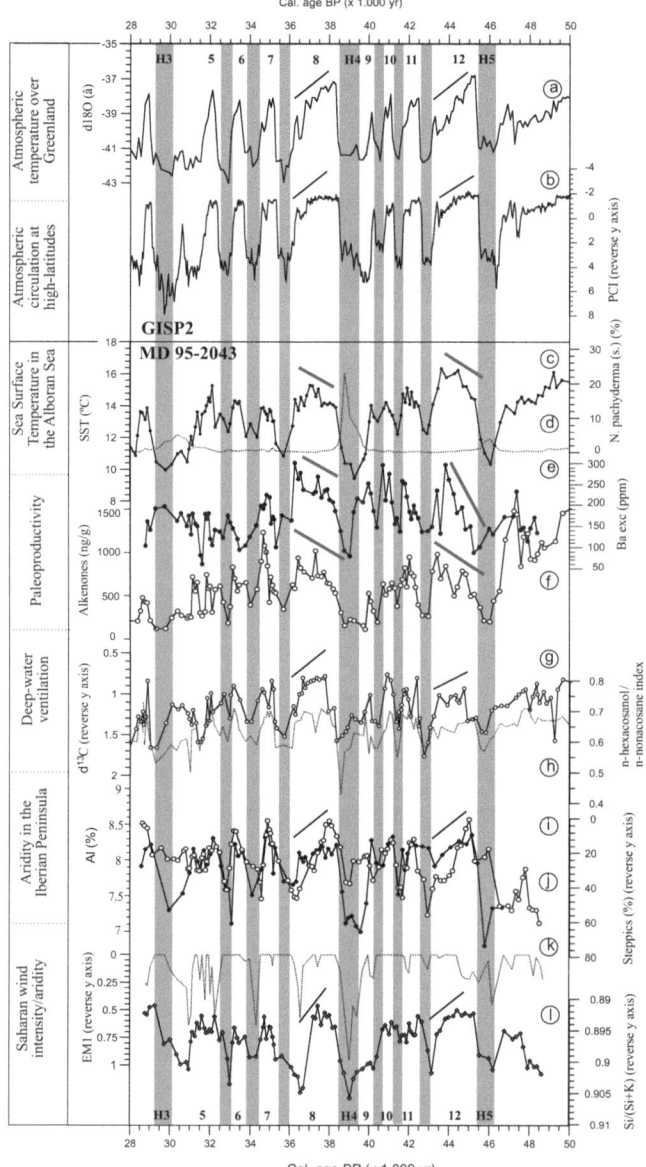

**Fig. 1.5** Comparison of different proxies selected from core MD95-2043 and ice core GISP2: **a** $\delta^{18}O$ and **b** PCI from GISP2 (Grootes and Stuiver 1997; Mayewski et al. 1997); **c** $Uk'_{37}$-SST (Cacho et al. 1999); **d** % of *N. pachyderma* (s.) (*dashed line*) (Cacho et al. 1999); **e** $Ba_{excess}$ (Moreno et al. 2004); **f** Alkenone total concentration (ng/g) (*white dots*) (Cacho et al. 2000); **g** $\delta^{13}C$ (benthics) (reverse y-axis) (Moreno et al. 2004); **h** n-hexacosanol/n-nonacosane index (*dashed line*) (Cacho et al. 2000); **i** steppic vegetation (*black dots*, reverse y-axis) (Sánchez-Goñi et al. 2002); **j** Aluminium percentage (*white dots*) (Moreno et al. 2004); **k** EM1 relative abundance (Moreno et al. 2002a), and **l** Si/(Si+K) (Moreno et al. 2002a) from MD95-2043 core. Processes represented are indicated in the *left-hand side boxes*. *Black* and *gray* lines are plotted to illustrate the differences in interstadial evolution (D/O interstadials 8 and 12) between atmospheric and marine systems. This figure was previously published in Moreno et al. (2005)

to the following two scenarios: (1) as related to cold stadial periods (lower SST), and (2) as associated with a warm interstadial period (higher SST) of a hypothetic D/O cycle

(Fig. 1.6). The definition of these two scenarios is based on several present-day mechanisms, such as the NAO, and on modelling results obtained with millennial-resolution (Ganopolski and Rahmstorf 2001). We therefore suggest that an atmospheric pressure gradient seesaw similar to today's NAO system may explain the records we obtained for the Alboran Sea. This glacial atmospheric oscillator operated on a millennial timescale, causing prolonged climate states similar to the much shorter periods of today's NAO extremes. Thus, in one mode (Fig. 1.6a), atmospheric pressure gradients in the North Atlantic region were high so that northwesterly wind intensity over the Mediterranean area was increased, as pointed out by some previous studies in which the intensity of the Mediterranean overturning was analyzed by means of terrestrial biomarkers and $\delta^{13}C$ measured in benthic forams (Cacho 2000). This mode would arise in response to a decreased North Atlantic SST driven by a deceleration in thermohaline overturn and decreased northward marine heat transport during D/O stadial periods and Heinrich events. This would have favored both drier conditions and more intense Saharan winds, which would have ultimately resulted in increased meridional transport of Saharan dust. Records of Si/Si+K, grain-size, and abundance of steppic vegetation are coherent with this reasoning (Fig. 1.5). In contrast, the interstadial periods (Fig. 1.6b) would have been characterized by weak atmospheric pressure gradients in the North Atlantic that may have favored the southward displacement of northwesterly winds. This situation implies an enhancement of rainfall in the Mediterranean region (Sánchez-Goñi et al. 2002). The southward location of the westerly winds and the stronger input of fluvially transported nutrients may have been the triggering mechanisms for the increases in paleoproductivity, as highlighted by the maximum values of barium and alkenones concentration (Fig. 1.5). In this way, the obtained results in the Alboran core can be integrated into the climate context of the North Atlantic, and at the same time demonstrate the close connection with subtropical latitudes.

Previous studies have suggested that, at a climatic millennial timescale, the tropics responded with an increase in the strength of the wind system (Porter and Zhisheng 1995; Schulz et al. 1998; An 2000; Leuschner and Sirocko 2000). However, in order to assign a role to the tropical latitudes in these observed abrupt changes, the temporal leads and lags between wind systems should be determined. From the Alboran record presented in Fig. 1.5, it seems that the maxima in the Saharan winds lead to maxima in the northwesterly wind system. Phase analyses carried out with this data set indicate that this lead is 320 years (Moreno et al. 2005). Although further studies will need to confirm that tropical processes lead those from higher latitudes, our results highlight the potential importance of dust supply

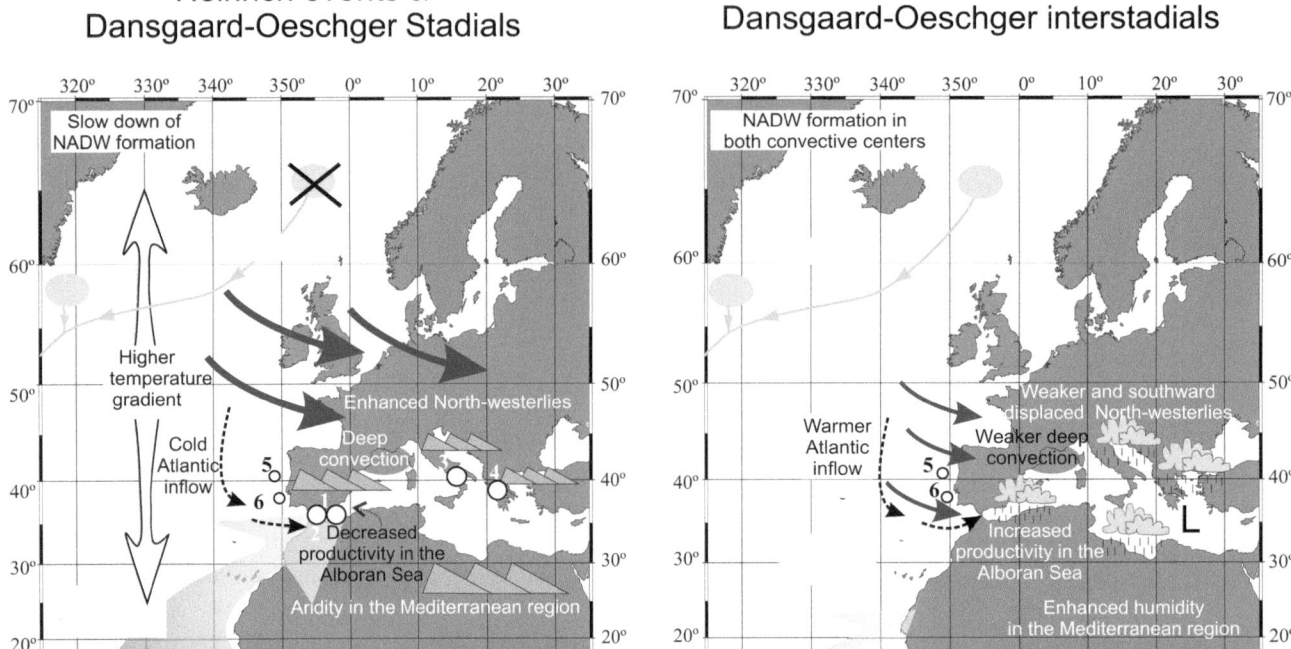

**a  Heinrich events & Dansgaard-Oeschger Stadials**

**b  Dansgaard-Oeschger interstadials**

1 [*Cacho et al.*, 1999, 2000; *Sánchez-Goñi et al.*, 2002; *Moreno et al.*, 2002]
2 [*Combourieu Nebout et al.*, 2002]
3 [*Allen et al.*, 1999]
4 [*Tzedakis*, 1999]
5 [*Roucoux et al.*, 2001]
6 [*Sánchez-Goñi et al.*, 2000]

**Fig. 1.6** D/O scenarios that summarize the main processes and features that controlled the Alboran record during HE and D/O stadial (**a**) and D/O interstadial periods (**b**). Numbers from 1 to 4 in the figure correspond to the following cores: *1*: MD95-2043 (Cacho et al. 1999); *2*: ODP976 (Combourieu Nebout et al. 2002); *3*: Lago Grande di Monticchio (Allen et al. 1999); *4*: Kopais basin (Tzedakis 1999). *This figure was previously published in* (Moreno et al., 2005)

from low latitudes in the global rapid climatic variability. In addition, this high-resolution study may help in our understanding of paleoecological and paleoclimatic conditions in northern Africa and the western Mediterranean when modern humans dispersed.

## Conclusions

Paleoclimate variability was inferred from marine sediment cores off the coast of North Africa (NW African margin and the Alboran Sea) in order to place investigations about the evolution and dispersal of modern humans over the last 250 kyr in a broad climatic context. The study of sediment cores from the North Canary Basin has shown that precession and eccentricity cycles play a role in driving changes in dust input

to the North Canary Basin. The record of dust can be interpreted in terms of a summer and a winter scenario during a minimum in the precessional index. During summers, trade winds were intensified, which resulted in higher productivity and the transport of coarser grains from a NW African source. Meanwhile, in the Sahel area, conditions were suitable for increased dust generation. Superimposed on the overall pattern of a trade wind enhancement during glacial periods, maxima in productivity and grain-size both appear at Terminations I, II, and III. At these periods of higher insolation but lower SST over the North Atlantic, the subtropical anticyclonic circulation may have intensified. Then, the strengthened trade winds forced upwelling and had the ability to carry coarser particles at glacial terminations. This mechanism thus explains the reinforcement of the coastal upwelling and associated filaments, and the productivity pulses recorded at terminations. In addition, the Canary

Current may play a role in transmitting cold meltwaters and nutrients from higher latitudes to the North Canary Basin.

The climatic pattern driven by monsoonal variations seems to be modulated after about 60–70 ka by the influence of high latitude processes due to the lower amplitude of the precessional cycle. The IMAGES Core MD95-2043 from the Alboran Sea allows us to evaluate the variation of fluvial versus eolian inputs along the OIS 3, a key interval in the history of abrupt climate changes. The records display a down-core variability that is similar in structure to the D/O climatic variability seen in the Greenland ice core records. The data indicate an increase of Saharan wind intensity during D/O stadial periods. Existing pollen records along the same core, notably increased abundances in steppic pollen taxa, indicate enhanced aridity on the southern Iberian Peninsula during the same intervals. This combined proxy pattern provides compelling evidence for a highly sensitive response of the low latitude atmospheric system to the D/O climatic cycles.

**Acknowledgments** I gratefully acknowledge the important role of Jean-Jacques Hublin and Shannon McPherron as editors of this volume and organizers of the conference "Modern Origins: A North African Perspective" in Leipzig, August 2007. I am indebted to my co-authors of the previous papers on which this manuscript is based: Jordi Targarona, Isabel Cacho, Anna Sanchez-Vidal, and Miquel Canals from the GRC Geociències Marines at the University of Barcelona (Barcelona, Spain); Jorijntje Henderiks from the Swiss Federal Institute of Technology (Zurich, Switzerland); Tim Freudenthal, Helge Meggers, and Holger Kuhlmann from the Universität Bremen (Bremen, Germany); Silvia Nave and Fatima Abrantes from the Instituto Geologico e Mineiro (Alfragide, Portugal); Maarten A. Prins from the Vrije Universiteit (Amsterdam, Netherlands), María-Fernanda Sánchez-Goñi from the University Bordeaux 1 (Bordeaux, France); Joan O. Grimalt from the ICER-CSIC (Barcelona, Spain); and Francisco J. Sierro from the University of Salamanca (Salamanca, Spain). The following journals gave the necessary permission to reproduce figures and main ideas: *Quaternary Science Reviews, Earth and Planetary Science Letters, Quaternary Research, Palaeogeography, Palaeoclimatology, Palaeoecology,* and *Contributions to Science*. J. Villanueva (CSIC, Barcelona), L. Dupont, R. Davenport, and S. Kasten (U. of Bremen), P. Bertrand, P. Martinez and F. Grousset (U. of Bordeaux), M. A. Bárcena (U. de Salamanca), and R. Zahn (U. Autònoma, Barcelona) for useful comments and discussions throughout my Ph.D. This study was mainly supported by the CANIGO project (MAS3-CT9-0060) and the IMAGES program. The data presented here were produced and elaborated during A. Moreno's Ph.D. thesis at the GRC Geociencies Marines (University of Barcelona, Spain) thanks to a Comissionat d'Universitats i Recerca fellowship.

# References

Allen, J. R. M., Brandt, U., Brauer, A., Hubberten, H. W., Huntley, B., Keller, J., et al. (1999). Rapid environmental changes in southern Europe during the last glacial period. *Nature, 400,* 740–743.

An, Z. (2000). The history and variability of the East Asian paleomonsoon climate. *Quaternary Science Reviews, 19,* 171–187.

Balsam, W. L., Otto-Bliesner, B. L., & Deaton, B. C. (1995). Modern and last glacial maximum eolian sedimentation patterns in the Atlantic Ocean interpreted from sediment iron oxide content. *Paleoceanography, 10,* 493–507.

Bergametti, G., Gomes, L., Coudé-Gaussen, G., Rognon, P., & Le Coustumer, M. N. (1989a). African dust observed over Canary Islands: Source-regions identification and transport pattern for some summer situations. *Journal of Geophysical Research, 94,* 14.855–14.864.

Bergametti, G., Gomes, L., Remoudaki, E., Desbois, M., Martin, D., & Buat-Menard, P. (1989b). Present transport and deposition patterns of African dusts to the Northwestern Mediterranean. In M. Leinen & M. Sarnthein (Eds.), *Paleoclimatology and paleometeorology: Modern and past patterns of global atmospheric transport* (pp. 227-252). Dordrecht: Kluwer Academic Publishers.

Bond, G., Showers, W., Cheseby, M., Lotti, R., Almasi, P., de Menocal, P., et al. (1997). A pervasive millenial-scale cycle in North Atlantic Holocene and glacial climates. *Science, 278,* 1257–1266.

Bozzano, G., Kuhlmann, H., & Alonso, B. (2002). Storminess control over African dust input to the Moroccan Atlantic margin (NW Africa) at the time of maxima boreal summer insolation: A record of the last 220 kyr. *Palaeogeography, Palaeoclimatology, Palaeoecology, 183,* 155–168.

Broecker, W. S., Peteet, D., & Rind, D. (1985). Does the ocean-atmosphere system have more than one stable mode of operation? *Nature, 315,* 21–26.

Brown, E. T., Johnson, T. C., Scholz, A., Cohen, A. S., & King, J. W. (2007). Abrupt change in tropical African climate linked to the bipolar seesaw over the past 55,000 years. *Geophysical Research Letters, 34.* doi:10.1029/2007GL031240.

Cacho, I. (2000). Respuesta del Mediterráneo Occidental a los cambios climáticos rápidos de los últimos 50.000 años. University of Barcelona: Análisis de biomarcadores moleculares.

Cacho, I., Grimalt, J. O., Pelejero, C., Canals, M., Sierro, F. J., Flores, J. A., et al. (1999). Dansgaard-Oeschger and Heinrich event imprints in Alboran Sea temperatures. *Paleoceanography, 14,* 698–705.

Cacho, I., Grimalt, J. O., Sierro, F. J., Shackleton, N. J., & Canals, M. (2000). Evidence for enhanced Mediterranean thermohaline circulation during rapid climatic coolings. *Earth and Planetary Science Letters, 183,* 417–429.

Cacho, I., Shackleton, N., Elderfield, H., Sierro, F. J., & Grimalt, J. O. (2006). Glacial rapid variability in deep-water temperature and d$^{18}$O from the Western Mediterranean Sea. *Quaternary Science Reviews, 25,* 3294–3311.

Chiapello, I., Bergametti, G., & Chatenet, B. (1997). Origins of African dust transported over the northeastern tropical Atlantic. *Journal of Geophysical Research, 102,* 13.701–713.709.

Clement, A. C., & Peterson, L. C. (2008). Mechanisms of abrupt climate change of the last glacial period. *Reviews of Geophysics, 46.* doi:10.1029/2006RG000204.

Combourieu Nebout, N., Turon, J. L., Zahn, R., Capotondi, L., Londeix, L., & Pahnke, K. (2002). Enhanced aridity and atmospheric high-pressure stability over the western Mediterranean during the North Atlantic cold events of the past 50 k.y. *Geology, 30,* 863–866.

Dansgaard, W., Johnsen, S. J., Clausen, H. B., Dahl-Jensen, D., Gundestrup, N. S., Hammer, C. U., et al. (1993). Evidence for general instability of past climate from a 250-kyr ice-core record. *Nature, 364,* 218–220.

Davenport, R., Neuer, S., Helmke, P., Pérez-Marrero, J., & Llinas, O. (2002). Primary productivity in the northern Canary Islands region as inferred from SeaWiFS imagery. *Deep Sea Research II, 19,* 3481–3496.

de Menocal, P. (1995). Plio-Pleistocene African climate. *Science, 270,* 53–59.

de Menocal, P., Ortiz, J., Guilderson, T. P., Adkins, J. F., Sarnthein, M., Baker, L., et al. (2000a). Abrupt onset and termination of the African Humid Period: Rapid climate responses to gradual insolation forcing. *Quaternary Science Reviews, 19,* 347–361.

de Menocal, P., Ortiz, J., Guilderson, T. P., & Sarnthein, M. (2000b). Coherent high- and low-latitude climate variability during the Holocene warm period. *Science, 288,* 2198–2202.

Dunbar, R. B. (2003). Leads, lags and the tropics. *Nature, 421,* 121–122.

Dupont, L. M., Jahns, S., Marret, F., & Ning, S. (2000). Vegetation change in equatorial West Africa: Time-slices for the last 150 ka. *Palaeogeography, Palaeoclimatology, Palaeoecology, 155,* 95–122.

Fabrés, J., Calafat, A., Sanchez-Vidal, A., Canals, M., & Heussner, S. (2002). Composition and spatio-temporal variability of particle fluxes in the Western Alboran Gyre, Mediterranean Sea. *Journal of Marine Systems, 33–34,* 431–456.

Fisher, D., & Wefer, G. (Eds.). (1999). *Use of proxies in paleoceanography. Examples from the South Atlantic.* Berlin: Springer.

Fletcher, W. J., & Sánchez Goñi, M. F. (2008). Orbital- and sub-orbital-scale climate impacts on vegetation of the western Mediterranean basin over the last 48,000 yr. *Quaternary Research, 70,* 451–464.

Freudenthal, T., Meggers, H., Henderiks, J., Kuhlmann, H., Moreno, A., & Wefer, G. (2002). Upwelling intensity and filament activity off Morocco during the last 250,000 years. *Deep Sea Research II, 19,* 3655–3674.

Ganopolski, A., & Rahmstorf, S. (2001). Rapid changes of glacial climate simulated in a coupled climate model. *Nature, 409,* 153–158.

Grootes, P., & Stuiver, M. (1997). Oxygen 18/16 variability in Greenland snow and ice with 103- to 105-year time resolution. *Journal of Geophysical Research, 102,* 26455–26470.

Grousset, F. E., Parra, M., Bory, A., Martinez, P., Bertrand, P., Shimmield, G. B., et al. (1998). Saharan wind regimes traced by the Sr-Nd isotopic composition of subtropical Atlantic sediments: Last Glacial maximum vs today. *Quaternary Science Reviews, 17,* 395–409.

Guieu, C., & Thomas, J. (1996). Saharan aerosols: From the soil to the ocean. In S. Guerzoni & R. Chester (Eds.), *The impact of desert dust across the Mediterranean* (pp. 207–216). Dordrecht: Kluwer Academic Publishers.

Holz, C., Stuut, J.-B. W., Henrich, R., & Meggers, H. (2007). Variability in terrigenous sedimentation processes off Northwest Africa and its relation to climate changes: Inferences from grain-size distributions of a Holocene marine sediment record. *Sedimentary Geology, 202,* 499–508.

Hooghiemstra, H., Stalling, H., Agwu, C. O. C., & Dupont, L. M. (1992). Vegetational and climatic changes at the northern fringe of the Sahara 250.000–5.000 years BP: Evidence from 4 marine pollen records located between Portugal and the Canary Islands. *Review of Palaeobotany and Palynology, 74,* 1–53.

Hughen, K. A., Overpeck, J. T., Peterson, L. C., & Trumbore, S. (1996). Rapid climate changes in the tropical Atlantic region during the last deglaciation. *Nature, 380,* 51–54.

Hurrell, J. W. (1995). Decadal trends in the North Atlantic oscillation: Regional temperatures and precipitation. *Science, 269,* 676–679.

Itambi, A. C., von Dobeneck, T., Mulitza, S., Bickert, T., & Heslop, D. (2009). Millennial-scale Northwest African droughts related to Heinrich events and Dansgaard-Oeschger cycles: Evidence in marine sediments from offshore Senegal. *Paleoceanography, 24,* PA1205. doi:10.1029/2007PA001570.

Jullien, E., Grousset, F. E., Malaize, B., Duprat, J., Sánchez-Goñi, M. F., Eynaud, F., et al. (2007). Low-latitude "dusty events" vs. high-latitude "icy Heinrich events". *Quaternary Research, 68,* 379–386.

Kasten, S., Haese, R., Zabel, M., Rühlemann, C., & Schulz, H. (2001). Barium peaks at glacial terminations in sediments of the equatorial Atlantic Ocean—relics of deglacial productivity pulses? *Chemical Geology, 175,* 635–651.

Kuhlmann, H. (2003). *Reconstruction of the sedimentary history offshore NW Africa: Application of core-logging tools, compilation of papers.* Ph.D. Dissertation, University of Bremen.

Kuhlmann, H., Freudenthal, T., Helmke, P., & Meggers, H. (2004a). Reconstruction of paleoceanography off NW Africa during the last 40,000 years: Influence of local and regional factors on sediment accumulation. *Marine Geology, 207,* 209–224.

Kuhlmann, H., Meggers, H., Freudenthal, T., & Wefer, G. (2004b). The transition of the monsoonal and the N Atlantic climate system off NW Africa during the Holocene. *Geophysical Research Letters, 31,* L22204. doi:22210.21029/22004GL021267.

Larrasoaña, J. C. (2012). A Northeast Saharan perspective on environmental variability in North Africa and its implications for modern human origins. In J.-J. Hublin & S. P. McPherron (Eds.), *Modern origins: A North African perspective.* Dordrecht: Springer.

Laskar, J., Robutel, P., Joutel, F., Gastineau, M., Correia, A. C. M., & Levrard, B. (2004). A long-term numerical solution for the insolation quantities of the Earth. *Astronomy and Astrophysics, 428,* 261–285.

Leuschner, D. C., & Sirocko, F. (2000). The low-latitude monsoon climate during Dansgaard-Oeschger cycles and Heinrich Events. *Quaternary Science Reviews, 19,* 243–254.

Lèzine, A. M., & Denèfle, M. (1997). Enhanced anticyclonic circulation in the eastern North Atlantic during cold intervals of the last deglaciation inferred from deep-sea pollen records. *Geology, 25,* 119–122.

Marret, F., & Turon, J. L. (1994). Paleohydrology and paleoclimatology off Northwest Africa during the last glacial-interglacial transition and the Holocene: Palynological evidence. *Marine Geology, 118,* 107–117.

Martinez, P., Bertrand, P., Shimmield, G. B., Cochrane, K., Jorissen, F., Foster, J. M., et al. (1999). Upwelling intensity and ocean productivity changes off Cape Blanc (northwest Africa) during the last 70,000 years: Geochemical and micropalaeontological evidence. *Marine Geology, 158,* 57–74.

Martinson, D. G., Pisias, N. G., Hays, J. D., Imbrie, J., Moore, T. C., & Shackleton, N. J. (1987). Age dating and the orbital theory of the Ice Ages: Development of a high-resolution 0 to 300,000-year chronostratigraphy. *Quaternary Research, 27,* 1–29.

Matthewson, A. P., Shimmield, G. B., Kroon, D., & Fallick, A. E. (1995). A 300 kyr high-resolution aridity record of the North African continent. *Paleoceanography, 10,* 677–692.

Mayewski, P. A., Meeker, L. D., Twickler, M. S., Whitlow, S., Yang, Q., Lyons, W. B., et al. (1997). Major features and forcing of high-latitude northern hemisphere atmospheric circulation using a 110,000-year-long glaciochemical series. *Journal of Geophysical Research, 102,* 26345–26366.

McIntyre, A., Ruddiman, W., Karlin, K., & Mix, A. C. (1989). Surface water response of the equatorial Atlantic Ocean to orbital forcing. *Paleoceanography, 4,* 19–55.

Meese, D. A., Gow, A. J., Alley, R. B., Zielinski, G. A., Grootes, P., Ram, M., et al. (1997). The Greenland Ice Sheet Project 2 depth-age scale: Methods and results. *Journal of Geophysical Research, 102,* 26411–26423.

Meggers, H., Freudenthal, T., Nave, S., Targarona, J., Abrantes, F., & Helmke, P. (2002). Assessment of geochemical and micropaleontological sedimentary parameters as proxies of surface water properties in the Canary Islands region. *Deep Sea Research II, 19,* 3631–3654.

Millot, C. (1999). Circulation in the western Mediterranean Sea. *Journal of Marine Systems, 20,* 423–442.

Moreno, A., & Canals, M. (2004). The role of dust in abrupt climate change: Insights from offshore Northwest Africa and Alboran Sea sediment records. *Contributions to Science, 2*, 485–497.

Moreno, A., Targarona, J., Henderiks, J., Canals, M., Freudenthal, T., & Meggers, H. (2001). Orbital forcing of dust supply to the North Canary Basin over the last 250 kyrs. *Quaternary Science Reviews, 20*, 1327–1339.

Moreno, A., Cacho, I., Canals, M., Prins, M. A., Sánchez-Goñi, M. F., Grimalt, J. O., et al. (2002a). Saharan dust transport and high latitude glacial climatic variability: The Alboran Sea record. *Quaternary Research, 58*, 318–328.

Moreno, A., Nave, S., Kuhlmann, H., Canals, M., Targarona, J., Freudenthal, T., et al. (2002b). Productivity response in the North Canary Basin to climate changes during the last 250,000 years: A multi-proxy approach. *Earth and Planetary Science Letters, 196*, 147–159.

Moreno, A., Cacho, I., Canals, M., & Grimalt, J. O. (2004). Millennial-scale variability in the productivity signal from the Alboran Sea record (western Mediterranean). *Palaeogeography, Palaeoclimatology, Palaeoecology, 211*, 205–219.

Moreno, A., Cacho, I., Canals, M., Grimalt, J. O., Sánchez-Goñi, M. F., Shackleton, N. J., et al. (2005). Links between marine and atmospheric processes oscillating at millennial time-scale. A multi-proxy study of the last 50,000 yr from the Alboran Sea (western Mediterranean Sea). *Quaternary Science Reviews, 24*, 1623–1636.

Moulin, C., Lambert, C. E., Dulac, F., & Dayan, U. (1997). Control of atmospheric export of dust from North Africa by the North Atlantic oscillation. *Nature, 387*, 691–694.

Mulitza, S., Prange, M., Stuut, J.-B., Zabel, M., von Dobeneck, T., Itambi, A. C., et al. (2008). Sahel megadroughts triggered by glacial slowdowns of Atlantic meridional overturning. *Paleoceanography, 23*. doi:10.1029/2008PA001637.

Nave, S., Freitas, P., & Abrantes, F. (2001). Coastal upwelling in the Canary Island region: Spatial variability reflected by the surface sediment diatom record. *Marine Micropaleontology, 42*, 1–23.

Nykjaer, L., & Van Camp, L. (1994). Seasonal and interannual variability of coastal upwelling along Northwest Africa and Portugal from 1981 to 1991. *Journal of Geophysical Research, 99*, 14197–14207.

Overpeck, J. T., Peterson, L. C., Kipp, N., Imbrie, J., & Rind, D. (1989). Climate change in the circum-north Atlantic region during the last deglaciation. *Nature, 338*, 553–557.

Paillard, D., Labeyrie, L., & Yiou, P. (1996). Macintosh program performs time-series analysis. *Eos Transactions, 77*, 379.

Parrilla, G., & Kinder, T. H. (1987). Oceanografía física del mar de Alborán. *Boletín del Instituto Español de Oceanografía, 4*, 133–165.

Perkins, H., Kinder, T., & La-Violette, P. (1990). The Atlantic inflow in the western Alboran Sea. *Journal of Physical Oceanography, 20*, 242–263.

Plaza, A. M. (2001). *Estudio paleoceanográfico de los testigos TG-7 (dorsal de Nazca-Pacífico) y MD95-2043 (mar de Alborán-Mediterráneo)*. Barcelona: CSIC.

Porter, S. C., & Zhisheng, A. (1995). Correlation between climate events in the North Atlantic and China during the last glaciation. *Nature, 375*, 305–308.

Prins, M. A., & Weltje, G. J. (1999). End-member modeling of siliciclastic grain-size distributions: The late Quaternary record of aeolian and fluvial sediment supply to the Arabian Sea and its paleoclimatic significance. In J. Harbaugh, W.L. Watney, E. Rankey, R. Slingerland, R. Goldstein & E. Franseen (Eds.), *Numerical experiments in stratigraphy: Recent advances in stratigraphic and sedimentologic computer simulations* (pp. 91–111). SEPM (Society for Sedimentary Geology) Special Publication 62.

Prospero, J. M. (1996). Saharan dust transport over the North Atlantic Ocean and Mediterranean: An overview. In S. Guerzoni & R.

Chester (Eds.), *The impact of desert dust across the Mediterranean* (pp. 133–151). Dordrecht: Kluwer Academic Publisher.

Rea, D. (1994). The paleoclimatic record provided by eolian deposition in the deep sea: The geologic history of wind. *Reviews of Geophysics, 32*, 159–195.

Rodriguez, S., Querol, X., Alastuey, A., Kallos, G., & Kakaliagou, O. (2001). Saharan dust contributions to PM10 and TSP levels in southern and eastern Spain. *Atmospheric Environment, 35*, 2433–2447.

Rousseau, D.-D., Sima, A., Antoine, P., Hatté, C., Lang, A., & Zöller, L. (2007). Link between European and North Atlantic abrupt climate changes over the last glaciation. *Geophysical Research Letters, 34*. doi:10.1029/2007GL031716.

Rühlemann, C., Müller, P., & Schneider, R. (1999). Organic carbon and carbonate as paleoproductivity proxies: Examples from high and low productivity areas of the tropical Atlantic. In G. Fischer & G. Wefer (Eds.), *Use of proxies in paleoceanography: Examples from the South Atlantic* (pp. 1–31). Berlin: Springer.

Sánchez-Goñi, M. F., Cacho, I., Turon, J. L., Guiot, J., Sierro, F. J., Peypouquet, J.-P., et al. (2002). Synchroneity between marine and terrestrial responses to millennial scale climatic variability during the last glacial period in the Mediterranean region. *Climate Dynamics, 19*, 95–105.

Schulz, H., von Rad, U., & Erlenkeuser, H. (1998). Correlation between Arabian Sea and Greenland climate oscillations of the past 110,000 years. *Nature, 393*, 54–57.

Schuster, M., Roquin, C., Duringer, P., Brunet, M., Caugy, M., Fontugne, M., et al. (2005). Holocene Lake Mega-Chad palaeo-shorelines from space. *Quaternary Science Reviews, 24*, 1821–1827.

Smith, J. R., Hawkins, A. L., Asmerom, Y., Polyak, V., & Giegengack, R. (2007). New age constraints on the Middle Stone Age occupations of Kharga Oasis, Western Desert, Egypt. *Journal of Human Evolution, 52*, 690–701.

Stuut, J.-B., Zabel, M., Ratmeyer, V., Helmke, P., Schefub, E., Lavik, G., et al. (2005). Provenance of present-day eolian dust collected off NW Africa. *Journal of Geophysical Research, 110*, D04202.

Thomson, J., Nixon, S., Summerhayes, C., Rohling, E. J., Schönfeld, J., Zahn, R., et al. (2000). Enhanced productivity on the Iberian margin during glacial/interglacial transitions revealed by barium and diatoms. *Journal of Geological Society of London, 157*, 667–677.

Tjallingii, R., Claussen, M., Stuut, J.-B. W., Fohlmeister, J., Jahn, A., Bickert, T., et al. (2008). Coherent high- and low-latitude control of the Northwest African hydrological balance. *Nature Geoscience, 1*, 670–675.

Torres-Padrón, M. E., Gelado-Caballero, M. D., Collado-Sánchez, C., Siruela-Matos, V. F., Cardona-Castellano, P. J., & Hernández-Brito, J. J. (2002). Variability of dust inputs to the CANIGO zone. *Deep Sea Research II, 19*, 3455–3464.

Tyson, R. V. (2001). Sedimentation rate, dilution, preservation and total organic carbon: Some results of a modelling study. *Organic Geochemistry, 32*, 333–339.

Tzedakis, C. (1999). The last climatic cycle at Kopais, Central Greece. *Journal of Geological Society of London, 156*, 425–434.

Voelker, A. (2002). Global distribution of centennial-scale records for marine isotope stage (MIS) 3: A database. *Quaternary Science Reviews, 21*, 1185–1212.

Wang, Y., Cheng, H., Edwards, L. R., Kong, X., Shao, X., Chen, S., et al. (2008). Millennial- and orbital-scale changes in the East Asian monsoon over the past 224,000 years. *Nature, 451*, 1090–1093.

Weldeab, S., Lea, D. W., Schneider, R., & Andersen, N. (2007). 155,000 years of West African monsoon and ocean thermal evolution. *Science, 316*, 1303–1307.

Weltje, G. J. (1997). End-member modeling of compositional data: Numerical-statistical algorithms for solving the explicit mixing problem. *Journal of Mathematical Geology, 29*, 503–549.

# Chapter 2
# A Northeast Saharan Perspective on Environmental Variability in North Africa and its Implications for Modern Human Origins

J. C. Larrasoaña

**Abstract** In this chapter, we recall a record of Saharan dust supply into the eastern Mediterranean Sea (ODP Site 967) to document Middle-Late Pleistocene environmental variations in the Northeast Sahara (NES). Distinctive dust flux minima ca. 330, 285, 240, 215, 195, 170, 125, 100, and 80 ka attest to the expansion of subtropical savannah landscapes throughout the NES during boreal summer insolation maxima, which drove penetration of the West African summer monsoon front up to 25–27°N. Such "green Sahara" periods broadly correlate with U-series ages of lacustrine and spring carbonates scattered throughout the NES, which are often associated with Acheulean and Mousterian archaeological sites that attest to widespread occupation of the area during pluvial episodes. In contrast, Aterian sites are linked to spring deposits and mountain areas during a prolonged period of hyperarid climate, which suggests adaptation to desert conditions. The Site 967 dust record has important implications for understanding the evolution and population dynamics of modern humans in Africa. Thus, the monsoon-driven alternation of "green Sahara" and hyperarid desert conditions throughout North Africa, combined with similarly paced environmental variations within tropical Africa, provides a favorable scenario for the speciation of *H. sapiens*, for a gradual accumulation of African modern behaviors as a whole, and for frequent out of Africa dispersals of modern human populations.

**Keywords** Dust record • Eastern Mediterranean • Environmental magnetism • Green Sahara • Modern human dispersals • North Africa • Ocean Drilling Program • Pleistocene

## Introduction

The origin of modern humans has led to one of the major debates in paleoanthropology over the last two decades (see Lahr and Foley 1998; McBrearty and Brooks 2000; Hovers and Kuhn 2006). At present, most genetic, paleoanthropological, and archaeological evidence point to a single origin of *Homo sapiens* in Africa (Stringer 2002; Mellars 2006), although the specific location and mechanism that gave birth to our species are the subject of a contested debate that is often taken beyond normal scientific inquiry into the realm of paradigmatic discussion (Lahr and Foley 1998; McBrearty and Brooks 2000).

One of the elements that is always considered to be a major factor conditioning human evolution is climatic variability, through its effect on landscape composition (Lahr and Foley 1998). Although temperate-cold climates have long been recognized as influencing recent human evolution in western Asia and Europe, it is still unclear what drove the evolution of modern humans in Africa (see Stringer 2002). Yet, and perhaps from a Eurocentric perspective, it is often assumed that evolution and behavioral development of modern humans in Africa was influenced by glacial-interglacial changes driven by climate variability at the high-latitudes (e.g., Lahr and Foley 1998; Mithen and Reed 2002; Stringer 2002; Mellars 2006). Thus, glacial periods would have conditioned the expansion of the Sahara at the expense of subtropical savannahs and equatorial rainforest. This situation would have been reversed during interglacial epochs in such a way that the Sahara might have nearly disappeared due to the expansion of subtropical savannahs (e.g., Lahr and Foley 1998). Although these dramatic expansions and contractions of tropical African landscapes are evidenced by paleoclimatic (Szabo et al. 1995; Jolly et al. 1998; Gasse 2000; Hooghiemstra et al. 2006; Weldeab et al. 2007; Tjallingii et al. 2008) and climate modeling data (Brovkin et al. 1998; Gasse 2000), a growing body of evidence accumulated over the last decade indicates that such contractions and expansions were driven primarily by internal dynamics of the monsoon system

J. C. Larrasoaña (✉)
Instituto Geológico y Minero de España,
Unidad de Zaragoza, C/Manuel Lasala 44 9B,
50006 Zaragoza, Spain
e-mail: jc.larra@igme.es

J.-J. Hublin and S. P. McPherron (eds.), *Modern Origins: A North African Perspective*,
Vertebrate Paleobiology and Paleoanthropology, DOI: 10.1007/978-94-007-2929-2_2,
© Springer Science+Business Media B.V. 2012

(Brovkin et al. 1998; Jolly et al. 1998; Gasse 2000), which is driven by incoming solar radiation in the low latitudes, rather than by high-latitude climate variability.

The debate on the African origin of modern humans has been further influenced by the fact that most researchers have focused their investigations on East and South Africa, but have largely ignored North Africa. This is surprising, especially in the case of Northeast Africa, because ample evidence attests to past Middle-Late Pleistocene pluvial episodes that fostered the recurrent occupation of what is a key location, now characterized by a hyperarid climate, in the gateway to Eurasia (Wendorf et al. 1993; Szabo et al. 1995; Smith et al. 2004, 2007; Kleindienst et al. 2008). Unfortunately, due to the intensive aeolian deflation in the Sahara Desert, continental records of North African climate variability are scarce, short, discontinuous, irregularly distributed, difficult to date, and provide only a fragmentary view of North African climate. Marine records of Saharan dust deposition into neighboring ocean basins (Tiedemann et al. 1994; de Menocal 1995, 2004; Matthewson et al. 1995; Moreno et al. 2001, 2002; Bozzano et al. 2002; Dinarès-Turell et al. 2003; Hamann et al. 2008; Itambi et al. 2009) might partly overcome these problems, but they are difficult to link with environmental variations at specific portions of the vast Saharo-Arabian Desert (Goudie and Middleton 2001; Prospero et al. 2002). As a result of these shortcomings, the paleoenvironmental and paleoclimatic scenario that framed the origin of modern humans in North Africa remains largely ignored.

In this chapter, we recall a record of Saharan dust deposition in the eastern Mediterranean Sea produced from Ocean Drilling Program (ODP) Site 967 (Larrasoaña et al. 2003). This record has been claimed to provide a proxy for the astronomically-controlled penetration of the West African monsoon into the northeastern Sahara (NES), which, in turn, conditioned variations in dust production in response to the expansion and retraction of savannah landscapes throughout the NES. We use the Site 967 record because, contrary to other marine dust records of North African climate variability from the Mediterranean Sea (Calvert and Fontugne 2001; Moreno et al. 2002; Dinarès-Turell et al. 2003; Hamann et al. 2008), the Atlantic Ocean (Tiedemann et al. 1994; de Menocal 1995, 2004; Matthewson et al. 1995; Moreno et al. 2001; Bozzano et al. 2002; Itambi et al. 2009) and the Arabian Sea (de Menocal 1995, 2004), it relates variations in dust production to environmental changes in a specific region within the Saharo-Arabian Desert belt, taking into account its complex physiography and interactions with climatic processes. The aim of this study is to use the Middle-Late Pleistocene (350–20 ka) record of Site 967 to: (1) develop a robust paleoclimatic and paleoenvironmental framework

for human occupation of the NES; and (2) examine its implications in the origin of modern humans.

## The ODP Site 967 Dust Record

ODP Site 967 was recovered at a water depth of 2,553 m on the northern slope of the Eratosthenes Seamount (34°04'N, 32°43'E) (Fig. 2.1). The studied sedimentary sequence of Site 967 consists of 90 m of Pliocene-Holocene hemipelagic bioturbated nannofossil oozes and nannofossil clays, and includes 79 visible sapropels (Kroon et al. 1998; Emeis et al. 2000). Sapropels are dark-colored layers that usually vary from 1 to 60 cm in thickness and contain up to 25% organic carbon (by weight) (Hilgen 1991; Lourens et al. 1996; Emeis et al. 2000). Sapropels are important because they mark the pace of an orbitally-driven climatic system that was exceptionally amplified due to the semi-enclosed nature of the Mediterranean basin. Formation of sapropels was controlled by ca. 21 kyr periodic changes in the amount of solar energy received in the northern low- and mid-latitudes during boreal summer insolation maxima (Hilgen 1991; Lourens et al. 1996; Emeis et al. 2000). At these times, intensification and enhanced northward penetration of the West African monsoon led to an increase in the freshwater discharge into the eastern Mediterranean (Rossignol-Strick 1983; Lourens et al. 2001; Rohling et al. 2002; Larrasoaña et al. 2003).

The Site 967 dust record is based on the high-resolution (1 cm) measurement of a laboratory-induced magnetization that was later demagnetized with an alternating magnetic field. The intensity of this so-called IRM@AF parameter is proportional to the content of hematite, which constitutes about a 6.5% (in weight) of the Saharan dust transported into the eastern Mediterranean (Tomadini et al. 1984). The correspondence between hematite contents and Saharan dust supply is evident in a Pliocene interval of the Site 967 record, which provides an exceptional view of African paleoclimate variability (Fig. 2.2) (Lourens et al. 2001). A distinctive cyclic pattern is evident in the Ti/Al ratio, with minimum values in the sapropels and highest values in the intercalated marls. Ti is linked to aeolian transport of heavy minerals in the distal marine sediments of Site 967 (Lourens et al. 2001), whereas Al is related to both aeolian (e.g., kaolinite; Foucault and Mélières 2000) and fluvial (e.g., smectite; Lourens et al. 2001) sources. Variations in Ti/Al can therefore be interpreted in terms of the relative contributions of aeolian (Saharan dust) and fluvial (Nile) sources. The Ti/Al curve strikingly parallels boreal summer insolation, which attests to a link between supply of Saharan dust and paleoclimate variability via the influence of solar radiation on monsoon dynamics in tropical Africa (Lourens et al. 2001). The amount of hematite, which is a mineral typically found in Saharan dust

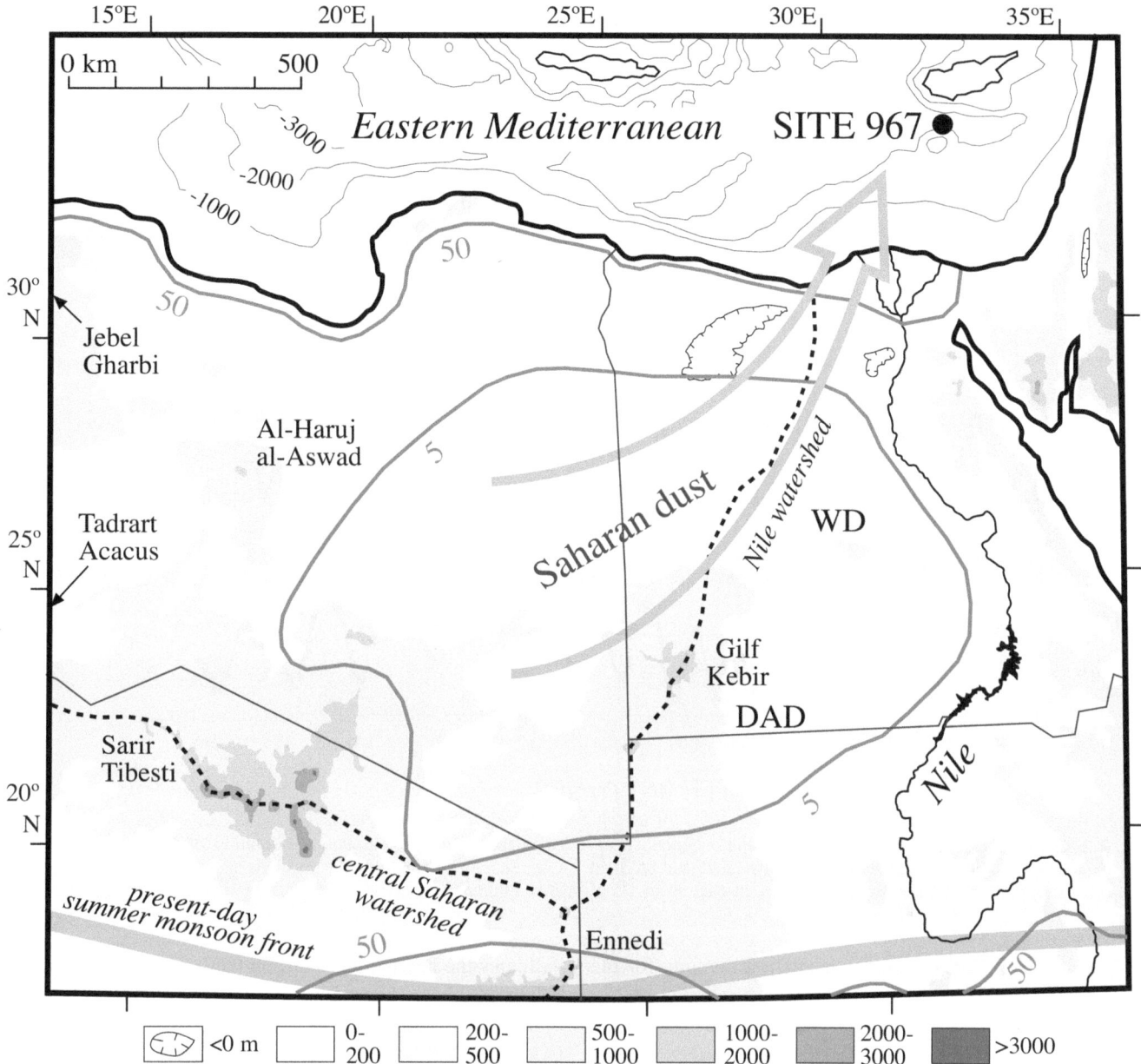

**Fig. 2.1** Present-day physiography of the northeastern Sahara and the eastern Mediterranean Sea, with location of ODP Site 967 and the main trajectory of northeastern Sahara dust transport (Dayan et al. 1991; Goudie and Middleton 2001). Thin *gray lines* indicate mean annual isohyets in mm per year (after Petit-Maire 2002). *WD* Western Desert, *DAD* Darb el Arba'in Desert

(e.g., Goudie and Middleton 2001; Prospero et al. 2002), closely mimics the Ti/Al curve. This demonstrates that the IRM@AF parameter can be used as a proxy for Saharan dust supply into the eastern Mediterranean.

## The Middle-Late Pleistocene Record

The interval of Site 967 between 2 and 16 revised meter composite depth (rmcd) includes sapropels S3 to S10 (Fig. 2.3). The age model for this interval is based on the characteristic sapropel pattern, which can be tuned to a summer insolation target curve by correlating sapropels to their corresponding insolation maxima (Hilgen 1991; Lourens et al. 1996; Emeis et al. 2000). IRM@AF values are larger than 0.2 A/m throughout the Middle-Late Pleistocene except within and around sapropels, where the IRM@AF parameter shows marked drops well below 0.1 A/m. This indicates that deposition of sapropels coincided with relatively short periods of decreased dust supply. As boreal summer insolation is driven by changes in the Earth's orbital precession (ca. 21 kyr cycles), and this, in turn, is modulated by the Earth's eccentricity (ca. 100 and

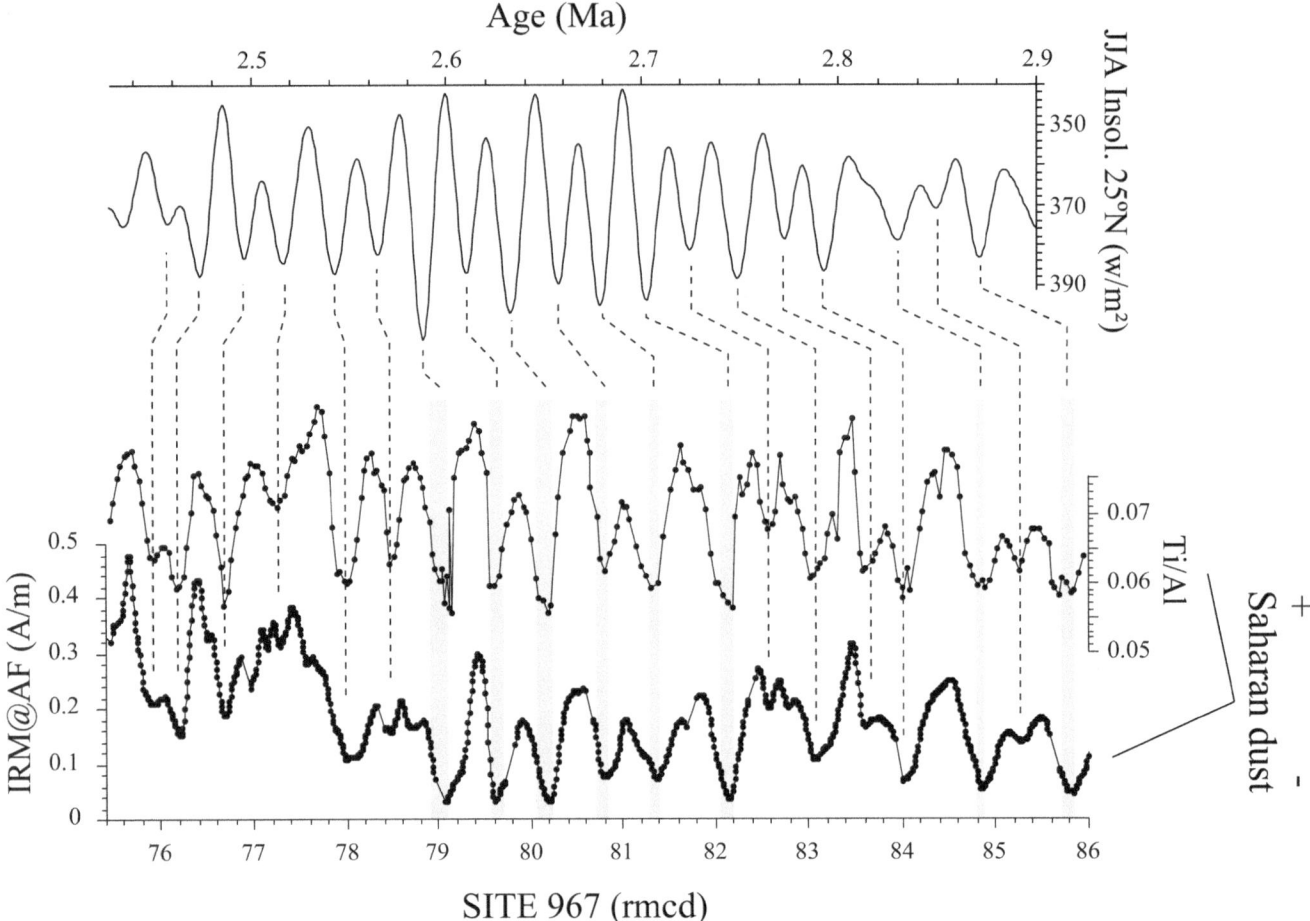

**Fig. 2.2** Geochemical (Lourens et al. 2001) and magnetic (Larrasoaña et al. 2003) data from ODP Site 967 between 75 and 86 rmcd (2.4–2.9 Ma), plotted against boreal summer insolation at 25°N calculated after the astronomical solution of Laskar et al. (2004). *Gray shaded bars* indicate the positions of sapropels. The age model is after Lourens et al. (2001)

400 kyr cycles), boreal summer insolation maxima for the last 350 kyr, and hence sapropels with dust minima, cluster near 100 ka eccentricity maxima (Fig. 2.3b). In contrast to sapropels, high IRM@AF values of intercalated sediments indicate that the supply of Saharan dust was significantly higher during their deposition. This is especially clear for the latest Pleistocene interval between 75 and 20 ka, where IRM@AF shows the highest values of up to 1 A/m. The $\delta^{18}O$ record of the studied interval is characterized by oscillations (>4‰) that are mostly marked by distinctive negative isotopic excursions within sapropels (Kroon et al. 1998). These excursions are larger than simultaneous glacial-interglacial variations (<1.2‰) in the Atlantic Ocean, and have been classically explained by massive drainage of isotopically-light monsoon rainfall via the Nile River into the eastern Mediterranean (Rossignol-Strick 1983). Despite these isotopic excursions, the characteristic glacial-interglacial pattern of the Middle-Late Pleistocene interval between marine isotopic stages 10 to 2 (350 to 20 ka) is clearly recognizable in the $\delta^{18}O$ record (Fig. 2.3), which

validates the age model based on the tuning of the sapropel pattern to the summer insolation target curve. It is worth noting that sapropels formed during interglacial stages (e.g., S3–S5 with stages 5a to 5e, S7 to S9 with stages 7 to 7e, and S10 to stage 9), but also during glacial stages such as S6 (stage 6). In addition, glacial stage 8 includes an interval characterized by a negative isotopic excursion and distinctively low IRM@AF values, typical for sapropels, and can therefore be interpreted as a sapropel (indicated by an arrow in Fig. 2.3) that has been erased by postdepositional oxidation (Larrasoaña et al. 2003, 2006).

In order to unravel the paleoclimatic and paleoenvironmental significance of the Site 967 dust record in the role of modern humans origins, we need to: (1) understand the climatic processes that control production and transport of Saharan dust into the eastern Mediterranean, and (2) integrate this knowledge with the paleoclimatic and archaeological evidence that portrays the response of landscape evolution and human adaptation to climate variability in the NES during the Middle-Late Pleistocene.

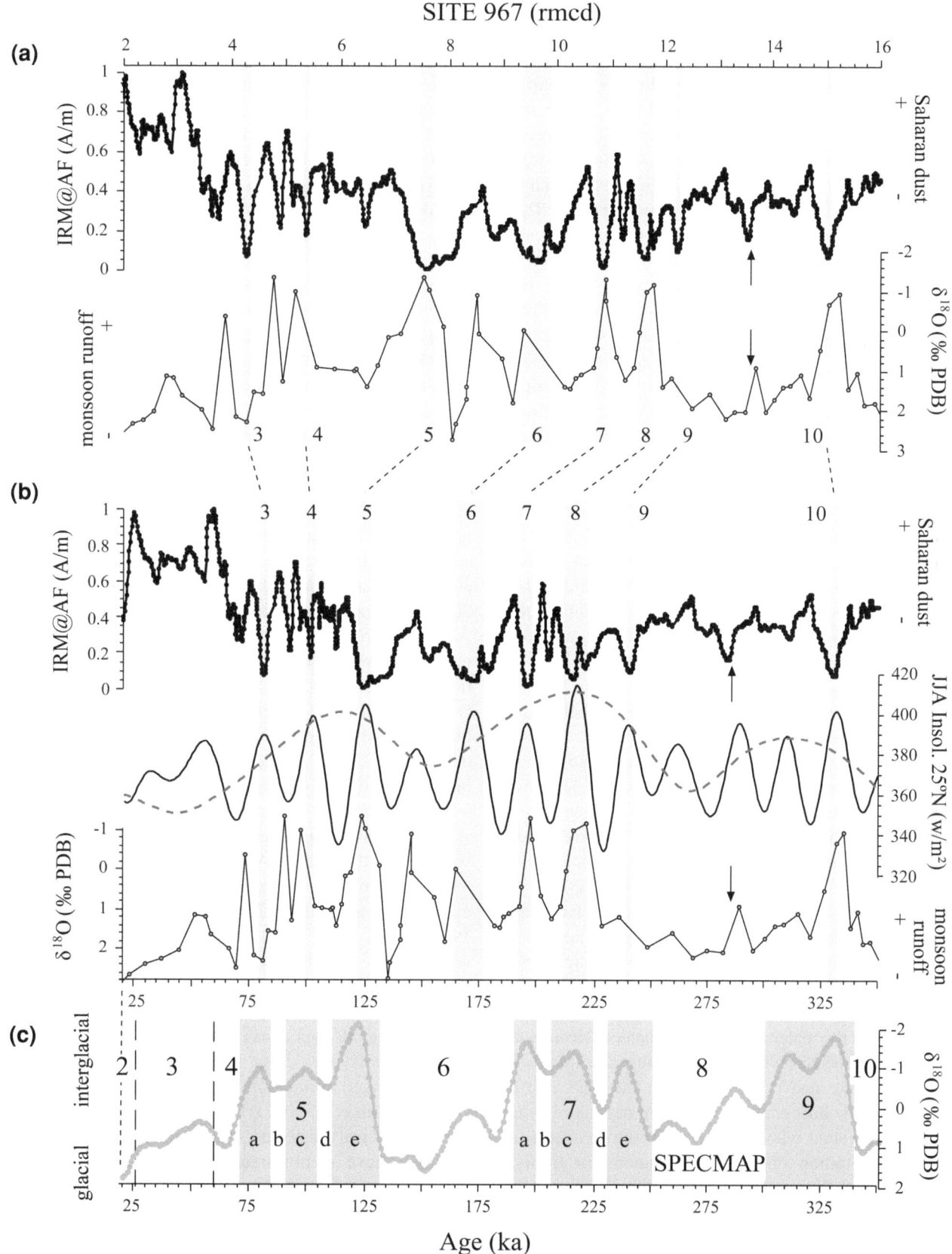

**Fig. 2.3 a** IRM@AF (Larrasoaña et al. 2003), oxygen isotopes (Kroon et al. 1998) and sapropels (*light gray shaded bars* 3–10) from ODP Site 967 between 2 and 16 rmcd. **b** IRM@AF and oxygen isotope data from the same interval plotted, together with sapropels and the boreal summer insolation curve (at 25°N) calculated after the astronomical solution of Laskar et al. (2004), against age. The *dashed lines* plotted with the insolation curve is the Earth's eccentricity parameter calculated after Laskar et al. (2004). The arrows indicate the position of an oxidized sapropel. **c** Oxygen isotope record of the SPECMAP (Imbrie et al. 1984). *White* and *gray bars* indicate glacial and interglacial periods, respectively

Satellite TOMS analyses (Prospero et al. 2002), geochemical data (Krom et al. 1999; Foucault and Mélières 2000; Weldeab et al. 2002), and back-trajectories of dust outbreaks (Dayan et al. 1991) indicate that the present-day dust source areas for dust transported into the eastern Mediterranean are the lowlands of eastern Libya, western Egypt, northeastern Chad and northwestern Sudan located between the Al-Haruj al-Aswad hill range, the Sarir Tibesti and Ennedi massifs, and the Nile River (Fig. 2.1). These areas broadly correspond to the driest part of the NES, which currently receives less than 5 mm of precipitation per year (Petit-Maire 2002). The hyperarid core of the NES contains fossil lacustrine and fluvial deposits and bedforms (Szabo et al. 1995; Pachur and Hoelzmann 2000; Rohling et al. 2002), and hosts a system of wadis that transport sediments from surrounding mountain areas into terminal alluvial fans, playas, and saline lakes (Pachur and Hoelzmann 2000). The easily weathered and deflated silt-rich sediments accumulated in these areas fuel the bulk of modern dust production (Goudie and Middleton 2001; Prospero et al. 2002).

The occurrence of lacustrine, palustrine, fluvial, and spring-related (tufa) deposits scattered throughout the NES attests to previous pluvial episodes during the Middle-Late Pleistocene (Szabo et al. 1995; Crombie et al. 1997; Sultan et al. 1997; Smith et al. 2004, 2007; Kieniewicz and Smith 2007; Kleindienst et al. 2008). Development of lake and fluvial systems and activation of springs has been linked to a poleward expansion of the tropical rainfall belt during periods of increased boreal summer insolation (Brovkin et al. 1998; Jolly et al. 1998; Gasse 2000). At those times, increased sensible heating over North Africa led to an intensification of the West African monsoon, which resulted in a northward shift of its summer front. Positive vegetation-albedo feedbacks pushed the summer monsoon front as far north as $\sim 25°$ (Brovkin et al. 1998; Gasse 2000), and conditioned the expansion of savannah landscapes throughout the whole Sahara Desert as far north and east as the NES (e.g., the "green Sahara" state of the climate modeling community; Brovkin et al. 1998; Gasse 2000).

This "greening of the Sahara" constitutes a key concept for unraveling the paleoclimatic and paleoenvironmental significance of the Site 967 record because it explains the formation of lake and river systems in the NES and the simultaneous low dust contents and sharp negative isotopic excursion associated with sapropels. On the one hand, the enhanced penetration of the summer monsoon front up to $\sim 25°N$, well beyond the central Saharan watershed, accounts for the massive drainage of isotopically-light monsoon rainfall not only via the Nile, but along the whole North African margin, into the eastern Mediterranean. This explains why the lightest $\delta^{18}O$ values associated with sapropels are typically found between Libya and southwest

Crete (Fontugne et al. 1994; Emeis et al. 2003) rather than off the Nile River, which, with its huge catchment, including both northern and southern hemisphere regions, drains monsoon rainfall with a relatively smaller range of isotopical variability throughout the year (Rohling et al. 2002). On the other hand, an increase in precipitation and vegetation cover would account for the stabilization of surface sediments in such a way that the production of dust would be severely dampened (Goudie and Middleton 2001; Prospero et al. 2002). The genetic link between "green Sahara" periods, sapropel formation, and insolation maxima is evidenced by the correspondence of lowest dust contents in sapropels with distinctive (>390 W/m$^2$) peaks in boreal summer insolation (Fig. 2.3). During periods of boreal summer insolation minima, which cluster at around 100 ka eccentricity minima, the weakened summer monsoon front would have remained south of the central Saharan watershed, which would have converted the NES into the barren hyperarid dust factory that it is today (e.g., the "desert Sahara" state of the climate modeling community, Brovkin et al. 1998). The dramatic hydrological changes in the NES, reported here on the basis of the Site 967 dust record, are consistent with in-phase changes in the Sahel (Tjallingii et al. 2008) and around the Gulf of Guinea (Weldeab et al. 2007). This points to a simultaneous response of the North African hydrological cycle to monsoon dynamics. In this regard, it is worth noting the key location of the NES in the farthest possible position away from the equatorial Atlantic Ocean, which is the source of moisture for the West African monsoon both along N–S and W–E transects. Identification of "green Sahara" periods in the NES therefore gives information on the occurrence of dramatic landscape variations that affected North Africa as a whole.

Figure 2.4 shows a comparison of the Site 967 dust record with the SPECMAP curve (Imbrie et al. 1984) and with uranium-series ages of lacustrine carbonates (Szabo et al. 1995) and tufas (Crombie et al. 1997; Sultan et al. 1997; Smith et al. 2004, 2007; Kleindienst et al. 2008) accumulated at different parts of the NES during past Middle-Late Pleistocene pluvial periods. The broad distribution of dates for the carbonates might result from the often large (>10 kyr) errors associated with most published uranium-series ages from lacustrine carbonates, but also from prolonged spring activity related to groundwater discharge (Smith et al. 2007). Despite the broad distribution of ages, statistical analyses based on probability density functions have demonstrated the clustering of carbonate deposition not only during most interglacial periods (e.g., during stages 5a, 5c, 5e, 7a, 7c, 7e), but also during glacial stages 6 and 8 (Fig. 2.4) (Smith 2012). Similar to what happens with sapropels, these data demonstrate that deposition of carbonates occurred more frequently than the 100 kyr glacial-interglacial cycles (Smith et al. 2007;

Smith 2012). We propose that such frequency corresponds with the ca. 21 kyr insolation cycles that are responsible for recurrent "green Sahara" periods. This interpretation, which might eventually be confirmed by improved uranium-series ages of NES carbonates, is further supported by the similar ca. 21 kyr pacing of increased groundwater movement in the NES, which is also driven by monsoon-fed aquifer recharge (Osmond and Dabous, 2004). These observations confirm previous claims suggesting that climate and landscape variability over North Africa are mainly driven by changes in incoming solar radiation via its influence on monsoon dynamics (Trauth et al. 2009), rather than by glacial-interglacial cycles linked to climatic variability at high latitudes, as has often been assumed (e.g., Lahr and Foley 1998; Mithen and Reed 2002; Mellars 2006). This does not imply that glacial-interglacial cycles do not have an effect on North African climate; this, in fact, has been demonstrated from several marine dust records in which glacial-interglacial oscillations are mainly reflected by changes in wind intensity and/or atmospheric circulation patterns (Matthewson et al. 1995; Moreno et al. 2001, 2002; Hamann et al. 2008; Itambi et al. 2009). It merely suggests that glacial-interglacial cycles exert a secondary imprint on a primary low-latitude climate mechanism in such a way that the influence of high-latitude climate variability becomes important at periods of lowest boreal summer insolation, when the monsoon system is severely weakened (Weldeab et al. 2007). Based on these results, and as has been proposed for sapropels (Lourens et al. 1996), we recommend that the "green Sahara" periods be named with the number of their correlative insolation peak (Fig. 2.4).

## Developing a Paleoclimatic and Paleoenvironmental Framework for Human Occupation of the NES

The Site 967 dust record sheds new light on the paleoclimatic and paleoenvironmental context that framed human occupation in the NES during the Middle-Late Pleistocene. Many of the lacustrine, fluvial, and tufa deposits scattered throughout the NES are associated with archaeological remains attributed to the Acheulean culture of the Early Stone Age (ESA) and to the Mousterian culture of the Middle Stone Age (MSA) (McHugh et al. 1988; Wendorf et al. 1993; Szabo et al. 1995; Haynes et al. 1997; Hill 2001; Mandel and Simmons 2001; Smith et al. 2004, 2007; Kleindienst et al. 2008) (Fig. 2.4). Such archaeological remains are typically found within silts that directly overlay and/or underlay lacustrine, fluvial, and tufa deposits.

Acheulean sites are associated with spring deposits from the oasis depressions of the Western Desert of Egypt at 24°–28°N (Smith et al. 2004), and with fluvial, lacustrine, and spring deposits from the Darb al-Arba'in Desert between Egypt and Sudan at 21–23°N (McHugh et al. 1988; Szabo et al. 1995; Haynes et al. 1997; Hill 2001; Mandel and Simmons 2001). Mousterian sites are mainly found in association with spring deposits from the Western Desert (Smith et al. 2004, 2007; Kleindienst et al. 2008), and with lacustrine carbonates from the Darb al-Arba'in Desert (Wendorf et al. 1993; Szabo et al. 1995). Previous studies have demonstrated that Acheulean and Mousterian sites attest to the recurrent human reoccupation of the NES during Middle-Late Pleistocene pluvial episodes. However, the precise timing and duration of these occupation events remains elusive, due to the discontinuity of the archaeological record and the often large errors associated with uranium-series dating of spring and lacustrine carbonates (Smith et al. 2007). The Site 967 dust record sheds light on these questions because it shows that conditions suitable for Acheulean and Mousterian occupation of the NES occurred during "green Sahara" periods ca. 330, 285, 240, 215, 195, 170, 125, 100, and 80 ka. "Green Sahara" periods paced by boreal summer insolation maxima also occurred before 350 ka (Larrasoaña et al. 2003), so they account for the occurrence of lacustrine and spring carbonates, often associated with Acheulean remains, whose ages are beyond the range of U-series dating (i.e., >350 ka) (Szabo et al. 1995; Crombie et al. 1997; Sultan et al. 1997; Hill 2001; Smith et al. 2004). Sedimentological, geochemical, and faunal evidence from lacustrine sediments of the Darb al-Arba'in Desert demonstrates that this area of the NES received at least 500 mm of annual rainfall during the "green Sahara" period ca. 125 ka, which enabled the widespread occurrence of wooded savannah landscapes inhabited by subtropical faunas (Kowalski et al. 1989; Wendorf et al. 1993). This 125 ka "green Sahara" period is associated with one of the highest boreal summer insolation maxima of the last 350 kyr, so it is likely that other "green Sahara" periods might have been characterized by relatively drier climates. In any case, paleoclimate evidence demonstrates that Middle-Late Pleistocene pluvial episodes were wetter than the Early-Middle Holocene "green Sahara" period between 6 and 10 ka (Szabo et al. 1995; Hoelzmann et al. 2000; Geyh and Thiedig 2008). At that time, rainfall in the Darb al-Arba'in Desert was <300 mm/year and oscillated between 50 and 150 mm/year in the Western Desert (Küper and Kröpelin 2006), so the whole NES then was covered by sparsely wooded grasslands and was inhabited by savannah to semi-desert dwellers (Nicoll 2004; Küper and Kröpelin 2006). Moreover, lacustrine deposits attest to the widespread occurrence of wetland area as far north as 26° (Szabo et al. 1995; Hoelzmann et al. 2000; Pachur and Hoelzmann 2000;

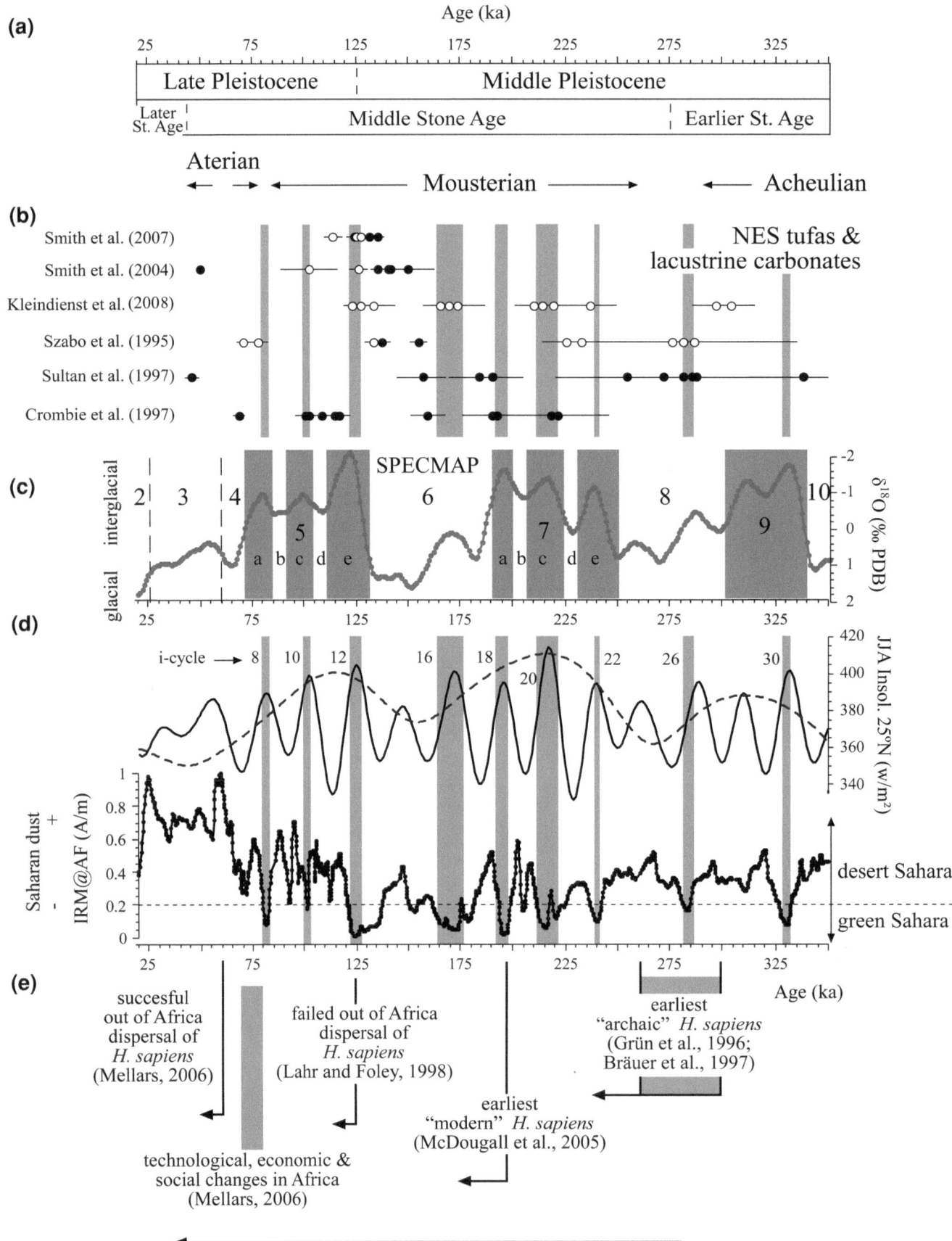

**Fig. 2.4** Comparison of: **a** Age range of archaeological industries found in the northeastern Sahara (see text); **b** Uranium-series ages of lacustrine and spring carbonates from the northeastern Sahara (after Smith et al. 2007). *White* (*black*) symbols indicate carbonates with (without) associated archaeological remains; **c** SPECMAP curve (Imbrie et al. 1984); **d** Boreal summer insolation curve (at 25°N) calculated after the astronomical solution of Laskar et al. (2004) (*dashed line* is the Earth's eccentricity parameter), plotted along with the IRM@AF record of ODP Site 967 (Larrasoaña et al. 2003); **e** Significant events in the evolution of *H. sapiens* (after Grün et al. 1996; Bräuer et al. 1997; Lahr and Foley 1998; McBrearty and Brooks 2000; McDougall et al. 2005; Mellars 2006). *Gray bars* indicate the positions of "green Sahara" periods as identified from lowest (<0.2 A/m) IRM@AF values. Numbers denote correlative insolation cycles (Lourens et al. 1996)

Leblanc et al. 2006). These data strongly suggest that during Middle-Late Pleistocene wetter "green Sahara" periods, a subtropical climate with a N–S gradient in increased humidity enabled expansion of wetland-spotted savannah landscapes throughout the NES and even farther north, as demonstrated by the occurrence of freshwater lakes up to 28°N (Wendorf et al. 1993; Szabo et al. 1995; Kieniewicz and Smith 2007; Geyh and Thiedig 2008). It is important to note that, according to the Site 967 dust record, these savannah landscapes prevailed in the NES for less than 5–10 kyr, which is the time span within each insolation cycle where the summer monsoon front penetrated well beyond the central Saharan watershed. This suggests that Acheulean and Mousterian reoccupation of the NES during Middle-Late Pleistocene "green Sahara" periods was restricted to short (<5–10 kyr) intervals separated by intervening hyperarid periods devoid of human occupation.

It might be argued that the proposed link between "green Sahara" periods and human reoccupation of the NES is too speculative, considering the errors associated with the uranium-series ages of lacustrine carbonates and tufas and the fact that not all lacustrine carbonates and tufas are associated with archaeological remains. New archaeological and geological surveys are clearly necessary to improve our knowledge on the link between human populations and climate variability in the NES. In the meantime, we can gain some insights on this topic by examining the latest Pleistocene-Holocene record in the NES. This period is characterized by a single "greening-yellowing of the Sahara" cycle, which has conditioned the preservation of a significant number of archaeological sites (nearly 300) for which accurate radiocarbon-based ages and environmental constraints are available (Nicoll 2004; Küper and Kröpelin 2006).

Geological and paleoclimatic data provide evidence for extensive eolian deflation under a hyperarid climate when the NES was devoid of human occupation between 20 and 10 $^{14}$C ka (Stokes et al. 1998; Maxwell and Haynes 2001; Swezey 2001), as demonstrated by the pervasive lack of

Late Paleolithic archaeological remains (Szabo et al. 1995; Nicoll 2004; Küper and Kröpelin 2006) (Fig. 2.5). Between 9.8 and 9.5 $^{14}$C ka, the NES rapidly shifted toward a semi-arid subtropical climate that was driven by the intensification and ~800 km northward shift of the West African summer monsoon front in response to a maxima in boreal summer insolation (Brovkin et al. 1998; Jolly et al. 1998; Gasse 2000). Geological, paleontological, and archaeological data indicate that the NES then hosted widespread wetland areas, was covered by sparsely wooded grasslands, and was inhabited by savannah to semi-desert dwellers that included hunter-gatherers (Nicoll 2004; Küper and Kröpelin 2006). The stabilization of this semi-arid climate by 8–7 $^{14}$C ka led to the development and widespread practice of sedentarism, pottery production, and domestic livestock keeping (Nicoll 2004; Küper and Kröpelin 2006). At ~6.3 $^{14}$C ka, a decrease in boreal summer insolation led to the rapid southward retreat of the West African summer monsoon front (Brovkin et al. 1998; Jolly et al. 1998; Gasse 2000) and conditioned the rapid return of hyperarid desert conditions to the NES (Stokes et al. 1998; Maxwell and Haynes 2001; Swezey 2001; Nicoll 2004; Küper and Kröpelin 2006) (Fig. 2.2d). This led to the exodus of human populations into neighboring, previously inhabited areas located to the south and east. Humans migrated to either the Sudanese plains, following the retreating summer monsoon rains, or to the Nile Valley, where they found stable refugium along the vegetated banks of the Nile. These migrations caused dramatic changes in population density and associated social structures, and ultimately led to the spread of pastoralism into tropical Africa and the emergence of the pharaonic civilization, respectively (Fig. 2.5) (Nicoll 2004; Küper and Kröpelin 2006). Human occupation of the NES after ~6.3 $^{14}$C ka became restricted to ecological refugia such as the Gilf Kebir plateau. By ~4.7 $^{14}$C ka, permanent settlements had disappeared, with the exception of the groundwater-fed oasis of the Western Desert of Egypt (Nicoll 2004; Küper and Kröpelin 2006).

The ODP Site 967 dust record captures the response of NES landscapes and human populations to climate variations strikingly well (Fig. 2.5a). On the one hand, the high dust contents characterizing the hyperarid latest Pleistocene display a sharp decrease that coincides with the initial spread of wetland spotted grasslands and the onset of human occupation at 9.8–9.5 $^{14}$C ka. On the other hand, low dust content throughout most of the Early-Middle Holocene, which coincided with the deposition of sapropel S1 in the eastern Mediterranean (Mercone et al. 2001), underwent a sharp increase at around 6 $^{14}$C ka that marks the return of hyperarid desert conditions and the beginning of the exodus from the desert.

We are aware that the archaeological record is still too scarce and its chronology too coarse to demonstrate a

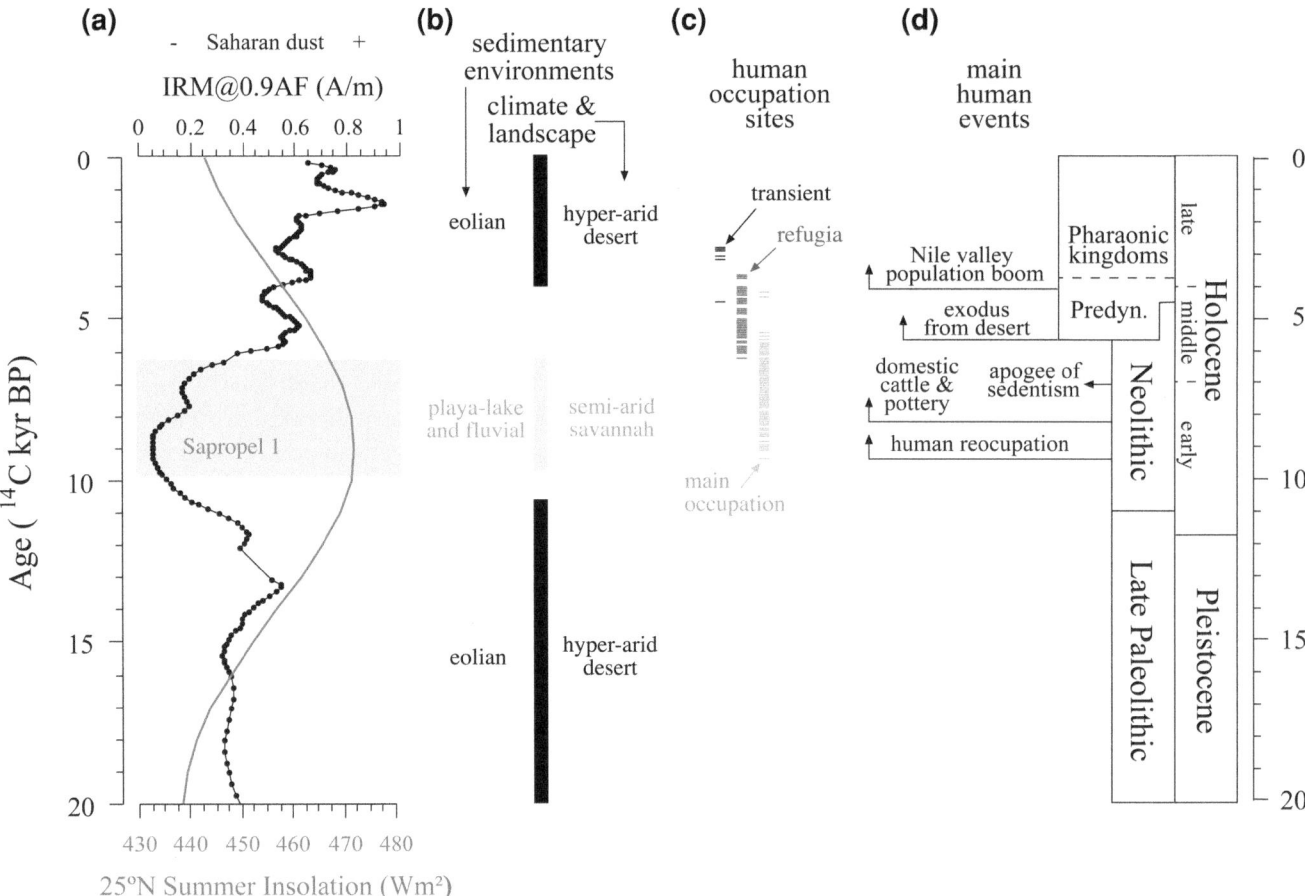

**Fig. 2.5** **a** IRM@AF values (Larrasoaña et al. 2003) from the latest Pleistocene-Holocene record of ODP Site 967, plotted along with the summer insolation curve at 25°N calculated after the astronomical solution of Laskar et al. (2004). The chronology for the interval is based on linear interpolation of ages between the top of the core, [14]C ages of the top and bottom of sapropel 1 (*gray shaded bar*) (Mercone et al. 2001), and the age of the previous sapropel (Kroon et al. 1998).

**b** Prevailing sedimentary environments, climatic conditions, and landscapes in the northeastern Sahara (after Stokes et al. 1998; Jolly et al. 1998; Gasse 2000; Maxwell and Haynes 2001; Swezey 2001; Nicoll 2004; Küper and Kröpelin 2006). **c** [14]C ages of archaeological remains in the northeastern Sahara (from Küper and Kröpelin 2006). **d** Chronology of main events in human history in the northeastern Sahara (after Nicoll 2004; Küper and Kröpelin 2006)

similar link between "green Sahara" periods and human reoccupation of the NES during the Middle-Late Pleistocene. However, we consider that the Holocene provides a valid "proof of concept" for relating human occupation and climate variability in the NES at previous times, especially when one considers that all Middle-Late Pleistocene "green Sahara" periods are linked to higher insolation peaks (and hence to more humid conditions) than those that prevailed during the Early-Middle Holocene.

Aterian sites in the NES, which range between 40 and 80 ka (Cremaschi et al. 1998; Smith et al. 2004; Barich et al. 2006; Garcea and Giraudi 2006), are associated with either tufa deposits from the Western Desert of Egypt and the Jebel Gharbi region of Northwest Libya (Smith et al. 2004; Barich et al. 2006; Garcea and Giraudi 2006) or with aeolian sands in the Tadrart Acacus range in Southwest Libya (Garcea 2004). The Site 967 dust record indicates that, between 75 and 20 ka, the NES was dominated by an

especially severe hyperarid climate that was conditioned by a weakened monsoon circulation at the time of lowest boreal insolation maxima. The presence of Aterian remains only near tufa deposits and in mountain areas under a prevailing hyperarid climate supports the view that Aterian groups were especially well-adapted to desert landscapes (Garcea 2004), provided that water was available in isolated areas that functioned as ecological refugia.

## Implications for the Origin and Population Dynamics of Modern Humans in North Africa

One of the key issues regarding the origin of modern humans is the evolution of *H. sapiens* as a distinctive species. Climate variability has been considered as one of the factors that might have influenced the speciation of

*H. sapiens*, but to date, no detailed causal paleoenviron-mental scenario has been proposed. Here, we argue that environmental variability in the Sahara, and in the NES in particular, might have played a critical role in driving the speciation of *H. sapiens*, because it is the region that has witnessed the most dramatic environmental variations within tropical Africa.

*H. sapiens* is often divided into "modern *H. sapiens*" (also known as "recent *H. sapiens*" or "*H. sapiens sensu stricto*"), which includes all living *H. sapiens* and their closest past relatives, and "archaic *H. sapiens*," which includes members of the stem group that are more closely related to "recent *H. sapiens*" than to *H. heidelbergensis* (McBrearty and Brooks 2000; Stringer 2002). The earliest fossil remains of "modern *H. sapiens*" are those from the Kibish Formation in southern Ethiopia. They date back to ca. 195 ka, and are related to an insolation-driven wet period around Lake Turkana that has been tentatively cor-related to sapropel 7 (McDougall et al. 2005). The earliest fossil remains of "archaic *H. sapiens*," formerly attributed to *H. helmei* (e.g., Lahr and Foley 1998), might date back to ca. 300 ka (Bräuer et al. 1997), with the earliest well-dated fossil being those of Florisbad (South Africa) ca. 260 ka (Grün et al. 1996). If, as suggested by Stringer (2002) on anatomical grounds, "archaic *H. sapiens*" is cladistically included within *H. sapiens*, the origin of our own species (*H. sapiens* hereafter) was around 260–300 ka, coinciding with the appearance of MSA technology (McBrearty and Brooks 2000; Stringer 2002).

According to the Site 967 dust record, "green Sahara" periods cluster at around 100 kyr eccentricity maxima centered at 315, 215, and 115 ka (Fig. 2.4). Within each of these clusters, short (<5–10 kyr) "green Sahara" periods in the NES were separated by relatively longer (10–15 kyr) hyperarid periods. On the contrary, during 100 kyr eccen-tricity minima at around 270, 150, and 45 ka, hyperarid desert conditions prevailed uninterrupted for longer time periods (>40 kyr). Especially significant is the fact that the time span between 260 and 300 ka, when *H. sapiens* most likely speciated, is centered around a long (ca. 90 kyr) period of hyperarid conditions in the NES that was only interrupted at ca. 285 ka by a brief "green Sahara" period that is related to the oxidized sapropel at 13.5 rmcd (Fig. 2.3). Prolonged hyperarid conditions in the NES in response to a long-term weakened monsoon, and simulta-neous relatively drier conditions in sub-Saharan Africa (see McDougall et al. 2005; Basell 2008), might have driven the fragmentation of habitats suitable for human occupation throughout most of tropical Africa. Between 325 and 290 ka, and between 280 and 225 ka, the NES (and the whole Sahara Desert) was most likely inhospitable except near ecological refugia associated with aquifer-related spring activity, mountain areas, and permanent rivers such

as the Nile. Any human population inhabiting the Sahara during the "green Sahara" episodes either ca. 330 or 285 ka might have been subsequently forced to migrate into these ecological refugia, to the Mediterranean or Red Sea coastal areas, or to the South, in search of more hospitable habitats (Fig. 2.6a). In sub-Saharan Africa, an insolation-driven change to relatively drier conditions (see McDougall et al. 2005) would have conditioned the contraction of autoch-tonous human populations around favorable habitats such as lake basin and mountain areas (Basell 2008), where they might have had to compete with immigrant populations. Prolonged habitat fragmentation, coupled with isolation in ecological refugia, adaptation to overall drier habitats, and competition between different human groups provide the optimum conditions for the accentuation of any genetic difference between separate groups, which might be pushed to the point of an effective reproductive isolation and thus lead to the emergence of *H. sapiens* as a distinct species. Since habitat fragmentation and isolation in ecological refugia was common throughout North and East Africa, it is difficult at present to make inferences on the specific region where the emergence of *H. sapiens* took place. Subsequent changes towards wetter conditions during the "green Sahara" episodes ca. 330 and 285 ka might account for an eventual rapid expansion of the most successful human species, *H. sapiens*, together with its technological inno-vation (e.g., MSA industries), through tropical Africa (Fig. 2.6b). Subsequent repeated expansions and contrac-tions of landscapes suitable for human occupation, driven by monsoon dynamics, would have favored recurrent con-tact between different African human populations, which provides support for models of modern human origins that advocate for a coalescence of genetic attributes within Africa (see Stringer 2002).

In addition to its potential implications for the emer-gence of *H. sapiens*, climate variability might have also had an important impact on *H. sapiens* population dynamics. Previous studies have demonstrated that an initial dispersal of *H. sapiens* into Southwest Asia through the NES and the Levant occurred ca. 125 ka (Bar-Yosef 1998; Lahr and Foley 1998; Grün et al. 2005). The Site 967 dust record suggests that climatic conditions favorable for the expan-sion of *H. sapiens* throughout the NES occurred more fre-quently, e.g., during at least 8 "green Sahara" episodes after 300 ka (Fig. 2.4). At these times, the subtropical savannah was connected with the narrow band of mild climate around the Mediterranean coast by a fringe of semiarid climate in such a way that sub-Saharan human populations could reach the Mediterranean coast of North Africa not only by fol-lowing the Nile River or the Red Sea coast, but also by crossing the Sahara. Although trans-Saharan crossing was possible throughout any location (Fig. 2.6b), preferential routes were the then-active rivers that drained the Tadrat

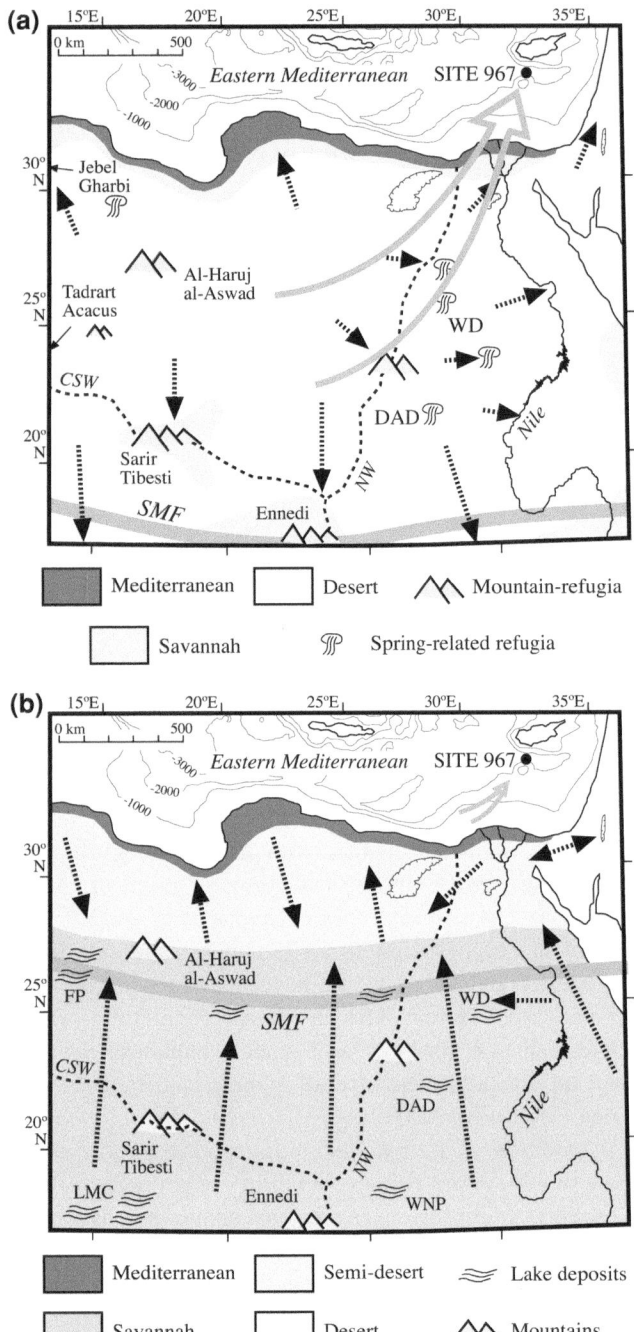

Fig. 2.6 Schematic reconstruction of environmental scenarios in the northeastern Sahara during: **a** "desert Sahara" state; **b** "green Sahara" state (e.g., 215, 195, 170, 125, 100, and 80 ka). SMF denotes the position of the boreal summer monsoon front (Gasse 2000). CSW and NW denote the central Saharan and Nile watersheds, respectively. The thickness of the *gray arrows* indicates the relative importance of dust supply into the eastern Mediterranean. *Dashed arrows* indicate potential dispersal routes for human populations. *WD* Western Desert, *DAD* Darb el Arba'in Desert, *WNP* West Nubian paleolake (Hoelzmann et al. 2000), *LMC* lake Mega Chad (Leblanc et al. 2006), *FP* Fezzan paleolake (Geyh and Thiedig 2008)

Acacus, Sarir Tibesti, Ennedi and Gilf Kebir massifs into the eastern Mediterranean (Fig. 2.1) (Drake et al. 2008; Osborne et al. 2008). The reason why only the "green Sahara" period around 125 ka witnessed a human expansion a step further across the Levant might lie in the fact that, only at that time, fully interglacial climate conditions operating in the Levant coincided with intensified monsoon conditions in tropical Africa in such a way that a "climatic window" was opened between Africa and Asia (Vaks et al. 2007). If so, similarly suitable conditions for previous "out of Africa" dispersals of humans also occurred at least at ca. 330 and 195 ka, coinciding with interglacial 9/green Sahara 30 and interglacial 7a-green Sahara 18, respectively (Fig. 2.4).

Through its impact on population dynamics, climate variability over the Sahara might also be behind the successful "out of Africa" migration of *H. sapiens* ca. 60 ka (Mellars 2006). Thus, massive depopulation of the Sahara following the end of the "green Sahara" period ca. 80 ka might have conditioned a rapid migration of Saharan populations into desert ecological refugia, coastal areas along the Mediterranean and the Red Seas, and sub-Saharan Africa. This, in turn, might have triggered the important technological, economic, and social changes observed in Africa between 80 and 70 ka (Mellars 2006) through fierce competition for the most favorable, yet far less abundant, habitats and resources. These changes ultimately made possible the successful colonization of Eurasia, which involved crossing the southern Red Sea through the Bab el-Mandeb Strait, seafaring along the Red Sea coast of the Sinai Peninsula, or a continental route through the Sahara and the Sinai Peninsula (see Derricourt 2005; Mellars 2006). Although speculative, this idea of a Saharan-forced "out of Africa" migration of *H. sapiens* should not be dismissed without a careful scrutiny of the paleoclimatic and archaeological Holocene record of the NES. Thus, during the Holocene, all the archaeological evidence demonstrates that a single "green Sahara" period, in which a rapid repopulation of the NES was followed by an "exodus from the desert," is primarily responsible for key events in human history, such as the emergence of the pharaonic civilization along the Nile River and the spread of pastoralism throughout tropical Africa (Fig. 2.5) (Nicoll 2004; Küper and Kröpelin 2006).

A last issue that might be also linked to climate variability through its impact on human population dynamics is that of modern human behavior. According to McBrearty and Brooks (2000), modern behavior resulted from a gradual accumulation of individual behavioral, economic, and technological innovations that are found scattered throughout most African regions, albeit sometimes at different times. Our results suggest that relatively fast (<20–40 kyr) expansions and contractions of human populations within tropical Africa in response to monsoon dynamics and the concomitant changes in landscape composition are a common element of the African Middle-Late Pleistocene. Such repeated expansions and contractions favored recurrent contact between different African populations, which is, in turn, necessary to explain the widespread occurrence of most behavioral, economic, and technological innovations throughout most African locations, from the Sahara to South Africa (McBrearty and Brooks 2000).

## Conclusions

The Middle-Late Pleistocene dust record from Site 967 presented here documents distinctive dust flux minima at ca. 330, 285, 240, 215, 195, 170, 125, 100, and 80 ka. These dust minima are linked to the insolation-driven penetration of the West African summer monsoon front over the NES, which, in turn, resulted in the expansion of subtropical savannah landscapes and suppressed dust production. These so-called green Sahara periods broadly correlate with U-series ages of lacustrine and spring carbonates scattered throughout the NES, which are often associated with Acheulean and Mousterian archaeological sites that attest to widespread occupation of the area during pluvial episodes. In contrast, Aterian sites are linked to spring deposits and mountain areas during a prolonged period of hyperarid climate in the NES, which suggests adaptation to desert conditions. The Site 967 dust record has important implications for understanding the evolution and population dynamics of modern humans in Africa. Alternation of "green Sahara" periods with hyperarid desert conditions in North Africa, coupled with simultaneous climate variability in sub-Saharan Africa, provide a scenario of fragmented habitats, isolation in ecological refugia, and competition for resources that would have favored speciation of *H. sapiens* 260–300 ka. Subsequent repeated expansion and contraction of subtropical savannah landscapes within Africa would have favored the coalescence of genetic and behavioral attributes of different human populations, providing the context for a gradual accumulation of African modern behaviors as a whole. Moreover, the disappearance of the Sahara during "green Sahara" periods might have made possible frequent out of Africa dispersals of human populations, whose success might have been ultimately controlled by climate conditions in the Levant. Finally, a sudden expansion of the Sahara in the latest Pleistocene ca. 75 ka, and a concomitant "exodus from the desert," might have triggered the important technological, economic, and

social changes that made the successful out of Africa dispersal of modern humans ca. 60 ka possible.

**Acknowledgments** I thank Jean-Jacques Hublin and Shannon McPherron, organizers of the meeting "Modern Origins: A North African Perspective" (Leipzig, August 2007) and editors of this volume, for making my attendance to the meeting and the publication of this contribution possible. I am also very grateful to Ana Moreno, who also facilitated my participation in the meeting, and to three anonymous reviewers, whose constructive comments and criticism greatly improved this manuscript. Samples were provided by the ODP, which is sponsored by the U.S. National Science Foundation and participating countries (including the U.K.) under management of Joint Oceanographic Institutions, Inc.

## References

Barich, B. A., Garcea, E. A. A., & Giraudi, C. (2006). Between the Mediterranean and the Sahara: Geoarchaeological reconnaissance in the Jebel Gharbi, Libya. *Antiquity, 80,* 567–582.

Bar-Yosef, O. (1998). On the nature of transitions: The Middle to Upper Palaeolithic and the Neolithic revolution. *Cambridge Archaeological Journal, 8,* 141–163.

Basell, L. S. (2008). Middle Stone Age (MSA) distributions in eastern Africa and their relationship to Quaternary environmental change, refugia and the evolution of *Homo sapiens. Quaternary Science Reviews, 27,* 2484–2498.

Bozzano, G., Kuhlmann, H., & Alonso, B. (2002). Storminess control over African dust input to the Moroccan Atlantic margin (NW Africa) at the time of maximal boreal summer insolation: A record of the last 220 kyr. *Palaeogeography, Palaeoclimatology, Palaeoecology, 183,* 155–168.

Bräuer, G., Yokoyama, Y., Falguères, C., & Mbua, E. (1997). Modern human origins backdated. *Nature, 386,* 337–338.

Brovkin, V., Claussen, M., Petoukhov, V., & Ganopolski, A. (1998). On the stability of the atmosphere-vegetation system in the Sahara/Sahel region. *Journal of Geophysical Research, 103,* 31613–31624.

Calvert, S. E., & Fontugne, M. R. (2001). On the late Pleistocene-Holocene sapropel record of climatic and oceanographic variability in the eastern Mediterranean. *Paleoceanography, 16,* 78–94.

Cremaschi, M., Di Lernia, S., & Garcea, E. A. A. (1998). Some insights on the Aterian in the Libyan Sahara: Chronology, environment, and archaeology. *African Archaeological Review, 15,* 261–286.

Crombie, M. K., Arvidson, R. E., Syurchio, N. C., Alfy, Z. E., & Zeid, K. A. (1997). Age and isotopic constraints on Pleistocene pluvial episodes in the Western Desert, Egypt. *Palaeogeography, Palaeoclimatology, Palaeoecology, 130,* 337–355.

Dayan, U., Heffter, J., Miller, J., & Gutman, G. (1991). Dust intrusions into the Mediterranean basin. *Journal of Applied Meteorology, 30,* 1185–1199.

de Menocal, P. B. (1995). Plio-Pleistocene African climate. *Science, 270,* 53–59.

de Menocal, P. B. (2004). African climate change and faunal evolution during the Pliocene-Pleistocene. *Earth and Planetary Science Letters, 220,* 3–24.

Derricourt, R. (2005). Getting "Out of Africa": Sea crossings, land crossings and culture in the Hominin migrations. *Journal of World Prehistory, 19,* 119–132.

Dinarès-Turell, J., Hoogakker, B. A. A., Roberts, A. P., Rohling, E. J., & Sagnotti, L. (2003). Quaternary climatic control of biogenic magnetite production and eolian dust input in cores from the Mediterranean Sea. *Palaeogeography, Palaeoclimatology, Palaeoecology, 195,* 195–209.

Drake, N. A., El-Hawat, A. S., Turner, P., Armitage, S. J., Salem, M. J., White, K. H., et al. (2008). Palaeohydrology of the Fazzan Basin and surrounding regions: The last 7 million years. *Palaeogeography, Palaeoclimatology, Palaeoecology, 263*, 131–145.

Emeis, K. C., Sakamoto, T., Wehausen, R., & Brumsack, H. J. (2000). The sapropel record of the eastern Mediterranean Sea—results of Ocean Drilling Program Leg 160. *Palaeogeography, Palaeoclimatology, Palaeoecology, 158*, 371–395.

Emeis, K. C., Schulz, H., Struck, U., Rossignol-Strick, M., Erlenkeuser, H., Howell, M. W., et al. (2003). Eastern Mediterranean surface water temperatures and $\delta^{18}O$ composition during deposition of sapropels in the late Quaternary. *Paleoceanography, 18.* doi: 10.1029/2000PA000617

Fontugne, M., Arnold, M., Labeyrie, L., Calvert, S. E., Paterne, M., & Duplessy, J. C. (1994). Palaeoenvironment, sapropel chronology and Nile River discharge during the last 20,000 years as indicated by deep sea sediment records in the eastern Mediterranean. *Radiocarbon, 34*, 75–88.

Foucault, A., & Mélières, F. (2000). Palaeoclimatic cyclicity in central Mediterranean Pliocene sediments: The mineralogical signal. *Palaeogeography, Palaeoclimatology, Palaeoecology, 158*, 311–323.

Garcea, E. A. A. (2004). Crossing desserts and avoiding seas: Aterian North Africa-European relations. *Journal of Anthropological Research, 60*, 27–53.

Garcea, E. A. A., & Giraudi, C. (2006). Late Quaternary human settlement patterning in the Jebel Gharbi. *Journal of Human Evolution, 51*, 411–421.

Gasse, F. (2000). Hydrological changes in the African tropics since the Last Glacial Maximum. *Quaternary Science Reviews, 19*, 189–211.

Geyh, M. A., & Thiedig, F. (2008). The Middle Pleistocene Al Mahrúqah Formation in the Murzub Basin, northern Sahara, Libya; evidence for orbitally-forced humid episodes during the last 500,000 years. *Palaeogeography, Palaeoclimatology, Palaeoecology, 257*, 1–21.

Goudie, A. S., & Middleton, N. J. (2001). Saharan dust storms: Nature and consequences. *Earth Science Reviews, 56*, 179–204.

Grün, R., Brink, J. S., Spooner, N. A., Taylor, L., Stringer, C. B., Franciscus, R. G., et al. (1996). Direct dating of Florisbad hominid. *Nature, 382*, 500–501.

Grün, R., Stringer, C., McDermott, F., Nathan, R., Porat, N., Robertson, S., et al. (2005). U-series and ESR analyses of bones and teeth relating to the human burials from Skhul. *Journal of Human Evolution, 49*, 316–334.

Hamann, Y., Ehrmann, W., Schmiedl, G., Krüger, S., Stuut, J. B., & Kuhnt, T. (2008). Sedimentation processes in the eastern Mediterranean Sea during the Late Glacial and Holocene revealed by end-member modelling of the terrigenous fraction in marine sediments. *Marine Geology, 248*, 97–114.

Haynes, C. V., Jr., Maxwell, T. A., El Hawary, A., Nicoll, K. A., & Stokes, S. (1997). An Acheulian site near Bir Kiseiba in the Darb el Arba'in Desert, Egypt. *Geoarchaeology—An Intenational Journal, 12*, 819–832.

Hilgen, F. J. (1991). Astronomical calibration of Gauss to Matuyama sapropels in the Mediterranean and implications for the geomagnetic polarity time scale. *Earth and Planetary Science Letters, 104*, 226–244.

Hill, C. L. (2001). Geologic context of the Acheulian (Middle Pleistocene) in the eastern Sahara. *Geoarchaeology—An Intenational Journal, 16*, 65–94.

Hoelzmann, P., Kruse, H. J., & Rottinger, F. (2000). Precipitation estimates for the eastern Sahara paleomonsoon based on a water balance model of the West Nubian Paleolake Basin. *Global and Planetary Change, 26*, 105–120.

Hooghiemstra, H., Lézine, A. M., Leroy, S. A. G., Dupont, L., & Marret, F. (2006). Late Quaternary palynology in marine sediments: A synthesis of the understanding of pollen distribution patterns in the NW African setting. *Quaternary International, 148*, 29–44.

Hovers, E., & Kuhn, S. (Eds.). (2006). *Transitions before the transitions: Evolution and stability in the Middle Paleolithic and Middle Stone Age.* New York: Springer.

Imbrie, J., Hays, J. D., Martinson, D. G., McIntyre, A., Mix, A. C., Morley, J. J., et al. (1984). The orbital theory of Pleistocene climate: Support from a revised chronology of the marine 6180 record. In A. Berger, J. Imbrie, J. Hays, G. Kukla, & B. Saltzman (Eds.), *Milankovitch and climate, Part 1* (pp. 269–305). Dordrecht: Plenum Reidel.

Itambi, A. C., von Dobeneck, T., Mulitza, S., Bickert, T., & Heslop, D. (2009). Millennial-scale Northwest African droughts related to Heinrich events and Dansgaard-Oeschger cycles: Evidence in marine sediments from offshore Senegal. *Paleoceanography, 24.* doi:10.1029/2007PA001570

Jolly, D., Prentice, I. C., Bonnefille, R., Ballouche, A., Bengo, M., Brenac, P., et al. (1998). Biome reconstruction from pollen and plant macrofossils data for Africa and the Arabian Peninsula at 0 and 6000 years. *Journal of Biogeography, 25*, 1007–1027.

Kieniewicz, J. M., & Smith, J. R. (2007). Hydrologic and climatic implications of stable isotope and minor element analyses of authigenic calcite silts and gastropod shells from a mid-Pleistocene pluvial lake, Western Desert, Egypt. *Quaternary Research, 68*, 431–444.

Kleindienst, M. R., Schwarcz, H. P., Nicoll, K., Churcher, C. S., Frizano, J., Giegengack, R., et al. (2008). Water in the desert: First report on Uranium-series dating of Caton-Thompson's and Gardner's "classic" Pleistocene sequence at Refuf Pass, Kharga Oasis. In M.F. Wiseman (Ed.), *Oasis Papers II: Proceedings of the Second Dakhleh Oasis Project Research Seminar* (pp. 25–54). Oxford: Oxbow Books.

Kowalski, K., Vanneer, W., Bochenski, Z., Mlynarski, M., Rzebikkowalska, B., Szyndlar, Z., et al. (1989). A last interglacial fauna from the eastern Sahara. *Quaternary Research, 32*, 335–341.

Krom, M. D., Cliff, R. A., Eijsink, L. M., Herut, B., & Chester, R. (1999). The characterisation of Saharan dust and Nile particulate matter in surface sediments from the Levantine basin using Sr isotopes. *Marine Geology, 155*, 319–330.

Kroon, D., Alexander, I., Little, M., Lourens, L.J., Matthewson, A., Robertson, A.H.F., et al. (1998). Oxygen isotope and sapropel stratigraphy in the eastern Mediterranean during the last 3.2 million years. In A.H.F. Robertson, K.C. Emeis, C. Richter & A. Camerlenghi, A. (Eds.), *Proceedings of the Ocean Drilling Program, Scientific Results* (vol. 160, pp. 181–190). College Station, TX: Ocean Drilling Program.

Küper, R., & Kröpelin, S. (2006). Climate-controlled Holocene occupation in the Sahara: Motor of Africa's evolution. *Science, 313*, 803–807.

Lahr, M. M., & Foley, R. A. (1998). Towards a theory of modern humans origins: Geography, demography, and diversity in recent human evolution. *Yearbook of Physical Anthropology, 41*, 137–176.

Larrasoaña, J. C., Roberts, A. P., Rohling, E. J., Winklhofer, M., & Wehausen, R. (2003). Three million years of monsoon variability over the northern Sahara. *Climate Dynamics, 21*, 689–698.

Larrasoaña, J. C., Roberts, A. P., Hayes, A., Wehausen, R., & Rohling, E. J. (2006). Detecting missing beats in the Mediterranean climate rhythm from magnetic identification of oxidized sapropels (Ocean Drilling Program Leg 160). *Physics of the Earth and Planetary Interiors, 156*, 283–293.

Laskar, J., Robutel, P., Joutel, F., Gastineau, M., Correia, A. C. M., & Levrard, B. (2004). A long-term numerical solution for the insolation quantities of the Earth. *Astronomy and Astrophysics, 428*, 261–285.

Leblanc, M. L., Leduc, C., Stagnitti, F., van Oevelen, P. J., Jones, C., Mofor, L. A., et al. (2006). Evidence for Megalake Chad, North-Central Africa, during the late Quaternary from satellite data. *Palaeogeography, Palaeoclimatology, Palaeoecology, 230*, 230–242.

Lourens, L. J., Antonarakou, A., Hilgen, F. J., Van Hoof, A. A. M., Vergnaud Grazzini, C., & Zachariasse, W. J. (1996). Evaluation of the Plio-Pleistocene astronomical timescale. *Paleoceanography, 11*, 391–413.

Lourens, L. J., Wehausen, R., & Brumsack, H. J. (2001). Geological constraints on tidal dissipation and dynamical ellipticity of the earth over the past three million years. *Nature, 409*, 1029–1033.

Mandel, R. D., & Simmons, A. H. (2001). Prehistoric occupation of Late Quaternary landscapes near Kharga Oasis, Western Desert of Egypt. *Geoarchaeology - An Intenational Journal, 16*, 95–117.

Matthewson, A. P., Shimmield, G. B., Kroon, D., & Fallick, A. E. (1995). A 300 kyr high-resolution aridity record of the North Africa continent. *Paleoceanography, 10*, 677–692.

Maxwell, T. A., & Haynes, C. V., Jr. (2001). Sand sheets dynamics and Quaternary landscape evolution of the Selima Sand Sheet, southern Egypt. *Quaternary Science Reviews, 20*, 1623–1647.

McBrearty, S., & Brooks, A. S. (2000). The revolution that wasn't: A new interpretation of the origin of modern human behaviour. *Journal of Human Evolution, 39*, 453–563.

McDougall, I., Brown, F. H., & Fleagle, J. G. (2005). Stratigraphic placement and age of modern humans from Kibish, Ethiopia. *Nature, 433*, 733–736.

McHugh, P. M., Breed, C. S., Schaber, G. S., McCauley, J. F., & Szabo, B. J. (1988). Acheulian sites along the "radar rivers", southern Egyptian Sahara. *Journal of Field Archaeology, 15*, 361–379.

Mellars, P. (2006). Why did human populations disperse from Africa ca. 60,000 years ago? A new model. *Proceedings of the National Academy of Sciences of the USA, 103*, 9381–9386.

Mercone, D., Thomson, J., Croudance, I. M., Siani, G., Paterne, M., & Troelstra, S. (2001). Duration of S1, the most recent sapropel in the eastern Mediterranean Sea, as indicated by accelerator mass spectrometry radiocarbon and geochemical evidence. *Paleoceanography, 15*, 336–347.

Mithen, S., & Reed, M. (2002). Stepping out: A computer simulation of hominid dispersal from Africa. *Journal of Human Evolution, 43*, 433–462.

Moreno, A., Targarona, J., Henderiks, J., Canals, M., Freudenthal, T., & Meggers, H. (2001). Orbital forcing of dust supply to the North Canary Basin over the last 250 kyr. *Quaternary Science Reviews, 20*, 1327–1339.

Moreno, A., Cacho, I., Canals, M., Prins, M. A., Sanchez-Goñi, M. F., Grimalt, J. O., et al. (2002). Saharan dust transport and high-latitude glacial climatic instability: The Alboran Sea record. *Quaternary Research, 58*, 318–328.

Nicoll, K. (2004). Recent environmental change and prehistoric human activity in Egypt and northern Sudan. *Quaternary Science Reviews, 23*, 561–580.

Osborne, A. H., Vance, D., Rohling, E. J., Barton, N., Rogerson, M., & Fello, N. (2008). A humid corridor across the Sahara for the migration "Out of Africa" of early modern humans 120,000 years ago. *Proceedings of the National Academy of Sciences of the USA, 105*, 16444–16447.

Osmond, J. K., & Dabous, A. A. (2004). Timing and intensity of groundwater movement during Egyptian Sahara pluvial periods by U-series analysis of secondary U in ores and carbonates. *Quaternary Research, 61*, 85–94.

Pachur, H. J., & Hoelzmann, P. (2000). Late Quaternary paleoecology and paleoclimates of the eastern Sahara. *Journal of African Earth Sciences, 30*, 929–939.

Petit-Maire, N. (2002). *Sous le Sable… des lacs; un voyage dans le temps*. Paris: CNRS Éditions.

Prospero, J.M., Ginoux, P., Torres, O., Nicholson, S.E., & Gill, T.E. (2002). Environmental characterization of global sources of atmospheric soil dust identified with the Nimbus 7 Total Ozone Mapping Spectrometer (TOMS) absorbing aerosol product. *Reviews of Geophysics, 40*, Art. No. 1002.

Rohling, E. J., Cane, T. R., Cooke, S., Sprovieri, M., Bouloubassi, I., Emeis, K. C., et al. (2002). African monsoon variability during the previous interglacial maximum. *Earth and Planetary Science Letters, 202*, 61–75.

Rossignol-Strick, M. (1983). African monsoons, an immediate climate response to orbital insolation. *Nature, 304*, 46–49.

Smith, J.R. (2012). Spatial and temporal variation in the nature of Pleistocene pluvial phase environments across North Africa. In J.-J. Hublin & S.P. McPherron (Eds.), *Modern origins: A North African perspective*. Dordrecht: Springer.

Smith, J. R., Giegengack, R., Schwarcz, H. P., McDonald, M. M. A., Kleindienst, M. R., Hawkins, A. L., et al. (2004). A reconstruction of Quaternary pluvial environments and human occupations using stratigraphy and geochronology of fossil-spring tufas, Kharga Oasis, Egypt. *Geoarchaeology, 19*, 1–34.

Smith, J. R., Hawkins, A. L., Asmerom, Y., Polyac, V., & Giegengack, R. (2007). New age constraints on the Middle Stone Age occupations of Kharga Oasis, Western Desert, Egypt. *Journal of Human Evolution, 52*, 690–701.

Stokes, S., Maxwell, T. A., Haynes, C. V., Jr., & Horrocks, J. L. (1998). Latests Pleistocene and Holocene sand-sheet construction in the Selima Sand Sea, Eastern Sahara. In A. S. Alsharhan, K. W. Glennie, G. L. Whittle, & C. G. S. C. Kendall (Eds.), *Quaternary deserts and climate change* (pp. 175–183). Rotterdam: AA Balkema.

Stringer, C. (2002). Modern human origins: Progress and prospects. *Philosophical Transactions of the Royal Society of London Series B - Biological Sciences, 357*, 563–579.

Sultan, M., Sturchio, N., Hassan, F. A., Hamdan, M. A. R., Mahmood, A. M., Alfy, Z. E., et al. (1997). Precipitation source inferred from stable isotopic composition of Pleistocene groundwater and carbonate deposits in the Western Desert of Egypt. *Quaternary Research, 48*, 29–37.

Swezey, C. (2001). Eolian sediment responses to late Quaternary climate changes: Temporal and spatial patterns in the Sahara. *Palaeogeography, Palaeoclimatology, Palaeoecology, 167*, 119–155.

Szabo, B. J., Haynes, C. V., Jr., & Maxwell, T. A. (1995). Ages of Quaternary pluvial episodes determined by uranium-series and radiocarbon dating of lacustrine deposits of the eastern Sahara. *Palaeogeography, Palaeoclimatology, Palaeoecology, 113*, 227–242.

Tiedemann, R., Sarnthein, M., & Shackleton, N. J. (1994). Astronomic timescale for the Pliocene Atlantic $\delta^{18}O$ and dust flux records of Ocean Drilling Program site 659. *Paleoceanography, 9*, 619–638.

Tjallingii, R., Claussen, M., Stuut, J. B. W., Fohlmeister, J., Jahn, A., Bickert, T., et al. (2008). Coherent high- and low-latitude control of the northwest African hydrological balance. *Nature Geosciences, 1*, 670–675.

Tomadini, L., Lenaz, R., Landuzzi, V., Mazzucotelli, A., & Vannucci, R. (1984). Wind-blown dust over the Central Mediterranean. *Oceanologica Acta, 7*, 13–23.

Trauth, M. H., Larrasoaña, J. C., & Mudelsee, M. (2009). Trends, rhythms and events in Plio-Pleistocene African climate. *Quaternary Science Reviews, 28*, 399–411.

Vaks, A., Bar-Matthews, M., Ayalon, A., Matthews, A., Halicz, L., & Frumkin, A. (2007). Desert speleothems reveal climatic window for African exodus of modern humans. *Geology, 35*, 831–834.

Weldeab, S., Emeis, K. C., Hemleben, C., & Siebel, W. (2002). Provenance of lithogenic surface sediments and pathways of riverine suspended matter in the eastern Mediterranean Sea: Evidence from $^{143}$Nd/$^{144}$Nd and $^{87}$Sr/$^{86}$Sr ratios. *Chemical Geology, 186*, 139–149.

Weldeab, S., Lea, D. W., Schneider, R. R., & Andersen, N. (2007). 155,000 years of West African monsoon and thermal ocean evolution. *Science, 316*, 1303–1307.

Wendorf, F., Schild, R., & Close, A. E. (Eds.). (1993). *Egypt during the last Interglacial: The Middle Paleolithic of Bir Tarwafi and Bir Sahara East*. New York: Plenum Press.

# Chapter 3
# Spatial and Temporal Variation in the Nature of Pleistocene Pluvial Phase Environments Across North Africa

J. R. Smith

**Abstract** Pleistocene-aged fluvial, lacustrine, and spring sediments across North Africa record times of enhanced rainfall relative to the present. Much, though not all of the preserved record, is found near modern oases or extant (sometimes seasonal) bodies of water. Pluvial phase indicators point to a variety of environmental conditions during humid events; reconstructed Pleistocene lakes range in size from several square kilometers to tens of thousands of square kilometers, and from long-lived perennial freshwater lakes, to large brackish lakes, to seasonal playas. Spring-fed wetlands would also have been a relatively common feature of pluvial phase environments. In most localities that record more than one pluvial phase, sedimentological and geochemical evidence suggests humid events following ~70 ka were less intense than those associated with Marine Isotope Stage (MIS) 5, though distributions of dates on these humid phases indicate they were still significant and continent-wide. Comparison of the timing of humid phases across the region is made possible through compilation of U-series, TL, and OSL dates on sediments indicative of relatively wet conditions. Available geochronological data suggest that the western Sahara and Mediterranean North Africa exhibit different climatic variation than the rest of the region, perhaps tied to the influence of Mediterranean rainfall.

**Keywords** Climate • Fluvial environment • Lacustrine environment • Monsoon • Precession • Spatial variation • Water resources

## Introduction

Although it has been known for over a century that North Africa periodically received significantly more rainfall during the Pleistocene than it does today (von Zittel 1883; Ball 1900; Beadnell 1909; Caton-Thompson 1952), we have not yet reached a detailed understanding of the pace or amplitude of Saharan climate change. Unlike marine environmental archives from the Mediterranean, Atlantic, or Red Seas (e.g., Rossignol-Strick 1985; de Menocal et al. 1991; Moreno et al. 2001; Rohling et al. 2002; Arz et al. 2003; Larrasoaña et al. 2003), terrestrial sedimentary records of Pleistocene environments in North Africa are generally discontinuous, and as such, provide only snapshots of climatic variation in the region. However, though they are generally less well-dated than marine archives, they do have the advantage of directly recording the environments available in particular locations. Furthermore, when directly associated with archaeological materials, terrestrial records can provide an immediate link between landscape and human activity (e.g., Caton-Thompson 1952; Wendorf et al. 1993; Kleindienst 1999; Smith et al. 2004; Barich et al. 2006). The sedimentology and geochemistry of pluvial (humid) phase deposits can be used to reconstruct these environments and landscapes, and geochronological analyses can establish the timing of humid events (e.g., Szabo et al. 1989; Wendorf et al. 1993; Szabo et al. 1995; Haynes et al. 1997; Sultan et al. 1997; Ouda et al. 1998; Schwarcz and Morawska 1998; Zouari et al. 1998; Churcher et al. 1999; Nicoll et al. 1999; Mandel and Simmons 2001; Wengler et al. 2002; Smith et al. 2004; Weisrock et al. 2006; Geyh and Thiedig 2008). An examination of terrestrial paleoenvironmental archives can also provide insight into spatial variation in the nature of available water resources within North Africa, as well as temporal variation in environmental conditions at individual locations. Developing an understanding of the timing of humid conditions throughout North Africa would allow for

J. R. Smith (✉)
Department of Earth and Planetary Sciences,
Washington University, 1 Brookings Drive,
Campus Box 1169, St. Louis, MO 63130, USA
e-mail: jensmith@wustl.edu

J.-J. Hublin and S. P. McPherron (eds.), *Modern Origins: A North African Perspective*,
Vertebrate Paleobiology and Paleoanthropology, DOI: 10.1007/978-94-007-2929-2_3,
© Springer Science+Business Media B.V. 2012

the identification of time periods when environments would have been favorable to human occupation. If climate shifted synchronously throughout the region, then North Africa as a whole would have oscillated between being largely habitable and largely uninhabitable throughout the Pleistocene. If, however, environmental change occurred asynchronously, it may have been possible for occupants to find productive areas by migrating within North Africa. Similarly, if successive humid events varied significantly in mean annual precipitation, resource availability and thus human adaptation to Saharan environments may also have been expected to vary.

Here, I will first provide a brief overview of the principal known sedimentary archives of pluvial phases during the Pleistocene, with a focus on the last 300 kyr, then examine variation in the nature and timing of these humid events across North Africa. In order to enable a spatio-temporal comparison of North African environments over a time scale relevant to modern human origins, this discussion will primarily involve directly dated stratigraphic sequences from the Middle to Late Pleistocene. Furthermore, this examination of Pleistocene environmental change will focus on those regions that are currently desert or near desert, and where the effects of climatic changes on habitability would be substantial. Thus, the Nile Valley and coastal margins will not be discussed explicitly. Even if all published literature were to be evaluated here, the vast expanses of North Africa yet unexplored, as well as the likelihood of loss of pluvial phase sedimentary archives to erosion, ensure that our records are incomplete. Thus, it is nearly certain that surface water would have been available more frequently and more extensively throughout the Sahara than can be documented here. Nonetheless, such an analysis can provide an understanding of the minimum extent of water resources in space and time, and also a sense of the variation in landscape settings across the Sahara.

## Paleolandscapes of the Eastern Sahara

The eastern Sahara, while host to the desert's only major river, represents the hyperarid core of the modern desert, particularly in its southern region. Deflational depressions, often structurally controlled throughout Egypt's Western Desert (Said 1990), represent discharge locations for the regional Nubian Aquifer (Thorweihe 1990; Sturchio et al. 2004; Patterson et al. 2005), and as such, are often oases under modern conditions. During humid phases, however, these depressions were sometimes filled with large lakes (McKenzie 1993; Wendorf et al. 1993; Kieniewicz and

Smith 2007; Churcher and Kleindienst, in press), or the sites of intensive spring activity (Caton-Thompson 1952; Crombie et al. 1997; Sultan et al. 1997; Kleindienst 1999; Nicoll et al. 1999; Brook et al. 2003; Smith et al. 2004; Kleindienst et al. 2008; Adelsberger and Smith 2010). The extent to which these water bodies were connected by drainage systems within the region is still rather difficult to determine as fluvial channels outside of oases are frequently only detected by remote sensing (e.g., McCauley et al. 1982; Robinson 2002).

## Lakes of the Eastern Sahara

During the Middle Pleistocene, several large lakes existed in central and southern Egypt. The Bir Tarfawi/Bir Sahara East region (Fig. 3.1) contains sediments recording a series of discrete lake phases ca. 230–60 ka, with Marine Isotope Stage (MIS) 5 as the wettest interval (Wendorf et al. 1993). Sedimentology and microfaunal analyses of the lacustrine sediments record significant fluctuations in salinity and water depth throughout this time range (Wendorf et al. 1993), but the geochemistry of the lake sediments also supports the persistence of a perennial lake for periods of ~7 kyr (McKenzie 1993). The Dakhleh Oasis depression also hosted several large lakes (and perhaps, at times, one very large lake), with a potential combined area of ~1700 km$^2$, also most likely during MIS 5 (Osinski et al. 2007; Kieniewicz and Smith 2009; Churcher and Kleindienst, in press). While sediments indicating deep water conditions are recorded from the Dakhleh paleolakes, much of the preserved stratigraphy indicates relatively shallow water, supporting aquatic vegetation. The highly irregular, yardanged topography which made up the lake floor (as documented by DGPS elevation measurements on the contact between pre-Quaternary bedrock and lacustrine sediments) would have resulted in substantial spatial variations in lake depth (Kieniewicz and Smith 2009; Churcher and Kleindienst, in press).

There are significant similarities between the environments recorded at Dakhleh and those at Bir Tarfawi/Bir Sahara East. Both appear to have been spatially heterogeneous, with deeper water bodies separated by marshy regions and at times, isolated, emergent spring mounds. The large Dakhleh paleolake, though, shows somewhat less obvious temporal variation, without evidence for substantial, abrupt changes in water depth throughout its existence (Kieniewicz and Smith 2009; Churcher and Kleindienst, in press). The generally savannah-adapted faunas recovered from both lakes suggest a semi-arid climate (Wendorf et al. 1993; Churcher et al. 1999). Energy

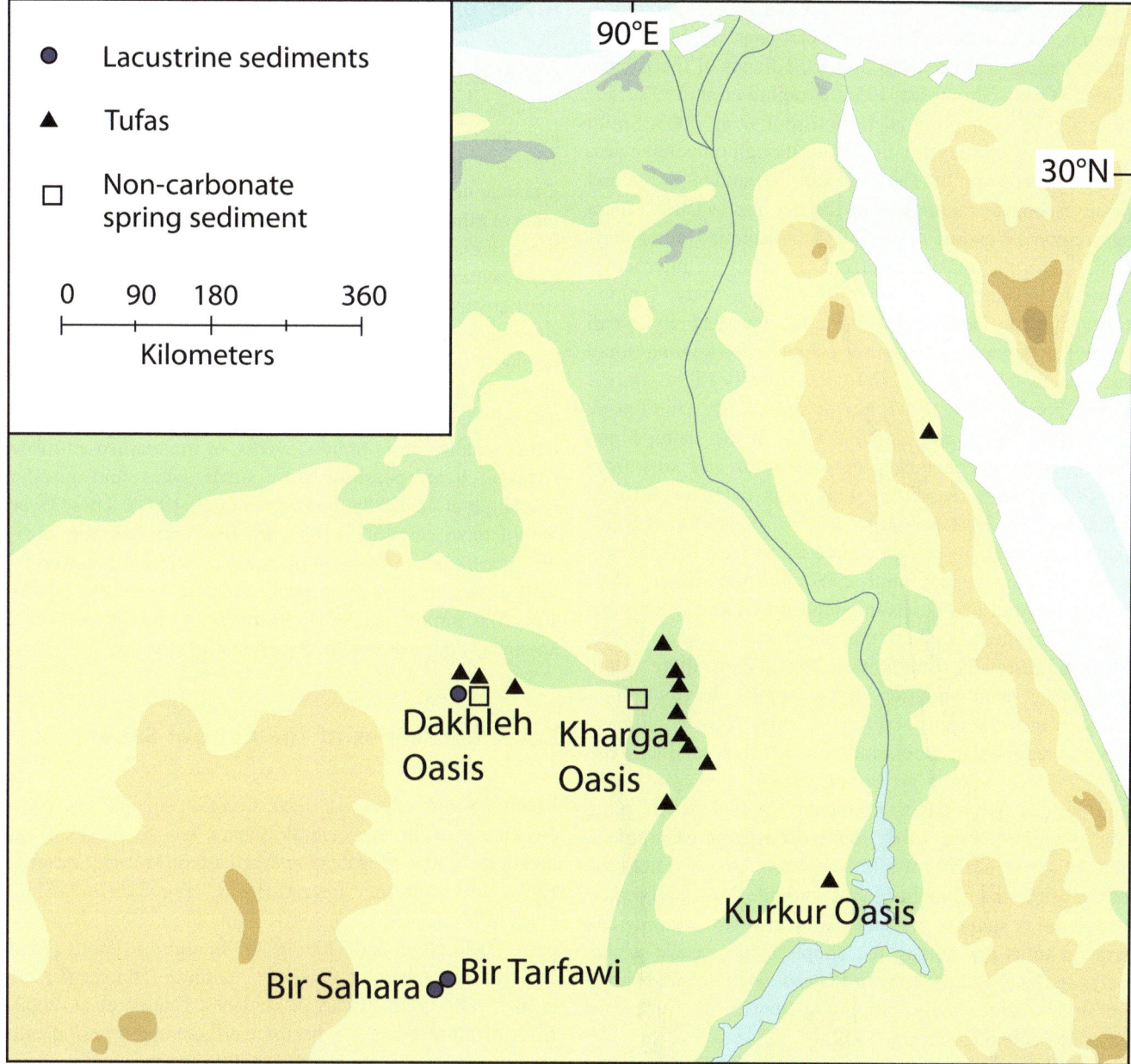

**Fig. 3.1** Eastern Saharan localities mentioned in the text

balance modeling for the Dakhleh paleolake suggests that at least 400 mm of rainfall would be required annually to support this lake, a number consistent with faunal evidence (Kieniewicz and Smith 2009).

Sedimentary evidence also exists in the Western Desert for ephemeral or relatively localized lacustrine settings; evaporite-rich, laminated sediments indicate the presence of most likely seasonal water bodies east of and within Dakhleh Oasis (Brookes 1993a, b). In southernmost Egypt and northern Sudan, in the vicinity of Bir Tarfawi/Bir Sahara East and the radar rivers (see below), in addition to fully lacustrine carbonates, Szabo et al. (1995) describe

carbonate sediments that formed from shallow water tables, or in localized ponds in sand sheets.

## Springs of the Eastern Sahara

The archaeological sequence developed by Caton-Thompson (1952) for the eastern Sahara was based on archaeological material associated with spring carbonates in Kharga Oasis. These carbonate sediments, at Kharga and elsewhere, have remained a focus of geological and geoarchaeological research since then. Fossil-spring carbonates, or tufas, are known largely from settings along the flanks of the Libyan Plateau (Caton-Thompson 1952; Crombie et al. 1997; Sultan et al. 1997; Nicoll et al. 1999; Brook et al. 2003; Smith et al. 2004; Kleindienst et al. 2008), though tufas have also been recorded from the Eastern Desert (Hamdan 2000), and spring carbonate mounds from the Bir Tarfawi/Bir Sahara East region (Wendorf et al. 1993). Substantial facies variations in spring tufas suggest that a variety of microenvironments existed during times of spring activity, ranging from stepped, cascading waterfalls separated by small (several m$^2$) pools, to possibly ephemerally-flowing small wadis, to marshy floodplains, to occasionally slightly larger lakes (Crombie et al. 1997; Nicoll et al. 1999; Smith et al. 2004; Kieniewicz and Smith 2007). Stable isotope and trace element geochemistry of gastropods and silts from an ~4 km$^2$ lake associated with tufas dated to MIS 5e indicate that at least one larger lake in Kharga Oasis was a stable, perennial freshwater source for at least several millennia (Kieniewicz and Smith 2007). Gastropod fauna from smaller (several m$^2$) ponds impounded by tufas at the same locality, however, suggest that these minor water bodies dried up seasonally (Smith et al. 2004). Thus, there would likely have been variation in the spatial distribution of reliable water resources at different scales. This would have occurred seasonally, within individual loci as smaller water sources dried up but larger ones remained, then over longer time periods (centuries to millennia), as regions of spring activity shifted. Particularly along the Kharga escarpment, areas of spring activity would have been attractive not only for the water and game they could provide, but also because the relatively shallow gradients associated with sloping and terraced tufas are among the simplest and easiest access routes to the top of the Libyan Plateau, where chert-bearing limestones additionally provide a source of lithic raw material (Caton-Thompson 1952).

In addition to the relatively extensive spring-deposited carbonate sediments along the margins of the Libyan Plateau, some oasis depressions in central Egypt also contain more spatially restricted iron-rich spring sediments (Caton-Thompson 1952; Brookes 1993a; Adelsberger and Smith 2010). At least in Dakhleh Oasis, these springs seem to have supported localized wetlands, as evidenced in part by reed casts preserved in spring mound goethite (Adelsberger and Smith 2010). These springs could have been active in relatively arid times, as they are supported by the extensive Nubian Aquifer (e.g., Sturchio et al. 2004), and some are still active today. In both the lacustrine and spring deposits, the frequency of aquatic vegetation suggests that a substantial portion of humid phase landscapes in the oases would have been marshy.

## Rivers of the Eastern Sahara

Gravel terraces along oasis margins record relatively short distance fluvial transport of sediments off the Libyan Plateau and into oasis depressions (Caton-Thompson 1952; Brookes 1993a; Kleindienst 1999), but larger, integrated regional drainage networks are often buried under recent eolian sediment or otherwise characterized by modest topography. Thus, these rivers have been mapped by shuttle-borne and satellite radar, which are capable of penetrating a relatively shallow sand cover (McCauley et al. 1982; Robinson 2002). These river valleys, predominantly studied in southern Egypt and northern Sudan (though also known from Libya; Rohling et al. 2002) were largely filled with sediment during the Tertiary. During the Pleistocene, they would have supported either smaller, inset braided rivers, or maintained a broader riparian zone, possibly with small lakes and marshes (McHugh et al. 1988a, b; Haynes et al. 1997). These rivers would have certainly been attractive corridors for longer distance travels throughout the region, and their discovery, as well as that of associated archaeological materials, emphasizes the importance of archaeological survey in what is apparently deep desert (McHugh et al. 1988a, b).

## Paleolandscapes of the Central Sahara

Much of the effort on reconstruction of pluvial phase environments in the central Sahara has focused on two locations: Lake Megafezzan/Wadi Shati/Murzuq Basin in Libya (hereafter, the "Fezzan Basin," Fig. 3.2; Petit-Maire et al. 1980; Gaven et al. 1981; Ouda et al. 1998; Armitage et al. 2007; Geyh and Thiedig 2008), and the Chotts (Chott el Jerid and Chott el Fejej) of southern Tunisia (Causse et al. 1988, 1989; Zouari et al. 1998; Causse et al. 2003). Both of these extensive basins were occupied periodically by large lakes throughout the Middle Pleistocene.

Geyh and Thiedig (2008) and Armitage et al. (2007) recognize four major episodes of Pleistocene lake formation in the Fezzan Basin; Armitage et al. (2007) found two episodes out of the range of OSL dating (i.e., >~400 ka), one at MIS 11, and another at MIS 5. Geyh and Thiedig (2008) recognize one lake phase out of range of U-series techniques (>425 ka), which they tentatively assign to MIS 11–12, and another at MIS 5, but unlike Armitage et al. (2007), they also find evidence for lacustrine sedimentation during MIS 9 and MIS 7–8. Reconstructed lake sizes for the Fezzan Basin also differ somewhat substantially. Geyh and Thiedig (2008) see a decrease in lake size during successive pluvial phases, from >100,000 km$^2$ in MIS 11–12 to 1,400 km$^2$ during MIS 5, and finally to <100 km$^2$ during

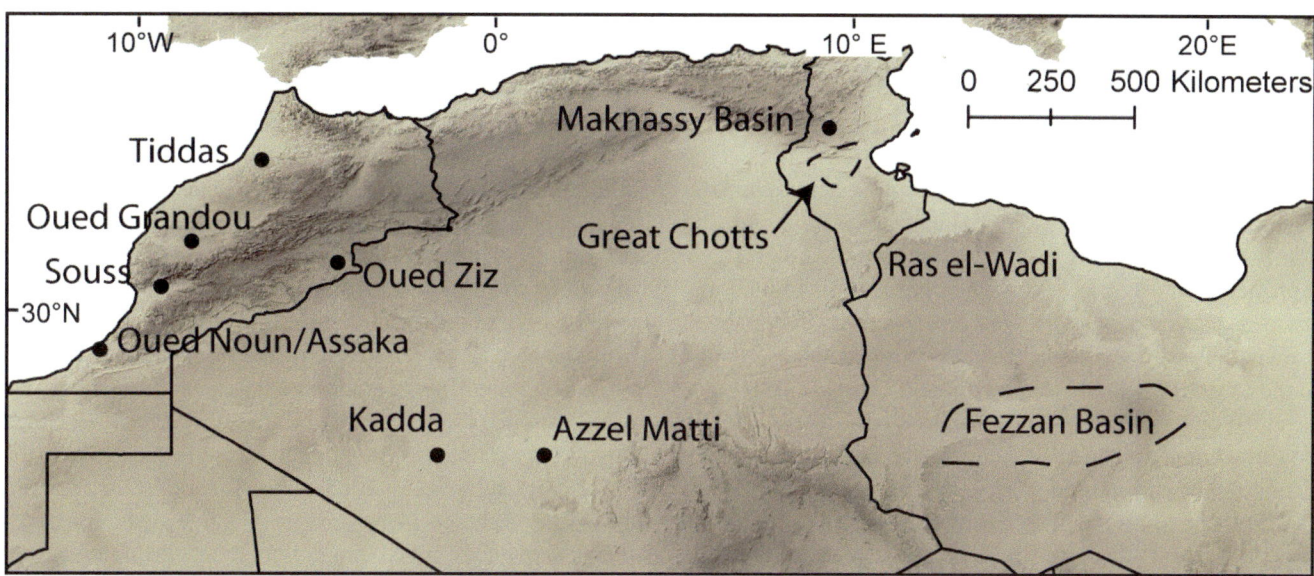

**Fig. 3.2** Central and western Saharan localities mentioned in the text

the Holocene humid event. Armitage et al. (2007) cite the estimate of Brooks et al. (2003), which puts the Holocene lake at 76,250 km$^2$. Thus, there is some disagreement as to the size and timing of lacustrine episodes. However, these studies, as well as prior work by Petit-Maire et al. (1980), present no evidence for significant lacustrine activity between MIS 5 and the Holocene. Based on the presence of *Cardium glaucum* (*Cerastoderma*) and the frequency of crossbedding and disarticulated shells in the deposits associated with the MIS 5 lake phase in Fezzan, this lake at this time is reconstructed as brackish, with strong currents (Geyh and Thiedig 2008).

Highstands of the saline Tunisian Great Chotts occurred due to influxes of continental waters during times of increased rainfall (Fontes and Gasse 1991; Zouari et al. 1998; Causse et al. 2003). Four lake phases were documented by U-series dating during the Pleistocene: at ∼200 ka, 150–130 ka, 110–90 ka, and 55–18 ka (Causse et al. 2003). The youngest phase, which the authors note may have instead consisted of several briefer events, appears to have been less intense than the previous ones, based on isotopic evidence from molluscs (Causse et al. 2003). Microfaunal analyses suggest significant fluctuations in the Chotts from fresh to slightly saline water to nearly marine salinities, though specific age constraints were not given for this variation (Zouari et al. 1998). Brackish water lacustrine environments are also indicated in the Maknassy Basin from 50 to 65 ka (Ouda et al. 1998).

Apart from the studies of large water bodies in the region, records of Pleistocene humidity are also contained in paleosols. A paleosol calcrete formed on loess indicates a slight increase in humidity at ∼30 ka in Northwest Libya (Barich et al. 2006). Paleosols in central western Libya also would have required a humid climate to form, but their age

is uncertain, with three distinct soil sequences attributed to the Tertiary, Middle Pleistocene, and Holocene (Cremaschi and Trombino 1998).

## Paleolandscapes of the Western Sahara

Dated records of Pleistocene climates in Algeria are relatively rare. Causse et al. (1988) describe lacustrine activity at Azzel Matti and Kadda (Fig. 3.2) between 80 and 100 ka. The mollusc shells dated (*Cerastoderma*) and the mineralogy (dolomite and aragonite as well as magnesium calcite, among others) of lacustrine carbonates (Causse et al. 1988) suggest that at least at some point during their duration, the lakes were brackish to saline.

In contrast to the large lake records of central and eastern North Africa, much of the record of the late Quaternary in Morocco seems to come from fluvial deposits. This could be the simple result of the ruggedness of Moroccan topography. However, the chronology of shifts in and the climatic significance of fluvial behavior has been, and remains, a matter of some debate (Weisrock 2003; Akdim and Julia 2006; Boudad et al. 2006; Weisrock et al. 2006). Thorp et al. (2002) document shifts in alluvial behavior on the southern margins of the Anti-Atlas Mountains from high energy, gravel-bearing rivers (drier climate), to lower energy, frequently flooding rivers carrying silts derived from weathering profiles (wetter climate). One of the silt phases is dated to 45 ± 3.7 ka. At Oued Noun/Oued Assaka, Weisrock et al. (2006) and Wengler et al. (2002) record a sequence of deposits indicating fluvial and spring activity dated tentatively to "wetter episodes" in late MIS 5–4, shifting to groundwater-fed swamps in MIS 3, with substantial accumulation of fine sediments between 45 and 30 ka. Based on paleontological and sedimentary evidence, Wengler et al. (2002) suggest that the region was covered by steppe-prairie during this time. High water tables and concomitant fluvial activity continued from 30 to 20 ka, and a shift to more variable and arid conditions seems to have happened post $\sim$20 ka (Weisrock et al. 2006). Fluvial and fluvio-lacustrine deposits of Quaternary age are documented elsewhere in Morocco (e.g., Tiddas; Naim et al. 1998; Oued Grandou and Oued M'tal; Ouadia and Aberkan 1999; the Middle Souss Valley; Bhiry and Occhietti 2004). Though they are often not directly dated, these sequences are not inconsistent with humid conditions during early MIS 5 ($\sim$120 ka), and less humid conditions (but still more humid than the present) during MIS 3, particularly $\sim$30–45 ka.

Much of this Middle Pleistocene activity goes undocumented in the travertine sequence of Boudad et al. (2003). They date travertine formations in Oued Ziz to $\sim$262 and 20 ka, with no indication of spring activity in the intervening time. Criticisms of the methodology and sampling density of this study have already been made (Akdim and Julia 2006). Nearby, travertine deposition was indicated by U-Th dates at $\sim$50–60, 135, 160, and 220 ka (Akdim and Julia 2005). However, it should be pointed out that while detailed field and geochronologic analyses certainly need to be carried out to assess the timing of travertine deposition in this region, a lack of authigenic carbonate precipitation does not necessarily indicate the absence of humid conditions. Thus, a hiatus in travertine deposition does not necessarily indicate unabated arid climates. In Egypt's Dakhleh and Kharga Oases, tufa formation has not yet been documented during the well-known Holocene humid phase, and only occasionally post 100 ka (Sultan et al. 1997; Brook et al. 2002, 2003; Smith et al. 2004, 2007; Kleindienst et al. 2008). Smaller scale pluvial events apparently neither recharge aquifers to the extent necessary to permit carbonate formation, nor allow for significant enough soil formation in aquifer recharge areas to result in groundwaters supersaturated with respect to calcium carbonate. Humid events during this later period of time are primarily recorded in clastic sedimentary deposits (Brookes 1993a).

## Timing of Humid Phases Across the Sahara

The following discussion will make use of a previously compiled database of 276 direct dates on sediments indicative of enhanced humidity (relative to the present) across the Sahara (Smith et al., unpublished). In order to avoid overwhelming the database with dates from the well-recognized

Holocene (Neolithic) pluvial phase (e.g., Fontes et al. 1983; Petit-Maire 1987; Kutzbach and Liu 1997; de Menocal et al. 2000; Gasse 2000; Hoelzmann et al. 2000), only dates >20 ka were collected. All dates published for a given locality were included in the database, except where the authors exhibited concerns that certain dates were erroneous, based on stratigraphic or geochemical evidence (e.g., Wendorf et al. 1993; Szabo et al. 1995; Geyh and Thiedig 2008), or where certain dates were on sediments indicative of arid climates, or of uncertain climatic affinity (Macklin et al. 2002; Armitage et al. 2007). Because of the frequency with which radiocarbon analyses of Saharan sediments have disagreed with chronology based on stratigraphy, archaeology, and other dating techniques (Fontes and Gasse 1983, 1991; Szabo et al. 1995; Causse et al. 2003), only U-series, OSL, and TL dates were included in the analyses. Some of the environmental sequences based on radiocarbon chronologies are undoubtedly accurate, but as this is exceedingly difficult to evaluate without access to dates on the same sequences by other techniques (which would be included in the database), radiocarbon analyses were simply left out. The period from 20 to 50 ka, then, may be more complicated climatically than can be addressed using this database. Probability density distributions for dates were plotted using the AgeDisplay Excel macro (Sircombe 2004).

Division of the Sahara into latitudinal and longitudinal zones for comparative analysis was done in part based on natural breaks in the data, particularly for examining longitudinal variation. Latitudinal zone boundaries, however, were shifted slightly from natural breaks in order to distribute data more evenly amongst the three zones, and also to reflect natural climatic zonation in North Africa, with the Mediterranean margin considered separately. As smaller subdivided datasets are considered, preservation of pluvial phase deposits (or lack thereof) becomes more of an issue in the interpretation of frequency distributions. However, due to an uneven geographic distribution of research, it was not possible to achieve equality in subsample sizes.

The compiled dates suggest that precession-scale control of monsoonal rainfall can be documented not only from marine records (e.g., Rossignol-Strick 1983; Moreno et al. 2001; Larrasoaña et al. 2003) but also directly from terrestrial climate archives. This $\sim$20 ka cyclicity of humid conditions is still evident from the regional compilations of dates presented here (Figs. 3.3 and 3.4). There are few extended (>20 ka) periods in which humid conditions cannot be documented somewhere in the Sahara. Similarly, despite the relative domination of young (<40 ka) dates in the small "West" sample, it can also be seen from these distributions that while across the Sahara conditions become generally more arid following $\sim$70 ka, they are not exclusively dry (Fig. 3.4). The decrease in frequency of dates older than $\sim$150 ka is more difficult to interpret climatically, as

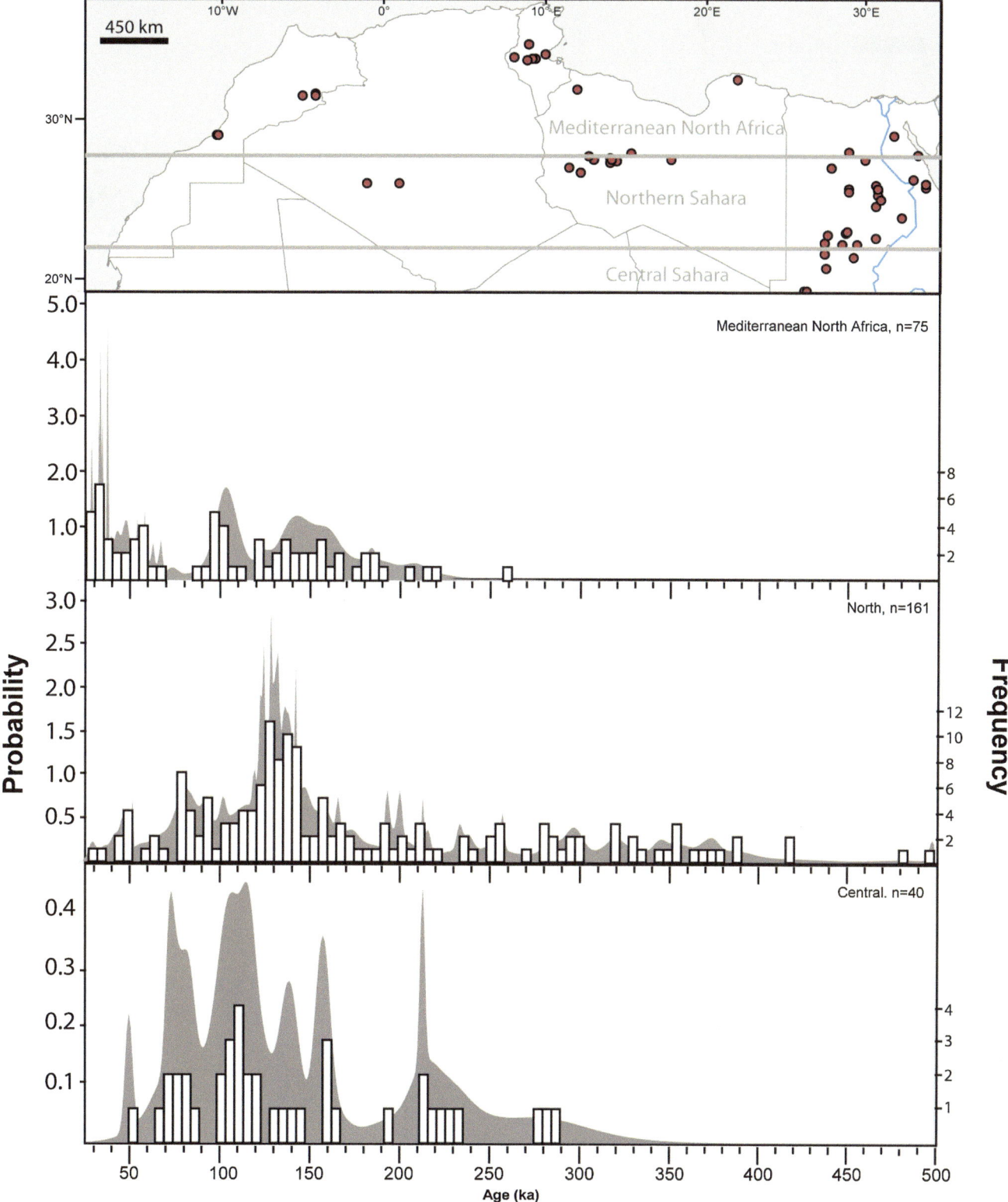

**Fig. 3.3**  Variation in pluvial phase timing between Mediterranean North Africa and the northern and central portions of the Sahara

older deposits may simply not be preserved as frequently. At least regionally, there is evidence that earlier pluvial events may have been more intense than the MIS 6–5 event; for the Fezzan Basin, Geyh and Thiedig (2008) reconstruct a decrease in lake level and lake extent through time, with the ∼130 ka lake smaller than older lake stages.

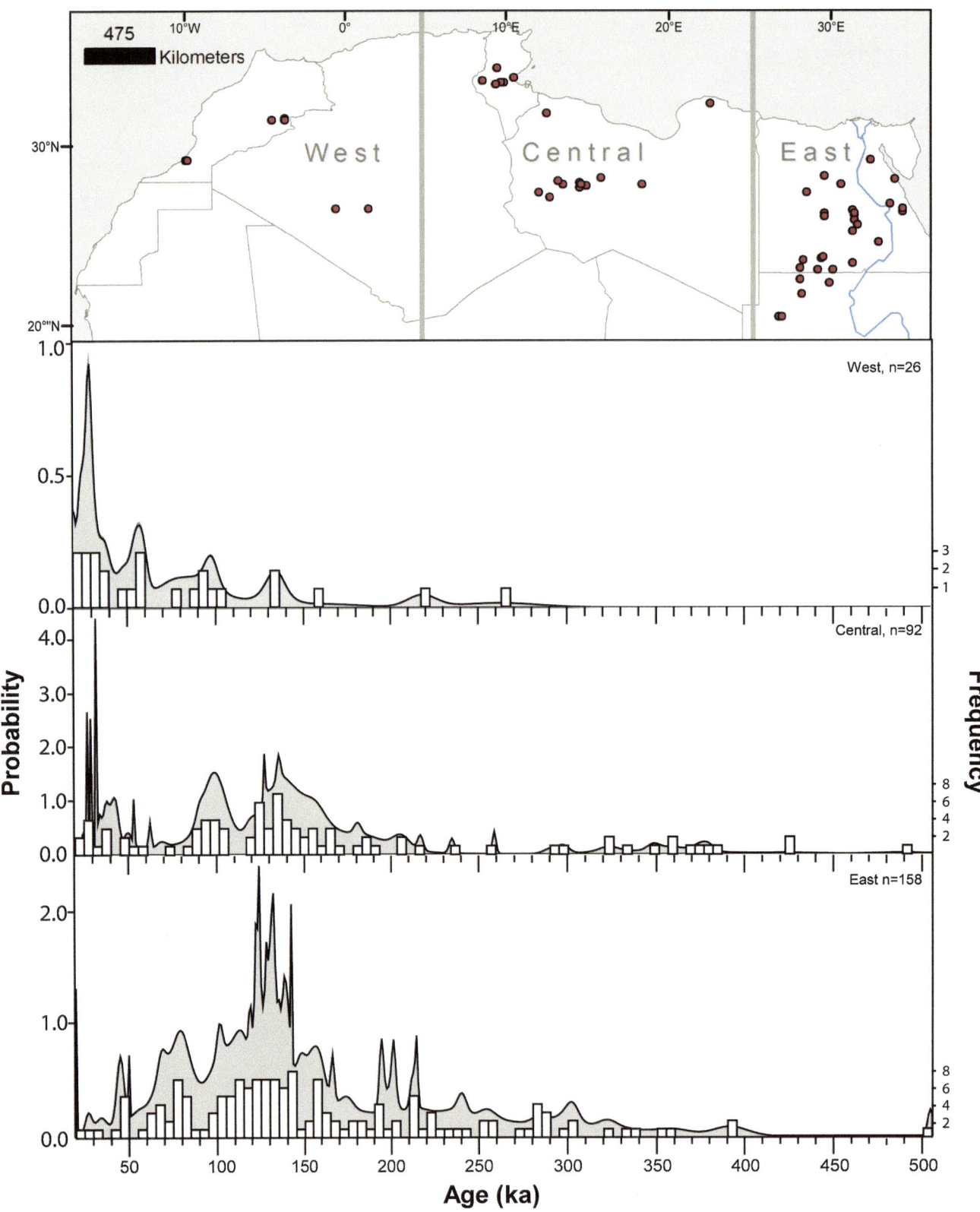

**Fig. 3.4** Variation in pluvial phase timing between eastern, central, and western portions of the Sahara

## North–South Variation in Pluvial Phase Timing

Based on the nature of the mechanisms bringing rain to the Sahara, latitudinal differences in the timing of pluvial phases could occur. The monsoon, which brings summer rain to North Africa, shifts in latitude with time (e.g., Kutzbach and Liu 1997). In addition, Mediterranean North Africa has the potential to receive winter rains associated with the Mediterranean rainfall belt, as seems to have occurred as far south as central Egypt during the Holocene humid phase (Arz et al. 2003; Brookes 2003; Kindermann et al. 2006).

The most striking difference in the regional compilation of pluvial phase dates is the relative frequency of <50 ka dates in the North, and their near absence in the South. This may imply a different climatic mechanism, perhaps the southward movement of Mediterranean winter rainfall, for enhanced humidity following ~45 ka. Causse et al. (2003) note that the stable carbon and oxygen isotope composition of molluscs from the Tunisian Chotts is distinctly different after 55 ka than it was in previous humid phases. They interpret this climatically as representative of less humid conditions, which is certainly consistent with climates across the Sahara at that time. However, the distinct oxygen isotope composition of mollusc shells from this time period could also reflect a change in moisture source from the Atlantic to the Mediterranean, although the nature of that change is difficult to constrain as it would be dependent on the relative geometries of water-bearing airmasses across the continent.

All regions show evidence for some humidity 120–140 ka, consistent with this being the wettest time across North Africa. The variation in regional records between 70 and 120 ka is somewhat more difficult to explain. The prominence of dates associated with MIS 5c, ~100 ka, and the absence of noticeable humid conditions during MIS 5a, ~80 ka, in the Mediterranean North Africa region is a distinct difference from the northern and central Sahara. Orbitally-based monsoon indices (Rossignol-Strick 1983) suggest progressive decreases in monsoonal strength from MIS 5e to 5c to 5a; a weaker MIS 5a monsoon that did not penetrate far enough north to substantially enhance rainfall over Mediterranean North Africa could explain the absence of dates associated with that period. However, this is not sufficient to explain the lack of MIS 5e dates. It may be that sediments related to an MIS 5e humid phase simply were not preserved in Mediterranean North Africa. Another difficulty with invoking a weaker monsoon to explain the absence of humid conditions around ~80 ka in Mediterranean North Africa is that subsequent monsoons should have been even weaker. Thus, it is necessary to invoke an alternate moisture source (i.e., the Mediterranean) or mechanism for increased rainfall other than the African monsoon to explain the prevalence of young dates in the North.

## East–West Variation in Pluvial Phase Timing

An examination of longitudinal variation in pluvial phase timing is somewhat less enlightening as the preponderance of dates in the eastern Sahara leads to a highly unbalanced data distribution. The dominance of relatively young humid conditions observed for the northern Sahara is strongly expressed in the very small western Saharan subsample. The pattern observed in western Saharan climate is intriguing: pluvial phase dates peak at 135, 100, 60, and 30 ka, with increasing numbers of dates with each younger humid phase. The latter is perhaps understandable as the result of a deliberate research focus on the Soltanian, or could be due to preservational biases resulting from the climate archives being predominantly fluvial. Older sediments may be destroyed completely in incisional settings, and buried in aggradational settings. While the timing of the older two (MIS 6–5) peaks in date distribution in the West is concurrent with that in the central and eastern Sahara, the younger two peaks do not seem to coincide with those in the other regions. The robustness, extent, and cause of the difference in humid phase timing between these areas are particularly interesting questions for future investigation; is the Atlantic margin of North Africa truly responding differently to glacial-interglacial oscillation than continental North Africa, or is it simply that the only records we currently have from that region are from northern localities, subject to Mediterranean-derived winter rainfall?

The variation in expression of MIS 5c and 5a between the regions noted above also occurs in this longitudinal distribution, with MIS 5a most strongly represented in the East, and less so in the central and western Sahara. If MIS 5a was the weakest monsoon event during MIS 5, it may be that only the region with the most extensive sampling was able to pick it up. Alternatively, perhaps there is some contribution of Indian Ocean sourced monsoonal rainfall to the eastern Sahara during an otherwise weaker MIS 5a; the Indian Ocean monsoon did penetrate as far as northern Sudan, even during the Holocene humid phase (Rodrigues et al. 2000).

## Spatial Variation Between Pluvial Phases

Further understanding of the spatial distribution of pluvial phase resources can be gained by examining the locations of sites recording humid conditions within each humid phase.

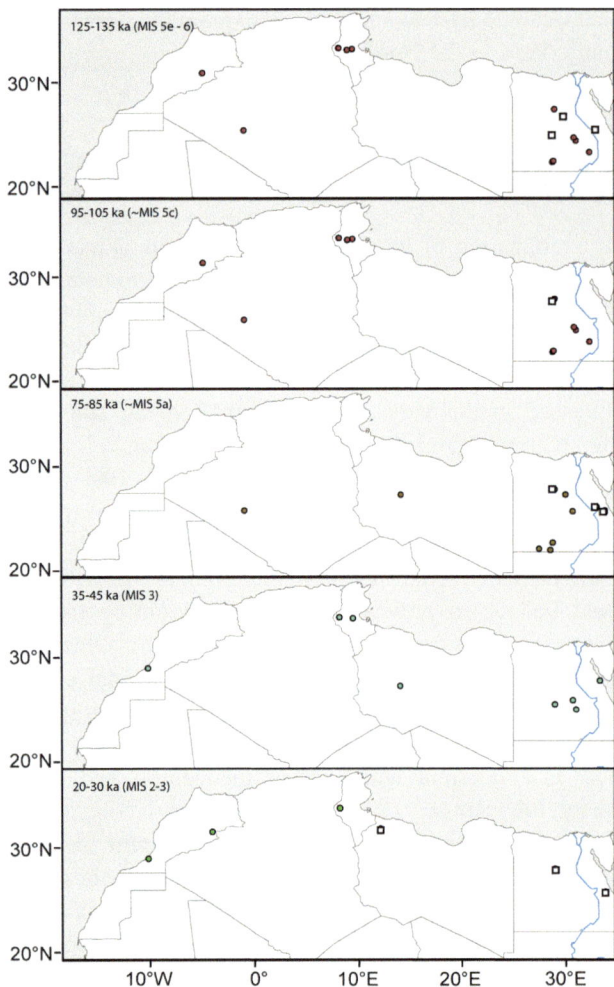

**Fig. 3.5** Localities recording humid conditions during peak pluvial phase conditions. Only localities with at least one mean date within the 10 ka range given are plotted for each pluvial phase; thus, this distribution underestimates the number of localities likely to be experiencing pluvial conditions during that time range. Open squares indicate pluvial phase indicators which require relatively little increase in rainfall, filled circles indicate lacustrine, spring, or fluvial deposits recording substantially enhanced rainfall

These phases can be recognized across North Africa at ~25, 40, 80, 100, and 130 ka (Fig. 3.5). The number of dates on sediments deposited prior to the large MIS 6-humid phase is comparatively low relative to those during or after. This is likely a result of the erosion associated with subsequent humid phases, and thus, the spatial distribution of these sediments is not addressed here. A distinction is made between sedimentary or geochemical indicators recording pluvial phases that require only modest increases in rainfall to form, such as paleosols or calcretes, and those that likely require larger increases, such as large lakes or extensive spring deposits. Included in these "minor" pluvial phase indicators are dates on remobilized uranium in Quaternary and non-Quaternary sediments in Egypt (Osmond

and Dabous 2004). This secondary mobilization of uranium would require enhanced groundwater flow. However, the magnitude of recharge needed is unknown, but presumed to be relatively minor.

Most of the trends described above are illustrated here; younger humid phases are recorded at fewer localities and more frequently by "minor" pluvial phase indicator. MIS 5a is particularly well-represented in the East, and the youngest humid phases are more strongly indicated in the West. It is clear, however, that despite using a relatively conservative method for plotting localities (localities were required to have a mean date fall within the given age range; thus 30.1 ± 10 ka would not be included in the 20–30 ka plot), all the identified "peak" times (Smith et al., unpublished) are represented throughout the Sahara.

## Conclusions

Pluvial phase environments were heterogeneous in space and time. The nature of water sources would have varied in size, quality (fresh or brackish), and reliability, both with location and with each successive pluvial phase. Large lakes seem to have primarily been a feature of MIS 5, though particularly in the western Sahara, notable subsequent humidity is recorded in fluvial sequences. Even these later, comparatively minor humid phases are recognizable across the continent, though their signature in the eastern Sahara is relatively subtle.

The slightly out-of-phase timing of humid conditions across the Sahara, if not an artifact of sampling and the sometimes substantial error associated with pluvial phase dates, brings up the possibility that somewhere within the Sahara, available water could be found throughout much of the last 150 kyr. This argument has been previously made for oasis locations (Kleindienst 1999) but could also hold true on a larger scale. Additional, well-dated environmental sequences from across the Sahara will be instrumental in testing the hypotheses presented here.

**Acknowledgments** This work was supported by NSF grant EAR-0447357 to Smith.

## References

Adelsberger, K. A., & Smith, J. R. (2010). Sedimentology, geomorphology and paleoenvironmental interpretation of spring-deposited ironstones and associated sediments, Dakhleh Oasis, Western Desert, Egypt. *Catena, 83*, 7–22.

Akdim, B., & Julia, R. (2005). The travertine mounds of Tafilalet (Morocco); morphology and genesis based on present-day analogues. *Zeitschrift fur Geomorphologie, 49*, 373–389.

Akdim, B., & Julia, R. (2006). Commentaire a la note intitulee Datation par la methode U/Th d'un travertin quaternaire du Sud-Est marocain : Implications paleoclimatiques pendant le Pleistocene moyen et superieur de L. Boudad et al. [C. R. Geoscience 335 (2003) 469–478]. *Comptes Rendus Geosciences*, 338, 581–582.

Armitage, S. J., Drake, N. A., Stokes, S., El-Hawat, A., Salem, M. J., White, K., et al. (2007). Multiple phases of North African humidity recorded in lacustrine sediments from the Fazzan Basin, Libyan Sahara. *Quaternary Geochronology, 2*, 181–186.

Arz, H. W., Lamy, F., Patzold, J., Muller, P. J., & Prins, M. (2003). Mediterranean moisture source for an Early-Holocene humid period in the Northern Red Sea. *Science, 300*, 118–121.

Ball, J. (1900). *Kharga Oasis, its topography and geology*. Cairo: Egyptian Survey Department.

Barich, B. A., Garcea, E. A. A., & Giraudi, C. (2006). Between the Mediterranean and the Sahara: Geoarchaeological reconnaissance in the Jebel Gharbi, Libya. *Antiquity, 80*, 567–582.

Beadnell, H. J. L. (1909). *An Egyptian oasis: An account of the oasis of Kharga in the Libyan Desert*. London: John Murray.

Bhiry, N., & Occhietti, S. (2004). Fluvial sedimentation in a semi-arid region; the fan and interfan system of the Middle Souss Valley, Morocco. *Proceedings of the Geologists' Association, 115*, 313–324.

Boudad, L., Kabiri, L., Farkh, S., Falgueres, C., Rousseau, L., Beauchamp, J., et al. (2003). Datation par la methode U/Th d'un travertin quaternaire du Sud-Est marocain : Implications paleoclimatiques pendant le Pleistocene moyen et superieur. U/Th dating of a Quaternary travertine from southern Morocco: Palaeoclimatic consequences during Middle and Upper Pleistocene. *Comptes Rendus Geosciences*, 335, 469–478

Boudad, L., Kabiri, L., Farkh, S., Falgueres, C., Rousseau, L., Beauchamp, J., et al. (2006). Réponse au commentaire de Brahim Akdim et Ramon Julia sur la note Datation par la methode U/Th d'un travertin quaternaire du Sud-Est marocain : Implications paleoclimatiques pendant le Pleistocene moyen et superieur [C. R. Geoscience 335 (2003) 469–478]. Comptes Rendus Geosciences, 338, 583

Brook, G. A., Embabi, N. S., Ashour, M. M., Edwards, R. L., Cheng, H., Cowart, J. B., et al. (2002). Djara Cave in the Western Desert of Egypt; morphology and evidence of Quaternary climatic change. *Cave and Karst Science, 29*, 57–66.

Brook, G. A., Embabi, N. S., Ashour, M. M., Edwards, R. L., Cheng, H., Cowart, J. B., et al. (2003). Quaternary environmental change in the Western Desert of Egypt: Evidence from cave speleothems, spring tufas, and playa sediments. *Zeitschrift fur Geomorphologie, 131*, 59–87.

Brookes, I. A. (1993a). Geomorphology and Quaternary geology of the Dakhla Oasis region, Egypt. *Quaternary Science Reviews, 12*, 529–552.

Brookes, I. A. (1993b). Late Pleistocene basinal sediments, Dakhla Oasis region, Egypt; a non-interglacial pluvial. In U. Thorweihe & H. Schandelmeier (Eds.), *Geoscientific research in Northeast Africa; Proceedings of the International Conference* (pp. 627–633). Rotterdam: A.A. Balkema.

Brookes, I. A. (2003). Geomorphic indicators of Holocene winds in Egypt's Western Desert: Geomorphology. *Geomorphology, 56*, 155–166.

Brooks, N., Drake, N., MacLaren, S., & White, K. H. (2003). Studies in geography, geomorphology, environment and climate. In D. J. Mattingly (Ed.), *The archaeology of Fazzan: Volume 1, synthesis* (pp. 33–74). Tripoli: Department of Antiquities.

Caton-Thompson, G. (1952). *Kharga Oasis in prehistory*. London: Athlone Press.

Causse, C., Conrad, G., Fontes, J.C., Gasse, F., Gibert, E., & Kassir, A. (1988). Le dernier "Humide" Pleistocene du Sahara nord-occidental daterait de 80-100 000 ans. *Comptes Rendus de l'Academie des Sciences, Serie 2, Mecanique, Physique, Chimie, Sciences de l'Univers, Sciences de la Terre, 306*, 1459–1464.

Causse, C., Coque, R., Fontes, J. C., Gasse, F., Gibert, E., Ben Ouezdou, H., et al. (1989). Two high levels of continental waters in the southern Tunisian chotts at about 90 and 150 ka. *Geology (Boulder), 17*, 922–925.

Causse, C., Ghaleb, B., Chkir, N., Zouari, K., Ben Ouezdou, H., & Mamou, A. (2003). Humidity changes in southern Tunisia during the late Pleistocene inferred from U-Th dating of mollusc shells. *Applied Geochemistry, 18*, 1691–1703.

Churcher, C. S., & Kleindienst, M. R. (in press). Great lakes in the Dakhleh Oasis: Mid-Pleistocene freshwater lakes in the Dakhleh Oasis Depressions, Western Desert Egypt. In A. J. Mills (Ed.), *The Oasis papers IV*. Oxford: Oxbow Books.

Churcher, C. S., Kleindienst, M. R., & Schwarcz, H. P. (1999). Faunal remains from a Middle Pleistocene lacustrine marl in Dakhleh Oasis, Egypt; Palaeoenvironmental reconstructions. *Palaeogeography, Palaeoclimatology, Palaeoecology, 154*, 301–312.

Cremaschi, M., & Trombino, L. (1998). The palaeoclimatic significance of Paleosols in southern Fezzan (Libyan Sahara); Morphological and micromorphological aspects. *Catena, 34*, 131–156.

Crombie, M. K., Arvidson, R. E., Sturchio, N. C., El Alfy, Z., & Abu Zeid, K. (1997). Age and isotopic constraints on Pleistocene pluvial episodes in the Western Desert, Egypt. *Palaeogeography, Palaeoclimatology, Palaeoecology, 130*, 337–355.

de Menocal, P. B., Bloemendal, J., & King, J. (1991). A rock magnetic record of monsoonal dust deposition to the Arabian Sea: Evidence for a shift in the mode of deposition at 2.4 Ma. *Proceedings of the Ocean Drilling Program, Scientific Results, 117*, 389–407.

de Menocal, P., Ortiz, J., Guilderson, T., Adkins, J., Sarnthein, M., Baker, L., et al. (2000). Abrupt onset and termination of the African Humid Period: Rapid climate responses to gradual insolation forcing. *Quaternary Science Reviews, 19*, 347–361.

Fontes, J. C., & Gasse, F. (1983). On the ages of humid Holocene and late Pleistocene phases in North Africa—remarks on late Quaternary climatic reconstruction for the Maghreb (North Africa) by P. Rognon. *Palaeogeography, Palaeoclimatology, Palaeoecology, 70*, 393–398.

Fontes, J. C., & Gasse, F. (1991). PALHYDAF (Palaeohydrology in Africa) program: Objectives, methods, major results. *Palaeogeography, Palaeoclimatology, Palaeoecology, 84*, 191–215.

Fontes, J. C., Coque, R., Dever, L., Filly, A., & Mamou, A. (1983). Paleohydrologie isotopique de l'Oued el Akarit (sud tunisie) au Pleistocene superieur et a l'Holocene. *Palaeogeography, Palaeoclimatology, Palaeoecology, 43*, 41–62.

Gasse, F. (2000). Hydrological changes in the African tropics since the Last Glacial Maximum. *Quaternary Science Reviews, 19*, 189–211.

Gaven, C., Hillaire-Marcel, C., & Petit-Maire, N. (1981). A Pleistocene lacustrine episode in southeastern Libya. *Nature, 290*, 131–133.

Geyh, M. A., & Thiedig, F. (2008). The Middle Pleistocene Al Mahruqah Formation in the Murzuq Basin, northern Sahara, Libya evidence for orbitally-forced humid episodes during the last 500,000 years. *Palaeogeography, Palaeoclimatology, Palaeoecology, 257*, 1–21.

Hamdan, M. A. E. (2000). Quaternary travertines of wadis Abu Had-Dib area, Eastern Desert, Egypt; paleoenvironment through field, sedimentology, age and isotopic study. *Sedimentology of Egypt, 8A*, 49–62.

Haynes, C. V., Maxwell, T. A., Hawary, A. E., Nicoll, K. A., & Stokes, S. (1997). An Acheulian site near Bir Kiseiba in the Darb el Arba'in Desert, Egypt. *Geoarchaeology, 12*, 819–832.

Hoelzmann, P., Kruse, H.-J., & Rottinger, F. (2000). Precipitation estimates for the eastern Saharan palaeomonsoon based on a water balance model of the West Nubian palaeolake basin. *Global and Planetary Change, 64*, 105–120.

Kieniewicz, J. M., & Smith, J. R. (2007). Hydrologic and climatic implications of stable isotope and minor element analyses of authigenic calcite silts and gastropod shells from a mid-Pleistocene pluvial lake, Western Desert, Egypt. *Quaternary Research, 68,* 431–444.

Kieniewicz, J., & Smith, J. R. (2009). Paleoenvironmental reconstruction and water balance of a Mid-Pleistocene pluvial lake, Dakhleh Oasis, Egypt. *GSA Bulletin, 121,* 1154–1171.

Kindermann, K., Bubenzer, O., Nussbaum, S., Riemer, H., Darius, F., Pollath, N., et al. (2006). Palaeoenvironment and Holocene land use of Djara, Western Desert of Egypt. *Quaternary Science Reviews, 25,* 1619–1637.

Kleindienst, M. R. (1999). Pleistocene archaeology and geoarchaeology of the Dakhleh Oasis: A status report. In C. S. Churcher & A. J. Mills (Eds.), *Reports from the survey of the Dakhleh Oasis 1977–1987* (pp. 83–108). Oxford: Oxbow Books.

Kleindienst, M. R., Schwarcz, H. P., Nicoll, K., Churcher, C. S., Frizano, J., Giegengack, R., et al. (2008). Water in the desert: First report on Uranium-series dating of Caton-Thompson's and Gardner's "classic" Pleistocene sequence at Refuf Pass, Kharga Oasis. In M. F. Wiseman (Ed.), *Oasis Papers II: Proceedings of the second Dakhleh Oasis Project research seminar* (pp. 25–54). Oxford: Oxbow Books.

Kutzbach, J. E., & Liu, Z. (1997). Response of the African monsoon to orbital forcing and ocean feedbacks in the middle Holocene. *Science, 278,* 440–443.

Larrasoaña, J. C., Roberts, A. P., Rohling, E. J., Winkhofer, M., & Wehausen, R. (2003). Three million years of monsoon variability over the northern Sahara. *Climate Dynamics, 21,* 689–698.

Macklin, M. G., Fuller, I. C., Lewin, J., Maas, G. S., Passmore, D. G., Rose, J., et al. (2002). Correlation of fluvial sequences in the Mediterranean basin over the last 200 ka and their relationship to climate change. *Quaternary Science Reviews, 21,* 1633–1641.

Mandel, R. D., & Simmons, A. H. (2001). Prehistoric occupation of Late Quaternary landscapes near Kharga Oasis, Western Desert of Egypt. *Geoarchaeology, 16,* 95–117.

McCauley, J. F., Breed, C. S., Schaber, G. G., McHugh, W. P., Issawi, B., Haynes, C. V., et al. (1982). Subsurface valleys and geoarchaeology of the Eastern Sahara revealed by Shuttle radar. *Science, 218,* 1004–1020.

McHugh, W. P., Breed, C. S., Schaber, G. G., McCauley, J. F., & Szabo, B. J. (1988a). Acheulian sites along the "radar rivers", southern Egyptian Sahara. *Journal of Field Archaeology, 15,* 361–379

McHugh, W. P., McCauley, J. F., Haynes, C. V., Breed, C. S., & Schaber, G. G. (1988b). Paleorivers and geoarchaeology in the southern Egyptian Sahara. *Geoarchaeology, 3,* 1–40

McKenzie, J. A. (1993). Pluvial conditions in the eastern Sahara following the penultimate deglaciation: Implications for changes in atmospheric circulation patterns with global warming. *Palaeogeography, Palaeoclimatology, Palaeoecology, 103,* 95–105.

Moreno, A., Targarona, J., Henderiks, J., Canals, M., Freudenthal, T., & Meggers, H. (2001). Orbital forcing of dust supply to the North Canary Basin over the last 250 kyr. *Quaternary Science Reviews, 20,* 1327–1339.

Naim, H., Aberkan, M., Irochene, M., & Kogbe, C. A. (1998). Les formations fluvio-lacustres plio-quaternaires de la region SW de Tiddas (Maroc Central). *Africa Geoscience Review, 5,* 199–206.

Nicoll, K., Giegengack, R., & Kleindienst, M. (1999). Petrogenesis of artifact-bearing fossil-spring tufa deposits from Kharga Oasis, Egypt. *Geoarchaeology, 14,* 849–863.

Osinski, G. R., Schwarcz, H. P., Smith, J. R., Kleindienst, M. R., Haldemann, A. F. C., & Churcher, C. S. (2007). Evidence for an approximately 200–100 ka meteorite impact in the Western Desert of Egypt. *Earth and Planetary Science Letters, 253,* 378–388.

Osmond, J. K., & Dabous, A. A. (2004). Timing and intensity of groundwater movement during Egyptian Sahara pluvial periods by U-series analysis of secondary U in ores and carbonates. *Quaternary Research, 61,* 85–94.

Ouadia, M., & Aberkan, M. (1999). Les formations fluviatiles quaternaires des vallees de l'Oued Grandou et de l'Oued M'Tal (Meseta occidentale Marocaine); Mise en place, origine et nouvelles donnees chronostratigraphiques. *Annales de la Societe geologique du Nord, 6,* 137–142.

Ouda, B., Zouari, K., Ben Ouezdou, H., Chkir, N., & Causse, C. (1998). Nouvelles donnees paleoenvironnementales pour le Quaternaire recent en Tunisie centrale (bassin de Maknassy). *Comptes Rendus de l'Academie des Sciences, Serie II Sciences de la Terre et des Planetes, 326,* 855–861.

Patterson, L. J., Sturchio, N. C., Kennedy, B. M., van Soest, M. C., Sultan, M., Lu, Z.-T., et al. (2005). Cosmogenic, radiogenic, and stable isotopic constraints on groundwater residence time in the Nubian Aquifer, Western Desert of Egypt. *Geochemistry, Geophysics, Geosystems - G3, 6,* Q01005.

Petit-Maire, N. (1987). Holocene paleomonsoon in northern Mali (22 degrees -24 degrees N/3 degrees -4 degrees W). In Anonymous (Ed.), International Union for Quaternary Research; XIIth international congress; Union Internationale pour l'Etude du Quaternaire; XII (super e) congres international (p. 242). International Union for Quaternary Research

Petit-Maire, N., Delibrias, G., & Gaven, C. (1980). Pleistocene lakes in the Shati area, Fezzan (27 degrees 30'N). In M. Sarnthein, E. Seibold, & P. Rognon (Eds.), *Sahara and surrounding seas; Sediments and climatic changes* (pp. 289–295). Rotterdam: A.A. Balkema.

Robinson, C. (2002). Application of satellite radar data suggest that the Kharga Depression in south-western Egypt is a fracture rock aquifer. *International Journal of Remote Sensing, 23,* 4101–4113.

Rodrigues, D., Abell, P. I., & Kroepelin, S. (2000). Seasonality in the early Holocene climate of Northwest Sudan; Interpretation of Etheria elliptica shell isotopic data. *Global and Planetary Change, 26,* 181–187.

Rohling, E. J., Cane, T. R., Cooke, S., Sprovieri, M., Bouloubassi, I., Emeis, K. C., et al. (2002). African monsoon variability during the previous interglacial maximum. *Earth and Planetary Science Letters, 202,* 61–75.

Rossignol-Strick, M. (1983). African monsoons, an immediate climate response to orbital insolation. *Nature (London), 304,* 46–49.

Rossignol-Strick, M. (1985). Mediterranean Quaternary sapropels, and immediate response of the African monsoon to variation of insolation. *Palaeogeography, Palaeoclimatology, Palaeoecology, 49,* 237–263.

Said, R. (Ed.). (1990). *The geology of Egypt.* Rotterdam: A.A. Balkema.

Schwarcz, H. P., & Morawska, L. (1998). Uranium-series dating of carbonates from Bir Tarfawi and Bir Sahara East. In F. Wendorf, R. Schild, & A. E. Close (Eds.), *Egypt during the last interglacial* (pp. 205–217). New York: Plenum Press.

Sircombe, K. N. (2004). AgeDisplay: An EXCEL workbook to evaluate and display univariate geochronological data using binned frequency histograms and probability density distributions. *Computers and Geosciences, 30,* 21–31.

Smith, J. R., Giegengack, R., Schwarcz, H. P., McDonald, M. M. A., Kleindienst, M. R., Hawkins, A. L., et al. (2004). A reconstruction of Quaternary pluvial environments and human occupations using stratigraphy and geochronology of fossil-spring tufas, Kharga Oasis, Egypt. *Geoarchaeology, 19,* 407–439.

Smith, J. R., Hawkins, A. L., Asmerom, Y., Polyak, V., & Giegengack, R. (2007). New age constraints on the Middle Stone Age occupations of Kharga Oasis, Western Desert, Egypt. *Journal of Human Evolution, 52,* 690–701.

Sturchio, N. C., Du, X., Purtschert, R., Lehmann, B. E., Sultan, M., Patterson, L. J., et al. (2004). One million year old groundwater in the Sahara revealed by krypton-81 and chlorine-36. *Geophysical Research Letters, 31*, 1–4.

Sultan, M., Sturchio, N. C., Hassan, F. A., Hamdan, M. A. R., Mahmood, A. M., El Alfy, Z., et al. (1997). Precipitation source inferred from stable isotopic composition of Pleistocene groundwater and carbonate deposits in the Western Desert of Egypt. *Quaternary Research, 48*, 29–37.

Szabo, B. J., McHugh, W. P., Schaber, G. G., Haynes, C. V., Jr., & Breed, C. S. (1989). Uranium-series dated authigenic carbonates and Acheulian sites in southern Egypt. *Science, 243*, 1053–1056.

Szabo, B. J., Haynes, C. V., Jr., & Maxwell, T. A. (1995). Ages of Quaternary pluvial episodes determined by uranium-series and radiocarbon dating of lacustrine deposits of Eastern Sahara. *Palaeogeography, Palaeoclimatology, Palaeoecology, 113*, 227–241.

Thorp, M., Glanville, P., Stokes, S., & Bailey, R. (2002). Preliminary optical and radiocarbon age determinations for upper Pleistocene alluvial sediments in the southern Anti Atlas Mountains, Morocco. *Comptes Rendus Geoscience, 334*, 903–908.

Thorweihe, U. (1990). Nubian Aquifer system. In R. Said (Ed.), *The geology of Egypt* (pp. 601–614). Rotterdam: A.A. Balkema.

von Zittel, K. (1883). *Beiträge zur geologie und palaeontologie der Libyschen Wüste und der angrenzenden gebiete von Aegypten.* Cassel: Fischer.

Weisrock, A. L. (2003). About the dating of upper-Pleistocene fluvial deposits in the arid zone of Morocco; Comparative data of radiocarbon, optic stimulation luminescence and uranium/thorium methods. *Comptes Rendus—Academie des sciences Geoscience, 335*, 277–278.

Weisrock, A., Wengler, L., Mathieu, J., Ouammou, A., Fontugne, M., Mercier, N., et al. (2006). Upper Pleistocene comparative OSL, U/Th and (super 14) C datings of sedimentary sequences and correlative morphodynamical implications in the south-western Anti-Atlas (Oued Noun, 29 degrees N, Morocco). *Quaternaire, 17*, 45–59.

Wendorf, F., Schild, R., & Close, A. E. (Eds.). (1993). *Egypt during the last Interglacial: The Middle Paleolithic of Bir Tarfawi and Bir Sahara East.* New York: Plenum Press.

Wengler, L., Weisrock, A., Brochier, J.-E., Brugal, J.-P., Fontugne, M., Magnin, F., et al. (2002). Enregistrement fluviatile et paleoenvironnements au Pleistocene superieur sur la bordure atlantique de l'Anti-Atlas (Oued Assaka, S-O marocain). *Quaternaire, 13*, 179–192.

Zouari, K., Chkir, N., & Causse, C. (1998). Pleistocene humid episodes in southern Tunisian chotts. *IAEA: Isotopic Techniques in the Study of Environmental Change, 18*, 543–554.

# Chapter 4
# The Faunal Context of Human Evolution in the Late Middle/Late Pleistocene of Northwestern Africa

D. Geraads

**Abstract** Mammalian faunas from the late Middle to the Late Pleistocene of northwestern Africa are reviewed, with special reference to their ecological meaning, keeping in mind that a number of problems, the first of which being the imperfectness of chronology, hinder their use as a simple proxy for environmental conditions. The proportions of open country antelopes among bovids and of gerbillids among rodents give some rough indications about the landscape and climate. Some sites in the second half of the Middle Pleistocene seem to document some improvement over the harsh conditions that prevailed earlier, but in Morocco at least, the Late Pleistocene sees the return of increasingly open, if not arid, conditions, with an overwhelming dominance of gazelles among large mammals. More research, especially on non-anthropic sites, is needed before the role of human hunting preferences in these proportions can be evaluated, but the dominance of gerbillids, and the loss of rodent diversity largely confirm inferences drawn from the large mammals.

**Keywords** Faunal environment • Late Pleistocene • Middle Pleistocene • Morocco • Northwestern Africa

## Introduction

It is usually believed that environmental change drove, or at least had some influence upon, evolutionary change in the human lineages. Mammals were a major component of past human environments and by being adapted to specific environments and relying on various food resources, they also provide some information on the vegetation cover, even if direct inference cannot be expected. Thus, ideally, mammalian assemblages can be used as a proxy for climatic change, although a number of problems related to chronology, taphonomy, sampling, and collecting procedures hinder such a reconstruction.

The first mammalian fossils from North Africa (indeed, from Africa) were discovered about 160 years ago, and many paleontological and archaeological sites were excavated between 1850 and 1950, but often by people who were not archaeologists or even scientists. Stratigraphic control on these old excavations was usually very poor, and often there is no stratigraphy at all. For more recently excavated sites, the stratigraphy is better known, but there are very few reliable absolute dates, especially when compared to East Africa. Precision is often poor, and dates are sometimes inconsistent (within a single site, lower levels may yield later dates). For Early and Middle Pleistocene sites, mammalian biochronology is usually able to sort the sites by age, but its discriminant power does not reach the ca. 10,000 years that are required to sort Late Pleistocene sites. As a whole, the chronological framework for the North African Late Pleistocene remains sketchy, and this severely hinders the reconstruction of climatic evolution. Thus, the following review of the main northwestern African sites, listed in more or less chronological order (Tables 4.1, 4.2), must be regarded as imperfect. Most of the late Middle to Late Pleistocene sites are located in Morocco, and many are concentrated in a small section of the coast (Fig. 4.1). It is thus clear that the representation of the various Quaternary biomes is very uneven.

## North African Pleistocene Mammals

Mammals can conveniently be divided into two groups, large ones and small ones. Besides different collection procedures, these two groups have different taphonomic histories, and cannot be approached in the same way.

D. Geraads (✉)
UPR 2147—Centre National de la Recherche Scientifique,
44 rue de l'Amiral Mouchez, 75014 Paris, France
e-mail: denis.geraads@evolhum.cnrs.fr

J.-J. Hublin and S. P. McPherron (eds.), *Modern Origins: A North African Perspective*,
Vertebrate Paleobiology and Paleoanthropology, DOI: 10.1007/978-94-007-2929-2_4,
© Springer Science+Business Media B.V. 2012

**Table 4.1** List of the sites discussed in the text, with their location, and very rough estimates of their ages (in ka)

| | | | |
|---|---|---|---|
| Zouhra | M | C | 40 |
| Mugharet el 'Aliya | M | C | 50 |
| Oulad Hamida 2, Grotte des Félins | M | C | 50 |
| Dar Bou Azza, Grotte des Gazelles | M | C | 50 |
| Thomas Quarry, locus D | M | C | 100 |
| Jebel Irhoud | M | I | 160 |
| Salé | M | C | 300 |
| Sidi Abderrahmane D2 | M | C | 350 |
| Doukkala | M | C | 400 |
| Aïn Bahya | M | C | 400 |
| Oulad Hamida, Grotte des Rhinocéros | M | C | 450 |
| Thomas Quarry, Hominid level | M | C | 500 |
| Aïn Maarouf | M | I | 700 |
| Tighenif | A | I | 800 |
| Thomas Quarry, level L | M | C | 1200 |

*A* Algeria, *M* Morocco; *C* coast, *I* inland

Many of the Pleistocene North African large mammals are quite similar to the East African ones; most of the genera, and even some species, are identical. This means that there was no continuous Saharan barrier for most of the Pleistocene. These taxonomic similarities allow us to interpret North African large mammal faunas with the ecological criteria used further south. However, many of these mammals provide only vague paleoecological indications. Potential exceptions are the Equidae, whose metapodial and hoof (terminal phalanx) proportions have been used elsewhere (Eisenmann 1984), but which have not been studied in detail in the sites considered here. Other families may also provide indications in some cases, but by far the most useful one is the family Bovidae. They have adapted to a variety of environments and for this reason, have received more attention than other groups. Some of them favor open dry environments; they belong to the tribe Alcelaphini, represented in the Middle and Late Pleistocene of the Maghreb by the genera *Connochaetes* (wildebeest), *Alcelaphus* (hartebeest) and *Parmularius* (extinct), and to the tribe Antilopini (*Gazella*). The oryx (Hippotragini) should also be added to this group, but as incomplete specimens can be confused with *Hippotragus*, an antelope often found under denser cover, they will be left aside. The other bovids belong to the tribes Bovini, with the aurochs (*Bos*) and buffaloes (*Pelorovis*); Reduncini, with the waterbuck and reedbuck (*Kobus* and *Redunca*); and the rare Tragelaphini (*Tragelaphus* and *Taurotragus*). These three latter tribes live in diverse environments, including forest, woodland, bushland, grassland, and wetland. They can also be found in fairly open areas but, on the whole, they indicate more closed and/or humid environments than the Alcelaphini–Antilopini group. Thus, to estimate the openness of the environment, I use pie-diagrams (Fig. 4.2) to represent the proportion of open country antelopes versus other non-hippotragine bovids, using NISP counts, as was done elsewhere, especially by Vrba (1975). It should be remembered that the term "open" carries only a vague assumption about the physiognomy of the landscape. However, the similarities between the faunas of North Africa and those of the eastern or southern African savannahs are so striking that some savannah-like biome must have been largely represented in the Maghreb, at least in the Middle Pleistocene.

Interpreting the environment from the large mammal assemblage is not straightforward, especially when dealing with collections from old excavations. Published identifications are not always reliable, all the more because they were often based upon teeth. Detailed counts of specimen numbers were never provided, and the relative abundance of taxa can be roughly estimated from the publications in only a few cases. Since most of the material is now lost, mixed, or at best distributed among various institutes, a revision of the material is usually impossible. The result is that there are very few sites with reliable faunal lists and proportions; most of these were excavated recently, or are still being excavated. Still, even for these ones, the taphonomic bias, a major obstacle to paleoecological reconstruction, remains. Since most faunas are found in caves, death of the animals may be the result of their having fallen into natural traps, but usually the fossil assemblage is only that part of the fauna that was selected and brought into the cave by carnivores or humans. As no detailed analysis has been yet undertaken, we can only assume that this selection did not significantly alter the relative abundances, or at least that this bias was rather uniform through sites. This huge

**Table 4.2** Faunal lists of the main localities. Mostly from Amani and Geraads (1993), Aouraghe and Abbassi (2002), Biberson (1961), Cheddadi (1986), Jaeger (1975, 1988), Michel (1989), Tong (1986), and personal data

| | Thomas 1, level L | Tighenif | Aïn Maarouf | Thomas 1, Hominid level | Oulad Hamida, GDR | Aïn Bahya composite | Doukkala II | Sidi Abderrahmane D2 | Salé | Jebel Irhoud | Thomas Quarry, locus D | Dar Bou Azza, Gazelles | Oulad Hamida 2, Félins | Zouhra |
|---|---|---|---|---|---|---|---|---|---|---|---|---|---|---|
| Elephantidae | + | + | + | | | | + | + | | | | | | |
| Rhinocerotidae | | | | | | | | | | | | | | + |
| Ceratotherium | | + | | + | + | + | + | + | | + | | | | |
| Stephanorhinus | | | | | | | | | + | | | + | + | |
| Equus | + | + | + | + | + | + | + | + | + | + | | + | + | + |
| Hippopotamus | + | + | + | | + | | | + | | | | | | |
| Kolpochoerus | + | | | + | | | | | | | | | | |
| Metridiochoerus | | + | | | | | | | | | | | | |
| Phacochoerus | | | + | + | + | | | | | | | | + | + |
| Sus | | | | | | | ? | ? | | | | + | | + |
| Camelus | | + | | | + | | | | | | | | | |
| Cervidae | | | | | | | ? | | | | | | | |
| Tragelaphini | | + | ? | | | | | | | + | | | | + |
| Bos | | + | | | | + | + | + | | + | | + | | + |
| Pelorovis/Bos | | | + | + | | | + | | | | | | + | + |
| Pelorovis | | | | | + | | + | | | | | | | |
| Redunca | | | | | + | | + | + | | | | | | |
| Hippotragini | | + | | + | + | | + | + | | + | | | | + |
| Alcelaphini | | + | + | | + | | + | + | | + | | + | | + |
| Connochaetes | | + | + | + | + | | + | + | + | + | | + | + | + |
| Ammotragus | | | | | | | | | | ? | | | | |
| Gazella sp. | + | + | | | | | | | | | | | | |
| G. atlantica | | + | | + | + | + | + | + | | + | | + | + | + |
| G. cuvieri | | | | | | | + | + | | + | | + | | + |
| G. dorcas | | | | | | | + | + | | | | + | + | |
| Theropithecus | | + | | + | + | + | + | | | | | + | + | |
| Panthera | | + | | + | + | + | | | | + | | | + | + |
| Felis, small | | + | | | | | | + | | + | | + | + | + |
| Hyaena hyaena | | + | | | + | | + | + | | | | + | + | |
| Crocuta crocuta | | + | + | | + | | + | + | | | | + | | + |

**Table 4.2** (continued)

| | Thomas 1, level L | Tighenif | Aïn Maarouf | Thomas 1, Hominid level | Oulad Hamida, GDR | Aïn Bahya composite | Doukkala II | Sidi Abderrahmane D2 | Salé | Jebel Irhoud | Thomas Quarry, locus D | Dar Bou Azza, Gazelles | Oulad Hamida 2, Félins | Zouhra |
|---|---|---|---|---|---|---|---|---|---|---|---|---|---|---|
| Herpestes | | | | + | | | | | | | | | | + |
| Lupulella mohibi | | + | | + | + | | | + | | | | | | |
| Canis aureus | | | | | | + | + | | | + | | + | + | + |
| Vulpes | | + | | + | | + | + | | | + | | + | + | + |
| Lycaon | | + | | + | + | | + | | | | | | | |
| Mellivora | | + | | + | + | | | | | | | | | |
| Lutra | | + | | + | | | | | | | | | | |
| Ursus | | + | | + | + | + | | + | | | | | | |
| Crocidura | + | | | + | + | | | | | + | + | | + | |
| Lagomorpha | + | + | | + | + | + | + | | | + | + | + | + | |
| Hystrix | | | | + | + | + | + | | | + | + | + | + | + |
| Arvicanthis | | + | | | | | + | | | | | | | |
| Lemniscomys | | | | | | | | | | + | | | | |
| Mus spretus | | | | | | | | | | + | + | | + | |
| Mus sp. | | | | + | | + | + | | + | | | + | | |
| Mus hamidae | | | | + | + | | | | | | | | | |
| Paraethomys sp. | + | | | | | | | | | | | | | |
| P.tighennifae | + | + | | + | + | | | | | | | | | |
| P.darelbeidae | | | | | | + | + | + | + | | | | | |
| P.ras | | | | | | | | | | + | | | | |
| P.filfilae | | | | | | | | | | + | | | | |
| Praomys | + | + | | + | + | + | | + | + | | | | | |
| Meriones | + | + | | + | + | + | + | + | + | + | + | + | + | + |
| Gerbillus | + | + | | + | + | + | | + | + | + | + | + | + | + |
| Eliomys | | | | + | + | + | | | | | + | | | |
| Ellobius atlanticus | | | | + | + | + | | | | | | | | |
| Ellobius africanus | | + | | | | | | + | | | | | | |
| Ellobius barbarus | | | | | | | | | | | | | + | |
| Jaculus | | | | | | | | | | | | | | |

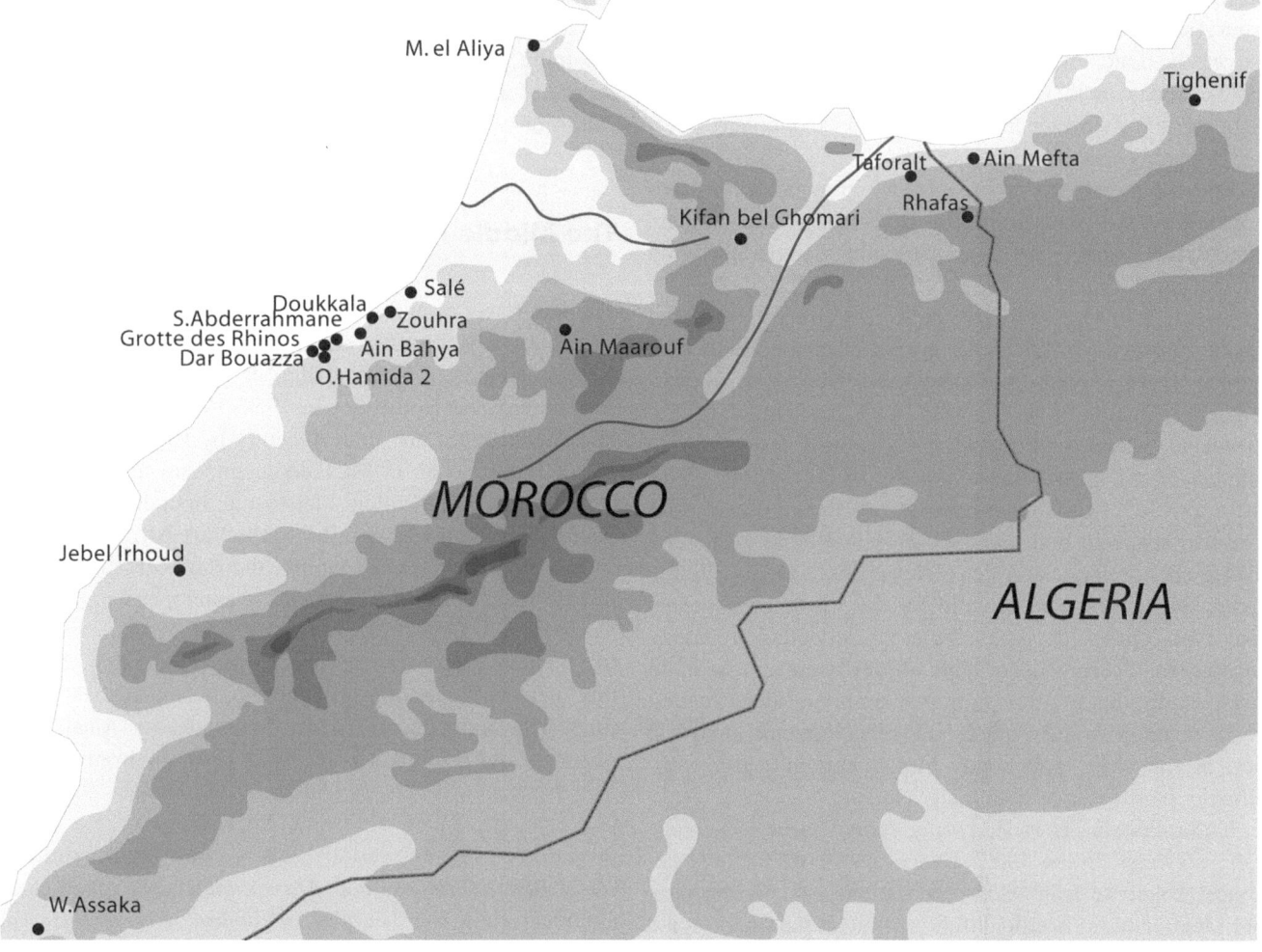

**Fig. 4.1** Map of northwestern Africa with the sites mentioned in the text

assumption emphasizes the need for future archaeozoolog-ical studies.

The small mammals include rodents, insectivores, and bats, but the paleoecological contribution of the latter groups has not yet been investigated in the Maghreb. Rodent diversity is remarkably low in the Pleistocene of North Africa, as there are only about 10 genera, many of which belong to long-lasting lineages. They are very useful for chronology because, as rodent teeth can be collected in large numbers, minor size changes can be detected, and their morphology evolves quickly. The main families are the Muridae and Gerbillidae.

The Gerbillidae now live in deserts or subdesertic areas, and can be considered a good indicator of aridity. They are represented by the living genera *Gerbillus*, known since the Late Pliocene (Geraads 1995) and *Meriones*, known since the Early/Middle Pleistocene (Tong 1986), to which an extinct form known only from Tighenif, *Mascaramys*, must be added (Tong 1986).

The Muridae (rats and mice of today) are found in a variety of environments, and provide no ecological infor-mation *per se* but a great abundance of murids is unlikely to be found in arid regions. These include:

- *Paraethomys*, a genus of large size known since the Late Miocene (Geraads 1998), but which became extinct in the Late Pleistocene;
- *Mus*, the mouse, which is the most common murid since the latest Miocene;
- a genus of intermediate size, referred to the comprehen-sive genus *Praomys*, or to a sub-genus endemic of North-Africa, *Berberomys* (Jaeger 1975);
- *Arvicanthis*, a widespread genus today, but common only at Tighenif in North Africa;
- the rare *Lemniscomys*.

Other rodents are much less abundant. The garden dor-mouse *Eliomys* (Gliridae) appears with the Middle Pleis-tocene, and is still living in North Africa today. The voles are represented by the single genus *Ellobius*, a large burrowing form with procumbent incisors, which emigrated from Asia perhaps 1 Ma, but disappeared from North Africa before the Late Pleistocene for unknown reasons (Jaeger 1988). Today, it is an inhabitant of the steppes of central Asia, so that its presence rules out heavy tree cover, but it was probably able to withstand cold.

Rodents have been much used for climatic reconstructions, but their poor diversity in North Africa hinders quantitative analysis such as estimating temperatures by comparing the numbers of murid and arvicolid species, as was done in Europe, for example. I have instead compared the numbers of gerbillids and murids, illustrated here with pie-diagrams; a high proportion of gerbillids can be taken as a good indicator of aridity (Fig. 4.2). Both numbers are estimated, whenever possible (i.e., in most cases) by the number of m1 s (first lower molars), or by the number of m1 s plus M1 s (first upper molars). As they are of similar size and shape in both families, preservation or collection biases are minimal, and since the genera are always the same, this comparison is valid for the whole Pleistocene. However, it has to be assumed that the collecting predators, such as small carnivores or birds of prey, were similar in all sites. Again, this is an unverified assumption (identification of the predator would require detailed taphonomic analyses) but, in any case, taphonomic bias is certainly less problematic than for large mammals.

Unfortunately, as rodents only started to be systematically collected in the 1970s (Jaeger 1973, 1975), they are almost unknown from older excavations, and in any case old identifications are doubtful. All rodent data here come from recently excavated sites.

Perhaps the most striking aspect of North African mammalian faunas from the Late Pliocene to the Late Pleistocene is their low diversity. For some groups, this is probably partly the result of their latitudinal position (some are intermediate between the intertropical and temperate zones) because diversity decreases from the equator to the poles, but this gradient primarily affects herbivores. For instance, only one site has more than one species of non-human primates, none have more than two Suidae, Equidae, or Proboscidea, none have more than one Giraffidae, and none have more than eight species of rodents. This contrasts with a normal or even great diversity of carnivores, which is not tightly linked to prey diversity. This poor diversity of primary consumers is an indicator of generally harsh conditions. However, it does not imply that they prevailed at every locality, because a temporary return to more favorable conditions for a few thousand years would probably have been unable to restore a greater diversity, simply because other taxa were unavailable there.

The large and small mammals together can be used to estimate the open versus closed nature of the vegetation. I assume that these are roughly equivalent to cold versus warm, allowing the tentative placement of each locality in an isotope stage, given its biochronological age, and assuming also that global changes had a significant impact on the Moroccan continental climate, despite various local conditions.

## The Middle Pleistocene

The late Early or early Middle Pleistocene site of Tighenif (also known as Ternifine) in Algeria is the earliest site with human remains in North Africa. It was first excavated by Arambourg and Hoffstetter in 1954–1956 (Arambourg and Hoffstetter 1963) and more recently by Hublin and colleagues (Geraads et al. 1986). Paleomagnetism suggests an age close to the Early/Middle Pleistocene boundary. The large mammal fauna is very similar to the East African one; taxa such as the elephant *Loxodonta*, the rhinoceros *Ceratotherium*, the suid *Metridiochoerus*, the antelopes *Tragelaphus*, *Hippotragus*, *Oryx*, *Connochaetes*, and *Parmularius*, and the baboon *Theropithecus* are African endemics showing that there was certainly no continuous Saharan belt at that time. The large mammals are dominated by Alcelaphini, gazelles, and oryx, and point to an open environment, but near permanent water, as shown by the abundance of hippopotamuses. There are some fluctuations of the rodent proportions in the various layers, but the Gerbillidae (*Gerbillus*, *Meriones*, and *Mascaramys*) are much more abundant than the Muridae (*Praomys*, *Arvicanthis*, and *Paraethomys*), confirming the openness of the environment. The immigration of the steppic form *Ellobius* also suggests some cooling compared to earlier sites.

A similar fauna, with *Loxodonta*, *Parmularius*, and many hippos, but without rodents, was found at Aïn Maarouf in Morocco, also with a hominid remain (Geraads et al. 1992; Geraads and Amani 1997), and a similar age and environment can be inferred for this site. Both Tighenif and Aïn Maarouf can best be assigned to a cold stage, perhaps around OIS (Oxygen Isotope Stage) 20.

Later sites of the Middle Pleistocene are concentrated on the Moroccan Coast, between Rabat and Casablanca, all within a distance of less than 100 km (Geraads 2010a, b; Raynal et al. 2010). The Thomas quarries are located at the southwestern border of Casablanca, not far from the coast. Quarry works have cut through a succession of marine and continental deposits with a complex stratigraphy. Several archaeological levels are being excavated in the frame of the "Programme Casablanca" of the "Institut National des Sciences de l'Archéologie et du Patrimoine." They have yielded several hominid remains and large faunal samples, ranging from the Early Pleistocene (level L) to the Late Pleistocene. The fauna from level L is too poor to be

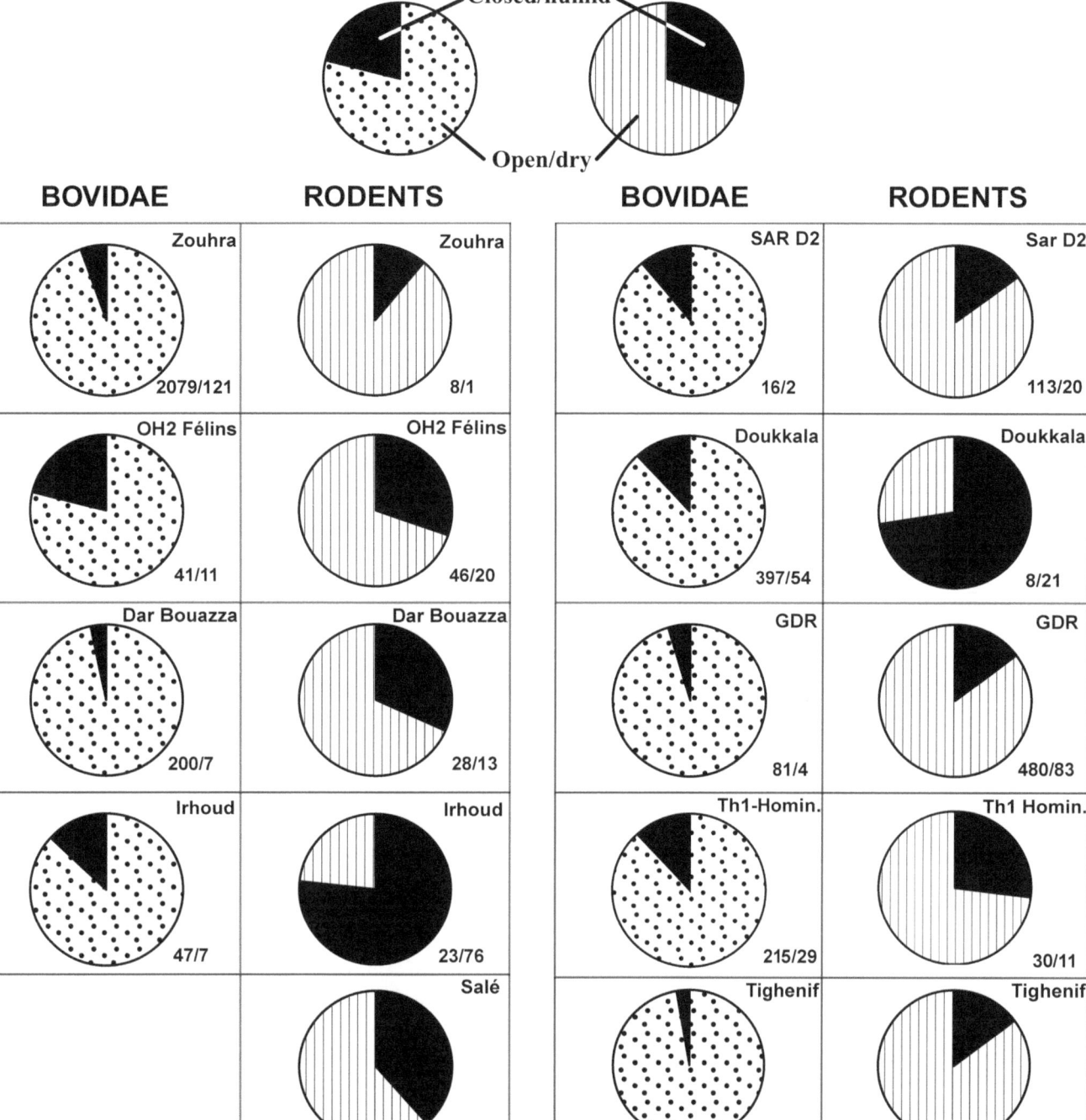

**Fig. 4.2** Pie-diagrams showing the relative abundances of Alcela-phini plus Antilopini versus other non-hippotragine bovids (*left columns*) and of Gerbillidae versus Muridae (*right columns*). Raw numbers of specimens are also given (number of identified specimens for bovids, number of first molars for rodents); the first number is that of the taxa of open-dry environments, the second one is that of those of closed-humid ones

ecologically meaningful. The fauna of the "Hominid level" at Thomas 1, which is probably younger than Tighenif, was mostly accumulated by carnivores, mainly Canidae, whose remains are quite common (Bernoussi 1997). The assemblage is similar to that of the Algerian site, but the dominance of gerbillids and of open country antelopes is less marked, and the rodent fauna shows some diversity, likely

reflecting some diversity in the environment. That this is not just the result of Thomas 1 being a cave near the coast is shown by the "Grotte des Rhinocéros" (Rhino Cave = GDR) in the nearby quarry of Oulad Hamida 1, which has also been excavated for more than 15 years by Raynal and colleagues (Raynal et al. 1993; Rhodes et al. 1994). It is biochronologically slightly younger than

Thomas 1–Hominid (Geraads 2002), but the environment is more similar to that of Tighenif, with very high proportions of Alcelaphini and gazelles among bovids, and of gerbils among rodents. The main difference with Tighenif is that *Mus* replaces *Arvicanthis* (Geraads 1994). The white rhinoceros *Ceratotherium*, which is very common in the site, is also a grazer that today lives in savannahs, and confirms the open dry physiognomy of the landscape, suggesting a cold stage, perhaps OIS 14 (Rhodes et al. 2006).

Thus, data from all these sites strongly suggest that the known early members of *Homo mauritanicus* (see Hublin 2001, for a discussion of the taxonomy of northwestern African hominids) lived in an environment that was fairly open and dry (perhaps slightly less so at Thomas 1-Hominid), but this is only an observation that cannot be falsified, as no other large mammal site is known in this period and area.

Aïn Bahya, between Rabat and Casablanca, is quite an interesting site, but is unfortunately hard to excavate because it was found in the railway cutting (Michel 1989). It has yielded some artifacts, a few large mammals, and good samples of rodents from various spots in the locality (Cheddadi 1986), which show that it is younger than GDR. All of the assemblages from Aïn Bahya are remarkable in their abundance of *Ellobius* and *Paraethomys*, while the gerbillids are always less common than the murids. The dominance of the murids certainly reflects relatively wet conditions, although the abundance of *Ellobius* argues against dense tree cover. Given its likely biochronological age, Aïn Bahya could belong to OIS 13 or 11.

The sites of Doukkala I and II between Rabat and Casablanca (Michel 1989; Michel and Wengler 1993) were both destroyed. There are no human remains, but some artifacts were found at Doukkala II. Biochronologically, the fauna clearly belongs to the Middle Pleistocene, with *Ceratotherium* (the mention of "*Dicerorhinus*" (recte: *Stephanorhinus*) *hemitoechus* by Michel 1989, is erroneous) and a *Paraethomys* close to *P. darelbeidae* (and not, as reported by Michel 1989, and Michel and Wengler 1993, to *P. ras*, a species known only at Irhoud). Unfortunately, although the infillings were said to be stratified, no data has ever been published about the stratigraphy and the published faunal lists (Michel 1989; Michel and Wengler 1993) are composite; even mixing between sites cannot be ruled out (Michel 1989). At Doukkala II, it may be that the rodents are from a different level than the bulk of the large mammal assemblage, which might help to explain the differences between the pie-diagrams. The large mammal fauna is not very characteristic, but the murids, which are abundant and rather diverse, despite the small sample size, with an *Arvicanthis* in addition to *Paraethomys* and *Mus*, and a single tooth of *Ellobius*, point to a temperate climate.

Sidi Abderrahmane near Casablanca is a reference site for the Pleistocene Moroccan stratigraphy (Biberson 1961). Biberson collected several large mammal assemblages from the various Amirian to Soltanian levels, but unfortunately most of them are now mixed in the collections of the Paris Muséum d'Histoire Naturelle and Faculté des Sciences of Rabat. However, I have been able to gather a small collection from level D2, which, fortunately, is also one of the levels from which Jaeger (1975: Sidi Abderrahmane 2) collected rodents. It is younger than the level that yielded the human mandible, and from the size and evolutionary stage of the rodents, D2 is certainly younger than Aïn Bahya. It is clear that we revert here to open conditions, with the gerbillids forming the bulk of the rodent fauna. This would indicate a cold climate, and we can tentatively refer this level to OIS 10.

A quarry near Salé, North of Rabat, yielded one of the most complete early hominid specimens of North Africa, probably assignable to a primitive *Homo sapiens* (see Hublin 1985). The associated large mammal fauna consists only of *Ceratotherium*, *Equus*, and *Connochaetes* (Jaeger 1973). These taxa again suggest open conditions, but the list is too short to draw firm conclusions. Unfortunately, the site is no longer within reach. Rodents were collected by Jaeger (1973, 1975) from a red breccia above the human skull, but the duration of the time gap between the deposition of each level is unknown. Rodents indicate an age younger than GDR or Aïn Bahya, but its age relative to Sidi Abderrahmane is doubtful, although stratigraphic data support a younger age for Salé (Raynal 2007, personal communication). The precise numbers of specimens of *Mus* and *Praomys* are unknown, but even assuming that these murids were each represented by a single m1, the proportion of Muridae is greater than in the previous sites. A similarly high proportion of murids is found in the roughly contemporary site of Aïn Mefta in Algeria (Jaeger 1975) at a relatively high elevation (the precise location of the site is unknown), and associated with *Macaca*, a rare taxon in North Africa, indicating wet and/or forested conditions. Thus, we can tentatively assign Salé to a temperate phase, perhaps a transitional phase before or after OIS 10, if not OIS 8. *Ellobius* became extinct afterwards. Since this genus certainly favored environments similar to the present-day continental steppes of Asia, we can guess that its extinction was caused by some warming.

The site of Jebel Irhoud has recently been dated to at least 160 ka (Smith et al. 2007). Ennouchi collected a good sample of large mammals from Jebel Irhoud, but it fully lacks stratigraphic data (Amani 1991; Amani and Geraads 1993; Amani and Geraads 1998). Conventional excavations were later conducted by Tixier and de Bayle des Hermens, but again no data have been published on the provenance of the fauna that they collected, and that Thomas (1981) studied. Among the bovids, the proportion of Bovini (a form close to *Bos primigenius*) is significant, but the most remarkable aspect of the assemblage is the overwhelming dominance of the gazelles; Jebel Irhoud is the first site in which they far outnumber all other bovids. As no archaeozoological analysis has been undertaken yet, we cannot tell whether this is because other bovids had become rare in the environment or because gazelles were selectively hunted, but in this regard, it is interesting to note that there is a significant difference in the collections made by Tixier and Ennouchi. *Gazella atlantica* is common in the collection made by Ennouchi, but absent in the collection made by Tixier, where *Gazella cuvieri* is the only gazelle. As the Ennouchi sample is richer in carnivore remains, it may be that the carnivores preferred *Gazella atlantica*, and humans preferred *Gazella cuvieri*. Another possibility is that Jebel Irhoud was near the limit of the range of one or both species, and that there were fluctuations of their relative abundances. In any case, it is clear that the faunal samples collected by Ennouchi and Tixier are not identical, suggesting possible heterochrony of the human remains.

The rodent sample collected by Ennouchi has a high proportion of murids, mostly *Paraethomys*, which are the last representatives of this genus in Morocco. Although the location of the site—it is about 50 km from the sea, at an elevation of more than 500 m—may bias comparisons with those of the coast, this assemblage points to a rather warm climate, perhaps OIS 5, and certainly not OIS 6, a stage that seems to be unrepresented in the mammalian record of North Africa.

The sharp contrast between the environments sampled at Aïn Bahya and Sidi Abderrahmane, for instance, demonstrates the ecological diversity of the environments colonized by late Middle Pleistocene hominids on the Atlantic Coast of Morocco. We may note that the hominid remains from Salé and Jebel Irhoud are less clearly associated with open environments than earlier hominids, but assuming that Moroccan early *Homo sapiens* lived in more forested areas than *H. mauritanicus* would be a rash conclusion.

## The Late Pleistocene

There were many changes at the beginning of the Late Pleistocene, with several immigration events, the chronology of which remain imprecise, from Eurasia. Among large mammals, these involve the rhinoceros *Stephanorhinus*, the wild boar *Sus*, and cervids. All these taxa have been reported from the earlier site of Doukkala (Michel 1989), but we have seen above that the rhino there is, in fact, *Ceratotherium*. The cervids, reported by Laquay (1986), have never been listed by Michel in his various faunal lists (Michel 1989; Michel and Wengler 1993), so that their provenance is doubtful. Only *Sus*

might indeed be present, but is represented by juvenile specimens only, which I could not access in Rabat.

With these new immigrants that put a European stamp on it, the North African large mammal fauna became more different than the East African one. However, these animals are never very common; as a whole, the fauna, at least in Morocco, loses diversity, with most sites now being primarily gazelle assemblages but their chronological succession is rather uncertain.

Above the speleothem that covers the Middle Pleistocene layers of Thomas 1, thick red infillings, called "Thomas 1-D," contain a rich microfauna, but no large mammals, and similar deposits, called "Thomas-rouge," occur in small pockets within level L of the same quarry (Raynal and Texier 1989). The rodent fauna is clearly impoverished, after the disappearance of *Ellobius*, *Paraethomys* (which had been present for 6 Myr), and *Praomys*, but it still has *Eliomys*.

Dar Bou Azza "Grotte des Gazelles" is a cave south of Casablanca that was discovered in 2005 during housing estate works, and soon destroyed. The stone-tool industry looks Aterian. The large mammal fauna consists almost exclusively of gazelles; among rodents, murids are not rare but all are mice. We do not know how much of the faunal assemblage is anthropic; it may be that human hunting bias increased the proportion of gazelles, but in any case, they certainly made up the bulk of the mammalian biomass.

The "Grotte des Félins" at Oulad Hamida 2 is another site in the same area that was destroyed before it could be properly excavated, also with a possible Aterian industry (Raynal et al. 2008). There are some buffaloes with the gazelles, but, again, all murids are mice. An unexpected occurrence is that of a jerboa, *Jaculus* sp. Today, this genus is represented in Morocco by two species living in deserts or subdeserts, *J. jaculus* and *J. orientalis*, the latter extending into the Atlas mountains (Aulagnier 1992), but the genus had never been reported, either living or in fossil form, in the coastal plains. Oulad Hamida 2 is perhaps not far from the climax of arid conditions on the seashore.

There are several other caves on the coast between Rabat and Casablanca, but not much has been published on the fauna. The levels with Aterian artifacts of the Zouhra Cave (also called El Harhoura 1) are the main exception. Again, gazelles form the bulk of the ungulate fauna (Aouraghe 2004, 2007, personal communication), and alcelaphines plus gazelles make up more than 90% of the bovid assemblage. The rodent sample is quite small but is again dominated by *Meriones* (Aouraghe and Abbassi 2002). All of this once more suggests open dry conditions.

More to the north, near Tangier, Mugharet el 'Aliya contains several levels dated to around 50 ka, in which the proportion of gazelles increases upwards, reaching 90% in

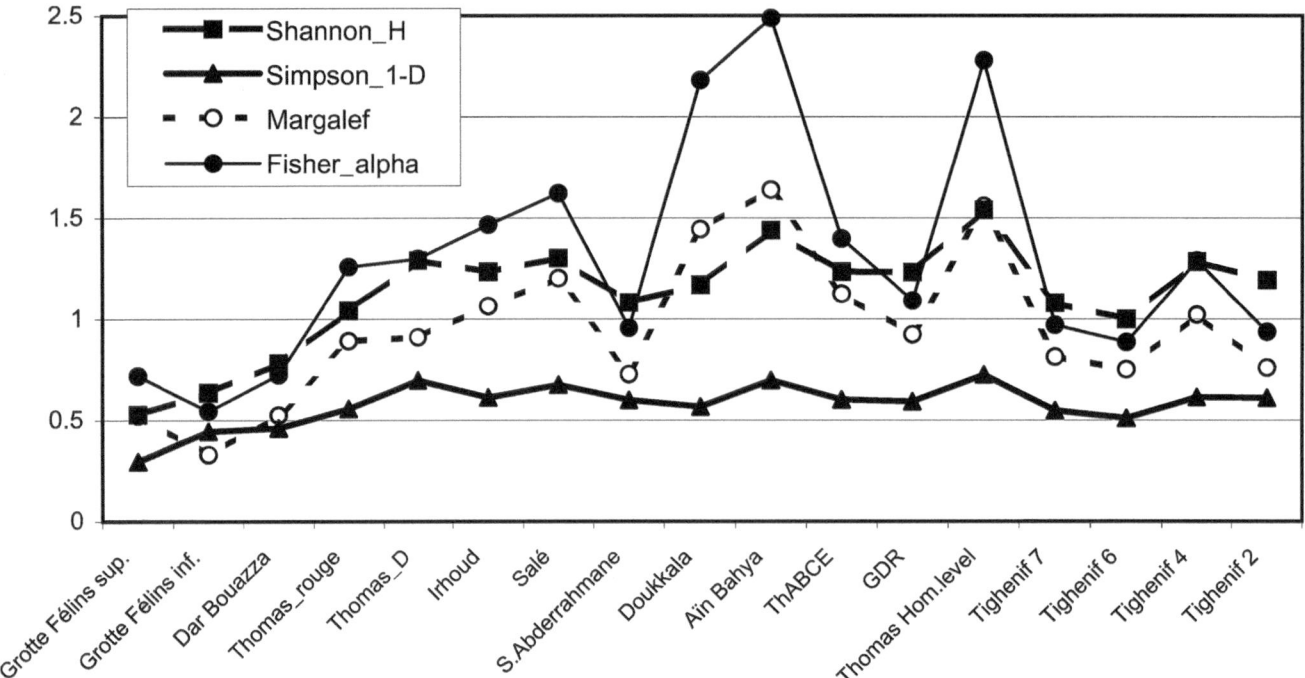

**Fig. 4.3** Graph showing the decrease in rodent diversity in the Late Pleistocene, using four commonly used indices, computed from the numbers of first molars. Sites are arranged according to their estimated age, decreasing from *right* to *left* (see Table 4.1)

the uppermost layer (Wrinn 2001; Wrinn and Rink 2003, personal communication).

Among large mammals, the main faunal change during the course of the Late Pleistocene of Morocco is the disappearance of the Cervidae and of *Stephanorhinus*, which are seemingly unknown above the Aterian levels; these local extinctions further decrease the large mammal diversity. As mentioned above, this could be partly due to human hunting preferences, but this decrease in diversity is also true of rodents, as shown in Fig. 4.3. Although it would be simplistic to directly relate high diversity to high vegetation productivity, it is very likely that the trend towards loss of diversity in the second part of the Late Pleistocene is significant, and that, after OIS 5, the coastal plain experienced a marked aridification, although not severe enough to force humans to migrate.

Other areas are less well-documented, but they show a somewhat different picture. More to the south, in the coastal part of the Anti-Atlas Mountains, Wengler et al. (2002) discovered some large mammals, primarily aurochs, dated to ca. 30–60 ka at Oued Assaka.

On the Algerian Coast, near Algiers, there are a number of cave fillings, but as most of them were excavated a long time ago, no radiometric data is available and thus their relative chronology is based upon cursory stratigraphic interpretations. They are loosely, if at all, associated with human presence, and their faunas are likely to provide a good picture of the environment. Unfortunately again, very few precise data are available about faunal proportions,

which can only be roughly estimated from the literature. *Hystrix* and *Mus* are the only reported rodents. At Pointe-Pescade (Pomel 1894), Bains Romains (Arambourg 1931), and Anglade (Arambourg 1935), the Bovini are probably dominant, as they definitely are in Grotte des Phacochères (Hadjouis 1985, 2002) and Sidi Said (Saoudi 1996), although *Gazella* is probably more common at Sintès (Arambourg 1932), but very rare at Anglade. Alcelaphines are rare or absent. Aïn Benian has the greatest diversity of ruminants, with nine species, and the gazelles make up only about one-third of the assemblage (Chaid-Saoudi, 2007, personal communication). Taphonomic and/or collecting biases are likely, but the overall picture of the Algerian Coast is definitely not that of a very open/dry or even sub-desertic landscape, in contrast to Morocco.

There are almost no data from the inland regions. At Rhafas, the only site that has been studied in detail (Michel 1989), there are only a few gazelles, although the sample size is very small. No data is available yet for the nearby site of Taforalt.

The fauna from the Aterian levels of Kifan bel Ghomari, excavated in the early twentieth century, is being revised by Chaid-Saoudi. The abundance of carnivores (mostly hyenas and canids) suggests that they were the main collectors, and that the high proportion of gazelles is not the result of human bias. However, several other bovid taxa are also present, and the fauna seems to have been more balanced than on the Atlantic Coast.

## Conclusions

For the few known sites of the first half of the Middle Pleistocene, the environment was clearly characterized by open landscapes, but some more humid and/or forested areas, perhaps inhabited by early *Homo sapiens*, are documented during the second half of this period. There is good evidence that aridity increased on the Atlantic Coast during the Late Pleistocene, leading to a loss of faunal diversity. Nonanthropic sites are badly needed there, but their absence at least does not falsify the hypothesis of a continuous human presence. In other areas, the picture is far less clear, and more excavations and archaeozoological studies would be welcome, but again, nothing speaks against a permanent occupancy of large parts of the Moroccan and northwestern Algerian inland regions, at least at some elevation.

**Acknowledgments** I am grateful to Jean-Jacques Hublin and Shannon McPherron for inviting me to the Leipzig symposium. Thanks also to H. Aouraghe, Y. Chaid-Saoudi, and P. Wrinn, who improved my data by providing unpublished results, and to J.-P. Raynal and three anonymous reviewers for helpful comments. Most of my research in Morocco is conducted within the "Programme Casablanca."

## References

Amani, F. (1991). *La faune du gisement à Hominidés du Jebel Irhoud. Contribution à l'étude de la chronologie et de l'environnement du Quaternaire marocain*. Université de Rabat: Thèse de troisième cycle.

Amani, F., & Geraads, D. (1993). Le gisement moustérien du Djebel Irhoud, Maroc : précisions sur la faune et la biochronologie, et description d'un nouveau reste humain. *Comptes-rendus de l'Académie des Sciences II, 316*, 847–852.

Amani, F., & Geraads, D. (1998). Le gisement moustérien du Djebel Irhoud, Maroc: précisions sur la faune et la paléoécologie. *Bulletin d'archéologie marocaine, 18*, 11–18.

Aouraghe, H. (2004). Les populations de mammifères atériens d'El Harhoura 1 (Témara, Maroc). *Bulletin d'Archéologie marocaine, 20*, 83–103.

Aouraghe, H., & Abbassi, M. (2002). Les rongeurs du site atérien d'El Harhoura 1 (Témara, Maroc). *Quaternaire, 13*, 125–136.

Arambourg, C. (1931). Observations sur une grotte à ossements des environs d'Alger. *Bulletin de la Société d'Histoire Naturelle d'Afrique du Nord, 22*, 169–176.

Arambourg, C. (1932). Note préliminaire sur une nouvelle grotte à ossements des environs d'Alger. *Bulletin de la Société d'Histoire Naturelle d'Afrique du Nord, 23*, 154–162.

Arambourg, C. (1935). La Grotte de la Carrière Anglade à Guyotville (Dpt d'Alger). *Bulletin de la Société d'Histoire Naturelle de l'Afrique du Nord, 26*, 15–22.

Arambourg, C., & Hoffstetter, R. (1963). Le gisement de Ternifine. I. Première partie—historique et géologie. *Archives de l'Institut de Paléontologie humaine, 32*, 9–36.

Aulagnier, S. (1992). *Zoogéographie des mammifères du Maroc : de l'analyse spécifique à la typologie de peuplement à l'échelle régionale*. Université de Montpellier II: Thèse d'état.

Bernoussi, R. (1997). *Contribution à l'étude paléontologique et observations archéozoologiques pour deux sites du Pléistocène moyen du Maroc Atlantique: la Grotte à Hominidés de la carrière Thomas 1 et la Grotte des Rhinocéros de la carrière Oulad Hamida 1 (Casablanca, Maroc)*. Thèse de doctorat, Université Bordeaux 1.

Biberson, P. (1961). Le cadre paléogéographique de la préhistoire du Maroc atlantique. *Publications du service des antiquités du Maroc, 16*, 1–235.

Cheddadi, A. (1986). *Les rongeurs d'age pléistocène moyen du site de l'Aïn Bahya (Skhirat, Maroc atlantique); Implications stratigraphiques et paléoécologiques*. Rabat: Thèse ENS Souissi.

Eisenmann, V. (1984). Sur quelques caractères adaptatifs du squelette d'*Equus* (Mammalia, Perissodactyla) et leurs implications paléoécologiques. *Bulletin du Muséum National d'Histoire Naturelle, 4ème série, C, 6*, 185–195.

Geraads, D. (1994). Rongeurs et Lagomorphes du Pléistocène moyen de la "Grotte des Rhinocéros", carrière Oulad Hamida 1, à Casablanca, Maroc. *Neues Jahrbuch für Geologie und Paläontologie Abhandlungen, 191*, 147–172.

Geraads, D. (1995). Rongeurs et Insectivores du Pliocène final de Ahl al Oughlam, Casablanca, Maroc. *Geobios, 28*, 99–115.

Geraads, D. (1998). Rongeurs du Mio-Pliocène de Lissasfa (Casablanca, Maroc). *Geobios, 31*, 229–245.

Geraads, D. (2002). Plio-Pleistocene mammalian biostratigraphy of atlantic Morocco. *Quaternaire, 13*, 43–53.

Geraads, D. (2010a). Biochronologie mammalienne du Quaternaire du Maroc atlantique, dans son cadre régional. *L'Anthropologie, 114*, 324–340.

Geraads, D. (2010b). Biogeographic relationships of Pliocene and Pleistocene North-western African mammals. *Quaternary International, 212*, 159–168.

Geraads, D., & Amani, F. (1997). La faune du gisement pléistocène moyen de l'Aïn Maarouf près de El Hajeb (Maroc). *L'Anthropologie, 101*, 522–530.

Geraads, D., Hublin, J.-J., Jaeger, J.-J., Sen, S., Tong, H., & Toubeau, P. (1986). The Pleistocene Hominid site of Ternifine, Algeria: new results on the environment, age, and human industries. *Quaternary Research, 225*, 380–386.

Geraads, D., Amani, F., & Hublin, J.-J. (1992). Le gisement pléistocène moyen de l'Aïn Maarouf près de El Hajeb, Maroc: présence d'un Hominidé. *Comptes-rendus de l'Académie des Sciences II, 314*, 319–323.

Hadjouis, D. (1985). Les bovidés (Artiodactyla, Mammalia) du gisement atérien des phacochères (Alger, Algérie). Interprétations paléoécologiques et phylogénétiques. *Comptes-rendus de l'Académie des Sciences II, 301*, 1251–1254.

Hadjouis, D. (2002). Un nouveau Bovini dans la faune du Pléistocène supérieur d'Algérie. *L'Anthropologie, 106*, 377–386.

Hublin, J.-J. (1985). Human fossils of the North African Middle Pleistocene and the origin of Homo sapiens. In E. Delson (Ed.), *Ancestors: The hard evidence* (pp. 283–288). New York: Alan R. Liss.

Hublin, J.-J. (2001). Northwestern African Middle Pleistocene hominids and their bearing on the emergence of *Homo sapiens*. In L. Barham & K. Robson-Brown (Eds.), *Human roots. Africa and Asia in the Middle Pleistocene* (pp. 99–121). Bristol: CHERUB, Western Academic and Specialist Press Ltd.

Jaeger, J.-J. (1973). Les faunes de Mammifères et les Hominidés fossiles du Pléistocène moyen du Maghreb. *Travaux de la R.C.P., 29*, 265–290.

Jaeger, J.-J. (1975). *Les Muridae (Mammalia, Rodentia) du Pliocène et du Pléistocène du Maghreb. Origine; Evolution; Données biogéographiques et paléoclimatiques*. Université de Montpellier: Thèse d'état.

Jaeger, J.-J. (1988). Origine et évolution du genre *Ellobius* au Maghreb. *Folia Quaternaria, 57*, 3–50.

Laquay, G. (1986). *Cervus elaphus* (Mammalia, Artiodactyla) du Pléistocène supérieur de la carrière Doukkala II (Rabat, Maroc). Sa comparaison avec le cerf würmien de France. *Revue de Paléobiologie, 4,* 143–147.

Michel, P. (1989). *Contribution à l'étude paléontologique des Vertébrés fossiles du Quaternaire marocain à partir de sites du Maroc atlantique, central et oriental.* Thèse de doctorat. Paris: Muséum National d'Histoire Naturelle.

Michel, P., & Wengler, L. (1993). Le site paléontologique et archéologique de Doukkala II (Maroc, Pleistocène moyen et supérieur): premier jalon en Afrique du Nord d'un comportement humain assimilable à un "charognage contrôlé et actif". *Comptes-rendus de l'Académie des Sciences II, 317,* 557–562.

Pomel, A. (1894). Sur une nouvelle grotte ossifère découverte à la Pointe-Pescade, à l'Ouest d'Alger-Saint-Eugène. *Comptes-rendus de l'Académie des Sciences, 119,* 986–989.

Raynal, J.-P., & Texier, J.-P. (1989). Découverte d'Acheuléen ancien dans la carrière Thomas 1 à Casablanca, et problème de l'ancienneté de la présence humaine au Maroc. *Comptes-rendus de l'Académie des Sciences II, 308,* 1743–1749.

Raynal, J.-P., Geraads, D., Magoga, L., El Hajraoui, A., Texier, J.-P., Lefèvre, D., et al. (1993). La grotte des Rhinocéros (carrière Oulad Hamida 1, anciennement Thomas III, Casablanca), nouveau site acheuléen du Maroc atlantique. *Comptes-rendus de l'Académie des Sciences II, 316,* 1477–1483.

Raynal, J.-P., Amani, F., Geraads, D., el Graoui, M., Magoga, L., Texier, J.-P., et al. (2008). Felids Cave, a new Upper Pleistocene Palaeolithic site at Casablanca (Morocco). *L'Anthropologie, 112,* 182–200.

Raynal, J.-P., Hublin, J.-J., & Geraads, D. (2010). Hominid Cave at Thomas Quarry I (Casablanca, Morocco): Recent findings and their context. *Quaternary International, 223–224,* 369–382.

Rhodes, E. J., Raynal, J.-P., Geraads, D., & Sbihi-Alaoui, F. Z. (1994). Premières dates RPE pour l'Acheuléen du Maroc atlantique (Grotte des Rhinocéros, Casablanca). *Comptes-rendus de l'Académie des Sciences II, 319,* 1109–1115.

Rhodes, E. J., Singarayer, J. S., Raynal, J.-P., Westaway, K. E., & Sbihi-Alaoui, F.-Z. (2006). New age estimates for the Palaeolithic assemblages and Pleistocene succession of Casablanca, Morocco. *Quaternary Science Reviews, 25,* 2569–2585.

Saoudi, Y. (1996). *Les vertébrés de Sidi Saïd (Tipaza). Résultats des premières campagnes de fouilles. In Peuplement et environnement naturel* (pp. 49–58). Alger: Journée d'étude du CNRPAH.

Smith, T. M., Tafforeau, P., Reid, D. J., Grün, R., Eggins, S., Boutakiout, M., et al. (2007). Earliest evidence of modern human life history in North African early *Homo sapiens. Proceedings of the National Academy of Sciences of the USA, 104,* 6128–6133.

Thomas, H. (1981). La faune de la Grotte à Néanderthaliens du Jebel Irhoud (Maroc). *Quaternaria, 23,* 191–217.

Tong, H. (1986). The Gerbillinae (Rodentia) from Tighennif (Pleistocene of Algeria) and their significance. *Modern Geology, 10,* 197–214.

Vrba, E. S. (1975). Some evidence of chronology and palaeoecology of Sterkfontein, Swartkrans and Kromdraai from the fossil Bovidae. *Nature, 254,* 301–304.

Wengler, L., Weisrock, A., Brochier, J.-E., Brugal, J.-P., Fontugne, M., Magnin, F., et al. (2002). Enregistrement fluviatile et paléoenvironnements au Pléistocène supérieur sur la bordure atlantique de l'Anti-Atlas (Oued Assaka, S-O marocain). *Quaternaire, 13,* 179–192.

Wrinn, P. J. (2001). *Reanalysis of the Pleistocene archaeofauna from Mugharet el 'Aliya, Tangier, Morocco: implications for the Aterian.* Comm. Liège: presented at the XIVth UISPP Congress.

Wrinn, P. J., & Rink, W. J. (2003). ESR dating of tooth enamel from Aterian levels at Mugharet el 'Aliya (Tangier, Morocco). *Journal of Archaeological Science, 30,* 123–133.

# Chapter 5
# New Data from the Site of Ifri n'Ammar (Morocco) and Some Remarks on the Chronometric Status of the Middle Paleolithic in the Maghreb

D. Richter, J. Moser and M. Nami

**Abstract** The definition and chronometric position of the Middle Paleolithic technocomplex of the "Aterian" are heavily debated. While for some, the presence of tanged lithics is sufficient to warrant the attribution of an assemblage to the "Aterian," for others there is more to the "Aterian," which is, however, defined in different ways. Here, we present thermoluminescence (TL) dating results on the multiple layer site of Ifri n'Ammar (Morocco), with an alternating double sequence of assemblages previously described as "Mousterian" and "Aterian" based on the presence/absence of tanged lithics. All the Middle Paleolithic industries at Ifri n'Ammar are technologically and typologically relatively similar to each other, as well as to the European Mousterian, with the most significant difference being the presence of tanged lithic artifacts in some layers. We prefer to use the term "Middle Paleolithic of Aterian facies/aspect," instead of assigning it the status of a distinct technocomplex which is used in a chronostratigraphical sense. TL data of $83.3 \pm 5.6$ ka (weighted; $n = 10$) on heated lithics is obtained for the uppermost Middle Paleolithic level (Upper OS), which contains tanged lithic tools, as well as personal ornaments, while the underlying layer (Lower OS), which lacks tanged pieces, dates to $130.0 \pm 7.8$ ka ($n = 9$). The latter age is also a minimum for its underlying layer, which again contains tanged items, and thus the first appearance of tanging is significantly older in comparison with chronometric data from other sites. The discrepancy between TL and radiocarbon dates from the same levels at Ifri n'Ammar brings into question the reliability of radiocarbon analysis at the limits of the method.

**Keywords** Aterian • Dating • Ifri n'Ammar • Maghreb • Middle Paleolithic • Middle Stone Age • Mousterian • Northwest Africa • Thermoluminescence

## The Maghrebine Middle Paleolithic and the Available Chronometric Data

The Maghreb is rich in Paleolithic sites, spanning from the Lower Paleolithic (e.g., Aïn Hanech or Carrière Thomas) to the Upper Paleolithic technocomplex of the Iberomaurusian (e.g., Taforalt or Ifri n'Ammar). While some authors prefer to include Northwest Africa into the African system of "Early/Middle/Late Stone Age" (e.g., McBrearty and Brooks 2000), others refer to the Middle Paleolithic from a European perspective, especially by using the term "Mousterian" (e.g., Carrière 1886; Nehren 1992; Debénath 2000).

However, the large areas and distance to the geographical areas where these terms were defined make the use of such terminologies for the Maghreb questionable. This is especially true for the Middle Paleolithic technocomplex of the "Aterian," which is closely associated, and, often exclusively, defined by tanged lithics as *fossil directeur*. However, any type fossil approach must consider equifinality, representativeness, and context, which, to a large extent, is lacking in the discussion on the Aterian. It is generally believed to have developed gradually out of the local Middle Paleolithic ("Mousterian") (e.g., Hahn 1984; Wengler 1997). However, it is important to underline that North Africa is approximately equal to Europe in area, and that the phenomenon of tanged lithic artifacts can be

D. Richter (✉)
Department of Human Evolution, Max Planck Institute of Evolutionary Anthropology, Deutscher Platz 6, 04103 Leipzig, Germany
e-mail: drichter@eva.mpg.de

J. Moser
Kommission für Archäologie Außereuropäischer Kulturen des Deutschen Archäologischen Instituts, Dürenstrasse 35-37, 53173 Bonn, Germany
e-mail: moser@kaak.dainst.de

M. Nami
Centre d'Inventaire et de Documentation du Patrimoine (CIDP), Ministère de la Culture, Rabat, Morocco
e-mail: m.nami@caramail.com

J.-J. Hublin and S. P. McPherron (eds.), *Modern Origins: A North African Perspective*, Vertebrate Paleobiology and Paleoanthropology, DOI: 10.1007/978-94-007-2929-2_5, © Springer Science+Business Media B.V. 2012

observed over an even larger geographical distribution, where such items can be found in Middle Paleolithic context on the Arabian Peninsula (McClure 1994) and even as far as India (James and Petraglia 2005). The definition or attribution of a technocomplex over such huge areas based on only one component (i.e., tanging) has to be questioned, as it would be difficult to explain by cultural theory.

The present day definition of the "Aterian" is unclear and little information has been provided in recent studies. It was originally defined by Reygasse (1921–1922, 1922a, b), was refined by Debénath et al. (1986), and was expanded upon recently by Wengler (2006). However, few authors provide information on their definition when an assemblage is attributed to the "Aterian." The problems with defining and using the term "Aterian" have received a great deal of attention. In a summary of the evidence for the "Aterian," Ferring (1975, p. 113) diplomatically notes that "…the problems are largely theoretical…" in delineating "Aterian" units from those preceding (or following) in time. But before this diplomatic description, he starts with the remark, "WHAT A MESS!" or, as Kleindienst (2001, p. 1) puts it, "The Aterian Industrial Complex, or Technocomplex, is poorly described, poorly dated, and suffers in its interpretations from a number of misconceptions imposed by prehistorians since it was first designated as a separate cultural stratigraphic unit by Reygasse (Reygasse 1921–1922, 1922b)." This point of view is also reflected in the title of her paper, "What is the Aterian?" (Kleindienst 2001).

The "Aterian" of the Maghreb is stratigraphically located above "Mousterian" in most sites and is always found below the Upper Paleolithic Iberomaurusian assemblages, if present (e.g., Bouzouggar et al. 2008), with the majority of sites located in Morocco. However, there is a growing body of evidence challenging the general validity of such a succession. Several sites have yielded assemblages with tanged items below "Mousterian," as, for example, in Wadi Gan (McBurney and Hey 1955, cited in Moyer 2003), Haua Fteah (Wendorf and Schild 1992), Uan Tabu and Uan Afuda (Cremaschi et al. 1998), Aïn El-Guettar (Aouadi-Abdeljaouad and Belhouchet 2008), Ifri n'Ammar (Mikdad et al. 2004), and even at Rhafas, where the presence of a single tanged piece did not lead to the interpretation of the assemblage as "Aterian" by Wengler (1997).

Even though these tanged pieces occur in small numbers in such contexts, the attempt to define a "culture" or technocomplex as "Aterian" by a certain percentage of tanged items (Bordes 1976–1977) is questionable, because of the dependencies of such criteria on assemblage size, site function, area excavated, etc. Despite the chronostratigraphically differentiated position of assemblages with tanged items, it is technically possible to apply the term technocomplex *sensu strictu*, but it is difficult to use the term "Aterian" in a cultural or ethnic sense, as a

chronometric marker, or in arguing for an "evolution" of lithic technologies with respect to tanging. Explanations other than a simple chronological succession have to be sought, like function, site function, ethnic groups, etc., for which more evidence from modern excavations is needed. Only then will it be possible to show links that will allow us to define and decide which system is appropriate. Here, we use the term Middle Paleolithic in a general sense, without the intention of invoking the European system.

## Chronometric Data on the Middle Paleolithic of the Maghreb

Until recently, the chronometric status of the "Aterian" in Northwest Africa was based mainly on radiocarbon data. In this study, we refer to radiocarbon data on the dimensionless $^{14}$C time scale ($^{14}$C years), unless it was converted to the calendric time scale (cal BP). Data from other chronometric dating methods is always referred to on the calendric time scale. More generalized radiocarbon data is not specifically marked when it is converted and is thus directly comparable on the calendric time scale.

The radiocarbon dating results led to the notion of placing the "Aterian" between 40 and 20 $^{14}$C ka (e.g., Debénath 1992). This time range has been expanded dramatically at its lower end by employing other dating techniques like Electron Spin Resonance (ESR) dating of tooth enamel at Mugharet el 'Aliya to 60 ka (Wrinn and Rink 2003), and TL dating of heated lithics at Rhafas to 80–60 ka (Mercier et al. 2007). Optically Stimulated Luminescence (OSL) dating results at Taforalt (Bouzouggar et al. 2007), and in Lybia at Uan Afuda and Uan Tabu (Cremaschi et al. 1998; Garcea 2004) suggest an age range of assemblages containing tanged "Aterian" lithics between 35 and 90 ka. The Middle Paleolithic is certainly rather ancient in the Maghreb, as evidenced by ESR/U-series dating of a tooth, which suggests an age of 160 ± 16 ka for the assemblages at Jebel Irhoud in Morocco (Smith et al. 2007), and its association with MIS 5 at Jebel Gharbi (Libya), which suggests an age between 128 and 71 ka (Barich et al. 2003, 2006).

Based on luminescence dating of the sequence at Rhafas, Mercier et al. (2007) placed the "transition from Mousterian to Aterian" between 70 and 80 ka. However, the underlying ("Mousterian") level 6d at Rhafas also contains a tanged piece (Fig. 332/12; Wengler 1993). The upper end of the Middle Paleolithic, with its supposedly final technocomplex of the "Aterian," has been defined by dosimetric dating to occur at about 35 ka (Wrinn and Rink 2003; Mercier et al. 2007), as well as by a number of radiocarbon data that point to an end of the "Aterian" at around 20 $^{14}$C ka (Debénath 1992).

**Table 5.1** Radiocarbon data of the Maghrebine Middle Paleolithic (MP) used for calibration with CalPal 2007 Hulu (Weninger and Jöris 2008) in Fig. 5.1 in addition to Table 5.2 data. Additional Upper Paleolithic (UP) radiocarbon data is mentioned in the text. Note that infinite radiocarbon data, or data lacking published uncertainties are not listed here. In case of uneven radiocarbon uncertainties the larger value was used in calibration

| Site | Age ($^{14}$C years) | Lab-# | Age (Cal BP a) | Material | Reference |
|---|---|---|---|---|---|
| **MP assemblages containing tanged tools** | | | | | |
| Ain Maarouf | 31,950 ± 600 | GrN-1965 | 36,370 ± 1,070 | Shell | Choubert et al. (1967) |
| Ain Shakshuk | 44,600 ± 2,430 | Beta-167097 | 48,200 ± 2,670 | Soil (total organic) | Garcea and Giraudi (2006) |
| Ain Shakshuk | 43,530 ± 2,110 | Beta-167098 | 47,040 ± 2,210 | Charred material | Garcea and Giraudi (2006) |
| Berard | 31,800 ± 1,900 | I-3951 | 36,730 ± 2,200 | Gastropode | Hébrard (1970[a]) |
| Bir Tarfawa 14 | 43,300 ± 3,000 | SMU-177 | 47,230 ± 2,960 | | Debénath et al. (1986) |
| Bou Hadid | 33,900 ± 1,900 | DF-143, T428 | 38,460 ± 2,180 | Mollusc | Alimen et al. (1966[a]) |
| Bou Hadid | 32,700 ± 1,700 | DF-143, T429 | 37,650 ± 2,190 | Mollusc | Alimen et al. 1966[a] |
| Bou Hadid | 35,950 ± 2,050 | | 39,910 ± 2,130 | | Alimen et al. (1966[b]) |
| Contrebandiers | 22,630 ± 500 | GIF-2576 | 27,140 ± 680 | Marine shell | Delibrias et al. (1982) |
| Contrebandiers | 12,500 ± 170 | GIF-2577 | 14,800 ± 360 | Bone | Delibrias et al. (1982) |
| Contrebandiers | 35,200 ± 2,100 | GIF-2578 | 39,260 ± 2,250 | Marine shell | Delibrias et al. (1982) |
| Contrebandiers | 14,460 ± 200 | GIF-2579 | 17,560 ± 220 | Bone | Delibrias et al. (1982) |
| Contrebandiers | 24,500 ± 600 | GIF-2582 | 29,260 ± 650 | Bone | Delibrias et al. (1982) |
| Contrebandiers | 12,170 ± 160 | GIF-2583 | 14,290 ± 310 | Bone | Delibrias et al. (1982) |
| Contrebandiers | 23,700 ± 1,000 | GIF-2585 | 28,340 ± 1,190 | Carbonaceous sed. | Delibrias et al. (1982) |
| Dar es-Soltan | 37,220 ± 290 | TO-2045 | 42,010 ± 310 | Marine shell | Occhietti et al. (1993) |
| Dar es-Soltan | 16,090 ± 90 | TO-2046 | 19,200 ± 150 | Continent. *Helix* sp. | Occhietti et al. (1993) |
| El Harhoura I | 25,580 ± 130 | TO-2047 | 30,390 ± 190 | Continent. *Helix* sp. | Occhietti et al. (1993) |
| Haua Fteah | 47,000$_{-3,200}^{+2,300}$ | GrN-2023 | 50,850 ± 3,720 | Burnt bone | McBurney (1967) |
| Wadi Sel | 43,530 ± 2,110 | Beta-167098 | 47,040 ± 2,210 | | Garcea and Gaudi n.d.[c] |
| Meteo | 15,100 ± 180 | GIF-7627 | 18,260 ± 240 | Charcoal | Wengler (1993) |
| Taforalt | 19,080 ± 250 | GIF-2278 | 22,960 ± 260 | Gastropod shell | Delibrias et al. (1982) |
| Taforalt | 21,860 ± 330 | GIF-2280 | 26,160 ± 530 | Gastropod shell | Delibrias et al. (1982) |
| Taforalt | 19,400 ± 250 | GIF-2280 | 23,240 ± 270 | Same sample | Delibrias et al. (1982) |
| Wadi El Hay | 24,300 ± 500 | Ly-4209 | 29,100 ± 550 | | Wengler (1993) |
| Wadi Sel | 44,600 ± 2,430 | Beta-167097 | 48,200 ± 2,670 | Organic sediment | Barich and Garcea (2008) |
| Shashuk | 43,530 ± 2,110 | Beta-167098 | 47,040 ± 2,210 | Organic sediment | Giraudi (2005) |
| **MP assemblages without tanged tools** | | | | | |
| Chetaibi-West | 25,850 ± 1,000 | Mc-630 | 30,630 ± 890 | | Nehren 1992 |
| Chetaibi-West | 27,870 ± 800 | Ny-170 | 32,560 ± 690 | | Nehren (1992) |
| Chetaibi-West | 30,330 ± 1,870 | Ny-172 | 35,120 ± 2,100 | | Nehren (1992) |
| El Kala | 23,450 ± 900 | Mc-633 | 28,120 ± 1,140 | | Nehren (1992) |
| Haua Fteah | 43,400 ± 1,300 | GrN-2564 | 46,610 ± 1,500 | Burnt bone | McBurney (1967) |
| **Selected UP sites** | | | | | |
| Jebel Gharbi | 27,310 ± 320 | | 31,980 ± 230 | | Barich et al. (2006) |
| Shakshuk | 27,740 ± 140 | | 32,240 ± 190 | | Garcea (2004) |
| Shakshuk | 30,870 ± 200 | | 34,920 ± 290 | | Garcea (2004) |

[a] cited in Wengler (1997)
[b] cited in Debénath et al. (1986)
[c] cited in Garcea (2004)

In the following study on the chronometric position of the "Aterian," we follow the respective authors' attributions of Maghrebine Paleolithic assemblages to specific technocomplexes because of the lack of a commonly agreed upon definition for the "Aterian" (Table 5.1). We restrict the analysis to the Maghreb and do not include Northeast

African data because of the large distances involved. While the lower limit of the "Aterian" is certainly much older than the limits of the radiocarbon method, it has been shown by other dating techniques to occur between 60 and 90 ka (Cremaschi et al. 1998), or might be even considerably older than 80 ka if tanged items from El Akarit are included (Reyss et al. 2007). Thus, the upper limit is not well-defined either. Given the problems with defining the "Aterian," it is necessary to assess each individual assemblage on a comparable basis. This not only leads to problems agreeing on a common definition, but also results in the huge task of reanalyzing all these assemblages in order to obtain comparable data. We will, therefore, abandon this rather ambitious task and will instead refer to renewed ongoing or recent excavations (e.g., Rhafas, Contrebandiers, El Haroura, El Mnasra, Taforalt, etc.), which will provide us with a wealth of data in the future. Therefore, we are basing our analysis on the data currently available. Few Northwest African Middle Paleolithic sites containing tanged lithics have radiocarbon age information available for analysis (n = 21), and even fewer without such items (n = 9) have this information available. This is because, in both cases, about two-thirds of all available radiocarbon data have either no uncertainties reported or infinite ages.

A fundamental problem with the application of chronometric dating methods in archaeology is the association of the sample dated (or the chemical extract of it) with a given archaeologically defined or identified event. For example, a number of radiocarbon determinations for assemblages containing tanged lithics were obtained on gastropods. Their association with the occupation must be questioned because gastropods dig deep into sediment. Similarly, charcoal can be highly mobile and unless a distinct hearth feature is sampled, the association with a respective occupation is questionable. An inherent problem with radiocarbon dating methods is the issue of contamination, where a very small amount of contamination can lead to erroneous ages (e.g., Aitken 1990) and infinite radiocarbon ages can therefore become finite. It is virtually impossible to prove the absence of contamination. It should also be mentioned that there are a number of radiocarbon records that indicate the possible presence of extreme and complex variation of radiocarbon levels before 30 ka (e.g., Voelker et al. 2000; Beck et al. 2001; Hughen et al. 2004). If correct, radiocarbon data could thus underestimate the "true" age by several thousands of years in the lower range of the method. But perhaps even more important is the lack of a commonly agreed upon procedure to calibrate radiocarbon ages in order to convert the dimensionless radiocarbon data to the calendric units that are fundamentally needed to interpret any archaeological processes. Here, we use the calibration software CalPal, which employs the HULU2007 calibration curve, as one possible way to convert radiocarbon data to a linear timescale (Weninger and Jöris 2008). When such an approach is used, it has to be kept in mind that the original radiocarbon data should always be reported (Table 5.1) in order to allow the calibration with future improved datasets, analogous to the history of the dendrochronological calibration approach, which received its last major revision in 2004 for IntCal04 (Friedrich et al. 2004). A problem that will not even be solved by a commonly agreed upon calibration procedure is the presence of plateaus in the calibration record, which leads to difficulties with the interpretation of the archaeological record.

Additional dating methods are therefore necessary for the verification of radiocarbon data. However, the other available methods are not without problems. In uranium series dating, the initial thorium has to be estimated and corrections are needed for the detritus, in addition to the sampling of a specific mineral phase, for example, in the dating of travertine (Mallick and Frank 2002). Dosimetric dating methods such as TL, OSL, and ESR dating all suffer from the general assumption of the constancy of the external dose rate, a parameter that is difficult to verify. In addition, in ESR dating, the uptake of uranium has to be modeled or determined by U-series. The events dated are not identical between the different methods and samples. For example, in OSL dating, the sedimentation is the event dated; in contrast, TL dating on heated lithics can be directly related to a prehistoric event.

Given the problems with samples and procedures in radiocarbon dating listed above, it seems appropriate to evaluate any radiocarbon data using thorough quality criteria, as suggested by Pettitt et al. (2003), for example. Furthermore, recent advances in radiocarbon sample preparation procedures (Brown et al. 1988; Bird et al. 1999 but also see Hüls et al. 2007) have shown the need to remeasure many samples older than 30 ka $^{14}$C (e.g., Higham et al. 2009). However, the published radiocarbon data on the "Aterian" is too poor to stand the scrutiny of such an approach where, in many cases, not even a vague association of a sample with human occupation can be postulated (e.g., gastropods). The amount of data that would result from a qualitative evaluation would be too small for any meaningful analysis. We therefore analyze the available data as it is in order to check if any meaningful pattern can be obtained from this data. When doing so with the currently-available radiocarbon data, no distributions or trends in the non-calibrated or in the calibrated radiocarbon data for the Maghrebine Middle Paleolithic can be observed, which could be related to any theory or interpretation of distinct human occupations, range, and spread of technocomplexes, etc. (Fig. 5.1). While probability distributions are notoriously difficult to interpret and might suppress valuable information (Galbraith 1998), they are currently the only way to analyze these data, because Bayesian approaches would require a more rigorous

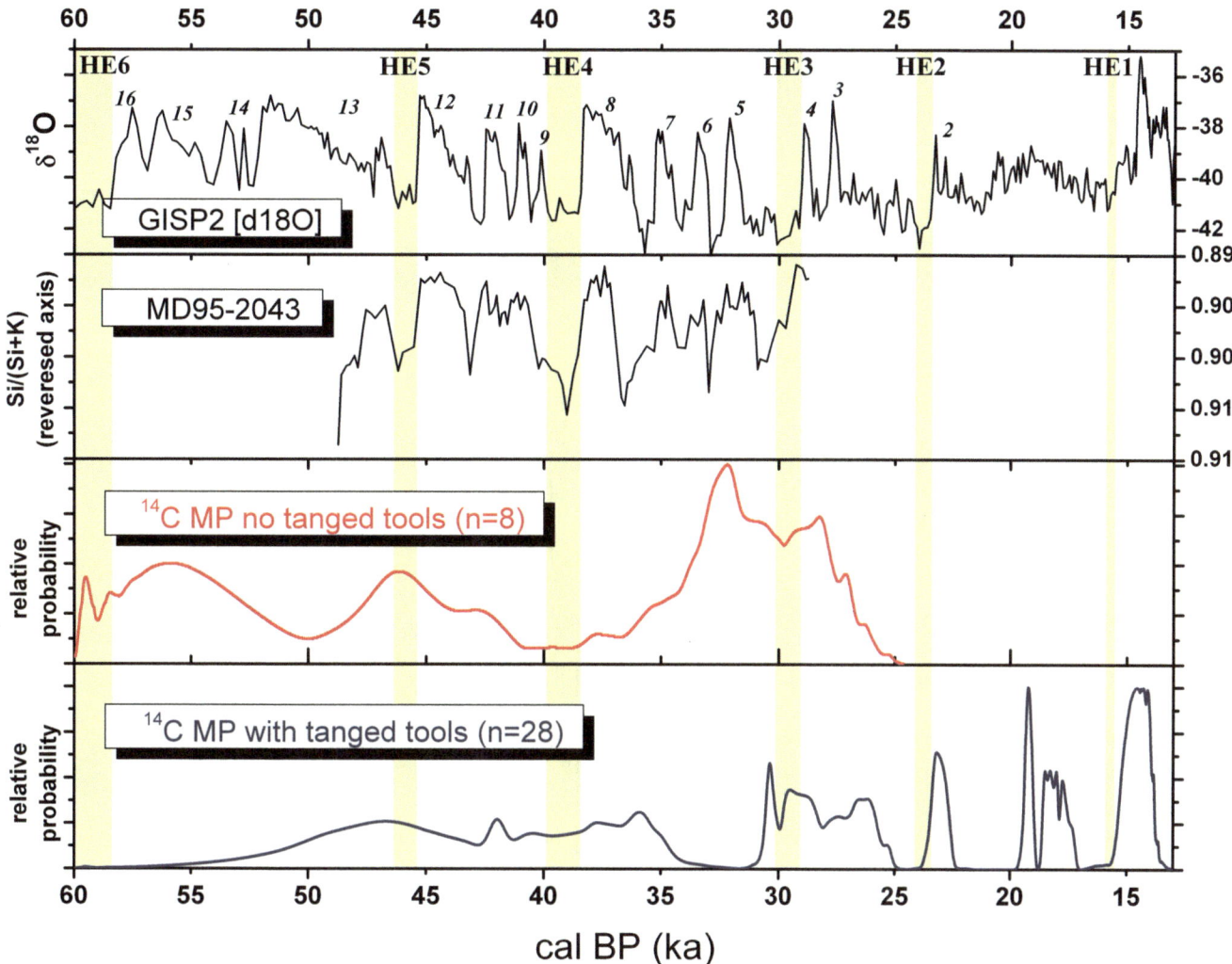

**Fig. 5.1** Proxy data and radiocarbon data for the Maghrebine Middle Paleolithic. $\delta^{18}O$ for GISP2 ice core ,is given with Dansgaard-Oeschger (D/O) events in italic and Heinrich events (HE) as shaded bars. Si/(Si + K) indicates the dust input in marine core MD95-2043 as a proxy for NW-African aridity (Moreno et al. 2002, 2005). Radiocarbon data for Maghrebine Middle Paleolithic assemblages containing and lacking tanged lithics (Table 5.1) were converted to the calendric time scale with CalPal 2007 Hulu (Weninger and Jöris 2008). Note that radiocarbon data from Ifri n'Ammar (Table 5.2) is included. For data with uneven uncertainties the larger value was used for calibration

treatment of the data than is currently possible (see discussion above). The data show a number of distinct distributions for the Middle Paleolithic assemblages with tanged lithics (Fig. 5.1), especially in its upper end, where there is a large overlap with radiocarbon data for the Iberomaurusian (earliest occurrence: 16–17 ka $^{14}C$ cal BP at Ifri n'Ammar (Moser 2003; employing IntCal04); probably 17–29 ka at Taforalt based on a variety of dating methods (Bouzouggar et al. 2007; Bouzouggar et al. 2008). The Iberomaurusian, however, is always located stratigraphically above Middle Paleolithic assemblages and it is suspected that the upper limits of radiocarbon results for assemblages with tanged lithics is probably represented by the upper end of the quasi continuous distribution. The separated peaks are likely to represent data on non- or badly associated material. All data for assemblages

with tanged lithics younger than 20 ka almost certainly have to be dismissed (Fig. 5.1). At the other end of the age distribution, the tail in radiocarbon data at the methodological limits is clearly visible and it is obvious from dating results by other techniques that this tail, or the lack of a more or less abrupt start, is not only the result of sample contamination, but because the "Aterian" is much older than the lower limits of the radiocarbon dating method. However, there is complete overlap with calibrated radiocarbon data from the Middle Paleolithic assemblages lacking tanged lithics, which actually show a rather similar distribution to the data for assemblages with tanged items, except the absence of age results younger than approximately 28 ka on the calendric time scale (CalPal 2007 Hulu). This coincides with the end of the more or less continuous distribution of calibrated radiocarbon data from Middle Paleolithic context with tanged lithics. It is peculiar that a low in the probability distribution for sites containing tanged items corresponds to a high for the ones not containing such tools between approximately 34 and 31 ka. However, given the low numbers and potentially poor quality of the results, a clear correlation, for whatever reason, cannot be expected.

Proxy data, and especially dust records as a direct proxy for Northwest African aridity (e.g., Giraudi 2005), can be synchronized with high latitude Heinrich events (Jullien et al. 2007). The upper ends of the continuous part of the two radiocarbon age distributions correspond to Heinrich event 2 (HE2) as recorded in ice cores, marine cores of the Atlantic Ocean and the Mediterranean (Moreno et al. 2001), and as well as in marine dust records from the Sahara, which much better reflect the climatic conditions of Northwest Africa (Fig. 5.1). This event coincides with an insolation minimum and the climate was probably drier and colder in the Maghreb than in present day (Moreno 2007, personal communication). However, a causal connection between the observed radiocarbon distribution and changes in human subsistence strategies, tool making, etc., can certainly not be established. The ending of Middle Paleolithic assemblages containing tanged tools appears to coincide with the upper limit of radiocarbon ages for assemblages not containing tanged material (Fig. 5.1). This might suggest a link to paleoclimatological changes, but does not quite fit the oldest data for Upper Paleolithic sites for the Maghreb, which have always been found to stratigraphically overlie Middle Paleolithic deposits. For example, ages of 27,310 ± 320 [14]C years (calendric years 31,980 ± 230 cal BP, according to CalPal 2007 HULU) or 30 ± 9 ka by U/Thorium dating are reported from blade industries at Jebel Gharbi (Barich et al. 2006; Barich and Garcea 2008), or from Shakshuk with ages of 27,740 ± 140 and 30,870 ± 200 [14]C years (Garcea 2004), which would calibrate to 32,240 ± 190 and 34,920 ± 290 years (CalPal 2007 HULU), respectively. This data agrees with the results from dating methods other than radiocarbon, where the youngest occurrences of assemblages with tanged items is about 30 ka for the "transition" of Middle to Upper Paleolithic. Taking the oldest and youngest chronometric age as proxies for first and last occurrence is, of course, problematic, especially when datasets are small, but there are currently no alternatives.

## The Site of Ifri n'Ammar

The site with the most convincing evidence against a simple succession of "Mousterian" and "Aterian" is Ifri n'Ammar, located 470 m above sea level in the Rif Oriental of Morocco. A section of almost 7 m was excavated by a

**Fig. 5.2** Photo of the excavation of the lower layers at Ifri n'Ammar

cooperative field project of the Institute National des Sciences de l'Archéologie et du Patrimoine (INSAP) and the German Archaeological Institute (DAI) between 1997 and 2005 (Fig. 5.2). Below an Iberomaurusian occupation, which occurred after a severe erosion (Moser 2003; Mikdad et al. 2004; Nami and Moser 2010), the sequence shows an alternating double sequence of "Aterian" (layers "Upper OS" and "Upper OI") and "Mousterian" (layers "Lower OS" and "Lower OI"), based on the presence/absence of tanged items (Figs. 5.3, 5.4, 5.5).

Most of the individual Middle Paleolithic layers (Fig. 5.3) are separated from each other by layers of secondary carbonates (Mikdad et al. 2004), some of which are of substantial thickness. No stratigraphical disturbances, except the easily recognizable intrusions by the Iberomaurusian occupation, were observed during excavation. Mixing between these units is therefore considered to be very unlikely, especially when the sterile deposits are taken into consideration.

The Middle Paleolithic sequence at Ifri n'Ammar can be divided in two major units of varying thicknesses ("occupation inférieure" and "occupation supérieure," Fig. 5.3). During excavation, it was possible to distinguish some of these layers in sub-units to some extent, but they have to be

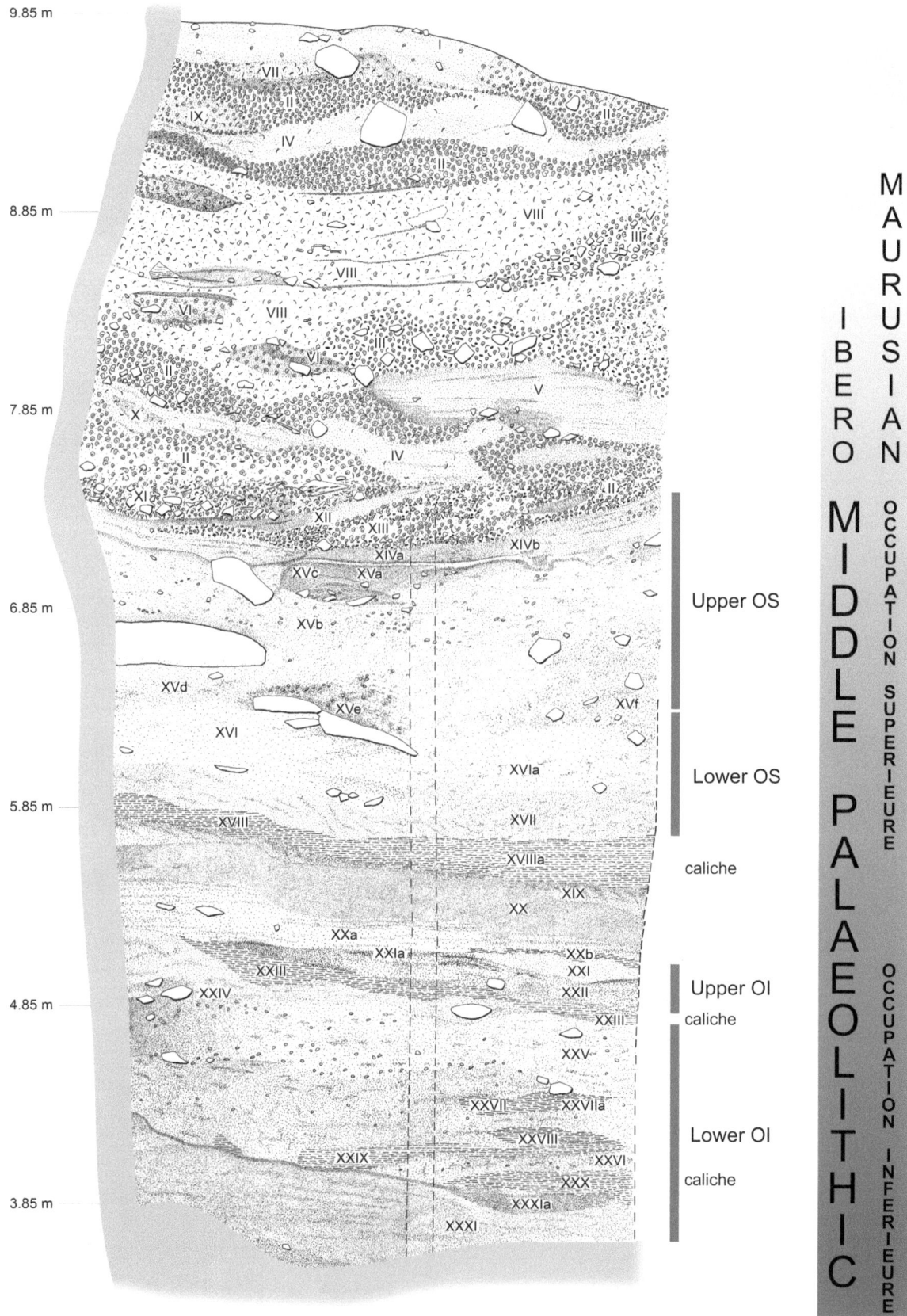

**Fig. 5.3** Synthetic stratigraphy of Ifri n'Ammar below the Iberomaurusian layers. Upper OS: XIVb-XV—*reddish–brownish* silt with gravel and stones. Lower OS: XVI—*brownish–yellowish* silt, XVII—*brownish–yellowish* silty sediment with gravels, stones and bands with charcoal. Almost sterile deposits: XVIII—caliche, XIX—*brownish–reddish* silty fine sand, XX—*orange* silty sediment. Upper OI: XXI—*orange* compact silty sediment, XXII—*brownish-black* silty sediment. Almost sterile deposits: XXIII—caliche, XXIV—*brownish-black* silty fine sand with gravel and stones. Lower OI: XXV—*orange–brownish* compact silty fine sand, XXVI—*grayish–brownish* silty fine sand with gravel and stones, XXVII—caliche, XXVIII—caliche, XXIX—caliche, XXX—caliche, XXXI *blackish–brownish* silty fine sand, XXXIa—lense of *black* silty fine sand

**Fig. 5.4** Examples of cores from Ifri n'Ammar: Levallois flake cores from Upper OS (*1–4*), Lower OI (*5*), Lower OS (*6*) and a recurrent Levallois flake core from Upper OS (*7*)

**Fig. 5.5** Examples of tools from Ifri n'Ammar: Foliates (*1–2*), tanged pieces (*3–6*) and an elongated Mousterian point (*8*) from Upper OS, convergent side scraper (*7*) and a denticulate (*9*) from Lower OS

regarded as palimpsests. The uppermost layer of the "occupation supérieure" (Upper OS, ~1 m thick) contains a rich lithic assemblage composed of various side scrapers, Mousterian points, scrapers, denticulates, and notched pieces. Furthermore, uni- and bi-facial foliates and tanged tools, which are typically referred to as components of the "Aterian," are common. Right below follows a ~0.7 m thick layer of the "occupation supérieure" (Lower OS) with a similar assemblage, but it completely lacks tanged tools and foliates, even though a similar area of 15 m² has been excavated of this unit. Separated by ~0.7 m of almost sterile sediment, which is mainly composed of caliche, the uppermost unit of the "occupation inférieure" (Upper OI, ~0.3 m thick) follows, which contains tanged tools of "Aterian type" beside the "typical" Middle Paleolithic industry. The oldest Middle Paleolithic assemblage so far at Ifri n'Ammar was excavated as the lower layer of the "occupation inférieure" (Lower OI, ~1 m thick), made up of side scrapers, denticulates, and notched pieces, is separated again by a 0.2 m thick, mostly sterile caliche from the overlying "Upper OI." Again, the "Lower OI" does not contain any tanged tools. The excavated area of approximately 12 m² is small, but still sufficient to warrant a good lithic sample when compared to many other sites. Therefore, given the current knowledge of the Maghrebine Middle Paleolithic, it is unlikely that tanged items were not present only in the excavated area. A variety of raw materials were used for tool production, including flint, chert, chalcedony, quartzite, and basalt.

A sample of 6,717 lithics from a 4 m² area, in which all archaeological layers are represented, were analyzed for their technology and typology, but all retouched tools were included. All components of the reduction sequences are present in the site, with the unmodified component dominating (>80%) throughout the sequence. Levallois blank production stays the same throughout the sequence with a decreased use of the centripetal reduction scheme and increased percentage of blades from bottom to top, with blades together with discoidal and Levallois cores.

Here, we report results on the lithic analysis of the two uppermost Middle Paleolithic layers ("Upper OS" with tanged lithics and "Lower OS" lacking tanged lithics), which are almost identical in terms of their respective typology and technology, as well as proportions of types and features. Results of the analysis are therefore reported together to some extent for the two assemblages. Levallois flakes make up 6 and 10% of the assemblages from layers "Upper OS" and "Lower OS," respectively. The type of approximately half of the Levallois cores (Boëda 1994) are "nucléus levallois préferentiel à éclat," while the type of about one-third are "nucléus récurrent à éclat" for both levels. "Nucléus Levallois preferential à Pointe" make up 3.7%, Levallois blade cores ("Nucléus Levallois récurrent à lame"), "Nucléus Levallois récurrent à Pointe," "Nucléus Levallois epuissés," and "Nucléus Levallois double sur les deux faces" are rare, representing only 0.9% of all Levallois cores.

The lithics of the upper two Middle Paleolithic layers represent a technocomplex on the basis of the Levallois technique, with side scrapers being a major component. But in contrast to Jebel Gharbi, for example, which is estimated to be of a similar age, new "modern" elements are present at Ifri n'Ammar (tanged pieces, bifacial foliates, etc.). There is little change in the composition of these assemblages. Throughout the sequence of these upper two layers, the number of lithics increases from layer "Lower OS," which contains no tanged pieces, up to layer "Upper OS," which contains a spectacular number of tanged items (n = 63). There appears to be changes in the proportions of certain typological elements from layer "Lower OS" to "Upper OS." Tanged pieces seem to replace certain types, although not entirely. The percentage of side scrapers and Mousterian points decreases significantly while these types were increasingly equipped with tangs. The only new element, therefore, is the tanging of artifacts, together with a less-pronounced increase in notched and denticulated pieces. This indicates that notching, whether Clactonian-type or retouched, is correlated with tanging and probably relates to the development of tanged points.

Tanged points can therefore be viewed as a technical phenomenon, rooted in a specific response of hunter-gatherer populations to achieve or improve certain aspects of subsistence and mobility. The apparent function of tangs as being exclusively related to hafting has to be questioned (see also Garcia 2012, for a discussion on micro-wear analysis of tangs) because tangs were produced in the Maghreb on all kinds of tool types, which include types like scrapers, where the often proposed use of scraping would inevitably result in an almost instant breakage at the base of the tang.

## Radiometric Dating at Ifri n'Ammar

Few radiocarbon data on charcoal are available from Ifri n'Ammar (Mikdad et al. 2002). At Ifri n'Ammar, a phenomena can be observed that is the opposite of "normal" contamination, where more recent carbon is detected through the radiocarbon measurement of the soluble (humic acid) fraction from the charcoal. In contrast, here, the soluble fraction produces ages older than the non-soluble fraction, which means that old, not young, carbon was deposited in the charcoal. The source of this old carbon from older deposits is evidenced by the presence of thick deposits of caliche at Ifri n'Ammar, especially between

**Table 5.2** Radiocarbon data (Mikdad et al. 2002) and calibration results on the soluble and non-soluble fractions of charcoal samples from the upper two Middle Paleolithic layers at Ifri n'Ammar

| Layer | Humic acid | Non-soluble | CalPal 2007 Hulu | Lab-# |
|---|---|---|---|---|
| Upper OS | 51,330 +1,990/−1,590 | 39,700 +1,320/−1,130 | 43,620 ± 980 | KIA-8822 |
| Lower OS | 57,390 +4,580/−2,900 | 38,740 +2,290/−1,780 | 42,850 ± 1,800 | KIA-8823 |
| Lower OS | > 53,630 | 51,480 +1,470/−1,240 | 55,560 ± 2,520 | KIA-8824 |
| ibid. | | 51,370 +2,490/−1,900 | 54,960 ± 3,140 | ibid. |

layers XVII and XIX, XXII and XXIV, and also between XXVI and XXXI. For caliche to form, water needs to percolate upwards through the sediments by evaporational processes. This percolation indicates the presence of high temperatures. Together with considerable amounts of water, this is probably related to the water table of the nearby SW-NE wadi having been much higher than it currently is and the present day NW–SE running larger wadi not being the main receiving water course.

The radiocarbon results on the non-solulable fraction thus suggest an age of about 40 ka for layer "Upper OS" (no tanged items), and 40–50 ka for layer "Lower OS," which lacks tanged pieces (Table 5.2). These results are at the fringes of the radiocarbon method and thus should be verified by other dating methods.

## Thermoluminescence (TL) Dating

TL dating of heated flint (or chert, hornstone, quartzite, etc.) determines the time elapsed since the last incidence of firing, which is usually associated with prehistoric activities. Naturally occurring fires are unlikely to be responsible for the heating of material in the vast majority of Paleolithic sites (see also Alperson-Afil et al. 2007). In any case, the penetration depth of fire in sediment is very low (Bellomo 1993) and burning roots do not produce high enough temperatures to zero the TL-signal ($\sim 400°C$), as is required for TL-dating. Aside from the possibility of guano spontaneously combusting in cave sites (Binford and Ho 1985), it is unlikely that artifacts were heated by natural fires, especially as there is no evidence of large amounts of guano in the small and shallow cave of Ifri n'Ammar.

Dosimetric dating methods are based on structural damages in the crystal lattice of minerals and an omnipresent ionizing radiation from radioactive elements from the surrounding sediment and the sample itself, as well as secondary cosmic rays. This causes a radiation dose (paleodose or P) to accumulate in the crystal in the form of electrons in excited states. For dating application, only electrons in metastable states are targeted, which are resident over periods of time much longer then the anticipated age (approximately 50 Ma; Wintle and Aitken 1977).

Detailed descriptions of the principles of luminescence dating methods can be found elsewhere (Aitken 1985, 1998; Wagner 1998; Bøtter-Jensen et al. 2003), and a general account of TL dating of lithics is given in Richter (2007).

The paleodose (P) is a function of the dose rate ($\dot{D}$; the ionizing radiation per time unit), which provides the clock for the dating application and the time the crystal was exposed to the omnipresent radiation. Exposure to light or temperature causes the electrons to relax to a ground state, sometimes by emitting a photon, which is the luminescence. If the temperature is high enough ($> \sim 400°C$), the drainage is sufficient to relax all electrons relevant to the luminescence method used, i.e., the clock is set to zero by this event. The completeness of the resetting of the TL-signal used for dating is verified with the "heating plateau" test (Aitken 1985). A flat ratio (Fig. 5.6, dotted line) of the TL-signal from unirradiated (natural) versus the TL emitted by additionally irradiated material (natural + dose) indicates the sufficiency of the prehistoric heating event (Fig. 5.6). The intensity of the luminescence signal (number of photons) increases with the total absorbed dose (P) in a crystal and is therefore a function of exposure time to radiation.

An age can be calculated with the following simplified formula,

$$age = \frac{paleodose}{dose\ rate} = \frac{P_{(Gy)}}{\dot{D}_{(Gy \cdot a^{-1})}},$$

where the paleodose (P) is expressed in Gy and the dose rate ($\dot{D}$) in Gy per time unit (usually in a or ka).

The denominator ($\dot{D}$) of the age formula consists of two parameters, the internal ($\dot{D}_{internal}$) and the external dose rate ($\dot{D}_{external}$).

$$age = \frac{P_{(Gy)}}{\dot{D}_{(Gy \cdot a^{-1})}} = \frac{P}{\dot{D}_{internal} + \dot{D}_{external}}$$

Any variability of one of the parameters of $\dot{D}$ through time makes it difficult to estimate the age of a heated flint (e.g., Richter 2007). All parts which are considered to be potentially geochemically unstable, like cortex or patinated portions, are carefully removed with a water cooled diamond saw from the flint samples prior to TL-dating. The internal dose-rate ($\dot{D}_{internal}$), which is measured on a subset

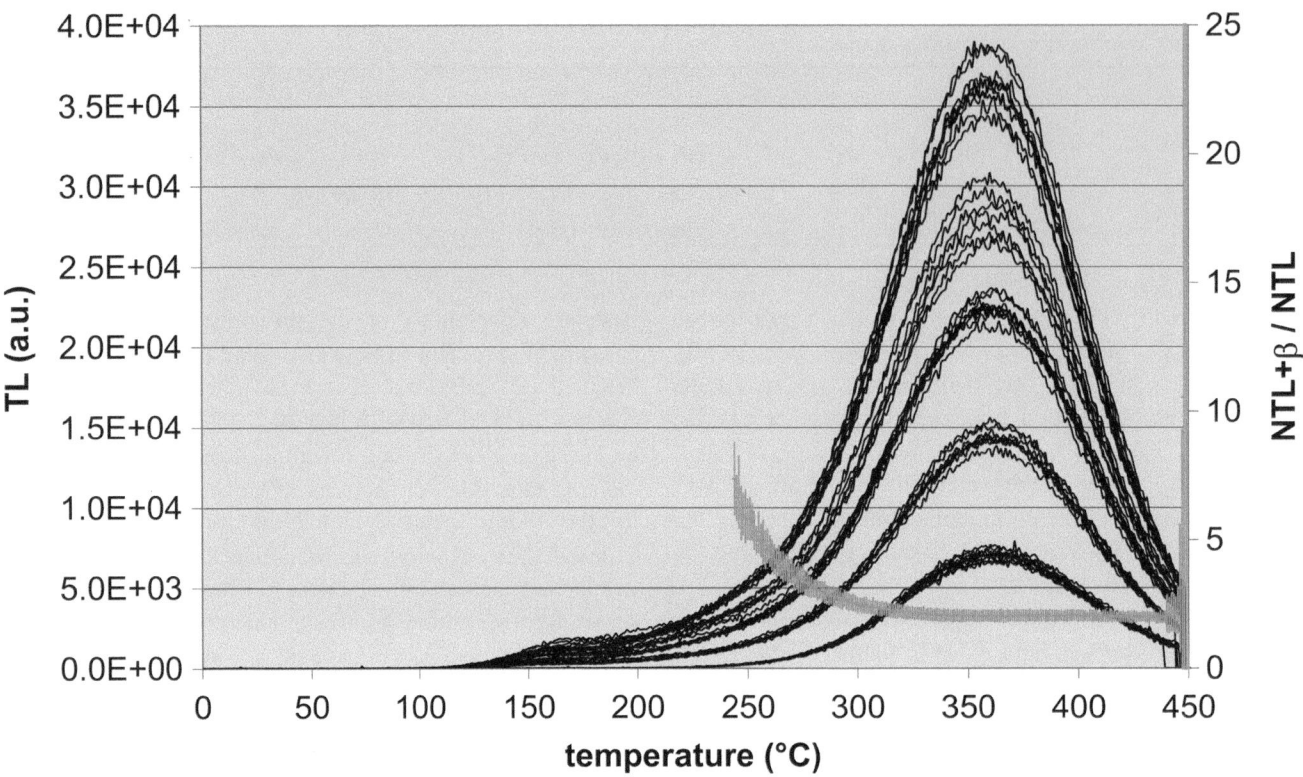

**Fig. 5.6** TL glow curves of sample EVA-LUM-05/07 and ratio of natural TL (NTL) to natural TL with an additive does (NTL + $\beta$) which indicates the heating plateau as a test of the sufficiency of the prehistoric heating for TL-dating purposes

of the sample by Neutron Activation Analysis (NAA), is thus considered to be constant over the time span of interest. This is an advantage of heated flint TL-dating over most other dosimetric dating methods, and reduces the uncertainty given for any age estimate. Most uncertainties in TL-dating of heated flint derive from the estimates of uncertainties associated with the ionizing radiation from the surrounding sediment ($\dot{D}_{external}$), which is measured by either gamma spectrometry or insertion of dosimeters in the sediments for a specified period of time. In order to simplify the estimation of $\dot{D}_{external}$, and thus reduce the uncertainties, each sample is carefully stripped of its outer 2 mm surface area (approximately the range of $\beta$-radiation from isotopes contained in the surrounding sediment) with a water cooled diamond saw prior to analysis. The paleodose (P) is determined from the TL signal (limited to UV-blue detection by a BG25 + HA50 in front of an EMI 9236QA photomultiplier), which is measured by heating sample aliquots at a constant rate of 5 K s$^{-1}$, producing the glow curves (Fig. 5.6). The standard approaches to determine P employ two series of aliquots with several mg of grains from the crushed "stripped core" of the sample (Multiple-Aliquot-Additive-Regeneration = MAAR). The sensitivity of the sample to ionizing radiation is determined by the luminescence yield after irradiation with increasing doses from calibrated radioactive sources. Some aliquots receive additive doses, while others get heated in the laboratory at 360°C for 90 min and then irradiated. Sensitivity changes due to laboratory heating are assumed to be dose independent and the quoted conditions are assumed to reflect the prehistoric heating conditions to a sufficient degree within the uncertainties of the approach. Here, we employed between 7 and 12 aliquots for 4 additive dose points and between 6 and 9 aliquots and 5 dose points for the regeneration growth curve. Various regression/fitting analyses are used to determine P (for an overview, see, e.g., Richter 2007). While the majority of growth curves of the samples from Ifri n'Ammar can be fitted with a straight line (Fig. 5.7), few show sublinear behavior, and linear regression analysis would yield erroneous results for these data sets. Comparisons of paleodose results by linear extrapolation of the appropriate samples and the application of a slide method (Valladas and Gillot 1978; Mercier 1991; Mercier et al. 1992) showed agreement well within 1$\sigma$ between the two methods while providing almost identical uncertainties. All data was therefore analyzed with the slide method in order to retain a common systematic error for all samples from this respect. The limits of integrals for analysis were defined by the overlap in temperatures of the heating as well as the paleodose-plateau, which is the range

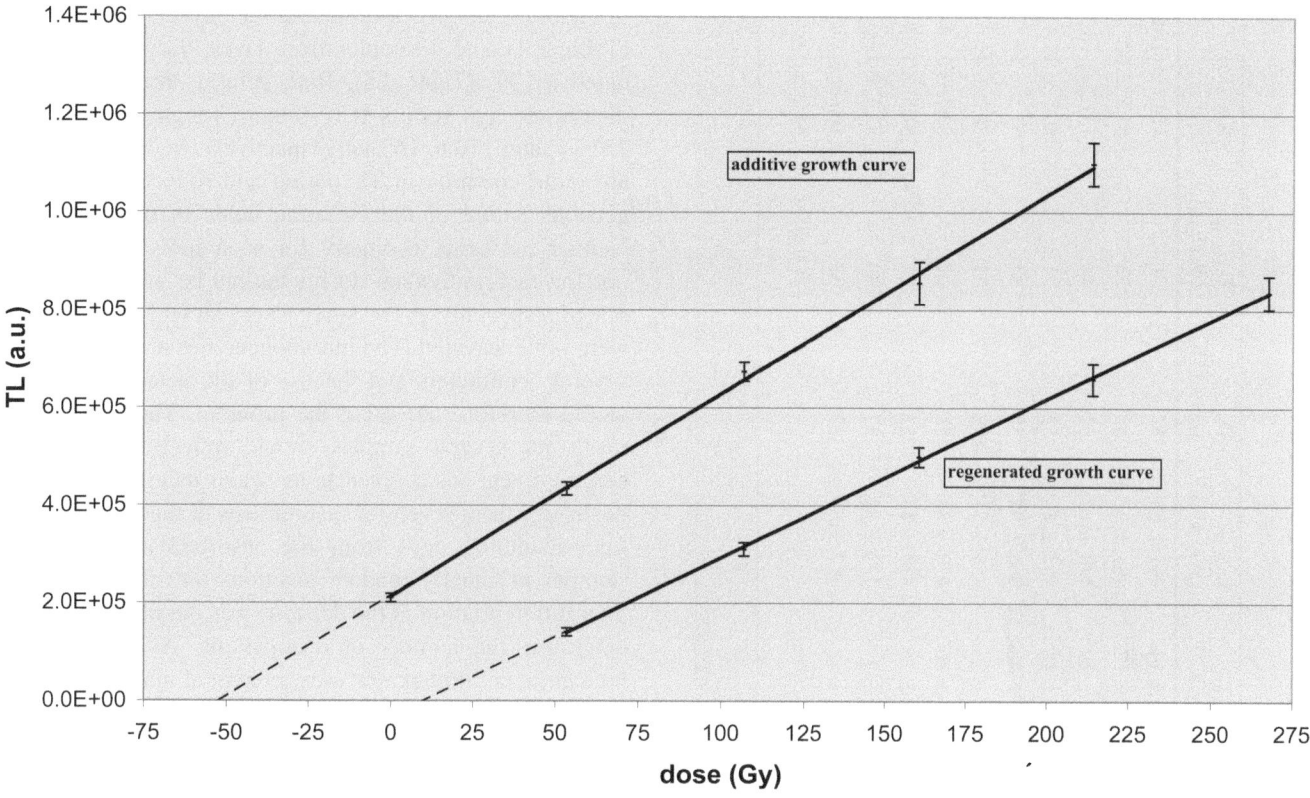

**Fig. 5.7** Additive and regeneration TL growth curves for sample EVA-LUM-05/07

**Table 5.3** $\gamma$-dose rates ($\dot{D}_{\gamma-external}$) of $Al_2O_3$:C dosimeters for layers "Upper OS" and "Lower OS" at their present day moisture for the latter, and corrected for the former (see text)

| Dosimeter | ($\mu Gy\ a^{-1}$) | Dosimeter | ($\mu Gy\ a^{-1}$) |
|---|---|---|---|
| V-1 | 359 | | |
| V-2 | 555 | VII-1 | 455 |
| V-3 | 421 | VII-2 | 733 |
| V-4 | 426 | VII-3 | 490 |
| Average | 440 | | 559 |

of temperatures yielding constant paleodose results. The sensitivity of each individual sample to alpha radiation was determined with the additive method to obtain b-values.

## Results of Luminescence Measurements on Heated Flint Samples From Ifri n'Ammar

The external $\gamma$-dose rate ($\dot{D}_{\gamma-external}$) was determined by $Al_2O_3$:C dosimeters, which were buried in the sediment through the course of 1 year. For layer "Upper OS" four dosimeters were placed at the only available profile at the mouth of the cave (5 m maximum distance to samples),

while three dosimeters could be placed along the main profile (Table 5.3) in layer "Lower OS" (2.5 m maximum distance to samples). The presence of (mainly Iberomaurusian) disturbances and the absence of extended profiles prevented measurements at more locations. The moisture of the sediments was determined in the laboratory and differences by a factor of two were observed between samples originating from inside the cave (6–10%) and at the drip line area (20%). The $\gamma$-dose rates for the dosimeters in layer "Upper OS" were therefore adjusted accordingly to a moisture value of 10%, because the samples dated all came from inside the cave and were not influenced by water from the drip line which, in addition, was located further north in the past. Otherwise the age results for layer "Upper OS" would be overestimated because of the reduced $\gamma$-dose rate due to the influence of dripping water at the area of the present day measurement. An overall uncertainty of 20% was used for the external $\gamma$-dose rate, which is roughly equivalent to the overall variation observed with the dosimeters. Despite the evidence of water movement, the U- and Th-series were found to be in equilibrium for sediment samples analyzed by low level (Ge) $\gamma$-spectrometry. A 5% uncertainty was employed for the cosmic dose rate after Barbouti and Rastin (1983).

**Table 5.4** TL results for samples from layer "Upper OS"

| EVA-LUM | Paleodose (Gy) | b-Value (Gy μm²) | U (ppm) | Th (ppm) | K (ppm) | $D_{int.eff}$ (μGy a⁻¹) | $D_{ext.eff}$ (μGy a⁻¹) | $D_{cosmic}$ (μGy a⁻¹) | $D_{internal}$ (% $D_{total}$) | Age (ka) |
|---|---|---|---|---|---|---|---|---|---|---|
| 05/01 | 75.9 ± 0.6 | 1.16 ± 0.03 | 0.99 ± 0.07 | 0.02 ± 0.01 | 497 ± 124 | 203 ± 14 | 461 ± 82 | 52 | 31 | 114.2 ± 17.8 |
| 05/02 | 45.9 ± 0.2 | 1.90 ± 0.02 | 0.62 ± 0.06 | 0.08 ± 0.05 | 1,530 ± 199 | 233 ± 18 | 468 ± 42 | 52 | 33 | 65.5 ± 6.1 |
| 05/03 | 109.9 ± 7.8 | 1.12 ± 0.01 | 2.08 ± 0.10 | 0.06 ± 0.02 | 291 ± 157 | 368 ± 19 | 470 ± 84 | 52 | 44 | 131.2 ± 18.3 |
| 05/04 | 52.8 ± 0.5 | 2.07 ± 0.01 | 0.66 ± 0.06 | 0.19 ± 0.03 | 998 ± 210 | 204 ± 18 | 452 ± 80 | 52 | 31 | 80.3 ± 12.4 |
| 05/05 | 69.0 ± 1.3 | 1.86 ± 0.03 | 0.87 ± 0.08 | 0.24 ± 0.04 | 1,010 ± 323 | 242 ± 28 | 466 ± 83 | 52 | 34 | 97.5 ± 14.7 |
| 05/06 | 53.3 ± 1.4 | 2.45 ± 0.02 | 0.60 ± 0.07 | 0.12 ± 0.03 | 1,200 ± 312 | 228 ± 26 | 455 ± 81 | 52 | 33 | 78.2 ± 11.8 |
| 05/07 | 62.3 ± 0.1 | 1.33 ± 0.02 | 0.90 ± 0.08 | 0.18 ± 0.04 | 1,490 ± 298 | 274 ± 26 | 470 ± 84 | 52 | 37 | 83.7 ± 12.4 |
| 05/08 | 53.9 ± 2.0 | 1.52 ± 0.01 | 0.46 ± 0.06 | 0.11 ± 0.04 | 1,280 ± 294 | 183 ± 25 | 470 ± 84 | 52 | 28 | 82.5 ± 13.2 |
| 05/12 | 55.3 ± 1.4 | 1.50 ± 0.02 | 0.51 ± 0.06 | 0.40 ± 0.08 | 1,260 ± 34 | 200 ± 9 | 472 ± 84 | 52 | 30 | 82.3 ± 13.0 |
| 06/18 | 71.2 ± 3.9 | 1.22 ± 0.03 | 0.68 ± 0.08 | 0.22 ± 0.04 | 1,420 ± 38 | 231 ± 12 | 477 ± 85 | 52 | 33 | 100.6 ± 15.5 |
| | | | | | | | | | Weighted average | 83.3 ± 5.6 |

A total of 10 samples from layer "Upper OS" (Table 5.4), and 9 samples from layer "Lower OS" were dated by TL (Table 5.5). Both datasets reveal normally distributed age results (Chi-square) ranging from 50 to 131 ka, and 110 to 163 ka, respectively. Such wide ranges are quite common in TL dating and reflect the inhomogeneous nature of the radiation fields in the sediments. Neither the exact geometry for each individual sample, nor for each individual dosimeter can be precisely determined, especially as the latter were placed in profiles that were not excavated. The inhomogeneities are at a scale of several centimeters and the use of the nearest dosimeter would therefore not solve the problem. The average TL result for several samples should reflect the age of a heating event for a given layer when using an averaged dosimetry from several measurements at random positions for calculating ages from the absorbed doses of flint samples at equally random positions (Richter 2007). No finer differentiation of the deposits was possible in order to determine the number of occupations. As the archaeological layer corresponds to a geological unit, all the data is treated as being contemporaneous within the uncertainties of the method. The best age estimates for the layers at Ifri n'Ammar are therefore provided by the weighted average ages (with the systematic uncertainties twice included) of 83.3 ± 5.6 ka for layer "Upper OS," which includes tanged pieces, and 130.0 ± 7.8 ka for layer "Lower OS," which lacks tanged pieces.

In fact, it could be argued that these are minimum ages, given the presence of the caliche as a strong indicator for the presence of a lot of water moving through the sediments, which reduces the gamma radiation significantly. However, this parameter is hard to estimate and we therefore prefer to use a more conservative approach by using the moisture content as described above.

These weighted TL data contrast with the age estimates provided by radiocarbon. They differ significantly and even when the older age estimates from the soluble fraction would be considered to reflect the "true" age, there is no overlap at the $2\sigma$ level of probability. In order to obtain TL ages in the range of the radiocarbon data, the moisture of the sediment would have to have been at 0% or even less, a level which is obviously not possible. This is also in contrast to the information provided by the presence of the caliche layers, which indicate that much more water was present at times, which means that the water content used for the TL data is probably a minimum estimate for the entire burial period. Assuming that the charcoals are truly associated with the prehistoric events in question here, the radiocarbon data of the non-soluble fraction therefore has to be suspected of being contaminated by a small amount of modern carbon, which produced finite ages instead of infinite ones.

**Table 5.5** TL results for samples from layer "Lower OS"

| EVA-LUM | Paleodose (Gy) | b-value (Gy μm²) | U (ppm) | Th (ppm) | K (ppm) | $D_{int.eff.}$ (μGy a⁻¹) | $D_{ext.eff.}$ (μGy a⁻¹) | $D_{cosmic}$ (μGy a⁻¹) | $D_{internal}$ (% $D_{total}$) | Age (ka) |
|---|---|---|---|---|---|---|---|---|---|---|
| 05/09 | 127.6 ± 2.5 | 1.26 ± 0.02 | 1.26 ± 0.11 | 0.26 ± 0.06 | 1,930 ± 560 | 370 ± 47 | 578 ± 106 | 47 | 39 | 134.6 ± 20.0 |
| 06/14 | 107.2 ± 3.8 | 1.29 ± 0.01 | 0.57 ± 0.02 | 0.07 ± 0.02 | 840 ± 235 | 164 ± 19 | 586 ± 108 | 47 | 22 | 142.9 ± 24.7 |
| 05/10 | 98.4 ± 4.3 | 1.62 ± 0.02 | 0.60 ± 0.06 | 0.07 ± 0.03 | 366 ± 77 | 135 ± 10 | 567 ± 104 | 47 | 19 | 140.2 ± 24.7 |
| 05/11 | 75.7 ± 0.6 | 1.22 ± 0.02 | 0.60 ± 0.05 | 0.06 ± 0.03 | 590 ± 159 | 148 ± 15 | 572 ± 105 | 47 | 21 | 105.0 ± 18.2 |
| 06/15 | 135.5 ± 2.5 | 1.93 ± 0.02 | 1.03 ± 0.06 | 0.11 ± 0.03 | 1,770 ± 336 | 326 ± 28 | 578 ± 106 | 47 | 36 | 149.9 ± 22.8 |
| 07/04 | 135.3 ± 1.8 | 1.53 ± 0.02 | 1.20 ± 0.09 | 0.09 ± 0.04 | 505 ± 17 | 249 ± 13 | 578 ± 106 | 47 | 30 | 163.7 ± 26.2 |
| 07/04 | 90.5 ± 0.9 | 1.05 ± 0.03 | 0.43 ± 0.06 | 0.09 ± 0.03 | 308 ± 11 | 97 ± 9 | 589 ± 108 | 47 | 14 | 131.9 ± 24.1 |
| 06/16 | 97.0 ± 3.9 | 0.96 ± 0.03 | 1.08 ± 0.07 | 0.18 ± 0.03 | 1,750 ± 298 | 318 ± 25 | 561 ± 103 | 47 | 36 | 110.3 ± 16.9 |
| 06/17 | 90.1 ± 3.4 | 0.97 ± 0.02 | 0.53 ± 0.04 | 0.13 ± 0.03 | 371 ± 12 | 119 ± 6 | 581 ± 107 | 47 | 17 | 128.7 ± 23.1 |
| | | | | | | | | | Weighted average | 130.0 ± 7.8 |

## Conclusions

Based on the stratigraphy of Ifri n'Ammar, with its sequence of Middle Paleolithic deposits containing and lacking tanged lithics, the technology of tanging cannot be related to a certain epoch, technocomplex, culture, etc., as it frequently is, but, rather, represents aspects of mobility, subsistence strategies, and a reflection of cognitive skills related to contact and exchange of groups. The technique of tanging could be seen as a component of the behavioral modernity of the anatomically modern humans that are responsible for the production of the Maghrebine Middle Paleolithic assemblages, as is evidenced at Jebel Irhoud. While further detailed analyses are required for explaining the presence and absence of the tanging technique, its early presence fits very well with a generally requested repertoire of "anatomically modern behavior," even though this repertoire is certainly not well-defined (see, e.g., Zilhão 2007). However, there is growing evidence for "modern behavior" in various forms from Middle Paleolithic assemblages containing tanged tools in the Maghreb. Examples include personal ornaments like pierced Nassarius shells at Taforalt (Bouzouggar et al. 2007), and hearth and other structures, as well as pigment use and bone tools at El Mnasra (El Hajraoui 1994; Nespoulet et al. 2008). The chronostratigraphic sequence of the Maghreb with "Mousterian" followed by "Aterian" must, therefore, be revised, based on the sequence of Ifri n'Ammar and other sites. It thus appears more appropriate to abandon the term "Aterian" and refer to assemblages containing tanged artifacts in Middle Paleolithic context as a "facies" or "aspects" of the Middle Paleolithic (without chronostratigraphical significance), somewhat analogous to southern France, where the different facies do not necessarily imply a chronological succession (e.g., Jaubert 1999).

According to combined U-series/ESR dating results for Jebel Irhoud (Smith et al. 2007), the Middle Paleolithic in the Maghreb is at least 160 ± 16 ka old, and probably even older. The dating results from Ifri n'Ammar show that the technique of tanging is much older than was previously thought. It certainly goes beyond the 130 ka age obtained by TL-dating for layer "Lower OS" (130.0 ± 7.8 ka), which overlies the tanged artifacts contained in layer XXI. The TL age of 83.3 ± 5.6 ka for the uppermost Middle Paleolithic layer containing tanged items and personal ornaments consisting of pierced Nassarius shells is in contrast to radiocarbon data from this level. This points to the problems that arise when using radiocarbon dating at the limits of this method. The lack of detailed publication (association arguments, sample material, method used, isotopic values, etc.) of radiocarbon data prevents an analysis of well evaluated datasets and, therefore, the interpretations based on radiocarbon data have to be viewed with caution. However, despite the questions underlying the radiocarbon data, there appears to be a pattern that might be of significance to the archaeological interpretation, and that could even be independent of the problems with this dataset. The upper end of the chronological position of the Maghrebine Middle Paleolithic at about 25 ka, based on calibrated radiocarbon ages, is identical for assemblages with and without tanged items. This provides support to the notion to regard tanging as part of the Middle Paleolithic repertoire, and not as a distinct entity related to a chronostratigraphical position, and better describes it as "Middle Paleolithic of Aterian facies" or as an "aspect" of the Maghrebine Middle Paleolithic without any chronological implications.

**Acknowledgments** The samples for TL dating were prepared and measured by S. Albert (MPI-EVA), Neutron Activation Analysis was done by T. Schifer (Curt-Engelhorn-Zentrum Archäometrie, Mannheim), and HpGe-γ-ray spectrometry by D. Degering (VKTA Rossendorf). We thank them profoundly for their help and commitment. Figures 5.2, 5.3, 5.4 and 5.5 are courtesy of DAI Bonn.

# References

Aitken, M. J. (1985). *Thermoluminescence dating*. London: Academic Press.
Aitken, M. J. (1990). *Science-based dating in archaeology*. London: Longmans.
Aitken, M. J. (1998). *An introduction to optical dating. The dating of Quaternary sediments by the use of photon-stimulated luminescence*. Oxford: Oxford University Press.
Alperson-Afil, N., Richter, D., & Goren-Inbar, N. (2007). Phantom hearths and the use of fire at Gesher Benot Ya'aqov, Israel. *Paleoanthropology, 1*, 1–15.
Aouadi-Abdeljaouad, N., & Belhouchet, L. (2008). Recent prehistoric field research in Central Tunisia: Prehistoric occupations in the Meknassy Basin. *African Archaeological Review, 25*, 75–85.
Barbouti, A. I., & Rastin, B. C. (1983). A study of the absolute intensity of muons at sea level and under various thickness absorber. *Journal of Physics G: Nuclear Physics, 9*, 1577–1595.
Barich, B., & Garcea, E. (2008). Ecological patterns in the Upper Pleistocene and Holocene in the Jebel Gharbi, Northern Libya: Chronology, climate and human occupation. *African Archaeological Review, 25*, 87–97.
Barich, B. E., Bodrato, G., Garcea, E. A. A., Barbaro, C. C., & Giraudi, C. (2003). Northern Libya in the final Pleistocene–The late hunting societies of Jebel Gharbi. *Quaderni di Archeologia della Libya, 18*, 259–265.
Barich, B. E., Garcea, E. A. A., & Giraudi, C. (2006). Between the Mediterranean and the Sahara: Geoarchaeological reconnaissance in the Jebel Gharbi, Libya. *Antiquity, 80*, 567–582.
Beck, J. W., Richards, D. A., Edwards, R. L., Silverman, B. W., Smart, P. L., Donahue, D. J., et al. (2001). Extremely large variations of atmospheric $^{14}$C concentration during the last Glacial Period. *Science, 292*, 2453–2458.
Bellomo, R. V. (1993). A methodological approach for identifying archaeological evidence of fire resulting from human activities. *Journal of Archaeological Science, 20*, 525–553.

Binford, L. R., & Ho, C. K. (1985). Taphonomy at a distance: Zhoukoudian, "The cave home of Beijing Man"? *Current Anthropology, 26*, 413–429.
Bird, M. I., Ayliffe, L. K., Fifield, L. K., Turney, C. S. M., Cresswell, R. G., Barrows, T. T., et al. (1999). Radiocarbon dating of "old" charcoal using a wet oxidation, stepped-combustion procedure. *Radiocarbon, 41*, 127–140.
Boëda, E. (1994). *Le concept Levallois: variabilité des méthodes*. CNRS, monographie du CRA: Paris. 9.
Bordes, F. (1976–1977). Moustérien et Atérien. *Quaternaria, XIX*, 19–34.
Bøtter-Jensen, L., McKeever, S. W. S., & Wintle, A. G. (2003). *Optically stimulated luminescence dosimetry*. Amsterdam: Elsevier.
Bouzouggar, A., Barton, N., Vanhaeren, M., d'Errico, F., Collcutt, S., Higham, T., et al. (2007). 82,000-year-old shell beads from North Africa and implications for the origins of modern human behavior. *Proceedings of the National Academy of Sciences of the USA, 104*, 9964–9969.
Bouzouggar, A., Barton, R., Blockley, S., Bronk-Ramsey, C., Collcutt, S., Gale, R., et al. (2008). Reevaluating the age of the Iberomaurusian in Morocco. *African Archaeological Review, 25*, 3–19.
Brown, T. A., Nelson, D. E., Vogel, J. S., & Southon, J. R. (1988). Improved collagen extraction by modified longin method. *Radiocarbon, 30*, 171–177.
Carrière, G. (1886). Quelques stations préhistorique de la province d'Oran. *Bulletin de la Société de Géographie et d'Archéologie de la Province d'Oran, VI*, 136–154.
Choubert, G., Faure-Mwet, A., & Maarleveld, G. C. (1967). Nouvelles dates isotopiques du Quaternaire marocain et leur signification. *Comptes Rendus de l'Académie des Sciences, 264*, 434–437.
Cremaschi, M., Lernia, S. D., & Garcea, E. A. A. (1998). Some insights on the Aterian in the Libyan Sahara: Chronology, environment, and archaeology. *African Archaeological Review, 15*, 261–286.
Debénath, A. (1992). Hommes et cultures matérielles de l'Atérien marocain. *L'Anthropologie, 96*, 711–719.
Debénath, A. (2000). Le peuplement préhistorique du Maroc: données récentes et problèmes. *L'Anthropologie, 104*, 131–145.
Debénath, A., Raynal, J. P., Roche, J., & Ferembach, D. (1986). Stratigraphie, habitat, typologie et devenir de l'Atérien marocain: données récentes. *L'Anthropologie, 90*, 233–246.
Delibrias, G., Guillier, M.-T., & Labeyrie, J. (1982). GIF Natural Radiocarbon Measurements IX. *Radiocarbon, 24*, 291–343.
El Hajraoui, M. A. (1994). L'industrie osseuse Atérienne de la Grotte d'El Mnasra (Région de Témara, Maroc). *Préhistoire Anthropologie Méditerranéennes, 3*, 91–94.
Ferring, C. R. (1975). Aterian in North African prehistory. In F. Wendorf & A. E. Marks (Eds.), *Problems in prehistory: North Africa and the Levant* (pp. 113–126). Dallas, TX: Southern Methodist University.
Friedrich, M., Remmele, S., Kromer, B., Hofmann, J., Spurk, M., Kaiser, K. F., et al. (2004). The 12, 460-year Hohenheim oak and pine tree-ring chronology from Central Europe—a unique annual record for radiocarbon calibration and paleoenvironment reconstructions. *Radiocarbon, 46*, 1111–1122.
Galbraith, R. F. (1998). The trouble with "probability density" plots of fission track ages. *Radiation Measurements, 29*, 125–131.
Garcea, E. A. A. (2004). Crossing deserts and avoiding seas: Aterian North African–European relations. *Journal of Anthropological Research, 60*, 27–54.
Garcea, E. A. A. (2012). Modern human desert adaptations: A Libyan perspective on the Aterian Complex. In J.-J. Hublin & S. McPherron (Eds.), *Modern origins: A North African perspective*. Dordrecht: Springer.

Garcea, E. A. A., & Giraudi, C. (2006). Late Quaternary human settlement patterning in the Jebel Gharbi. *Journal of Human Evolution, 51*, 411–421.

Giraudi, C. (2005). Eolian sand in peridesert northwestern Libya and implications for Late Pleistocene and Holocene Sahara expansions. *Palaeogeography, Palaeoclimatology, and Palaeoecology, 218*, 161–173.

Hahn, J. (1984). Südeuropa und Nordafrika. In O. Bar-Yosef, G. Corvinus, J. Hahn, H. H. Loofs-Wissowa, H. Müller-Beck, A. Ono, et al. (Eds.), Neue forschungen zur altsteinzeit. Forschungen zur allgemeinen und vergleichenden archäologie (pp. 1–231). München: C.H. Beck.

Higham, T. F. G., Barton, H., Turney, C. S. M., Barker, G., Ramsey, C. B., & Brock, F. (2009). Radiocarbon dating of charcoal from tropical sequences: Results from the Niah Great Cave, Sarawak, and their broader implications. *Journal of Quaternary Science, 24*, 189–197.

Hughen, K., Lehman, S., Southon, J. R., Overpeck, J., Marchal, O., Herring, C., et al. (2004). [14]C activity and global carbon cycle changes over the past 50,000 years. *Science, 303*, 202–207.

Hüls, M. C., Grootes, P. M., & Nadeau, M.-J. (2007). How clean is ultrafiltration cleaning of bone collagen? *Radiocarbon, 49*, 193–200.

James, H. V. A., & Petraglia, M. D. (2005). Modern human origins and the evolution of behavior in the Later Pleistocene record of South Asia. *Current Anthropology, 46*, 3–27.

Jaubert, J. (1999). *Chasseurs et artisans du Moustérien*. Paris: Maison des Roches-Le Seuil.

Jullien, E., Grousset, F., Malaize, B., Duprat, J., Sanchez-Goni, M. F., Eynaud, F., et al. (2007). Low-latitude "dusty events" vs. high-latitude "icy Heinrich events". *Quaternary Research, 68*, 379–386.

Kleindienst, M. R. (2001). What is the Aterian? The view from the Dakhleh Oasis and the Western Desert, Egypt. In C. A. Hope & G. E. Bowen (Eds.), *Dakhleh Oasis Project monograph* (pp. 1–14). Oxford: Oxbow Books.

Mallick, R., & Frank, N. (2002). A new technique for precise uranium-series dating of travertine micro-samples. *Geochimica et Cosmochimica Acta, 66*, 4261–4272.

McBrearty, S., & Brooks, A. S. (2000). The revolution that wasn't: A new interpretation of the origin of modern human behavior. *Journal of Human Evolution, 39*, 453–563.

McBurney, C. B. M. (1967). *The Haua Fteah (Cyrenaica) and the Stone Age of the South-East Mediterranean*. Cambridge: Cambridge University Press.

McBurney, C. B. M., & Hey, R. W. (1955) cited in Moyer (2003). Prehistory and Pleistocene geology in Cyrenaican Libya: A record of two seasons' geological and archaeological fieldwork in the Gebel Akhdar hills. Cambridge: Cambridge University Press.

McClure, H. A. (1994). A new Arabian stone tool assemblage and notes on the Aterian industry of North Africa. *Arabian Archaeology and Epigraphy, 5*, 1–16.

Mercier, N. (1991). Flint palaeodose determination at the onset of saturation. *Nuclear Tracks and Radiation Measurements, 18*, 77–79.

Mercier, N., Valladas, H., & Valladas, G. (1992). Observations on palaeodose determination with burnt flints. *Ancient TL, 10*, 28–32.

Mercier, N., Wengler, L., Valladas, H., Joron, J. L., Froget, L., & Reyss, J. L. (2007). The Rhafas Cave (Morocco): Chronology of the mousterian and aterian archaeological occupations and their implications for Quaternary geochronology based on luminescence (TL/OSL) age determinations. *Quaternary Geochronology, 2*, 309–313.

Mikdad, A., Moser, J., & Ben-Ncer, A. (2002). Recherche préhistoriques dans le gisement d'Ifri n'Ammar au Rif Oriental (Maroc). *Premiers résultats. Beiträge zur Allgemeinen und Vergleichenden Archäologie, 22*, 1–20.

Mikdad, A., Moser, J., Nami, M., & Eiwanger, J. (2004). La stratigraphie du site d'Ifri n'Ammar (Rif Oriental, Maroc): premiers résultats sur les dépôts du Paléolithique moyen. *Beiträge zur Allgemeinen und Vergleichenden Archäologie, 24*, 125–137.

Moreno, A., Targarona, J., Henderiks, J., Canals, M., Freudenthal, T., & Meggers, H. (2001). Orbital forcing of dust supply to the North Canary Basin over the last 250 kyr. *Quaternary Science Reviews, 20*, 1327–1339.

Moreno, A., Nave, S., Kuhlmann, H., Canals, M., Targarona, J., Freudenthal, T., et al. (2002). Productivity response in the North Canary Basin to climate changes during the last 250,000 yr: A multi-proxy approach. *Earth and Planetary Science Letters, 196*, 147–159.

Moreno, A., Cacho, I., Canals, M., Grimalt, J. O., Sanchez-Goni, M. F., Shackleton, N., et al. (2005). Links between marine and atmospheric processes oscillating on a millennial time-scale. A multi-proxy study of the last 50,000 yr from the Alboran Sea (Western Mediterranean Sea). *Quaternary Science Reviews, 24*, 1623–1636.

Moser, J. (2003). *La Grotte d'Ifri n'Ammar. Tome 1. L'Ibéromaurusien*. Köln: AVA-Forschungen 8. Linden-Soft.

Moyer, C. C. (2003). The organisation of lithic technology in the Middle and Early Upper Palaeolithic Industries at the Haua Fteah, Libya. Ph.D. Dissertation, Cambridge University.

Nami, M., & Moser, J. (2010). La Grotte d'Ifri n'Ammar. Le Paléolithique Moyen. Forschungen zur Archäologie Außereuropäischer Kulturen 9. Wiesbaden, Reichert Verlag.

Nehren, R. (1992). Zur Prähistorie der Maghrebländer (Marokko, Algerien, Tunesien). In Materialien zur allgemeinen und vergleichenden archäologie (Vol. 49). Mainz: Philipp von Zabern.

Nespoulet, R., El Hajraoui, M., Amani, F., Ben Ncer, A., Debénath, A., El Idrissi, A., et al. (2008). Palaeolithic and Neolithic occupations in the Témara Region (Rabat, Morocco): Recent data on hominin contexts and behavior. *African Archaeological Review, 25*, 21–39.

Occhietti, S., Raynal, J. P., Pichet, P., & Texier, J.-P. (1993). Aminostratigraphie du dernier cycle climatique au Maroc atlantique, de Casablanca a Tanger. *Comptes Rendus de l'Académie des Sciences Paris, Série II, 317*, 1625–1632.

Pettitt, P. B., Davies, W., Gamble, C. S., & Richards, M. B. (2003). Palaeolithic radiocarbon chronology: Quantifying our confidence beyond two half-lives. *Journal of Archaeological Science, 30*, 1685–1693.

Reygasse, M. (1921–1922). Études de palethnologie maghrébine (IIe série). *Notices et Mémoires de la Société archéologique de Constantin, LIII*, 159–204.

Reygasse, M. (1922a). Découverte d'un outillage moustérien à outils pédonculés atérien dans le Tidikelt, Oued Asriouel, Région d'Aoulef Chorfa. *XLVIe Congr. de l'A.F.A.S., Montpellier*, 471–472.

Reygasse, M. (1922b). Note au sujet de deux civilisations préhistoriques africaines pour lesquelles deux termes nouveaux me paraissent devoir être employés. *XLVIe Congr. de l'A.F.A.S., Montpellier*, 467–471.

Reyss, J.-L., Valladas, H., Mercier, N., Froget, L., & Joron, J.-L. (2007). Application des méthodes de la thermoluminescence et des déséquilibres dans la famille de l'uranium au gisement archéologique d'El Akarit. In J. P. Roset & M. Harbi-Riahi (Eds.), *El Akarit. Un site archéologique du paléolithique moyen dans le sud de la Tunisie. Éditions recherche sur les civilisations* (pp. 357–363). Paris: Culturefrance.

Richter, D. (2007). Advantages and limitations of thermoluminescence dating of heated flint from Palaeolithic sites. *Geoarchaeology, 22*, 671–683.

Smith, T. M., Tafforeau, P., Reid, D. J., Grün, R., Eggins, S., Boutakiout, M., et al. (2007). Earliest evidence of modern human life history in North African early Homo sapiens. *Proceedings of the National Academy of Sciences of the USA, 104*, 6128–6133.

Valladas, G., & Gillot, P. Y. (1978). Dating of the Olby lava flow using heated quartz pebbles: Some problems. *PACT, Revue du Groupe Européen d'Études pour les Techniques Physiques, Chimiques et Mathématiques Appliquées à l'Archéologie, 2,* 141–150.

Voelker, A. H. L., Grootes, P. M., Nadeau, M. J., & Sarnthein, M. (2000). Radiocarbon levels in the iceland sea from 25–53 kyr and their link to the earth's magnetic field intensity. *Radiocarbon, 42,* 437–452.

Wagner, G. A. (1998). *Age determination of young rocks and artifacts.* Berlin: Springer.

Wendorf, F., & Schild, R. (1992). The Middle Palaeolithic of North Africa: A status report. In F. Klees & R. Kuper (Eds.), *New light on the Northeast African past* (pp. 39–80). Köln: Heinrich Barth-Institut.

Wengler, L. (1993). *Cultures préhistoriques et formations quaternaires au Maroc oriental. Relations entre comportements et paléoenvironnements au Paléolithique moyen.* Université de Bordeaux I: Thèse de Doctorat d'Etat ès Sciences.

Wengler, L. (1997). La transition du Moustérien à l'Atérien. *L'Anthropologie, 101,* 448–481.

Wengler, L. (2006). Innovations et normes techniques dans le Paléolithique moyen et supérieur du Maghreb: Une alternative aux migrations? In L. Astruc, F. Bon, V. Léa, P.-Y. Milcent & S. Philibert (Eds.), Normes techniques et pratiques sociales: de la simplicité des outillages pré- et protohistoriques. Antibes, APDCA, 93–105.

Weninger, B., & Jöris, O. (2008). A [14]C age calibration curve for the last 60 ka: The Greenland-Hulu U/Th timescale and its impact on understanding the Middle to Upper Paleolithic transition in Western Eurasia. *Journal of Human Evolution, 55,* 772–781.

Wintle, A. G., & Aitken, M. J. (1977). Thermoluminescence dating of burnt flint: Application to a Lower Palaeolithic site, Terra Amata. *Archaeometry, 19,* 111–130.

Wrinn, P. J., & Rink, W. J. (2003). ESR dating of tooth enamel from Aterian levels at Mugharet el'Aliya (Tangier, Morocco). *Journal of Archaeological Science, 30,* 123–133.

Zilhão, J. (2007). The emergence of ornaments and art: An archaeological perspective on the origins of "behavioral modernity". *Journal of Archaeological Research, 15,* 1–54.

# Chapter 6
# Amino Chronology and an Earlier Age for the Moroccan Aterian

J.-P. Raynal and S. Occhietti

**Abstract** As part of a joint French-Moroccan research project, a dating program in collaboration with Canadian researchers from GEOTOP has been ongoing since 1987. Due to the regional mean temperature (18–19°C) and Morocco's intermediate-high thermal history, racemization ratios in molluscs are sufficiently high to establish a reliable zoned scale using late cycle sites. The lithostratigraphic and archaeological framework of the last climatic cycle has been established and is supported by other chronological data. Using this framework and additional AMS and $^{14}$C dates on the same sample units that were used for amino acid analysis, preliminary amino acid zones using whole shell protein for the last climatic cycle along the northern part of the Atlantic Coast of Morocco are proposed. An attempt to date archaeological layers beyond the $^{14}$C limit of application is presented for three Middle Paleolithic (Aterian) sites in the Rabat–Temara area. Estimated ages fall between 50 and 85 ka, according to the calibration curves established during the project.

**Keywords** Amino chronology • Aterian • Atlantic Coast • Caves • Middle Paleolithic • Molluscs • Morocco

This chapter uses a chronological scale developed over the last 15 years (Occhietti et al. 1993, 1999, 2002; Occhietti and Raynal 1996) from amino acid analysis on shells from several cave sites near Rabat and Temara to resolve questions regarding the origin and evolution of modern humans in Morocco during the transition between the Middle and Upper Stone Age, which are related to the Aterian in this part of Africa. Our specific objective is to propose age brackets for the various Aterian units sampled, based on D-alloisoleucine/L-isoleucine (Alle/Ile) epimerization ratios from several regional species.

The history of the last 125,000 years in Morocco remains poorly known despite much lithostratigraphic, chronological, and archaeological work. Since 1977, research based programs combining prehistoric archaeology[1] with inter-university co-operation[2] have produced a lithostratigraphic and cultural dating framework for the last climatic cycle ($^{14}$C, Optically Stimulated Luminescence (OSL), and Thermoluminescence (TL)) (Debénath et al. 1982, 1984, 1986; Texier et al. 1988, 1992; Daugas et al. 1989, 1999; Ousmoï 1989; Texier and Raynal 1989; Rhodes 1990; Smith et al. 1990; Raynal et al. 1992; Lefevre et al. 1994). In addition to these correlated data, conventional and accelerator radiocarbon ages have been obtained on some of the samples that were subjected to amino acid analysis. Moderately high thermal conditions along the northern Moroccan Coast (presently 18–19°C) are the likely cause of the relatively fast racemization ratios and these have favored the identification of several amino zones in the Late Pleistocene and Holocene. Analysis has allowed us to establish the following amino- chronological framework for

J.-P. Raynal (✉)
Centre National de la Recherche Scientifique,
Université Bordeaux 1, UMR 5199 PACEA, PPP,
Bâtiment B18, Avenue des Facultés,
33405 Talence Cedex, France
and
Department of Human Evolution, Max Planck Institute for
Evolutionary Anthropology, Deutscher Platz 6,
04103 Leipzig, Germany
e-mail: jpraynal@wanadoo.fr

S. Occhietti
Université du Québec à Montréal, GEOTOP, C.P. 8888,
Succ. Centre-Ville, Montréal, QC H3C 3P8, Canada
and
Université de Lorraine, CERPA, 23 Boulevard Albert 1er, BP
3397, 54015 Nancy Cedex, France
e-mail: serge.occhietti@gmail.com

---

[1] Studies performed by the *Mission préhistorique et paléontologique française au Maroc*, then by the *Mission "Littoral Maroc"* in collaboration with *the Institut National des Sciences de l'Archéologie et du Patrimoine du Royaume du Maroc* (INSAP), within the framework of the France-Morocco *Convention relative aux recherches archéologiques et anthropologiques* (January 19, 1971, revised in December 1979).

[2] Several theses prepared at the Universities of Bordeaux 1 and Paris VI.

J.-J. Hublin and S. P. McPherron (eds.), *Modern Origins: A North African Perspective*,
Vertebrate Paleobiology and Paleoanthropology, DOI: 10.1007/978-94-007-2929-2_6,
© Springer Science+Business Media B.V. 2012

**Table 6.1** Epimerization ratios for total amino acids in *Patella* and *Mytilus* marine shells (n = number of analyzed samples)

| Stratigraphy | Archaeology | Site[a] | $^{14}$C BP Age[b] | Species | AIle/Ile | n |
|---|---|---|---|---|---|---|
| MIS 1 Holocene | | El Kiffen beach 0–2 m | Around 3500 | *Patella* | 0.081 ± 0.019 | 10 |
| | Final Neolithic | Dar es-Soltan I Unit 2 | | *Patella* | 0.107 ± 0.021 | 15 |
| | | | | *Mytilus* | 0.153 ± 0.043 | 3 |
| | | Skhirat bar base | UQ-1557 : 4950 ± 150[a] 5702 ± 168 calBP[b] | *Patella* | 0.145 ± 0.016 | 6 |
| | | | | *Mytilus* | 0.251 ± 0.030 | 6 |
| MIS 4, 3 and 2 "Soltanian" | Iberomaurusian | Dar es-Soltan II Unit 3, cave | UQ-1558 : 16 500 ± 250[a] 19 788 ± 388 CalBP[b] | *Patella* | 0.101 ? | 1 |
| | Aterian | Dar es-Soltan II Unit 6, cave | TO-2045 : 37 220 ± 290[a] 41 966 ± 333 CalBP[b] | *Patella* | 0.229 ± 0.022 | 6 |
| | | | | *Mytilus* | 0.376 ± 0.017 | 3 |
| | ? | El Mnasra c.3 Cave | | *Patella* | 0.280 ± 0.051 | 3 |
| | Aterian | El Mnasra c.6 Cave | | *Patella* | 0.282 ± 0.018 | 3 |
| | Below Aterian | Dar es-Soltan I Unit M, cave | | *Patella* | 0.336 ± 0.097 | 3 |
| | Aterian ? *Homo* | Dar es-Soltan II Unit 7, cave | | *Patella* | 0.369 ± 0.020 | 2 |
| MIS 5 "Ouljian" | Intertidal deposits | Larache | | *Patella* | 0.533 ± 0.125 | 3 |
| | | Dar Reddad Ben Ali | | *Patella* | 0.519 ± 0.099 | 3 |
| | | Dar Bouazza | | *Patella* | 0.543 ± 0.068 | 7 |
| MIS 7 | | Oulad Aj Jmel | | *Patella* | 0.842 ± 0.208 | 5 |
| | | | | *Mytilus* | 1.251 ± 0.109 | 3 |
| | | Aïn Diab | | *Mytilus* | 1.21 (?) | 1 |
| MIS 9 "Harounian" | | Oulad Aj Jmel | | *Patella* | 0.950 ± 0.106 | 8 |

Stratigraphic units are labeled after the excavators' terminology
[a] The same sampling was used for $^{14}$C and amino acids
[b] Calibration data set used is CalPal-2007Hulu (Weninger and Jöris 2008), using CalPal-2007online (Danzeglocke et al. 2008; CalPal-2007online. http://www.calpal-online.de/)

the Moroccan Atlantic Coast between Casablanca and Tangier (Occhietti et al. 1993, 1999, 2002; Occhietti and Raynal 1996), extending as far as Agadir (Weisrock et al. 1999):

(1) Holocene amino group (Group H): characterized by a very rapid rate, a linear relationship and a significant epimerization- $^{14}$C age relationship (Tables 6.1 and 6.2).

(2) Marine Isotopic Stages (MIS) 4-3-2 amino group (Group S): related to the Soltanian—*sensu lato*, characterized by an inflection of the racemization rates (Tables 6.1 and 6.2).

(3) MIS 5 amino group (Group O): represented by marine units associated with the raised shorelines of the last interglacial, between 0 and +6 m placed in the Ouljian of Morocco. Group O is characterized by regular average epimerization ratios in the order of 0.500 to 0.550 (Table 6.1).

(4) Older amino groups beyond the scope of this study: Bir Feghloul (Table 6.1), OAJ, and Ancient Groups.

## Sampling

In general, three specimens of each pertinent species of marine and continental origin were analyzed for each sampled unit. To limit the effect of superficial-layer overheating, which can induce an excessive racemization in samples from sites other than in caves, shells were taken at 20 cm or more from the exposed surface of the deposit and in places sheltered from the sun. Marine shells originate from natural or anthropic thanatocoenoses, so particular attention was given to identifying and discarding introduced fossil shells and those that exhibited traces of burning. Continental gastropod shells (*Helix* and *Rumina*) were collected from cave fill and ancient bars within dunes. Because of the small number of sites containing terrestrial gastropods studied on the Atlantic Coast, samples from the Mediterranean sites of Kaf-Tat-El-Ghar Cave (Province of Tetouan) and from the dune bar at Nador were used as comparative data.

**Table 6.2** Epimerization ratios for total amino acids in *Helix* shells (n = number of analysed samples)

| Stratigraphy | Archaeology | Site[a] | Age $^{14}$C BP | Alle/Ile | n |
|---|---|---|---|---|---|
| | | Nador dune bar | UQ-1462 : 2950 ± 100[a] | 0.033 ± 0 | 3 |
| | | | 3120 ± 139 calBP[b] | | |
| Holocene | Contemporaneous of Cardial | Kaf Taht El Ghar cave | Ly-3821 : 6050 ± 120 charcoal | 0.062 ± 0.008 | 7 |
| | | | 6936 ± 163 calBP[b] | | |
| Soltanian | Migration into Iberomaurusian layer | Dar es-Soltan II cave | | 0.104 ± 0.015 | 9 |
| | Migration into Aterian layer | Dar es-Soltan II cave | TO-2046 : 16 090 ± 90[a] 19 214 ± 237 calBP[b] | 0.234 ± 0.010 | 7 |
| | Reworking of Aterian layers during Upper Paleolithic | El Mnasra bed 3 | | 0.335 ± 0.097 | 3 |
| | Migration into Aterian layer | El Mnasra, 350-370 | | 0.271 ± 0.033 | 3 |
| | Migration into Aterian layer | El Mnasra, 390-400 | | 0.241 ± 0.018 | 3 |
| | Contemporaneous of Aterian | El Harhoura I cave | TO-2047 : 25 580 ± 130[a] 30 597 ± 320 calBP[b] | 0.312 ± 0.046 | 4 |

Stratigraphic units are labelled after the excavators' terminology
[a] The same sampling was used for $^{14}$C and amino acids
[b] Calibration data set used is CalPal-2007Hulu (Weninger and Jöris 2008), using CalPal-2007online (Danzeglocke et al. 2008; CalPal-2007online. http://www.calpal-online.de/, accessed 2008–2009)

The sites sampled between Casablanca and Tangier were (Fig. 6.1):

• Beach formations and combined dune bars at Casablanca (El Kiffen, Dar Bouazza, Oulad Aj Jmel West, Aïn Diab, El Hank); Skhirat (Rouazi); Salé (Sidi Moussa); Larache; Tangier (Oued Tahadart, Achakar).
• Cave fill at Témara (El Mnasra, El Harhoura I and II); Rabat (Dar es-Soltan I and II); Tangier (Les Idoles and El Khril at Achakar).

Specific attention was focused on the prehistoric sites of the Rabat-Temara-Skhirat littoral, which is rich in Aterian human remains and Ibero-maurusian and Neolithic artifacts, but lacking in chronological markers. Except for the Rouazi necropolis at Skhirat (Daugas et al. 1984, 1989, 1999; Lacombe and Daugas 1988; Lacombe et al. 1990; Daugas 2002), these are cave sites. The cave fill of continental origin at Dar es-Soltan 1 (Ruhlmann 1951) was formerly used to define the Soltanian (Choubert et al. 1956). Excavations in the nearby cavity of Dar es-Soltan 2 (Debénath 1972, 1975, 1976, 1978) revealed a similar fill that probably represents only part of the time interval classically allotted to the Soltanian. From the base, these are: bioclastic sands (unit 7), at the top of which were human remains (probably Aterian) discovered by Debénath (1975); followed by continental deposits with Aterian and Ibero-maurusian remains (units 6–3); overlain by deposits that are primarily anthropic and Neolithic (unit 1). A similar stratigraphic arrangement occurs in the caves known as El Harhoura I and II, and El Mnasra (Debénath 1980, 1982; Debénath and Lacombe 1986; Lacombe et al. 1991; El Hajraoui 1993,

1994, 2001; Debénath and El Hajraoui 1999; Nespoulet et al. 2008a, b).

## HPLC Analysis: Analytical Procedure

Shell samples were taken at the apex (for *Patella* and *Helix*) or on the bulging area near the hinge on bivalves. This strategy limits the intra-specimen variability pointed out by Hearty et al. (1986). At the GEOTOP Laboratory in Montreal, Quebec, high-pressure liquid chromatography (HPLC) was used for amino acid separation, followed by fluorescence detection.

All analyses were performed without transferring sample solutions between the dissolution and hydrolysis manipulations. This protocol is similar to that used in the 1980s and 1990s by the laboratories at Boulder, Colorado (INSTAAR), and Aberystwyth (Great Britain), and also takes into account the experiments by Miller (1982).

The D-alloisoleucine/L-isoleucine ratios are measured using the areas of the relevant chromatogram spikes. All area measurements of these spike areas were simulated on a screen to guarantee homogeneous measurements and to take into account the variability of the chromatogram baseline. The Alle/Ile ratios measured at the GEOTOP Laboratory on the standard samples sent by Wehmiller to different amino chronology laboratories varied between 0 and 7% with respect to the interlaboratory values (Wehmiller 1984).

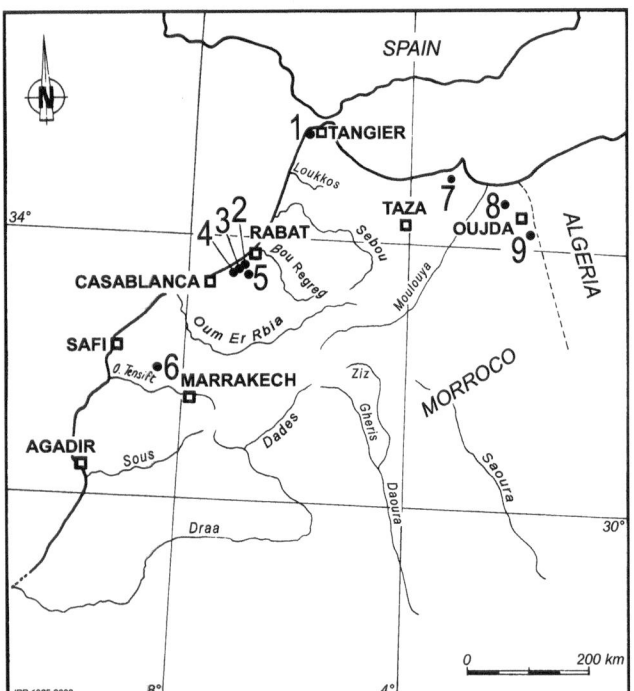

**Fig. 6.1** 1 to 9. Middle Stone Age sites of Atlantic Morocco studied and other main reference localities (all are caves, except 5). *1* Mugharet el 'Aliya; *2* Dar es-Soltan 1 and 2; *3* El Harhoura 1 and 2; *4* El Mnasra; *5* Chaperon Rouge I and II; *6* Jebel Irhoud; *7* Ifri n' Ammar; *8* Taforalt; *9* Rhafas

## Amino Acid Results

### Marine Shells

The genus *Patella* has the best distribution in space and time; it has a medium racemization rate and shows satisfactory intraspecific variability (Bowen and Sykes 1985), though less favorable than that of *Glycimeris*, for the Mediterranean Basin (Hearty 1986). The *Patella* racemization ratios offer a suitable reference for the amino-chronological framework of Morocco's Atlantic Coast, and are useful for identifying events that occurred during Marine Isotope Stage (MIS) 2 and 3. Thus, they provide an inexpensive absolute dating method. Although not as widely distributed as *Patella*, *Mytilus*, another marine genus, is also useful.

The epimerization ratios of these genera indicate significant differences with increasing values coherent with their stratigraphic position. Radiocarbon ages (Table 6.1, Fig. 6.2) confirm that the epimerization ratio of *Mytilus* is rapid (Miller and Mangerud 1985); the ratios of *Helix* are in concordance with the stratigraphic position.

The high value of the epimerization ratio of *Patella* in layer 3 of El Mnasra Cave shows that there was deep reworking of the subjacent Aterian layers during Neolithic occupation of the cavity. A radiocarbon date of 21,586 BP was recently

published for the Aterian of El Mnasra (El Hajraoui 2001), but lacked precise stratigraphic location and any indication of the material dated. Nevertheless, this recent date combined with our estimates suggests that, in this cavity, Aterian occupation probably spanned several thousands of years.

The high value of the epimerization ratio of *Patella* in unit M of Dar es-Soltan 1 (from sands at the bottom of the sequence and significantly below the lowest Aterian layer) and from unit 7 at Dar es-Soltan 2 (below the lowest Aterian layer) are consistent with their stratigraphic position at the very bottom of both sections. At Dar es-Soltan 1, radiocarbon dates already provided age estimates >27,000 BP for "Upper Aterian" and >30,000 BP for "Lower Aterian" (Berger et al. 1965), but we were unable to get suitable samples for amino acid dating from these layers.

### Terrestrial Gastropods

When compared to radiocarbon dates, the Alle/Ile average ratios obtained on the genus *Helix* show an epimerization rate that is nearly linear until at least around 25 ka. The intraspecific variability is zero for Nador and seems to increase in some older samples (Table 6.2, Fig. 6.3).

The average curve in Fig. 6.2 indicates homogeneity of the racemization rates between sites several hundred kilometers apart but with similar thermal histories. It underlines the fact that shells accumulated in caves are favorable materials for amino chronology because of an attenuation of thermal variations.

This primary series of measurements of the epimerization ratios of the genus *Helix* opens the way for dating archaeological material using such shell material. It will be particularly useful for sites and deposits where [14]C methods are not available because of the well known radiocarbon anomalies specific to terrestrial gastropods (Goodfriend 1987).

## Amino Acid Results: Discussion and Interpretation

### Discussion

The significance of variations in the measured ratios within each lithological unit must be addressed before interpreting amino chronology data. Variability can be due to epimerization, as well as external geological and archaeological processes. We have tried to eliminate most of the external factors that could have modified the thermal conditions of the sampled shells by carefully sampling away

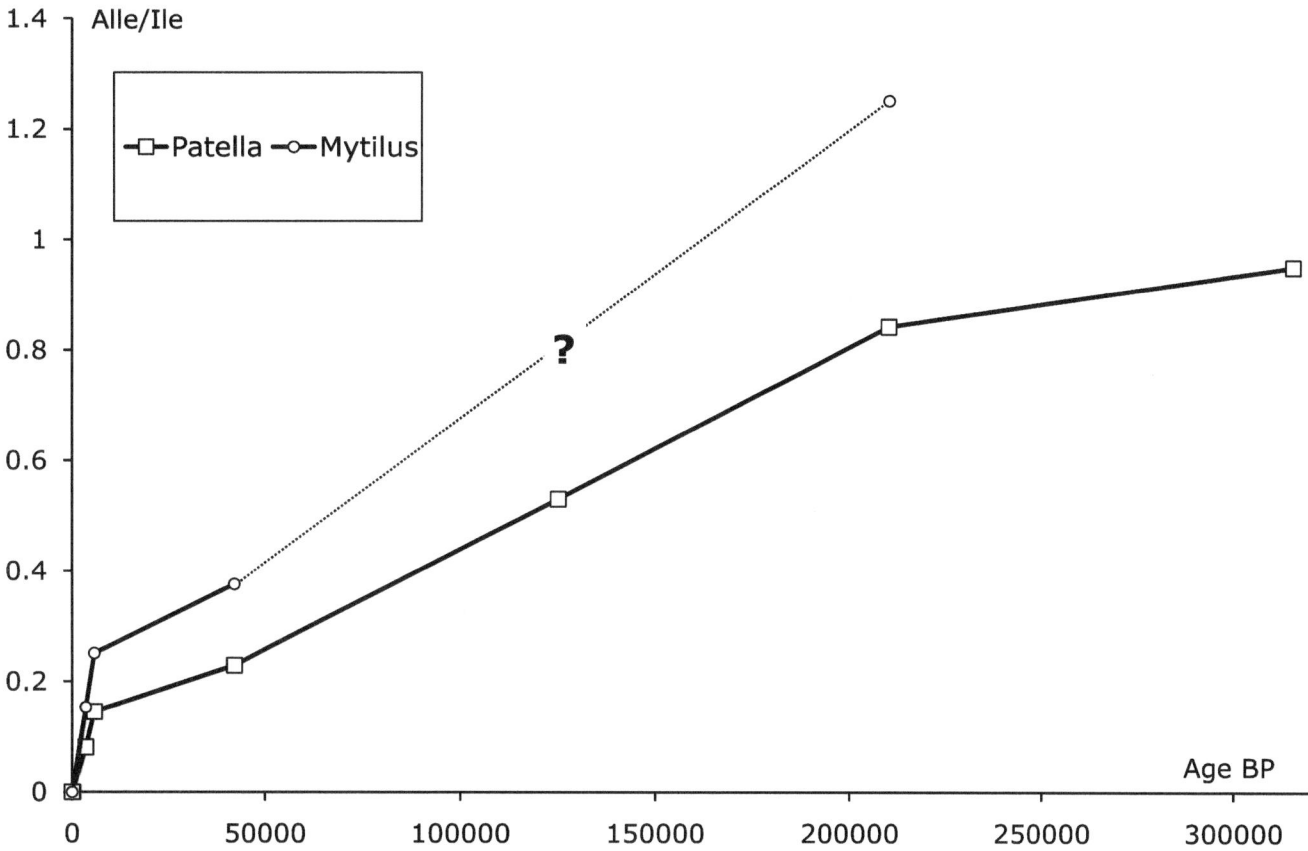

**Fig. 6.2** Epimerization ratios in *Patella and Mytilus* shells compared to $^{14}$C dates or to sea-level-high stands assessed ages (MIS 5e "Ouljian s.l." at 0–6 m, 125 y; MIS 7 at 7–8 m, 210 kya; MIS 9 "Harounian" at 8–12 m, 315 kya) after recent works at Casablanca

from highly exposed outcrops. Sampling in caves offers the best stratigraphic control, and reworked or introduced material is usually easily indentified. Of course, heat from human fires can influence areas larger than those which contain visible ashes and burnt shells. In such cases, marine and terrestrial shells would be overheated, leading to anomalous epimerization ratios in both.

We are aware of the limitations of amino acid dating. Most of the factors that interfere with the racemization processes and explain intra- and interspecific variability of D/L ratios are described by Wehmiller and Belknap (1978), Kriausakul and Mitterer (1978, 1983), Kimber et al. (1986), and Murray-Wallace (1993). We are also aware that the results obtained with the technical procedure we used at the GEOTOP Lab could differ from those obtained by reverse chromatography (Kaufman and Manley 1998), and that those using the inner part of *Patella* shells might give very precise values (Penkman et al. 2008). Nevertheless, our results are internally consistent. The powders extracted from the shells represent the equivalent of aliquots, and, consequently, the results between different samples are directly comparable.

## Age Estimates

The colonization of underlying units by continental gastropods over a depth of several tens of centimeters poses a major problem for dating. It appears that the Aterian unit at Dar es-Soltan II containing *Patella* dated to 41,966 ± 333 CalBP (TO-2045) and *Helix* dated to 19,214 ± 237 calBP (TO-2046) was colonized by gastropods 20,000 years after its deposition. A secondary colonization of the Aterian layer at El Harhoura 1 cave around 25 ka is also suspected because TL dates for this layer, which contains the "Moroccan point," have been older (BOR 56: 41,160 ± 3500, and BOR 57: 32,150 ± 4800 BP; Debénath 1992). Moreover, the Aterian layers were capped by a speleothem that has been dated by H.P. Schwarcz to 66.5 ± 5.2 (bottom) and 58.5 ± 5.2 (top) ka by the Uranium–Thorium method (*in litteris*). The same invasive colonization phenomenon probably explains the occurrence of gastropods in the lower units at El Mnasra, which are estimated to date between 20 and 25 ka (Fig. 6.4). Recent unpublished dates in a similar age bracket obtained in the Casablanca area also indicate

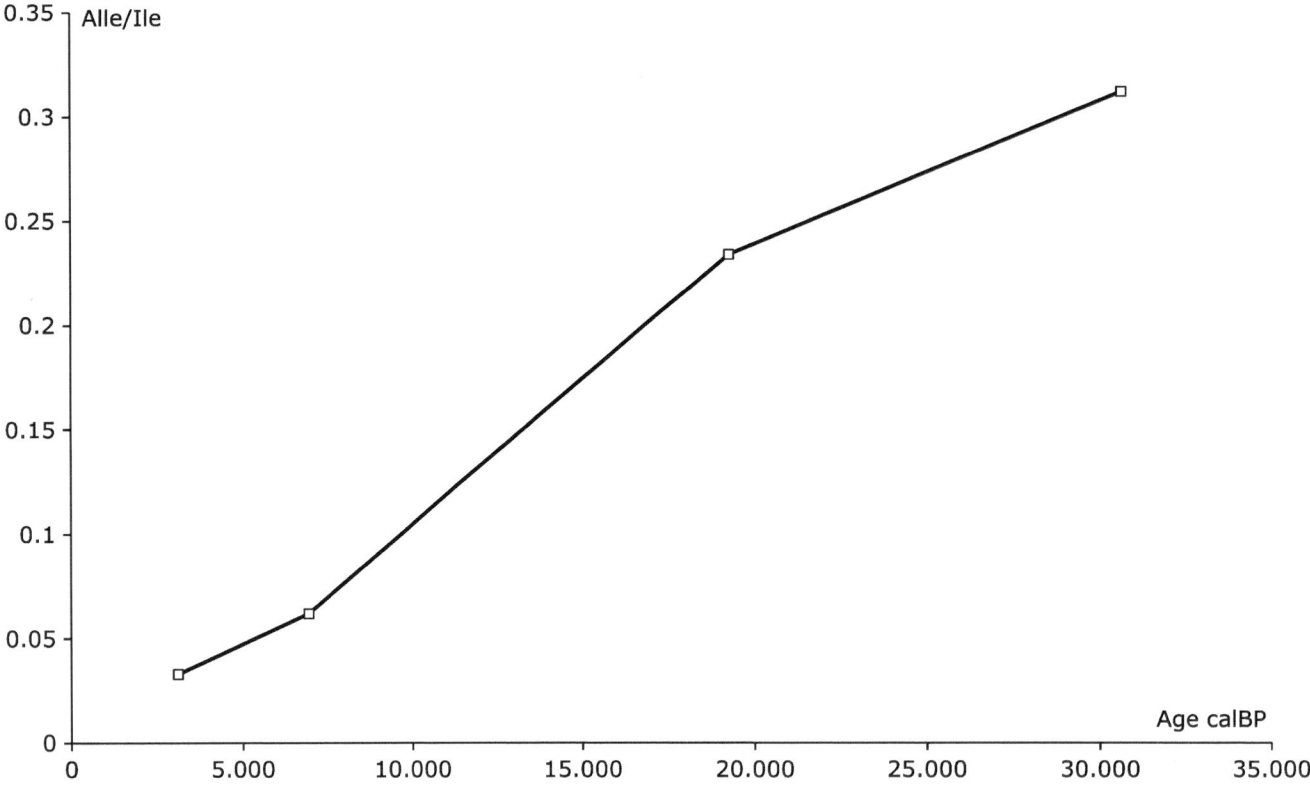

**Fig. 6.3** Epimerization mean ratios in continental gastropod shells (*Helix*) as compared to [14]C dates

several episodes of colonization of the underlying Middle Paleolithic units by *Helix*. Finally, at El Mnasra, gastropods from lower Aterian units have been reintroduced into younger units by the activity of humans or burrowing animals (Fig. 6.4).

For marine shells, the resulting calibration curves imply two hypotheses (Fig. 6.5). Age estimations have been established from measurements made on samples of *Patella* coming from Aterian layers at three archaeological sites: Dar es-Soltan I (unit M, Ruhlmann's excavations), Dar es-Soltan II (unit 7 with *Homo* remains), and El Mnasra.

The different layers we analyzed all lie between 55 and 85 ka. At Dar es-Soltan II, the sediment that contains the *Homo* remains thus occupies a chronological position around 80 ka, somewhat older than the lowest stratigraphic unit M of Dar es-Soltan I, which is placed around 70 ka. This occurrence provides a maximum estimated age for the human fossils.

## Discussion: The Aterian Time

Our relatively high estimation for the age and extent of the Aterian is not really surprising. Other Moroccan sites have recently provided similar estimates for Aterian assemblages.

Some faunal remains from the Mugharet el 'Aliya site collected during previous excavations of various Aterian layers have recently been dated by ESR on tooth enamel and have provided dates between 40 and 60 ka, or 55 and 80 ka, according to the uranium enrichment model (Wrinn and Rink 2003). More recently, TL dates of between 60 and 80 ka were obtained on heated flint at Rhafas Cave (Mercier et al. 2007), while TL and OSL dating at Taforalt provided results between 60 and 108 ka for stratigraphic unit E. This unit contains foliated artifacts of the MSA (Bouzouggar et al. 2007), which have been considered as Aterian since the earlier excavations of J. Roche (Raynal 1980). Moreover, older dates obtained by TL have been reported for the long sequence of Ifri n'Ammar Cave (Richter et al. 2012), in which assemblages described as Mousterian and Aterian alternate between MIS 6 and 5. Finally, detailed OSL results recently obtained after new excavations at Dar es-Soltan I Cave (Barton et al. 2009) clearly confirm our estimations.

It had long been thought that the Aterian progressively emerged from a Middle Paleolithic substratum, defined by various authors as a Mousterian-like assemblage dominated by Levallois flaking and an abundance of scrapers (Reygasse 1921; Ruhlmann 1948; Roche 1968; Bordes 1976; Wengler 1986, 1997; Wengler et al. 2001). The chronological position of this Mousterian within MIS 6 and 5 was deduced on the basis of bio-stratigraphy and some ESR dates from the classical

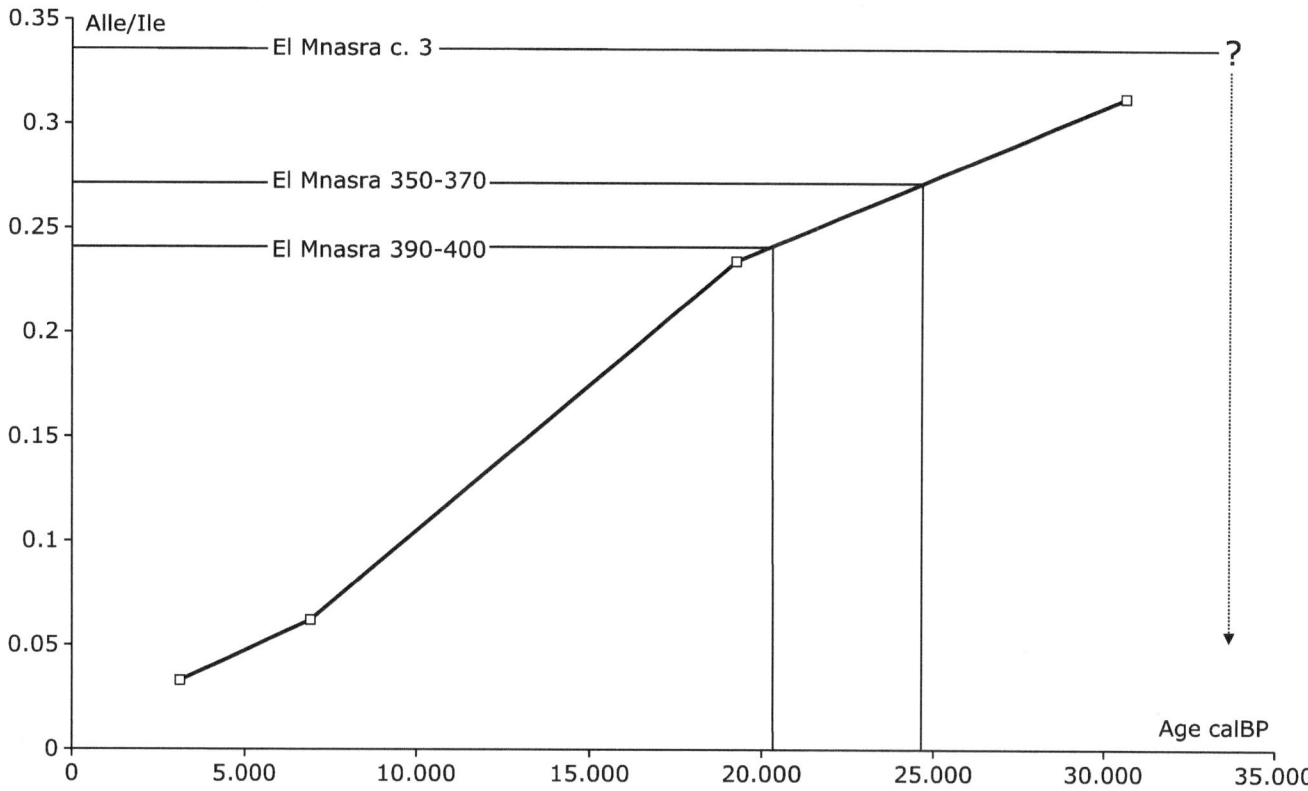

**Fig. 6.4** Calibration curve and age estimates for some *Helix* shells in El Mnasra archaeological units

hinterland site of Jebel Irhoud (3) (Hublin 1989; Amani 1991; Amani and Geraads 1993; Geraads and Amani 1998), and was recently confirmed by a $160 \pm 16$ ka ESR/U-series date obtained from a human tooth (Smith et al. 2007). After recent work on the Atlantic Coast (Raynal et al. 2002, 2004; Sbihi-Alaoui and Raynal 2004; Raynal 2005; Rhodes et al. 2006; Raynal et al. 2009), it is proposed that this Mousterian might derive from the late regional Acheulean, which is present at Casablanca during MIS 8 (and possibly 7) in the *Oulad Aj Jmel* member of the *Kef El Haroun* formation, for example at the Sidi Abderrahmane-Extension site, where the production of small Levallois flakes occurs along with a few hand-axes and a lesser number of cleavers. The Levallois technique developed at Casablanca from hand axe and cleaver technology (Biberson 1961), and was applied to the production of small flakes at Cap Chatelier during MIS 10, prior to 0.37 Ma. The Moroccan situation thus reasonably fits with what is known throughout most of Africa, where "*the Acheulian-MSA transition is marked by the disappearance of hand axes, their replacement by regionally distinct forms of points, and an increased reliance on Levallois and other methods of flake and blade production*" (Tryon et al. 2006, p. 201).

The age and processes of the disappearance of the Aterian and the possible contemporaneity of its terminal stages with Iberic Old Solutrean industries was underlined on the basis of young radiocarbon dates (Debénath et al. 1986; Otte et al. 2004). If many of these previous dates must now be revised in light of methodological progress, the OSL dates obtained on the Chaperon Rouge I and II sites (Texier et al. 1988; Rhodes 1990; Smith et al. 1990; Raynal et al. 1992) show very clearly that the Aterian occupations:

(1) postdate a yellow colluvial and lixiviated soil, dated by OSL to around 41 ka, which is the product of humid conditions, and,

(2) are overlain by sand sheets deposited during a major arid phase, occurring between 28 and 10 ka, which contain Upper Paleolithic assemblages (Ibero-maurusian) (Texier and Raynal 1989; Raynal et al. 1992; Texier et al. 1992). New data have recently confirmed the chronological position of the Ibero-maurusian in Morocco (Bouzouggar et al. 2008).

Besides an apparent monotony in the lithic assemblages, which fails to reflect any synchronic variability in tasks and types of settlements, whether in caves or open air, or the possible existence of regional hominid groups, the diachronic variability of the Moroccan Middle Paleolithic nevertheless occasionally exhibits the presence of tanged artifacts, foliated pieces, tanged darts, tanged points on blades, etc. These artifact types have even been used to

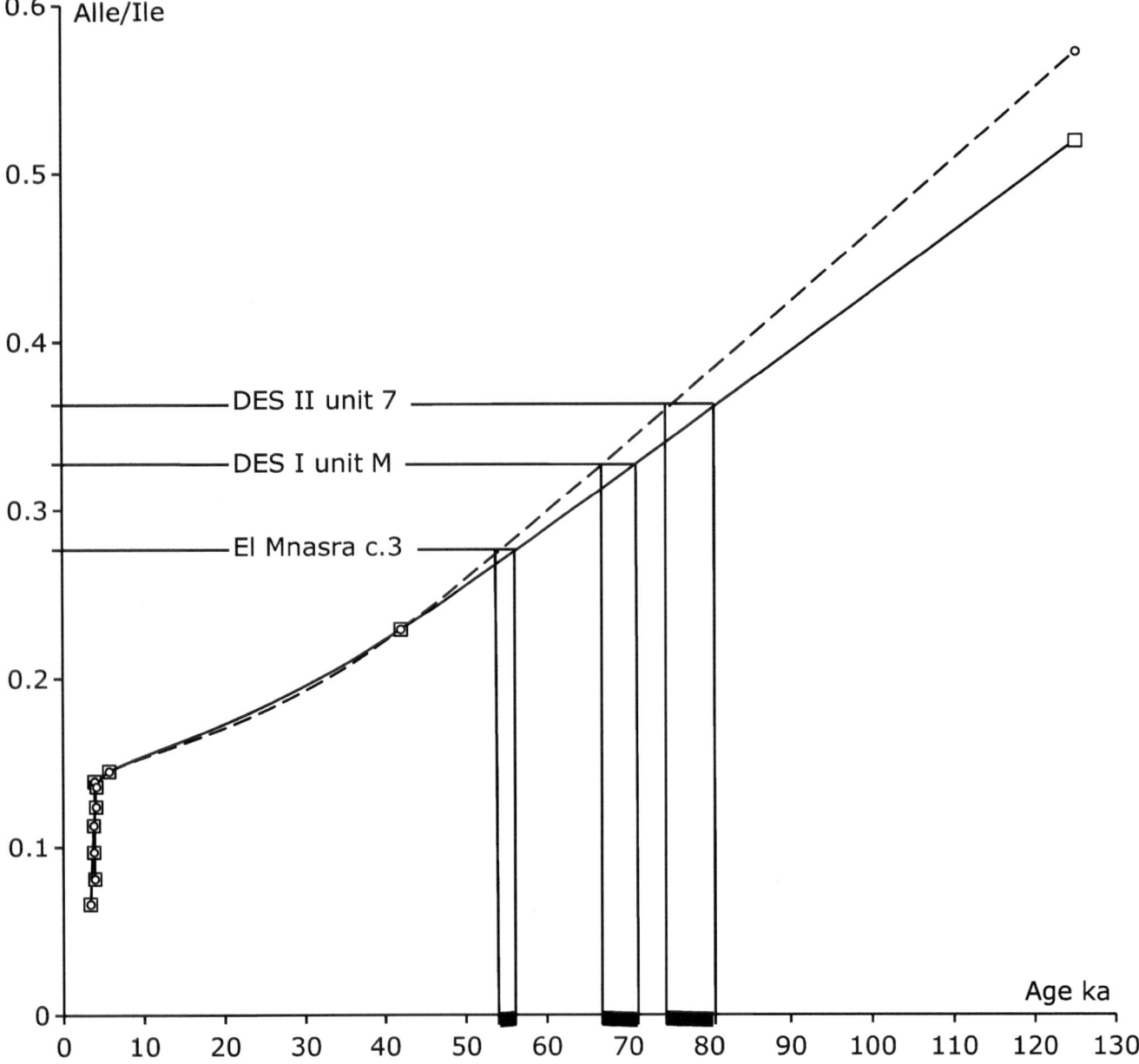

**Fig. 6.5** Calibration curves after original data (Lefevre et al. 1994; Occhietti et al. 1993, 2002)

define two facies (Ruhlmann 1948), namely a Lower Aterian, which looks very much like Mousterian, and an Upper Aterian when tanged pieces and foliates were abundant and accompanied by rarer types like pseudo-Saharian and Moroccan points. Undoubtedly, this regional complex now demands broad investigations that match its time dimension and likely origins. The Moroccan Middle Paleolithic is a model of continuous anthropological and cultural evolution whose roots may lie in the upper Acheulean, down to MIS 7–8, and which lasts until the end of MIS 3 (Fig. 6.6). However, it still lacks a detailed internal chronology, which must now be defined with the help of correlative dating methods applied to all types of materials, like we have demonstrated here.

## Conclusions

Continental gastropods like *Helix* are a useful amino-chronologic subject, considering the radiocarbon anomalies, *but their age is not necessarily the same as that of the unit containing them*. They usually colonized underlying units *below* their living surface and thus radiocarbon dates on

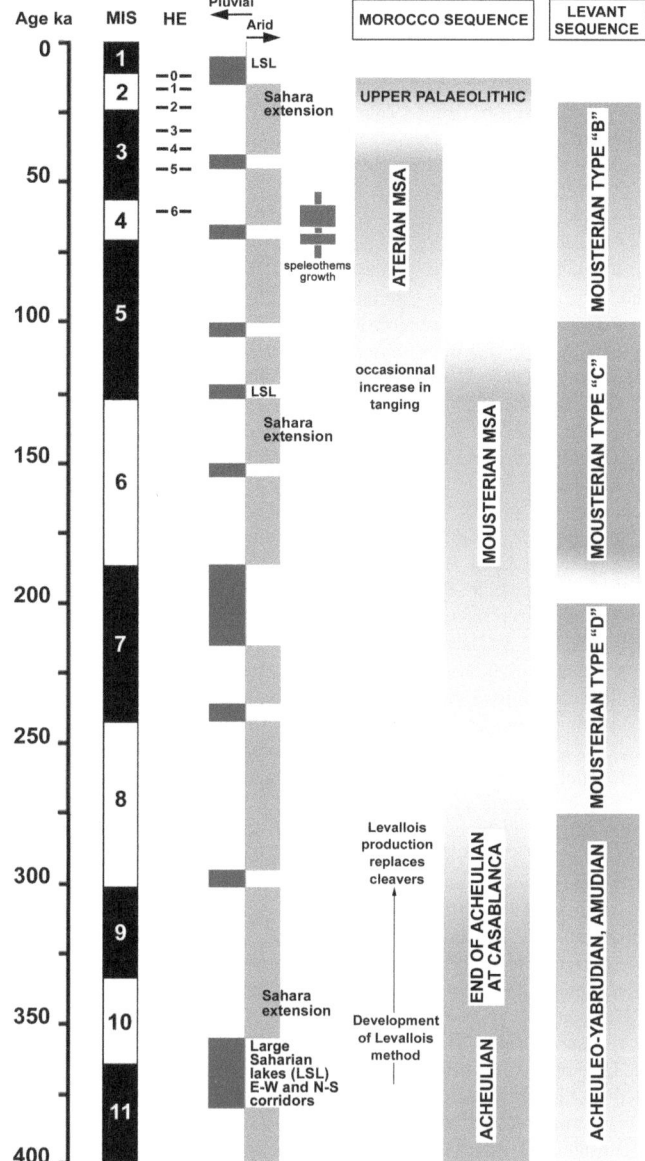

**Fig. 6.6** Chronological (after Bassinot et al. 1994) and environmental context (after Hooghiemstra et al. 1992; Matthewson et al. 1995; Gasse 2000, 2006; Moreno et al. 2002) of the prehistoric sequence of Atlantic Morocco for the last 400 kyr compared to the Levant sequence (after Valladas et al. 1998). Uranium/thorium dating of speleothems was performed by H.P. Schwarcz

their shells require careful interpretation. A 16 ka date obtained from a continental gastropod from an Aterian layer neither dates the Aterian nor makes it an Upper Paleolithic assemblage! Nevertheless, *Helix* epimerization, which is approximately linear until around 25 ka, provides an inexpensive absolute dating method when the reservations noted here are taken into account.

Archaeologists are greatly interested in marine shells. Inter-species variation between *P. safiana*, *P. intermedia*, and *P. aspera* have little significance and the racemization

ratios of *Patella* can be used as a reference for the amino-chronological framework of Morocco's Atlantic Coast. Age estimates from measurements made on samples of the genus *Patella* from Aterian layers (or from below Aterian layers) from three archaeological sites have provided results that range between 55 and 85 ka. It would be very interesting to check whether the maximum age of 75/85 ka at Dar es-Soltan 2, which is provided by marine shells from the sands surrounding the human remains, could be confirmed by directly dating the fossils. The existing estimates correspond with several recent data obtained in the same area by other dating methods.

The epimerization conditions of *Patella* and *Helix* are favorable for the identification of events occurring during MIS 4 and 3 ("Soltanian"), and finally, amino chronology allows for the rapid verification of the chronological homogeneity of archaeological assemblages and the selection of suitable samples for radiocarbon dating.

The Aterian assemblages of the Moroccan Atlantic Coast have previously been identified using the presence of tanged artifacts. In the studied area, their age lies between 40 and 80 ka at least. However, these peculiar artifacts may appear earlier, as indicated by the results obtained by TL at Ifri n'Ammar (Richter et al. 2012), and we already know that at least at Jebel Irhoud Cave (Smith et al. 2007), the Middle Paleolithic of Mousterian facies dates from MIS 6.

Ruhlmann (1948, p. 60) wrote about the Mousterian-Aterian complex of Morocco:

> *il faut s'appliquer à faire de nouvelles recherches afin d'éclaircir, avec les origines de l'Atérien, bien d'autres questions encore obscures se rapportant autant à son évolution et à ses faciès régionaux qu'à sa distribution et à sa chronologie relative, problèmes dont nos vues ne reposent que trop souvent sur des conjectures hardies, mais fragiles* (One must now undertake new research to enlighten both the question of the origin of the Aterian and other obscure matters dealing with its evolution, its regional fascies, its distribution, and its relative chronology, all problems that we consider on the grounds of audacious yet fragile hypotheses).

Despite being made in 1948, this statement still rings true!

**Acknowledgments** The authors are grateful to J.J. Hublin and S. McPherron for their invitation to participate in this meeting. They express their deep thanks to the Isotrace Laboratory at the University of Toronto and to J.P. Daugas, A. Debénath, M. A. El Hajraoui, and F.Z. Sbihi-Alaoui, who facilitated access to the sites and collections of Dar es-Soltan I and II, El Harhoura I, El Mnasra, Kaf Taht El Ghar, and Skhirat and allowed the sampling, to H.P. Shwarcz from McMaster University in Canada for providing U/TH dates, and to J. Hassar-Benslimane, Head of the National Institute of Archaeological Sciences and Heritage of the Kingdom of Morocco, for authorizing this project. The field surveys were supported by the Foreign Office of France and by the Natural Sciences and Engineering Research Council of Canada (NSERC), and were facilitated by the Ministry of Cultural Affairs of the Kingdom of Morocco and the Cultural, Scientific, and

Co-operative Office of the French Embassy at Rabat. The authors thank the anonymous readers whose comments allowed us to improve this chapter and P. Bindon for reviewing the English text.

# References

Amani, F. (1991). La faune du gisement à hominidés du Jebel Irhoud. Contribution à l'étude de la chronologie et de l'environnement du Quaternaire Marocain. Diplôme des Etudes Supérieures de 3è cycle de Géologie, Université Mohamed V, Faculté des Sciences de Rabat.

Amani, F., & Geraads, D. (1993). Le gisement moustérien du Djebel Irhoud, Maroc: Précisions sur la faune et la biochronologie, et description d'un nouveau reste humain. *Comptes rendus de l'Académie des sciences, 316*(II), 847–852.

Barton, N., Bouzouggar, A., Colcutt, S. N., Schwenninger, J.-L., & Clark-Balzan, L. (2009). OSL dating of the Aterian levels at Dar es-Soltan I (Rabat, Morocco) and implications for the dispersal of modern *Homo sapiens*. *Quaternary Science Reviews, 28*, 1914–1931.

Bassinot, F. C., Labeyrie, L. D., Vincent, E., Quidelleur, X., Shackleton, N. J., & Lancelot, Y. (1994). The astronomical theory of climate and the age of the Brunhes-Matuyama magnetic reversal. *Earth and Planetary Science Letters, 126*, 91–108.

Berger, R., Fergusson, G. J., & Libby, W. F. (1965). UCLA radiocarbon dates IV. *Radiocarbon, 7*, 356.

Biberson, P. (1961). *Le Paléolithique inférieur du Maroc atlantique.* Maroc: Publications du Service des Antiquités du fascicule 17.

Bordes, F. (1976). Moustérien et Atérien. *Quaternaria, Roma, XIX*, 19–34.

Bouzouggar, A., Barton, N., Van Haeren, M., D'Errico, F., Collcutt, S., Higham, T., et al. (2007). 82,000-year-old shell beads from North Africa and implications for the origins of modern human behavior. *Proceedings of the National Academy of Sciences of the USA, 104*, 9964–9969.

Bouzouggar, A., Barton, R. N. E., Blockley, S., Bronk-Ramsey, C., Collcutt, S., Gale, R., et al. (2008). Reevaluating the age of the Iberomaurusian in Morocco. *African Archaeological Review, 25*, 3–19.

Bowen, D. Q., & Sykes, G. A. (1985). Amino acid geochronology of raised beaches in south west Britain. *Quaternary Science Reviews, 4*, 279–318.

Choubert, G., Joly, F., Gigout, M., Marcais, J., Margat, J., & Raynal, R. (1956). Essai de classification du Quaternaire continental du Maroc. *Comptes Rendus de l'Académie des Sciences de Paris, 243*, 504–506.

Danzeglocke, U., Jöris, O., & Weninger, B., (2008). CalPal-2007online. Retrieved October 31, 2008, from http://www.calpal-online.de/

Daugas, J. P. (2002). Le Néolithique du Maroc: Pour un modèle d'évolution chronologique et culturelle. *Bulletin d'archéologie Marocaine, XIX*, 135–175.

Daugas, J.-P., Texier, J.-P., Raynal, J.-P., & Ballouche, A. (1984). *Nouvelles données sur le Néolithique marocain et ses paléoenvironnements : l'habitat cardial des grottes d'El Khril à Achakar (Province de Tanger) et la nécropole néolithique final de Rouazi à Skhirat (Province de Skhirat). 10° réunion annuelle des sciences de la Terre* (p. 167). Bordeaux, Paris: S.G.F.

Daugas, J.-P., Raynal, J. P., Ballouche, A., Occhietti, S., Pichet, P., Evin, J., et al. (1989). Le Néolithique nord-atlantique du Maroc: Premier essai de chronologie par le radiocarbone. *Comptes Rendus de l'Académie des Sciences de Paris, 308*(II), 681–687.

Daugas, J.-P., Raynal, J.P., El Idrissi, A., Ousmoi, M., Fain, J., Miallier, D., et al. (1999). Synthèse radiochronométrique concernant la séquence néolithique au Maroc. *Actes du 3e colloque International "¹⁴C et archéologie", Mémoires de la SPF, XXVI, and supplément à la Revue d'archéométrie*, 349–353.

Debénath, A. (1972). Nouvelles fouilles à Dar Es Soltane (champ de tir d'El Menzeh) près de Rabat, Maroc. *Bulletin de la Société Préhistorique française, 69*, 178–179.

Debénath, A. (1975). Découverte de restes humains probablement atériens à Dar Es Soltane (Maroc). *Comptes Rendus de l'Académie des Sciences de Paris, 281*, 875–876.

Debénath, A. (1976). Le site de Dar Es Soltan 2, à Rabat (Maroc). *Bulletin et Mémoires de la Société d'Anthropologie de Paris, 3*(13), 181–182.

Debénath, A. (1978). Le gisement préhistorique de Dar Es Soltane 2, champ de tir d'El Menzeh à Rabat (Maroc). Note préliminaire. 1: le site. *Bulletin d'Archéologie Marocaine, 11*, 9–23.

Debénath, A. (1980). Nouveaux restes humains atériens du Maroc. *Comptes Rendus de l'Académie des Sciences de Paris, 290*(D), 851–852.

Debénath, A. (1982). Découverte d'une mandibule humaine atérienne à El Harhoura, province de Rabat. *Bulletin d'Archéologie Marocaine, 12*, 1–2.

Debénath, A. (1992). Hommes et cultures matérielles de l'Atérien marocain. *L'Anthropologie, 96*, 711–720.

Debénath, A., & El Hajraoui, A. (1999). Les sites clés de Rabat et Témara. In J. P. Raynal, F. Z. Sbihi-Alaoui, & A. El Hajraoui (Eds.), *Maroc, terre d'origines* (pp. 39–44). Goudet: CDERAD.

Debénath, A., & Lacombe, J. P. (1986). Remarques sur la double sépulture néolithique du gisement d'El Harhoura II (Province de Témara, Maroc). *Arqueologia, 13*, 120–125.

Debénath, A., Raynal, J. P., Roche, J., Texier, J. P., & Laville, E. (1982). Mission préhistorique et paléontologique française au Maroc: Rapport d'activités pour l'année 1978. *Bulletin d'Archéologie Marocaine, XII*, 45–77.

Debénath, A., Daugas, J. P., Lefevre, D., Raynal, J. P., Roche, J., & Texier, J. P. (1984). Mission préhistorique et paléontologique française au Maroc. Rapport 1979. *Bulletin d'Archéologie Marocaine, 14*(1981–1982), 3–48.

Debénath, A., Raynal, J. P., Roche, J., & Texier, J. P. (1986). Stratigraphie, habitat, typologie de l'Atérien Marocain: Données récentes. *L'Anthropologie, 90*(2), 233–246.

El Hajraoui, M. A. (1993). Nouvelles découvertes néolithiques dans la région de Rabat. Grotte d'El Mnasra. *Méditerranéo, 2*, 105–121.

El Hajraoui, M. A. (1994). L'industrie osseuse atérienne de la grotte d'El Mnasra (Région de Témara, Maroc). *LAPMO, Université de Provence, CNRS, 3*, 91–94.

El Hajraoui, M. A. (2001). La Meseta côtière dans la Préhistoire. *Actes des 1ères Journées Nationales d'Archéologie et du Patrimoine, Volume 1: Préhistoire*, SMAP ed., 50–58.

Gasse, F. (2000). Hydrological changes in the African tropics since the Last Glacial Maximum. *Quaternary Science Reviews, 19*, 189–211.

Gasse, F. (2006). Climate and hydrological changes in tropical Africa during the past million years. *Comptes Rendu Palevol, 5*(1–2), 35–43.

Geraads, D., & Amani, F. (1998). Le gisement moustérien du Djebel Irhoud, Maroc : Précisions sur la faune et la paléoécologie. *Bulletin d'Archéologie Marocaine, Rabat, 18*, 11–18.

Goodfriend, G. A. (1987). Radiocarbon age anomalies in shell carbonate of land snails from semi-arid areas. *Radiocarbon, 21*(2), 159–167.

Hearty, P. J. (1986). An inventory of Last Interglacial (*sensu lato*) Age deposits from the Mediterranean Basin: A study of Isoleucine Epimerizationet U-Series dating. *Zeitschrift für Geomorphologie (Suppl.-Bd.), 62*, 51–69.

Hearty, P. J., Miller, G. H., Stearns, C. E., & Szabo, B. J. (1986). Aminostratigraphy of Quaternary shorelines in the Mediterranean Basin. *Geological Society of America Bulletin, 97*, 850–858.

Hooghiemstra, H., Stalling, H., Agwu, C. O. C., & Dupont, L. M. (1992). Vegetational and climatic changes at the northern fringe of the Sahara 250.000–5.000 years BP: Evidence from 4 marine pollen records located between Portugal and the Canary Islands. *Review of Palaeobotany and Palynology, 74,* 1–53.

Hublin, J.-J. (1989). Les origines de l'homme de type moderne: Europe occidentale et Afrique du Nord. In G. Giacobini (Ed.), *Hominidae. Actes du 2e Congrès intern. de Paléontologie humaine (Turin)* (pp. 223–230). Milan: Jaca Book.

Kaufman, D. S., & Manley, W. F. (1998). A new procedure for determining DL amino acid ratios in fossils using reverse phase liquid chromatography. *Quaternary Science Reviews, 17*(11), 987–1000.

Kimber, R. W. L., Griffin, C. V., & Milnes, A. R. (1986). Amino acid racemization dating: Evidence of apparent reversal in aspartic acid racemization with time in shells of Ostrea. *Geochimica et Cosmochimica Acta, 50,* 1159–1161.

Kriausakul, N., & Mitterer, R. M. (1978). Isoleucine epimerization in peptides and proteins: Kinetic factors and application to fossil proteins. *Science, 201,* 1011–1014.

Kriausakul, N., & Mitterer, R. M. (1983). Epimerization of COOH-terminal isoleucine in fossil dipeptides. *Geochimica et Cosmochimica Acta, 47,* 963–966.

Lacombe, J. P., & Daugas, J. P. (1988). La nécropole néolithique de Rouazi-Skhirat. *Bulletin et Mémoires de la Société d'Anthropologie de Paris, 5*(XVème série(4)), 308–309.

Lacombe, J. P., Daugas, J. P., & Sbihi-Alaou, F. Z. (1990). La nécropole néolithique de Rouazi-Skhirat (Maroc), Présentation de l'étude des sépultures. *Bulletin et Mémoires de la Société d'Anthropologie de Paris, 2*(3–4), 55–60.

Lacombe, J. P., El Hajraoui, A., & Daugas, J. P. (1991). Etude anthropologique préliminaire des sépultures néolithiques de la grotte d'El Mnasra (Témara, Maroc). *Bulletin de la Société d'Anthropologie du Sud-Ouest, XXVI,* 163–176.

Lefevre, D., Texier, J. P., Raynal, J. P., Occhietti, S., & Evin, J. (1994). Enregistrements-réponses des variations climatiques du Pleistocène supérieur et de l'Holocène sur le littoral de Casablanca (Maroc). *Quaternaire, 5*(3–4), 173–180.

Matthewson, A. P., Shimmield, G. B., Kroon, D., & Fallick, A. E. (1995). A 300 kyr high-resolution aridity record of the North African continent. *Paleoceanography, 10,* 677–692.

Mercier, N., Wengler, L., Valladas, H., Joron, J. L., Froget, L., & Reyss, J. L. (2007). The Rhafas Cave (Morocco): Chronology of the mousterian and aterian archaeological occupations and their implications for Quaternary geochronology based on luminescence (TL/OSL) age determinations. *Quaternary Geochronology, 2,* 309–313.

Miller, G. H. (1982). Amino Acid Geochronology Laboratory, January 1981 through May 1982. Report of current activities, Institute of Arctic and Alpine Research and Department of Geological Sciences. University of Colorado

Miller, G. H., & Mangerud, J. (1985). Aminostratigraphy of European marine interglacial deposits. *Quaternary Science Reviews, 4,* 215–278.

Moreno, A., Cacho, I., Canals, M., Prins, M. A., Sanchez-Goni, M. F., Grimaly, J. O., et al. (2002). Saharan dust transport and high-latitude glacial climatic variability: The Alboran Sea record. *Quaternary Research, 58,* 318–328.

Murray-Wallace, C. V. (1993). A review of the application of the amino acid racemisation reaction to archaeological dating. *The Artefact, 16,* 19–26.

Nespoulet, R., Debénath, A., El Hajraoui, A., Michel, P., Campmas, E., Oujaa, A., et al. (2008a). Le contexte archéologique des restes humains atériens de la région de Rabat-Témara (Maroc): Apport des fouilles des grottes d'El Mnasra et d'El Harhoura 2. In H. Aouraghe, H. Haddoumi & K. El Hammouti (Eds.), *Actes des quatrièmes rencontres des Quaternaristes Marocains (RQM4),*

*Le Quaternaire marocain dans son contexte méditerranéen* (pp. 356–375). Oujda

Nespoulet, R., El Hajraoui, M., Amani, F., Ben Ncer, A., Debénath, A., El Idrissi, A., et al. (2008b). Palaeolithic and Neolithic occupations in the Témara Region (Rabat, Morocco): Recent data on hominin contexts and behavior. *African Archaeological Review, 25,* 21–39.

Occhietti, S., Raynal J. P. (1996). La méthode de datation par les acides aminés appliquée à la Préhistoire du Maroc. *XIIIè Congès UISPP, Section 2, sous-section 2.2, Forli, 1996, Abstracts, 1,* 25.

Occhietti, S., Raynal, J. P., Pichet, P., & Texier, J. P. (1993). Aminostratigraphie du dernier cycle climatique au Maroc atlantique, de Casablanca à Tanger. *Comptes Rendus de l'Académie des Sciences de Paris, 317*(II), 1625–1632.

Occhietti, S., Raynal, J. P., Pichet, P., Daugas, J. P., El Hajraoui, A. (1999). Calibration du ratio épimerisation de l'isoleucine par le $^{14}$C: Exemple du Maroc. *Actes du 3e colloque International "$^{14}$C et archéologie", Mémoires de la SPF t. XXVI et supplément à la Revue d'archéométrie* (pp. 33–37).

Occhietti, S., Raynal, J. P., Pichet, P., & Lefevre, D. (2002). Aminostratigraphie des formations littorales pléistocènes et holocènes de la région de Casablanca, Maroc. *Quaternaire, 13*(1), 55–64.

Otte, M., Bouzouggar, A., & Kozlowski, J. (2004). *La Préhistoire de Tanger (Maroc)* (Vol. 105). Liège: ERAUL.

Ousmoï, M. (1989). *Application de la datation par thermoluminescence au Néolithique Marocain.* Thèse, Université de Clermont II.

Penkman, K. E. H., Kaufman, D. S., Maddy, D., & Collins, M. J. (2008). Closed-system behaviour of the intra-crystalline fraction of amino acids in mollusc shells. *Quaternary Geochronology, 3,* 2–25.

Raynal, J. P. (1980). Taforalt. *In* Mission préhistorique et paléontologique française au Maroc Rapport d'activité 1978. *Bulletin d'Archéologie Marocaine, XII,* 69–72.

Raynal, J. P. (2005). L'Afrique du Nord. Le Maghreb préhistorique. Origines et diversité du Néolithique au Maghreb. In *Archéologies. Vingt ans de recherches françaises dans le monde* (pp. 219–230). Paris: Maisonneuve et Larose et ADPF.ERC.

Raynal, J. P., Texier, J. P., Lefevre, D., Rhodes, E. (1992). Les Sables Beiges de Couverture et l'Atérien en Mamora, nouveaux éléments de chronologie numérique. Colloque international *L'homme préhistorique de Témara et ses contemporains du bassin méditerranéen depuis 100 000 ans,* Témara, 21–23 septembre 1992, pré-actes.

Raynal, J. P., Sbihi-Alaoui, F. Z., Magoga, L., Mohib, A., & Zouak, M. (2002). Casablanca and the early occupation of North-Atlantic Morocco. *Quaternaire, 13*(1), 65–77.

Raynal, J. P., Sbihi-Alaoui, F. Z., Magoga, L., Mohib, A., & Zouak, M. (2004). The lower palaeolithic sequence of Atlantic Morocco revisited after recent excavations at Casablanca. *Bulletin d'Archéologie Marocaine, 20,* 44–76.

Raynal, J. P., Sbihi-Alaoui, F. Z., Mohib, A., & Geraads, D. (2009). Préhistoire ancienne au Maroc atlantique: Bilan et perspectives régionales. *Bulletin d'Archéologie Marocaine, XXI,* 9–54.

Reygasse, M. (1921). *Nouvelles études de Palethnologie maghrébine.* Braham: Constantine.

Rhodes, E. J. (1990). Optical dating of quartz from sediments. Thesis, Oxford University.

Rhodes, E. J., Singarayer, J. S., Raynal, J. P., Westaway, K. E., & Sbihi-Alaoui, F. Z. (2006). New age estimates for the Palaeolithic assemblages and Pleistocene succession of Casablanca, Morocco. *Quaternary Science Reviews, 25*(19–20), 2569–2585.

Richter, D., Moser, J., Nami, M. (2012). New data from the site of Ifri n'Ammar (Morocco) and some remarks on the chronometric status of the Middle Paleolithic in the Maghreb. In J.-J. Hublin & S.

McPherron (Eds.), *Modern origins: North African perspective*. Dordrecht: Springer.

Roche, J. (1968). L'Atérien de la grotte de Taforalt (Maroc oriental). *Bulletin d'archéologie Marocaine, VII*(1967), 11–56.

Ruhlmann, A. (1948). A propos de la subdivision de l'Atérien marocain. *Publications du Service des Antiquités du Maroc, 8*, 9–68.

Ruhlmann, A. (1951). *La grotte préhistorique de Dar Es Soltan*. Hespéris, XI

Sbihi-Alaoui, F. Z., & Raynal, J. P. (2004). Casablanca: Un patrimoine préhistorique exceptionnel. *Bulletin d'Archéologie Marocaine, 20*, 17–43.

Smith, W., Rhodes, E. J., Stokes, S., Spooner, N. A., & Aitken, M. J. (1990). Optical dating of sediments: Initial quartz results from Oxford. *Archaeometry, 32*, 19–31.

Smith, T. M., Tafforeau, P., Reid, D. J., Grün, R., Eggins, S., Boutakiout, M., et al. (2007). Earliest evidence of modern human life history in North African early *Homo sapiens*. *Proceedings of the National Academy of Sciences of the USA, 104*, 6128–6133.

Texier, J. P., & Raynal, J. P. (1989). Les "sables beiges" du Nord-Ouest du Maroc: Nouvelles interprétations dynamiques, chronologiques et paléoclimatiques. *Comptes Rendus de l'Académie des Sciences de Paris, 309*(II), 1577–1582.

Texier, J. P., Huxtable, J., Rhodes, E. J., Miallier, D., & Ousmoi, M. (1988). Nouvelles données sur la situation chronologique de l'Atérien au Maroc et leurs implications. *Comptes Rendus de l'Académie des Sciences de Paris, 307*(II), 827–832.

Texier, J. P., Lefevre, D., & Raynal, J. P. (1992). La Formation de la Mamora. Le point sur la question du Moulouyen et du Salétien du Maroc Nord-Occidental. *Quaternaire, 3*(2), 63–73.

Tryon, C. A., Mcbrearty, S., & Texier, P. J. (2006). Levallois lithic technology from the Kapthurin Formation, Kenya: Acheulian origin and Middle Stone Age diversity. *African Archaeological Review, 22*(4), 199–229.

Valladas, H., Mercier, N., Joron, J. L., & Reyss, J. L. (1998). Gif Laboratory dates for Middle Paleolithic Levant. In T. Akazawa, K. Aoki, & O. Bar-Yosef (Eds.), *Neandertals and modern humans in western Asia* (pp. 69–75). New York: Plenum Press.

Wehmiller, J. F. (1984). Interlaboratory comparison of amino acid enantiomeric ratios in fossil Pleistocene mollusks. *Quaternary Research, 22*, 109–120.

Wehmiller, J. F., & Belknap, D. F. (1978). Alternative kinetic models for the interpretation of amino acid enantiomeric ratios in Pleistocene mollusks: Examples from California, Washington, and Florida. *Quaternary Research, 9*, 330–348.

Weisrock, A., Occhietti, S., Hoang, C. T., Lauriat-Rage, A., Brebion, P., & Pichet, P. (1999). Les séquences littorales pléistocènes de l'Atlas atlantique entre Cap Rhir et Agadir, Maroc. *Quaternaire, 10*(2–3), 227–244.

Wengler, L. (1986). Position géochronologique et modalités du passage Moustérien-Atérien en Afrique du Nord. L'exemple de la grotte du Rhafas au Maroc oriental. *Comptes Rendus de l'Académie des Sciences de Paris, 303*(série II(12)), 1153–1156.

Wengler, L. (1997). La transition du Moustérien à l'Atérien. *L'Anthropologie, 101*, 448–481.

Wengler, L., Wengler, B., Brochier, J., El Azouzzi, M., Margaa, A., Mercier, N., et al. (2001). La grotte du Rhafas (Maroc oriental) et les recherches sur le Paléolithique moyen. *Actes des 1ères Journées Nationales d'Archéologie et du Patrimoine, Volume 1: Préhistoire*, SMAP ed., 67–81.

Weninger, B., & Jöris, O. (2008). Towards an absolute chronology at the Middle to Upper Palaeolithic transition in western Eurasia: A new Greenland Hulu time-scale based on U/Th ages. *Journal of Human Evolution, 55*(5), 772–781.

Wrinn, P. J., & Rink, W. J. (2003). ESR dating of tooth enamel from Aterian levels at Mugharet el Aliya (Tangier, Morocco). *Journal of Archaeological Science, 30*, 123–133.

# Chapter 7
# The Identity and Timing of the Aterian in Morocco

A. Bouzouggar and R. N. E. Barton

**Abstract** Until recently, any discussions of the Aterian of Morocco were severely limited by the paucity of secure dating evidence. New stratigraphic studies in northern, eastern, and, most recently, Atlantic Morocco have begun to address this major gap in our knowledge. It is clear that the very late record for the Middle Paleolithic in Morocco of 40–20 ka can certainly now be extended as far back as more than 110 ka. Another area for discussion concerns the nature and identity of the Aterian technology through time, which is urgently in need of review and redefinition. Rather than concentrating on the very narrow issue of pedunculate tools, we feel that more consideration should be given to a wider set of variables.

**Keywords** Aterian • Chronology • Morocco • Redefinition

## Introduction

In the nineteenth century, the first pedunculate Aterian lithic artifacts were recorded at Eckmuhl Cave (Carrière 1886) in the Oran region of Algeria. After this discovery, many other Aterian sites were discovered, including Bir El Ater, south of Tébessa, in Algeria, which became the eponymous site for this technology. Early descriptions of the Aterian were provided by Reygasse (1922) and Pallary (1927), and these were followed by more detailed studies by other authors, especially in Morocco and Algeria (Antoine 1938; Hugot 1958; Bobo 1956).

Despite the acknowledged wealth of evidence for the Aterian in North Africa and its known existence in Morocco for over 50 years (Ruhlmann 1936, 1951; Antoine 1937, 1950a, 1950b), this "Middle Paleolithic" technology remains surprisingly poorly researched and ill-defined. A new opportunity to reassess the Aterian emerged from a wide program of research and exploration undertaken by a joint Morocco-UK team of archaeologists in northern, eastern, and most recently Atlantic Morocco. The aims of this continuing project include the investigation of the nature and chronology of the Aterian, including its environmental and paleoclimatic contexts, and the reconstruction of climatic change in this region during Marine Isotope Stages (MIS) 6–2. Here, we review some of the early results of this work as it applies to the timing of the Aterian and offer comments on the nature of the industry itself. Some of the most pressing questions concern the first appearance of the Aterian in Morocco, its relationship with other Middle Paleolithic finds, its evolutionary development, and its latest occurrences. One of the most contentious issues concerns its relationship with the "Mousterian," which is generally considered to have occurred earlier than the Aterian and is characterized by an absence of pedunculates (Balout 1965; Camps 1974). Despite the presence of Mousterian at Jebel Irhoud in association with early *Homo sapiens* (Hublin et al. 1987; Smith et al. 2007) and at Rhafas (Wengler 1993), the cultural origins and development of this industry remain disappointingly vague.

## The Persistent Problem of the Origin and Chronology of the Aterian

One of the problems of the early dating work on the Aterian was that it was largely dependent on the conventional radiocarbon method. As a consequence and despite the

A. Bouzouggar (✉)
Institut National des Sciences de l'Archéologie et du Patrimoine, Rabat-Instituts, Madinat Al Irfane Angle rues 5 et 7, Hay Riad, Rabat, Morocco
and
Department of Human Evolution, Max Planck Institute for Evolutionary Anthropology, Deutscher Platz 6, 04103 Leipzig, Germany
e-mail: bouzouggar@minculture.gov.ma

R. N. E. Barton
Institute of Archaeology, University of Oxford, 36 Beaumont Street, Oxford, OX1 2PG, UK
e-mail: nick.barton@arch.ox.ac.uk

J.-J. Hublin and S. P. McPherron (eds.), *Modern Origins: A North African Perspective*,
Vertebrate Paleobiology and Paleoanthropology, DOI: 10.1007/978-94-007-2929-2_7,
© Springer Science+Business Media B.V. 2012

**Table 7.1** New OSL dates and proposed correlation of stratigraphic units between different Aterian cave sites of the Témara region, Morocco

| MIS | Dar es-Soltan I | Dar es-Soltan II | Contrebandiers | El Mnasra |
|---|---|---|---|---|
| 5.5 | GI (144–123 ka) | Layer 7 (121 ka) | Layer 15 (129 ka) Layer 14 (main part) | ?Layer 13 subunits "A17" to "A15" |
| 5.4 | Strong erosion | (Disturbance) | (Poor exposure) | Layer 13 ?base of subunit "A14" |
| 5.3 | G2 (119–106 ka) | Layer 5 (101 ka) | Layer 14 (top) to Layer 11 (121–100 ka) | Top Layer 13 (118 ka) Layers 12 to 5a (111–105 ka) |
| 5.2 | (No obvious evidence) | (Disturbance) | ?Layer 9 (100–96 ka) | – |
| 5.1 | G3 (87–68 ka) | – | (No obvious evidence) | – |
| 4 | Erosion at base G4 (61–52 ka) | – | Layer 8 (59 ka) | Erosion at base (undated) ?Layer 4 |
| 3 | G5 (base unit 14) (33 ka) | Layer 3b2 (31 ka) | – | – |
| 2 | G5 (main unit 14) | Layer 3a (13 ka) | ?Upper Layer 7 etc. | – |
| 1 | G5 (units 15–16) (7–6 ka) | – | – | ?Layers 3–1 |

The unmodified central OSL age estimates are shown in brackets

publication of infinite ages for some samples, a relatively recent origin for this industry was proposed. Such an interpretation of the evidence allowed for the idea that the Aterian was introduced in Northwest Africa during a wetter climatic phase of MIS 3 (ca. 50–40 ka) (Debénath et al. 1986). The first serious challenge to this model came from the application of luminescence dating to sites in the eastern Sahara (Cremaschi et al. 1998; Garcea 2001, 2004) which suggested the existence of a much longer chronology for this industry. In Morocco, Electron Spin Resonance dating methods have subsequently shown that the Aterian in El 'Aliya Cave in the northern peninsula can be dated to between 42 (EU) and 56 ka (LU) (Wrinn and Rink 2003), though uncertainties persist due to the fact that the samples derive from "old" excavations. At Rhafas Cave, luminescence ages of between 80 and 70 ka have now been recorded for the earliest Aterian and to around 90–80 ka for the latest Mousterian (Mercier et al. 2007). At Grotte des Pigeons, Taforalt, a range of proxies (Optically Stimulated Luminescence, Thermoluminscence, and Uranium Series dating) indicate that the earlier Aterian levels lie between 73 and 91 ka (Bouzouggar et al. 2007a). Even older dates have been claimed for Ifri n'Ammar, eastern Morocco (Mikdad et al. 2004; Richter et al. 2012), while new research in Dar es-Soltan 1 demonstrates that the Aterian *sensu lato* in this cave dated to 110 ka and is of considerably greater antiquity than previously suspected, with a chronology extending back into MIS 5 (Barton et al. 2009). The Aterian in the Atlantic area between Rabat and Témara was once considered to be represented only by its middle and late phases (Roche 1969; Roche and Tixier 1976). New OSL chronology has clearly shown that this technology is considerably earlier in this part of Morocco than previously suspected and older too perhaps than in Tunisia and parts of the Libyan Sahara or on its northern fringes (Table 7.1)

(Schwenninger et al. 2010). In caves near Rabat, the Aterian *sensu lato* can be extended as far back as 100 ka at Dar es-Soltan II in layer 5, between 121 and 100 ka at Contrebandiers Cave in the top of layer 14 (top) and layer 11 and 111 ka at El Mnasra in layer 12 (Schwenninger et al. 2010). Based on all of these new data, it is clear that the known chronology has at least tripled in length since the original application of scientific dating techniques. It would now be of little surprise if its beginnings could not be traced back to MIS 6 or even earlier but this will require further testing and verification at individual sites using multiproxy dating techniques.

The same pattern seen in Morocco has been noted in Tunisia, where an Aterian-like industry at Oued El Akarit has recently been dated to more than 70 ka (Roset and Harbi-Riahi 2007). The excavators pointed out that the technology of "Niveau 1" is very similar to the Aterian (Roset and Harbi-Riahi 2007, p. 118 plate XXVII, n° 1 and 2, p. 225 plate LXV, plate LXXXVIII n° 4–7) since it contains pedunculates and includes examples with intensive thinning of the proximal parts of the blanks.

In terms of its origins, the Aterian in Morocco is still considered to have developed out of the local early Middle Paleolithic, also known as the Mousterian (Balout 1965; Camps 1974). However, since the term "Mousterian" was imposed by French prehistorians, it implies that this is the same technology as in Europe, or at least as in Southwest France. In the absence of alternative names for this industry, it is still the terminology most widely used in North Africa (Hublin et al. 1987; Wengler 1993; Mikdad and Einwanger 2000; Wengler 2006). Perhaps surprisingly, however, the North African Mousterian technology has rarely been described in detail and is thus not well-defined. Often it is regarded as a technology with a high Levallois component and an abundance of scrapers, which has invited

**Table 7.2** Different typological classifications of the Aterian in Morocco

| Ruhlmann 1945 | Antoine (1950a, b) | Balout (1955) | Roche (1969) | Bordes (1977) | Wengler (1973) |
|---|---|---|---|---|---|
| Upper Levalloisian | Aterian I | Early Aterian | Aterian (?) | Aterian with Mousterian Tradition | Early Aterian |
| Lower Aterian | Aterian II | Middle Aterian | Middle Aterian | Pre-Aterian | Proto-Aterian |
| Upper Aterian | Aterian III | Middle Aterian | Middle Aterian | | |
| Upper Aterian | Aterian IV | Late Aterian | Typical Aterian | Typical Aterian | Late Aterian |

**Table 7.3** The percentage of pedunculates from major Aterian sites

| Site | % of the pedunculates | Source of the data |
|---|---|---|
| Dar es-Soltan I, layer C2 | 30.2 | Ruhlmann (1951) |
| Taforalt, layer F | 1.3 | Roche (1969) |
| Rhafas, layer 2 | 12.3 | Wengler (1993) |
| Contrebandiers, layer 11 | 13.89 | Bouzouggar (1997a, b) |

comparisons with the European "*Moustérien de faciès Levallois*" (Balout 1965; Hublin et al. 1987) or "Ferrassie" (Wengler 2006) variant. Other authors would include industries with Mousterian blades and Levallois points in this definition, though these notably decrease in the later stages of its development (Camps 1974). While it seems that the human types associated with the North African Mousterian are earlier *Homo sapiens* (Smith et al. 2007), the associations with the Aterian are generally archaic forms whose affinities are still controversial (Trinkaus 2007). In Morocco, the Mousterian at Jebel Irhoud does not seem to contain pedunculates (Hublin et al. 1987), even if some doubts exist over the earlier collections (Balout 1965, p. 142 and plate II n° 2). The assemblage associated with the human remains in layer 18 is described as being a predominantly Levallois industry, with side scrapers, but only rarely blades, and a total absence of pedunculates and end scrapers (Hublin et al. 1987). The same can probably be said for the rest of the archaeological sequence at this site (Salih 1995).

## The Earliest Definitions of the Aterian

The first definitions and subdivisions of the Aterian in Morocco were mainly based on lithic typology (Antoine 1937, 1950a), with reference to four main lithic types in particular:
• Mousterian points
• pedunculate points also known as "Aterian points"
• bifacial and elongated retouched points
• small discoidal cores.

Included in this was the widely held view that the technology had a significant Levallois component (Caton-Thompson 1946). Bordes (1976–1977) subsequently

divided the Aterian into three major substages based on typological criteria. An almost identical scheme was later adopted by Roche for his analysis of Grotte des Pigeons at Taforalt, where he identified an *Upper Aterian* in layer D, a *Middle Aterian* in layer F, and a rich *Aterian of Mousterian tradition* without pedunculate pieces in layer H (Roche 1969). Despite difficulties concerning the original position of the Roche layers, the same subdivisions were employed for Rhafas Cave 24 years later (Wengler 1993) (Table 7.2).

The most widely accepted classification scheme is still that of Tixier (1967), which describes the Aterian as being made up of Levallois blades, end scrapers, and a high proportion of tanged pieces (around 25%). This definition is mainly based on the description of the industry present at Bir El-Ater (Oued Djebbana, Algeria).

Other authors would include a variant with pebble tools within the Aterian (Texier 1985–1986; Debénath 1992, 1994), but this seems to be restricted to a solitary example from the open air site at Chaperon Rouge, Morocco, so it might be regarded as a special case. Despite this broad consensus, there are, in fact, very few assemblages that meet all of the criteria outlined by Tixier. In particular, there is a marked variability in the percentage of tanged pieces, ranging from 1.3% in layer F at Taforalt to over 30% at Dar es-Soltan 1 (Table 7.3). The latter is almost certainly exaggerated because of the recovery methods employed by Ruhlmann in his excavations between 1937 and 1938 (Ruhlmann 1951).

## Aterian Technology

Much of the attention on the Aterian has tended to concentrate on the very narrow issue of pedunculate tools, rather than considering a wider set of typological and technological variables. In its extreme form, the *fossile*

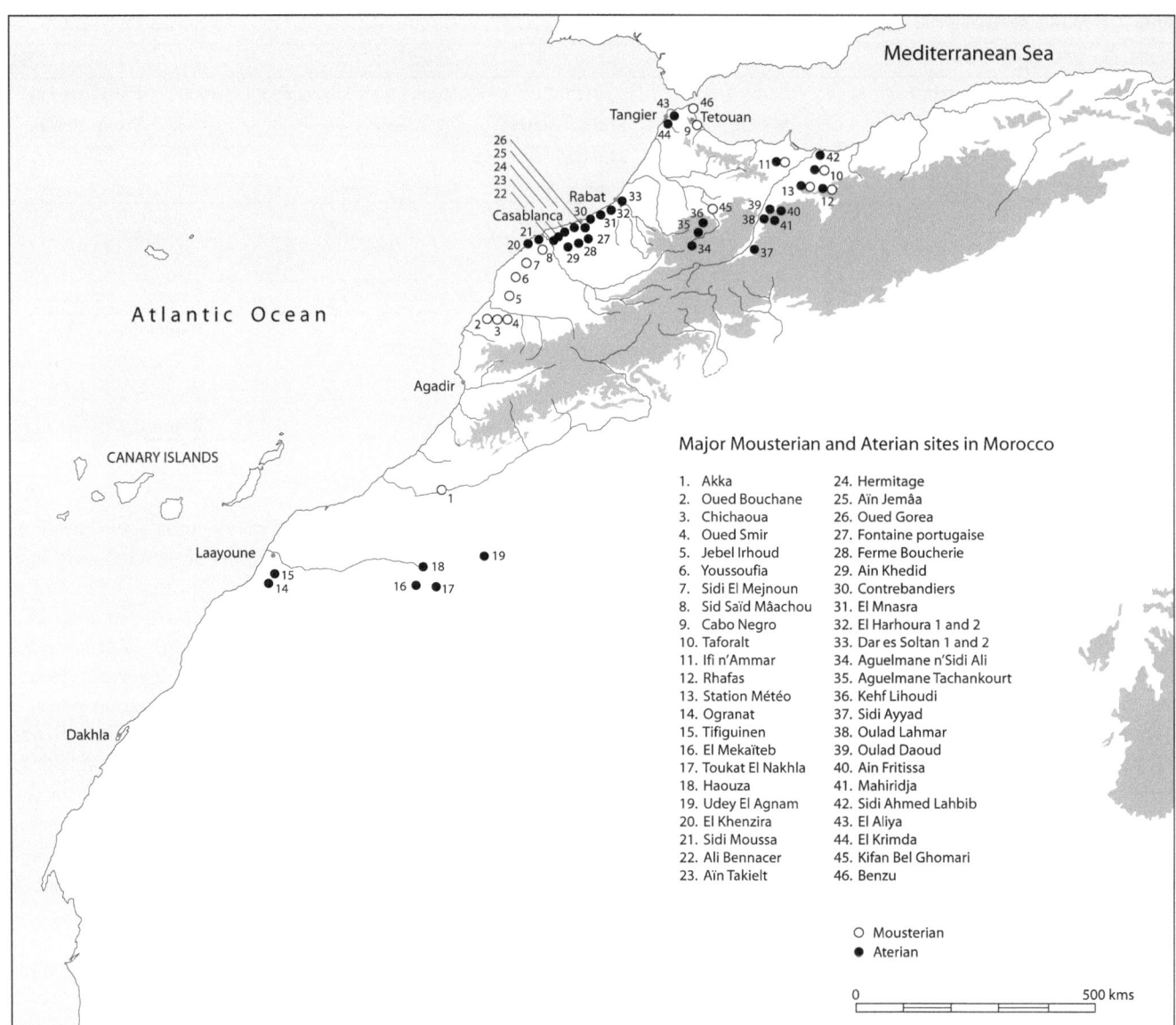

**Fig. 7.1** The distribution of Mousterian and Aterian sites in Morocco based only on the criteria of the presence/absence of pedunculates

*directeur* approach has led to the artificial subdivision of Middle Paleolithic/Middle Stone Age in Morocco into the local "Mousterian" and "Aterian," with only minor reference to the "early Middle Paleolithic" (Balout 1955, 1965; Camps 1974; Wengler 1993, 2006) (Fig. 7.1).

While it remains true that pedunculates can form an important component of Aterian tool assemblages (Tixier 1958–1959), it should not be forgotten that the same technique of making tangs is not unknown in the Late Acheulean, for example, in layer D2 at Cap Châtelier, Casablanca (Biberson 1961, p. 367, plate CLXXVII, piece n° 299), in the local Mousterian (Wengler 1993, 2006), as well as in the

Upper Paleolithic (Roche 1958–1959, p. 189) and the Neolithic (Camps 1974, p. 285, Fig. 84; p. 288, Fig. 87; and p. 307, Fig. 91).

## The Pedunculates

One of the problems with the typological approach is that it led to the assumption that pedunculates served solely or primarily as projectiles (Marchand and Aymé 1935;

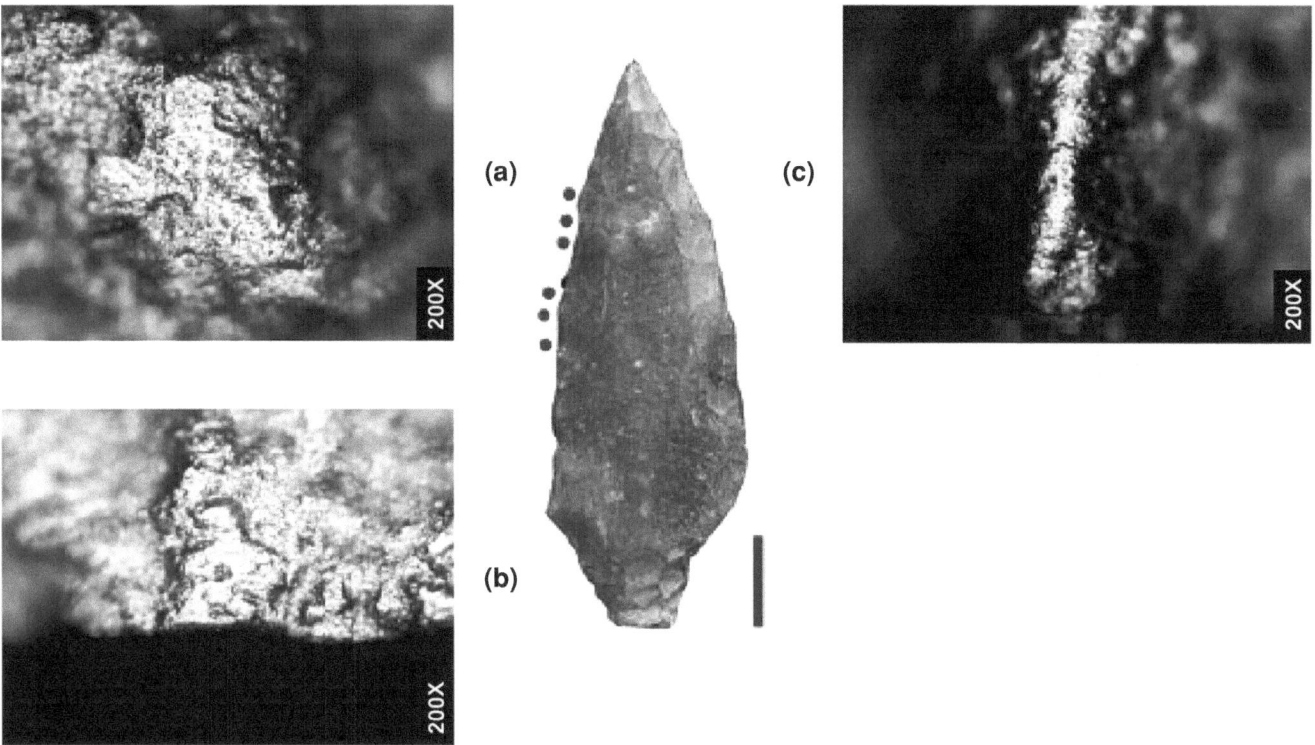

**Fig. 7.2** A pedunculate from an Aterian context at Taforalt that shows traces of cutting soft animal materials. The dots indicate where wear has occurred on the piece. Images **a** and **b** are close-up examples of that wear, and **c** is an on-edge view of the same

Cadenat 1939; but see Tixier 1958–1959, for an alternative point of view). However, contrary to this popular idea, our preliminary work on micro-wear traces has suggested that only a very few functioned in this way. Instead, they appear to have been used for a variety of tasks, including scraping and butchery. The micro-wear study of such pieces from Rhafas and Contrebandiers (Bouzouggar et al. 2007b) clearly showed that they were employed for working hard animal material (bones?) as well as softer materials, but rarely as projectiles. More recent work on finds from Taforalt would also appear to support this view (Fig. 7.2). During the 2008 season, pedunculates identified in an Aterian context were subjected to micro-wear analysis and, here again, despite the morphology of the points, the tools were clearly used for cutting soft animal materials (Igreja De Arojau, personal communication, 2008).

Partly on the basis of experimental studies (see Garcea 2012), but also from our preliminary micro-wear studies, which show that the artifacts were embedded in some kind of handle (Bouzouggar et al. 2007b), we suggest that many of the tools were also hafted.

## The Points

In our opinion, the tangs and other forms of preparation (e.g., bifacially thinning) on Aterian points and other tools are indicative of a developed *hafted* technology. The origins of this may not lie exclusively in the Aterian. For example, the thinning of points has also been described in the Mousterian assemblage at Jebel Irhoud (Balout 1965; Salih 1995) and elsewhere in North Africa, including at Oued El Akarit, Tunisia (Roset and Harbi-Riahi 2007). Indeed, this may imply connections between the Aterian and a local Mousterian, as has been suggested by other authors (Balout 1965; Camps 1974; Bordes 1976–1977; Hahn 1984), but these claims are difficult to substantiate and require further study.

An artifact of special interest is the so-called *Pointe Marocaine* (see Tixier 1958–1959 for a discussion of the term), a bifacial, thick and elongated pedunculate point (Fig. 7.3, n° 9). This point was first identified in 1931 in Aïn Takielt, an open air site near Casablanca (Antoine 1931), and has been described in different Aterian sites in Morocco, including at Oued Gorea and Tit Mellil, near Casablanca (Antoine 1934), at El Aliya Cave, Tangier

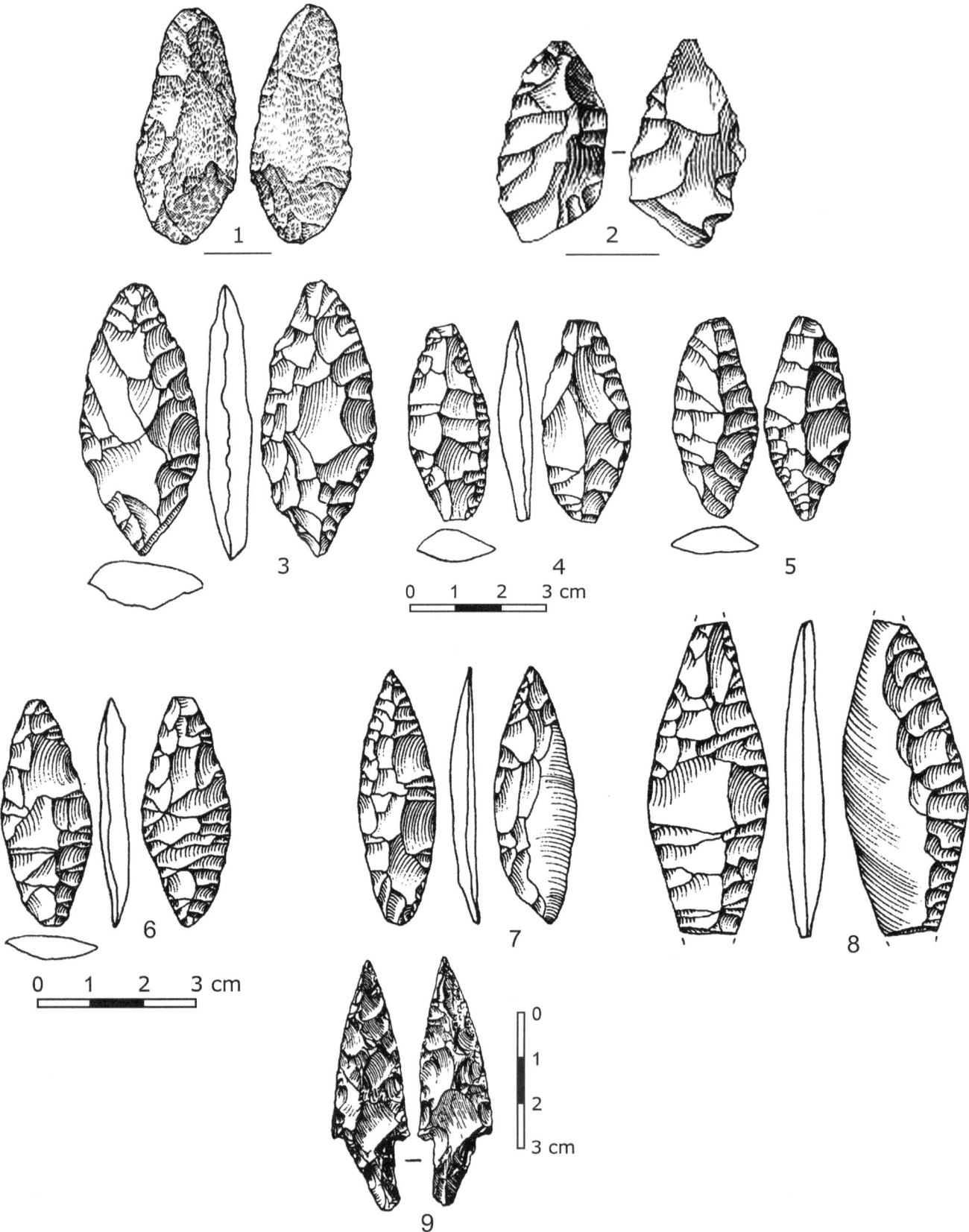

**Fig. 7.3** Bifacial foliates (*1–2* Grotte des Pigeons, Taforalt, layer 21; *3–5* El 'Aliya Cave, layer 6; *6–8* El 'Aliya Cave, layer 6; *9* Dar es-Soltan I, layer C2 (Ruhlmann 1951, and redrawn in Camps 1974)

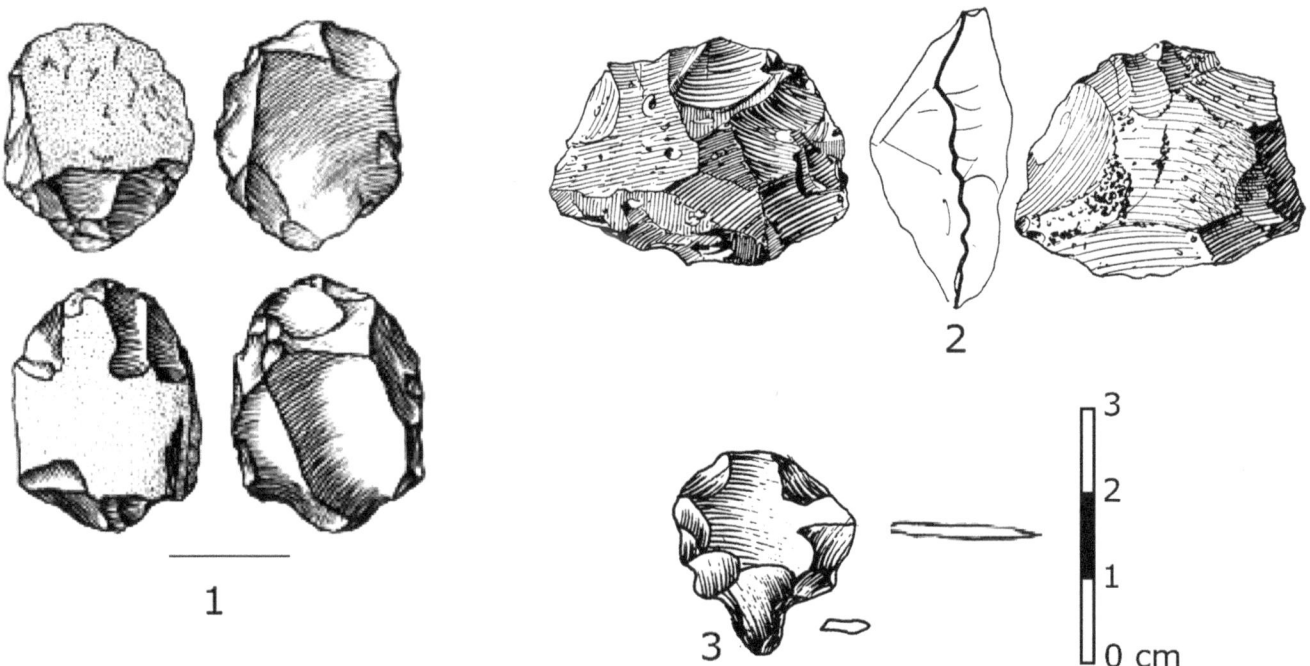

**Fig. 7.4** Small Levallois, discoidal cores and a pedunculate on a small flake (*1* Grotte des Pigeons, Taforalt, layer 21; *2* Dar es-Soltan 1, layer I; *3* Contrebandiers, layer 9)

(Howe and Movius 1947), and Dar es-Soltan 1, layer C2 (Ruhlmann 1951). This tool seems to exist only in Morocco. It was once considered to be a marker of the "Middle" or "Late" Aterian (Camps 1974), but with the re-dating of sites, it is at present difficult to see how it can be linked with such specific time intervals within the Aterian. Up until now, such tools have not been recognised in the local Mousterian.

Bifacial foliates provide another distinctive marker of Aterian technology. However, here we would differentiate between those with overall flaking, which are often bi-convex in section, and foliates in which the flat and invasive retouch is mainly confined to the dorsal surface (Fig. 7.3, 1–5). Both types are apparently absent in the Middle Paleolithic assemblages from Jebel Irhoud (Hublin et al. 1987). In contrast, they are relatively common in Aterian assemblages, though, for some interesting reason, their number appears to decrease whenever there is a rise in pedunculates. This is the case at Dar es-Soltan 1 (Roche 1956), at Taforalt in layer D (Roche 1967, 1969), and in Taforalt layer 21 of the recent excavations (Bouzouggar et al. 2007a). They also occur at El Aliya (Bouzouggar et al. 2002). The first category are generally made on thin blanks as at Dar es-Soltan 1 (Barton et al. 2009) and El Aliya (Bouzouggar et al. 2002), and more rarely on thick flakes as at Contrebandiers (Bouzouggar 1997a, b). Also of interest is the second category of foliate, which is made on blades or laminar flakes. The retouch on these is invasive and, on the ventral surface, is limited to near the proximal or distal

zones, possibly to both thin and straighten the tools. Two of the best examples are from El Aliya (Bouzouggar et al. 2002) (Fig. 7.3, 7–8). Although not exclusively so, these tools seem to be linked to blade production.

## The Small Discoidal and Levallois Cores

Apart from pedunculates and foliates, other representative components of the Aterian are small discoidal and Levallois cores. First identified in an Aterian context by Antoine (1938), and then at Tit Mellil, near Casablanca, and Aïn Fritissa by Tixier (1958–1959), these types of cores have also been recognized at Contrebandiers (Bouzouggar 1997a, b), El Aliya (Bouzouggar et al. 2002), and El Mnasra and El Harhoura (Nespoulet et al. 2008). The small Levallois cores are generally made on pebbles that were initially broken into two pieces using the anvil technique, before the preparation of the Levallois core surface (Bouzouggar 1997a, b). They were probably knapped to produce small flakes, which were sometimes transformed into pedunculates (Bouzouggar 1997a, b) (Figs. 7.3 and 7.4). Another question that needs to be tested is whether these small Levallois cores were themselves utilized as tools, because the scar of the preferential removal would provide a convenient thumb hold (Antoine 1938).

## The Blades

Another relevant and intriguing aspect of Aterian technology concerns the appearance of blades and bladelets. Although it is now commonly accepted that blades are one of the recurrent features of the Aterian industry (Tixier 1967; Wendorf and Schild 1992), this mainly relates to the production of elongated laminar artifacts from bipolar cores (Fig. 7.5, 1–3). Although none of the Aterian lithic collections contain genuine blade cores of the types described for the Upper Paleolithic (Roche 1963; Tixier 1963), the Aterians clearly had the technical skill to make them. Furthermore, in contrast to the widespread existence of small Levallois cores, the presence of blades and laminar flakes seems to vary much more from site to site and may show some degree of regional patterning. For example, these types are well-represented in sites in eastern and northern Morocco, at Rhafas Cave (Wengler 1993), at El Aliya (Bouzouggar et al. 2002), and probably at Taforalt (Bouzouggar et al. 2007a), but are much less common in the Atlantic coastal zone, at sites like Dar es-Soltan 1 (Ruhlmann 1951; Roche 1956) and at Contrebandiers from layers 10, 9 to 8 (Bouzouggar 1997a, b). It is possible that this may be chronologically related, since good quality raw materials were available in both areas.

The production of bladelets is a somewhat more complex issue (Fig. 7.5, 4–9). This debitage type has been recognized in Aterian contexts such as Contrebandiers (Roche 1963, 1976; Bouzouggar 1997a, b), but without true bladelet cores. Nevertheless, at Contrebandiers it was possible to reconstruct part of the chaîne opératoire, showing bladelets with negatives that appeared to cut the posterior crest and initiate a change in debitage direction (Bouzouggar 1997a, b). Examples of bladelet technology have also been recorded at Dar es-Soltan 1 in layer C1 (Ruhlmann 1951) and in eastern Morocco (Wengler 1993), and possibly also now at Taforalt.

## Other Aspects of Aterian Technology and Behavior

The presence of scrapers, including end scrapers, has often been discussed in relation to the Aterian in Morocco. However, it should be mentioned that their numbers are generally less significant than, for example, in the Mousterian of Southwest France. Among the most common forms are simple side scrapers and offset scrapers (déjetés), as in Rhafas Cave in layer 2 (Wengler 1993) and in Dar es-Soltan 1, layer C2 (Ruhlmann 1951; Roche 1956). And it is generally accepted that end scrapers become more important through time but with some regional and intersite variation. For example, at Dar es-Soltan 1, they are rare in layers I and C2 (Ruhlmann 1951).

The lithic artifacts in the Mousterian assemblages mainly reflect the use of the hard hammer technique (Hublin et al. 1987), but it seems that soft hammers were also used in the Aterian (Bouzouggar 1997a, b, 2001; Bouzouggar et al. 2002), even if this technique was probably already known from the late stages of the local Acheulean.

In addition to the stone artifacts, another potentially important signature of the Aterian may be found in the appearance of worked bone and ivory. At Dar es-Soltan 1, two ivory objects, one pointed, the other a small *plaquette*, were reported from layer I (Ruhlmann 1951). Subsequent examination has shown that the point was probably deliberately shaped and modified (Kaouane 2002). Bone tools have been described in El Mnasra (Hajraoui, 1994), though no detailed publication of them yet exists.

Two other aspects that merit attention concern the structuring of space within Aterian sites and the existence of symbolic artifacts. Strong evidence for the former occurs at El Mnasra Cave near Rabat, where a rich complex of apparently stone-lined hearths has been uncovered by excavation (El Hajraoui, 2004). Second, one of the most important discoveries to be made in recent years has been of personal bead ornaments in layer 21 at Taforalt (Bouzouggar et al. 2007a; d'Errico et al. 2009). These items are comprised of intentionally perforated *Nassarius* shells that were imported from marine shorelines and are similar to examples from sites like Blombos Cave in southern Africa (Henshilwood et al. 2002; D'Errico et al. 2005), but are of a slightly earlier age. Lately, more cases of *Nassarius* shell beads have been identified at Ifri n'Ammar (Richter et al. 2012), and at Rhafas and Contrebandiers, though the dating and precise contexts of these finds is still unclear (Fig. 7.6). At Taforalt, the use of red ochre is found in association with the shells and similar pigments are known from Rhafas (Wengler 1993) and El Mnasra, where small blocks of haematite are reported to exhibit traces of scraping (Nespoulet et al. 2008).

## Concluding Discussion

Any scope for discussing the Aterian of Morocco is still severely limited by the paucity of secure dating evidence. Although well-known Aterian sites, including those with human fossils, have been recorded on the Atlantic Coast at Dar es-Soltan 1 and 2, Contrebandiers, El Harhoura, and El Mnasra (Fig. 7.1), very little reliable dating evidence exists for the sequences in these caves. We hope new stratigraphic studies that are now in progress (Barton et al. 2009;

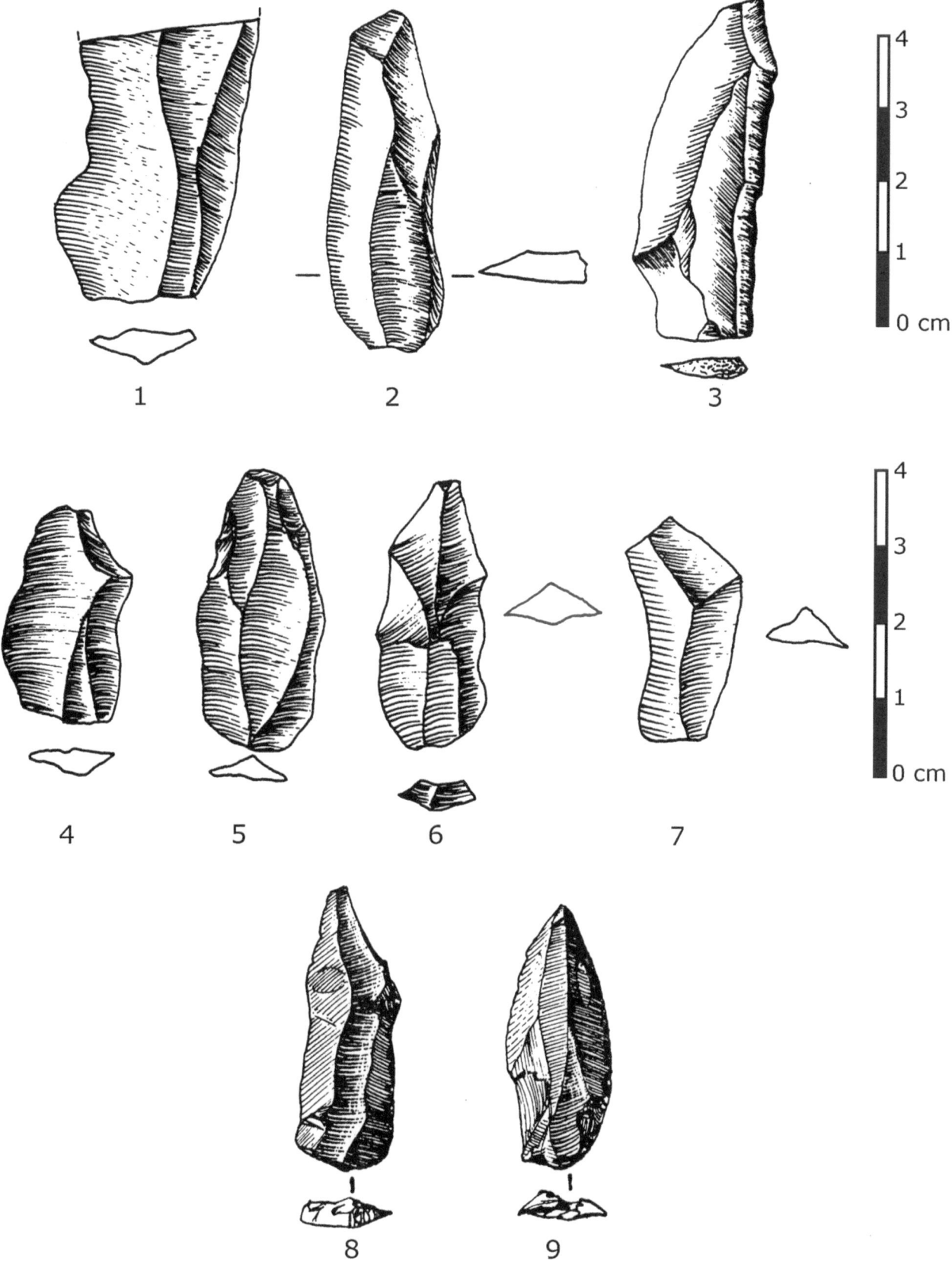

**Fig. 7.5** *1–3* Blades (Contrebandiers, layer 9); *4–7* Bladelets (Contrebandiers, layer 8); *8–9* Bladelets (Dar es-Soltan I, layer I after Ruhlmann 1951)

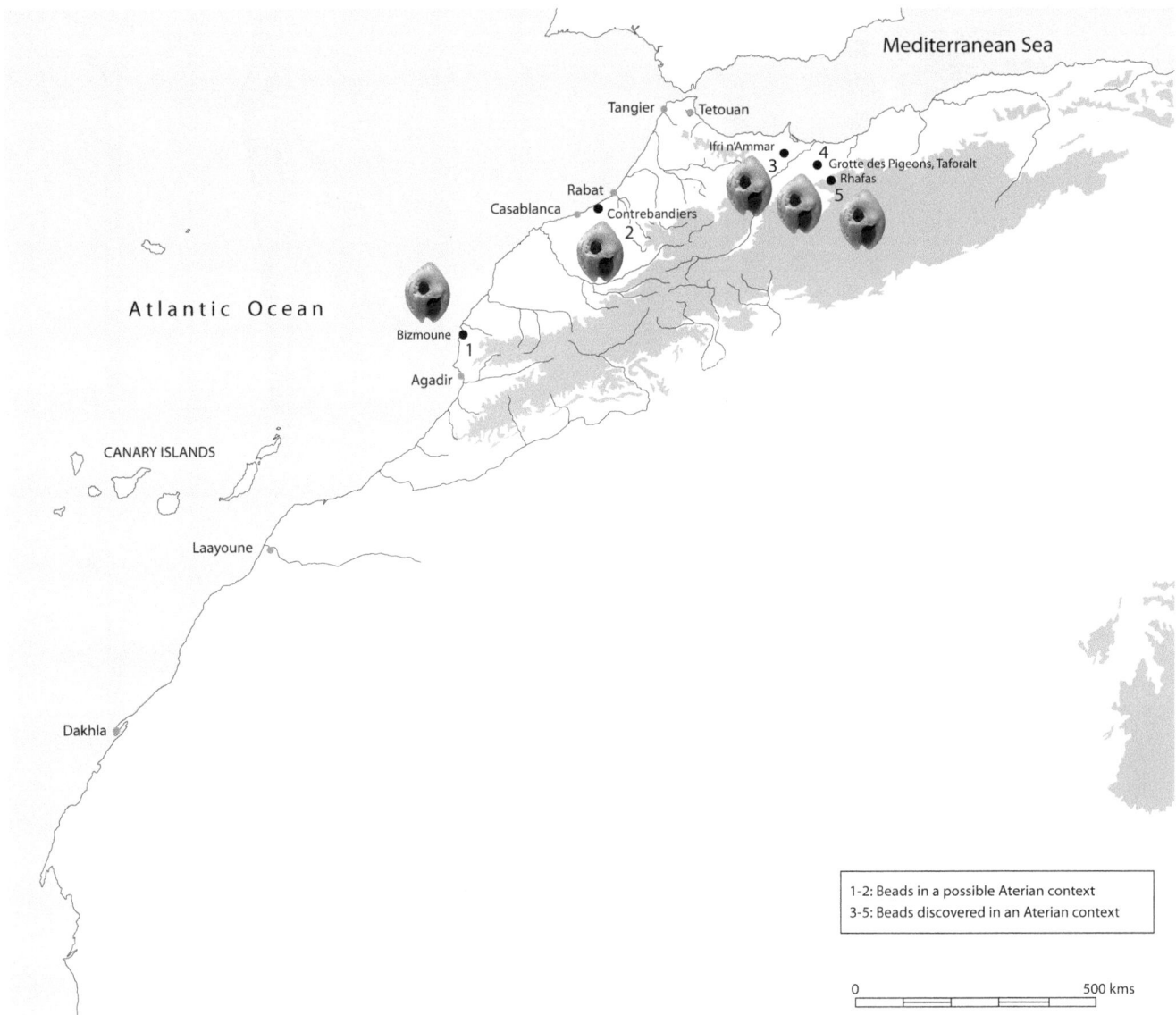

**Fig. 7.6** Shell ornaments discovered in an Aterian context in Morocco

Schwenninger et al. 2010) will rectify this major gap in our knowledge. If we accept the new dating results from Moroccan sites at face value, the age of the Aterian can certainly now be extended as far back as 100 ka if not earlier (Barton et al. 2009).

The existence of a longer chronology has significant implications for understanding the development of the Middle Paleolithic in Morocco and also challenges traditional views of the local "Mousterian" and its relationship to the Aterian. Recently, a Mousterian technology reported from Benzu rockshelter, near Sebta (northern Morocco) has been dated by U/Th to between 70 ka (U/Th IGM) and $173 \pm 10$ ka (U/Th IGM), though it is interesting that technological changes were already observed by the excavators in layer 2, which is dated by OSL to $254 \pm 17$ ka

(Shfd 020135) (Ramos et al. 2008). Apart from the analysis of raw materials, little other data are available concerning the lithic technology, except that the "Mousterian" from Benzu is dominated by side scrapers, rare blades, Mousterian points, notches, and denticulates. It should be noted that many of these components are also present in the Aterian. This could be relevant in the case of Ifri n'Ammar, where alternating Mousterian and Aterian occupations have been interpreted by the excavators largely according to the presence/absence of pedunculate tools (Mikdad et al. 2004), an idea that has gained support in other areas of Morocco (Wengler 2006).

Until now, it has largely been accepted that the Aterian lithic technology is characterized by a dominance of flake tools, and that the Upper Paleolithic is characterized by a dominance of tools on blades and bladelets. This distinction, which was already questioned by earlier researchers (Tixier 1967), has now come under renewed challenge following fresh studies of finds from Rhafas Cave (Wengler 1993), El Aliya (Bouzouggar et al. 2002), Taforalt (Bouzouggar et al. 2007a), El Mnasra (Nespoulet et al. 2008; Schwenninger et al. 2010), Dar es-Soltan 1 (Barton et al. 2009), Contrebandiers (Schwenninger et al. 2010), and El Khenzira (Ruhlmann 1936). Such claims of course need to be properly scrutinized and verified at sites with deep and continuous stratigraphies and supported by secure dating evidence. However, it is now becoming increasingly clear that the Aterian represents a very flexible technology, which, in addition to flakes, includes evidence for the use of "Middle Paleolithic" blades, bladelets, and flake tools, and even pebble tools occasionally. It is also apparent from an examination of the core morphology that Aterian knappers were perfectly capable of producing real blades of the type known in the Upper Paleolithic but, for some reason, did not consistently choose to do so or made bladelets instead. Much of the problem, in our opinion, is that too great an emphasis has been given to the pedunculate tools, which has resulted in other types being overlooked or missed. For example, the presence of small Levallois cores and manufacture of bifacial points, which form important components of the tool assemblages, may well be related to variations in the use of blade technology. Notwithstanding this observation, it is also apparent that laminar production and the manufacture of small Levallois cores are recurrent features of the Aterian technology as shown at Rhafas and Taforalt but less obviously at Ifri n'Ammar.

Finally, we feel that more consideration should be given in the future to non-lithic aspects of the technology, such as the manufacture and use of beads, the utilization of pigments, and the employment of hearths in structuring domestic space (Bouzouggar et al. 2007a; d'Errico et al. 2009). Contrary to received wisdom, some or most of the features described here are present both in Aterian and other "Middle Paleolithic" sites (Hublin et al. 1987; but see Balout 1965, for an opposing view). Therefore, we would suggest that key behavioral innovations seen in the Aterian may have been the result of cumulative development over a long period of time. We would further argue that defining what we mean by the *pre-Aterian* in Morocco is now just as important and necessary as characterizing the components that identify the Aterian.

**Acknowledgments** We would like to thank J.-J. Hublin and S. McPherron for their invitation to present a paper at the "Modern Origins: A North African Perspective" conference and to contribute to this volume. AB acknowledges the Moroccan CNRST for its support under the PROTARS P32/09 project and NB would like to thank the British Academy and NERC for funding excavation and post-excavation work. Both of the authors would like to pay special tribute to our colleagues and the large supporting team of specialists who are currently working on the UK-Moroccan team and, in particular, to Jean-Luc Schwenninger, Simon N. Collcutt, Louise Humphrey, Marian Vanhaeren, Francesco d'Errico, Tom Higham, Edward Hodge, Simon Parfitt, Edward Rhodes, Chris Stringer, Elaine Turner, Steven Ward, Abdelkrim Moutmir, and Abdelhamid Stambouli, who have contributed data incorporated in this chapter. Thanks to the three anonymous reviewers for their useful comments.

# References

Antoine, M. (1931). Notes de préhistoire marocaine : IV : Sur deux stations à outils pédonculés des environs de Casablanca. *Bulletin de la Société Préhistorique du Maroc, V*, 3–19.

Antoine, M. (1934). Un gisement atérien en place dans les alluvions de l'Oued Goréa près de Casablanca. *Bulletin de la Société de Préhistoire du Maroc, VIII*, 7–34.

Antoine, M. (1937). Notes de préhistoire marocaine : XIII—la question atéro-ibéromaurusienne au Maroc: Historique et mise au point. *Bulletin de la Société Préhistorique du Maroc, 11è année*, 45–58.

Antoine, M. (1938). Notes de préhistoire marocaine: XIV—Un cône de résurgence du Paléolithique moyen à Tit Mellil, près de Casablanca. *Bulletin de la Société Préhistorique du Maroc, 12*, 3–95.

Antoine, M. (1950a). Notes de préhistoire marocaine : XIX—L'Atérien du Maroc atlantique, sa place dans la chronologie Nord Africaine. *Bulletin de la Société Préhistorique du Maroc, Nouvelle Série, tome 1*, 4–47.

Antoine, M. (1950b). La chronologie de l'Atérien marocain et les fouilles américaines à Tanger. *Comptes Rendus des Séances Mensuelles de la Société des Sciences Naturelles du Maroc, 1*, 23–24.

Balout, L. (1955). *Préhistoire de l'Afrique du Nord : Essai de chronologie*. Paris: Paris Arts et Métiers Graphiques.

Balout, L. (1965). Données nouvelles sur le problème du moustérien en Afrique du Nord. Actas del V Congreso Panafricano de Prehistoria y de Estudio del Cuaternario. *Publicaciones del Museo Arqueologico Sant Cruz de Tenerife, tome 1*, 137–145.

Barton, R. N. E., Bouzouggar, A., Collcutt, S. N., Schwenninger, J.-L., & Clark-Balzan, L. (2009). OSL dating of the Aterian levels at Grotte de Dar es-Soltan I (Rabat, Morocco) and possible implications for the dispersal of modern *Homo sapiens*. *Quaternary Science Reviews, 28*, 1914–1931.

Biberson, P. (1961). *Le Paléolithique inférieur du Maroc atlantique*. Rabat: Publications du Service des Antiquités Marocaines.

Bobo, J. (1956). Un ensemble de stations moustéro-atériennes aux environs de Djanet (Tassili des Ajjer), notre Préliminaire. *Libyca, IV*, 263–268.

Bordes F. (1976-1977). Moustérien et Atérien. *Quaternaria, 1919–34*

Bouzouggar, A. (1997a). Economie des matières premières et du débitage dans la séquence atérienne de la grotte d'El Mnasra I (ancienne grotte des Contrebandiers-Maroc). *Préhistoire Anthropologie Méditerranéennes, 6*, 35–52.

Bouzouggar, A. (1997b). *Matières premières, processus de fabrication et de gestion des supports d'outils dans la séquence atérienne de la grotte d'El Mnasra I (ancienne grotte des Contrebandiers) à Témara (Maroc)*. Université Bordeaux I: Thèse de doctorat.

Bouzouggar, A. (2001). Technologie lithique de la séquence atérienne de la grotte des Contrebandiers à Témara. *Actes des Premières Journées Nationales de l'Archéologie et du Patrimoine, tome 1: Préhistoire*, 59–66.

Bouzouggar, A., Kozlowski, J. K., & Otte, M. (2002). Etude des ensembles lithiques atériens de la grotte d'El Aliya à Tanger (Maroc). *L'Anthropologie, 106*, 207–248.

Bouzouggar, A., Barton, N., Vanhaeren, M., d'Errico, F., Collcutt, S., Higham, T., et al. (2007a). 82,000 year-old shell beads from North

Africa and implications for the origins of modern human behavior. Proceedings of the National Academy of Sciences of the USA, 104, 9964–9969.

Bouzouggar, A., Barton, R.N.E., & De Araujo, I. (2007b). A brief overview of recent research into the Aterian and Upper Palaeolithic of Northern and Eastern Morocco. In B. Barich (Ed.), Tra il Sahara e il Mediterraneo: il Jebel Gharbi (Libia) e l'Archeologia del Maghreb (pp. 473–488). Edizioni Quasar. Scienze dell'Antichità, Storia Archeologia Antopologia.

Cadenat, P. (1939). Les objets pédonculés atériens du Koudiat bou Gherara. Bulletin de la Société de Géographie et d'Archéologie d'Oran, LX, 1939.

Camps, G. (1974). Les civilisations préhistoriques de l'Afrique du Nord et du Sahara. Paris: Doin.

Carrière, G. (1886). Quelques stations préhistoriques de la Province d'Oran. Bulletin de la Société de Géographie et d'Archéologie de la Province d'Oran, VI, 136–154.

Caton-Thompson, G. (1946). The Aterian Industry: Its place and significance in the Palaeolithic world. The Journal of the Royal Anthropological Institute of Great Britain and Ireland, 76, 87–130.

Cremaschi, M., Di Lernia, S., & Garcea, E. A. A. (1998). Some insights on the Aterian in the Libyan Sahara: Chronology, environment, and archaeology. African Archaeological Review, 15, 261–286.

D'Errico, F., Henshilwood, C., Vanhaeren, M., & Van Niekerk, K. (2005). Nassarius kraussianus shell beads from Blombos Cave: Evidence for symbolic behaviour in the Middle Stone Age. Journal of Human Evolution, 48, 3–24.

D'Errico, F., Vanhaeren, M., Barton, N., Bouzouggar, J., Mienis, H., Richter, D., et al. (2009). Additional evidence on the use of personal ornaments in the Middle Paleolithic of North Africa. Proceedings of the National Academy of Sciences of the USA, 106, 16051–16056.

Debénath, A. (1992). Hommes et cultures matérielles de l'Atérien marocain. L'Anthropologie, 96, 711–720.

Debénath, A. (1994). L'Atérien du Nord de l'Afrique et du Sahara. Sahara, 6, 21–30.

Debénath, A., Raynal, J.-P., Roche, J., Texier, J.-P., & Ferembach, D. (1986). Stratigraphie, habitat, typologie et devenir de l'Atérien Marocain: Données récentes. L'Anthropologie, 90, 233–246.

El Hajraoui, M. A. (2004). Le Paléolithique du domaine mésetien septentrional. Données récentes sur le littoral: Rabat, Témara et la Mamora, Thèse de doctorat d'Etat, Université Mohamed V.

Garcea, E. A. A. (2001). The Pleistocene and Holocene archaeological sequences. In E. A. A. Garcea (Ed.), Uan Tabu in the settlement history of the Libyan Sahara (pp. 25–49). Firenze: All'Insegna del Giglio.

Garcea, E. A. A. (2004). Crossing deserts and avoiding seas: Aterian North African–European relations. Journal of Anthropological Research, 60, 27–53.

Garcea, E. A. A. (2012). Modern human desert adaptations: A Libyan perspective on the Aterian Complex. In J.-J. Hublin & S. McPherron (Eds.), Modern origins: A North African perspective. Dordrecht: Springer.

Hahn, J. (1984). Südeuropa und Nordafrika. Neue forschungen zur altsteinzeit. Verlag C.H. Beck (Forshungen zur Allgemeinen und Vergleichenden Archäologie, 4, 1–231.

Hajraoui, M. A. (1994). L'industrie osseuse atérienne de la grotte d'el Mnasra. Préhistoire Anthropologie Méditerranéennes, 3, 91–94.

Henshilwood, C., D'Errico, F., Yates, R., Jacobs, Z., Tribolo, C., Duller, G. A., et al. (2002). Emergence of modern human behavior: Middle Stone Age engravings from South Africa. Science, 295, 1278–1280.

Howe, B., & Movius, H. L. (1947). A stone age cave site at Tangier. Preliminary report on the excavations at the Mughraret el-Aliya, or High Cave, in Tangier. Papers of the Peabody Museum, XXIII(1), 32.

Hublin, J.-J., Tillier, A.-M., & Tixier, J. (1987). L'humérus d'enfant moustérien (Homo 4) du Jebel Irhoud (Maroc) dans son contexte archéologique. Bulletin et Mémoires de la Société d'Anthropologie de Paris 4, série, XIV, 115–142.

Hugot, H. J. (1958). Préliminaires à une étude du Moustéro-Atérien du Tidikielt. Libyca, I, 87–102.

Kaouane, C. (2002). L'industrie sur matière dure animale de trois gisements néolithiques marocains : Kaf Taht el Ghar, Dar es Soltan I et la nécropole de Rouazi à Skhirat. Ph.D. Dissertation, Institut National des Sciences de l'Archéologie et du Patrimoine.

Marchand, H., & Aymé, A. (1935). Recherches stratigraphiques sur l'Atérien. Bulletin de la Société d'Histoire Naturelle d'Afrique du Nord, XXVI, 1935.

Mercier, N., Wengler, L., Valladas, H., Joron, J. L., Froget, L., & Reyss, J.-L. (2007). The Rhafas Cave (Morocco): Chronology of the Mousterian and Aterian archaeological occupations and their implications for Quaternary geochronology based on luminescence (TL/OSL) age determinations. Quaternary Geochronology, 2, 309–313.

Mikdad, A., & Einwanger, J. (2000). Recherches préhistoriques et protohistoriques dans le Rif oriental (Maroc): Rapport préliminaire. Beiträge zur Allgemeinen und Vergleichenden Archäologie, 20, 109–167.

Mikdad, A., Moser, J., Nami, M., & Einwanger, J. (2004). La stratigraphie du site d'Ifri n'Ammar (Rif Oriental, Maroc): Premiers résultats sur les dépôts du Paléolithique moyen. Beiträge zur Allgemeine und Vergleichende Archäologie Band, 24, 125–137.

Nespoulet, R., El Hajraoui, M. A., Amani, F., Ben-Ncer, A., Debénath, A., El Idrissi, A., et al. (2008). Palaeolithic and Neolithic occupations in the Témara region (Rabat, Morocco): Recent data on hominin contexts and behaviour. African Archaeological Review, 25, 21–40.

Pallary, P. (1927). Découvertes préhistoriques dans le Maroc oriental (1923–1926). l'Anthropologie, 37, 49–64.

Ramos, J., Bernal, D., Domínguez-Bella, S., Calado, D., Ruiz, B., Gil, M. J., et al. (2008). The Benzu' rockshelter: A Middle Palaeolithic site on the North African coast. Quaternary Science Reviews, 27, 2210–2218.

Reygasse, M. (1922). Note au sujet de deux civilizations préhistoriques africaines pour lesquelles deux termes nouveaux me paraissent devoir être employés (pp. 467–472). Montpellier: XLVIè Congrès de l'Association française pour l'avancement des sciences.

Richter, D., Moser, J., & Nami, M. (2012). New data from the site of Ifri n'Ammar (Morocco) and some remarks on the chronometric status of the Middle Paleolithic in the Maghreb. In J.-J. Hublin & S. McPherron (Eds.), Modern origins: A North African perspective. Dordrecht: Springer.

Roche, J. (1956). Etude sur l'industrie de la grotte de Dar-es-Soltane (Rabat). Bulletin d'Archéologie Marocaine, 1, 93–118.

Roche, J. (1958-1959). L'Epipaléolithique marocain. Libyca, VI-VII, 159–192.

Roche, J. (1963). L'Epipaléolithique Marocain. Lisbon: Fondation Calouste Gulbenkian.

Roche, J. (1967). L'Atérien de la grotte de Taforalt (Maroc oriental). Bulletin d'Archéologie Marocaine, 7, 11–56.

Roche, J. (1969). Les industries paléolithiques de la grotte de Taforalt (Maroc oriental). Quaternaria, 11, 89–100.

Roche, J. (1976). Chronostratigraphie des restes atériens de la grotte des Contrebandiers à Témara. Bulletin et Mémoires de la Société d'Anthropologie de Paris, 3, 165–173.

Roche, J., & Tixier, J.-P. (1976). Découverte de restes humains dans un niveau atérien supérieur de la grotte des Contrebandiers, à Témara (Maroc): Note. Comptes Rendu Académie des Sciences, 282(Série D), 45–47.

Roset, J.-P., & Harbi-Riahi, M. (2007). *El Akarit: Un site archéologique du Paléolithique Moyen dans le Sud de la Tunisie.* Paris: Editions Recherches sur les Civilisations.

Ruhlmann, A. (1936). Les grottes préhistoriques d' "El Khenzira" (région de Mazagan). *Publications du Service des Antiquités du Maroc, Rabat, 2,* 103–105.

Ruhlmann, A. (1951). *La Grotte préhistorique de Dar es-Soltan. Collection Hésperis* (Vol. 11, pp. 1–210). Paris: Institut des Hautes Études Marocaines.

Salih, A. (1995). *Le Moustérien de la grotte de J'bel Irhoud hominidés (J'Bilet, Maroc). L'homme Méditerranéen* (pp. 19–28). Provence: Mélanges offerts à Gabriel Camps. Publications de l'Université de Provence-LAPMO.

Schwenninger, J.-L., Collcutt, S. N., Barton, R. N. E., Bouzouggar, A., El Hajraoui, M. A., Nespoulet, R., et al. (2010). Luminescence chronology for Aterian cave sites on the Atlantic coast of Morocco. In E. A. A. Garcea (Ed.), *South-Eastern Mediterranean peoples between 130,000 and 10,000 years ago.* Oxford: Oxbow Books.

Smith, T. M., Tafforeau, P. T., Reid, D. J., Grün, R., Eggins, S., Boutakiout, M., et al. (2007). Earliest evidence of modern human life history in North African early *Homo sapiens. Proceedings of the National Academy of Sciences of the USA, 104,* 6128–6133.

Tixier, J. (1958–1959). Les pièces pédonculés de l'Atérien. *Libyca, VI–VII,* 127–158.

Tixier, J. (1963). *Typologie de l'Epipaléolithique du Maghreb, Mémoire n°2.* Alger: C.R.A.P.E.

Tixier, J. (1967). Procédés d'analyse et questions de terminologie concernant l'étude des ensembles industriels du Paléolithique récent et de l'Epipaléolithique dans l'Afrique du nord-ouest. In W. W. Bishop & J. D. Clark (Eds.), *Background to evolution in Africa* (pp. 771–820). Chicago: University of Chicago Press.

Texier, J.-P. (1985–1986). Le site atérien du Chaperon–Rouge I (Maroc) et son contexte géologique. *Bulletin d'Archéologie Marocaine, XVI,* 27–73.

Trinkaus, E. (2007). European early modern humans and the fate of the Neandertals. *Proceedings of the National Academy of Sciences of the USA, 104,* 7367–7372.

Wendorf, F., & Schild, R. (1992). The Middle Palaeolithic of North Africa: A status report. In F. Kless & R. Kuper (Eds.), *New light on the Northwest African past* (pp. 39–78). Köln: Heinrich Barth Institut.

Wengler, L. (1993). Formations quaternaires et cultures préhistoriques au Maroc oriental. Thèse de Doctorat d'État, l'Université Bordeaux I.

Wengler, L. (2006). *Innovations et normes techniques dans le Paléolithique moyen et supérieur du Maghreb: Une alternative aux migrations? XXVIè rencontres internationales d'archéologie et d'histoire d'Antibes (Astruc, Bon, Léa, Milcent et Philibert dir.)* (pp. 93–105). Antibes: Editions APDCA.

Wrinn, P. J., & Rink, W. J. (2003). ESR dating of tooth enamel from Aterian levels at Mugharet el 'Aliya (Tangier, Morocco). *Journal of Archaeological Science, 30,* 123–133.

# Chapter 8
# Late Pleistocene Human Subsistence in Northern Africa: The State of our Knowledge and Placement in a Continental Context

T. E. Steele

**Abstract** Zooarchaeological evidence has featured prominently in the debate of the mode and tempo of modern human origins during the Late Pleistocene of Africa. However, most of the assemblages included in the discussion are from southern Africa, and we have limited knowledge about human paleoecology from other parts of the continent. This chapter aims to review the zooarchaeological record for Late Pleistocene human subsistence in northwestern Africa. This region provides some of the earliest and most complete fossil evidence for anatomically modern humans, and it provides a unique aspect of human technological development in the Aterian stone tool industry. Understanding the subsistence of these early people is important for understanding human behavioral evolution. Until recently, most faunal analyses from the region have been focused on paleontology, biostratigraphy, and environmental reconstructions. Zooarchaeological studies have been limited by sample size, excavation methods, and the lack of good comparative data. What is known of the Northwest African record is placed in the context of our knowledge of Middle Stone Age subsistence from the rest of Africa, and future research directions are offered. The current projects in Morocco, as discussed in this volume, will provide valuable insights into North African subsistence during the Late Pleistocene.

**Keywords** Archaeozoology • Environment • Faunal analysis • *Gazella* • Hunting • Morocco • North Africa • Subsistence • Taphonomy • Zooarchaeology

## Introduction

Human fossil and genetic evidence point to Africa as the place where anatomically and behaviorally modern humans first emerged, either at the beginning of or during the Middle Stone Age (MSA, about 250–50 ka; Middle Paleolithic in northern Africa) or at its end with the transition to the Later Stone Age (LSA, about 50 ka to ethnohistoric times). Anatomically and behaviorally modern humans then expanded out of Africa after 50 ka to eventually populate the rest of the world; the goal of current research is to determine when, where, and how these anatomical and behavioral changes occurred (reviewed in McBrearty and Brooks 2000; Henshilwood and Marean 2003; Mellars 2006; Klein 2009). Paleoecological investigations using faunal remains from archaeological and paleontological sites are an essential part of this research. Reconstructing ancient human subsistence and environments will help to track behavioral changes and the timing of changes in human demography, including the population growth that occurred with the modern human expansion out of Africa.

Much of what we know about Late Pleistocene human subsistence and paleoecology and modern human origins is based on South African sites. This region possesses an abundance of sites with good bone preservation and long sequences, investigated through a rich history of paleoanthropological research. Unfortunately, few zooarchaeological analyses are available from sites in other parts of Africa (reviewed in Steele and Klein 2009). Fortunately, current work in North Africa has great potential to increase our coverage of Middle to Late Pleistocene African human subsistence and paleoecology.

As in South Africa, northwestern Africa has many cave sites with good bone preservation and a history of paleoanthropological and paleontological research. However, few of these assemblages have been studied from a zooarchaeological perspective. A zooarchaeological study not only

T. E. Steele (✉)
Department of Anthropology, University of California-Davis,
One Shields Ave, Davis, CA 95616-8522, USA
and
Department of Human Evolution,
Max Planck Institute for Evolutionary Anthropology,
Deutscher Platz 6, 04103 Leipzig, Germany
e-mail: testeele@ucdavis.edu

J.-J. Hublin and S. P. McPherron (eds.), *Modern Origins: A North African Perspective*,
Vertebrate Paleobiology and Paleoanthropology, DOI: 10.1007/978-94-007-2929-2_8,
© Springer Science+Business Media B.V. 2012

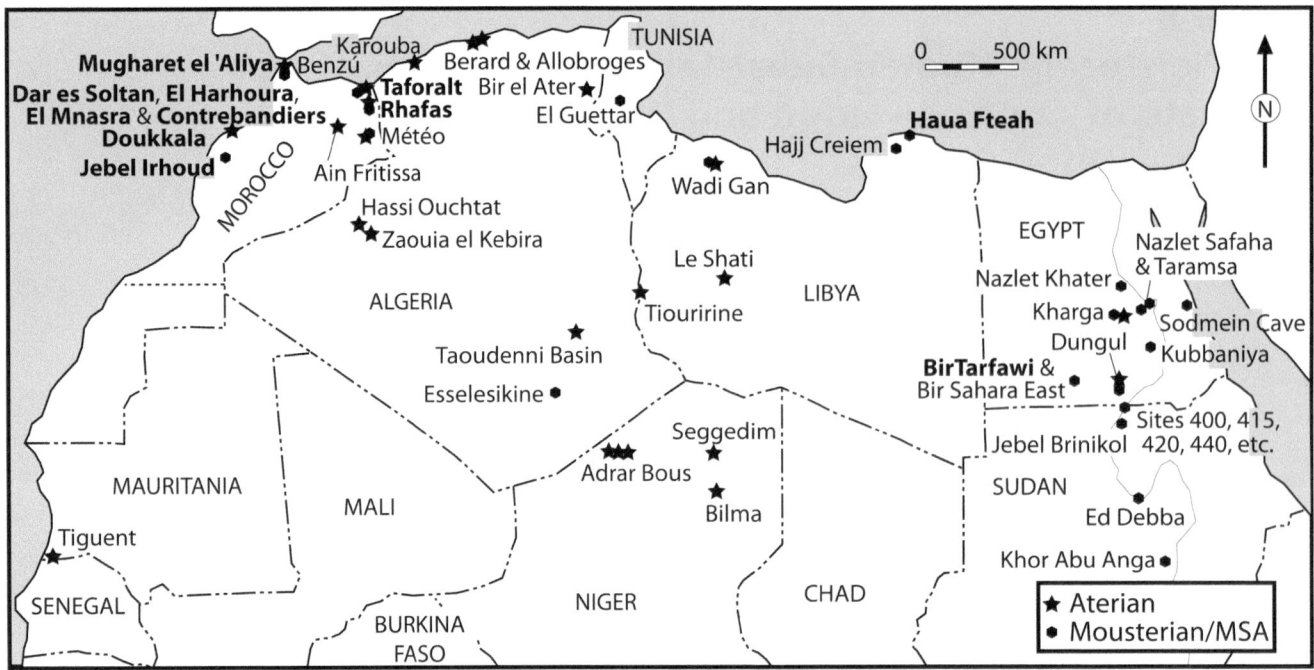

**Fig. 8.1** Map of the North African sites discussed in the text (*bold*), as well as additional Aterian and Mousterian sites (modified from Klein 2009). Key references are provided in Table 8.1

considers the species present and their relative abundance, but also cut-marks and breakage patterns, skeletal part representation, prey mortality profiles, carnivore impacts, etc. Zooarchaeologists quantify these measures within stratigraphic samples and make comparisons with other sites and actualistic studies. Importantly, they also consider the taphonomic history of the sample, such as the pre- and post-depositional processes that may bias a sample as it moves from a living animal to a zooarchaeological assemblage. These types of investigations allow researchers to separate human, carnivore, and natural contributions to an assemblage, so that we can more confidently investigate ancient human behavior to understand human prey choice and how humans reacted to past environmental and technological changes.

As the chapters in this volume demonstrate, North Africa has much to contribute to our understanding of modern human origins. Jebel Irhoud, Morocco, has yielded the largest human fossil assemblage that indicates that modern human morphology emerged from Africa (Bräuer 1989; Rightmire 1989; Hublin 2001; Harvati and Hublin 2012), and it is important to understand the archaeological context of these individuals. In addition, both Mousterian and Aterian lithic assemblages (both are part of the Middle Paleolithic) are found in the region (Aouadi-Abdeljaouad and Belhouchet 2012; Bouzouggar and Barton 2012; Garcea 2012; Hawkins 2012), and they may represent chronological, spatial, functional, stylistic, or environmental variation; faunal analysis can help address these issues.

The goal of this chapter is to discuss what we know about Late Pleistocene human ecology in Africa, which is primarily known from South Africa, to review the current status of zooarchaeological research in North Africa (Fig. 8.1 and Table 8.1) and compare it to what is known from elsewhere, and to make suggestions for future research.

## What is Known About Human Subsistence During the African MSA

The best information about human subsistence and paleoecology during the MSA comes from southern Africa. This region has benefited from a long history of archaeological research so that more than 70 MSA sites are known, and the local geology encourages good bone preservation in more than half of these sites (Klein 2009). Building on a rich paleontological tradition, zooarchaeological analyses have steadily increased in abundance since they began in the early 1970s. MSA assemblages contain a diversity of larger mammals, ranging from the very small blue duiker (*Cephalophus monticola*) to the very large extinct long-horned buffalo (*Pelorovis antiquus*). Suids (pigs) are present where the environment supports them, and equids (zebras) are consistently present, but never in large numbers. In coastal sites, fur seals (*Arctocephalus pusillus*) can be quite abundant. The representation of carnivores, both large and small, varies with each assemblage. The relative abundances of

**Table 8.1** The sites that are discussed in the text, along with their associated industries and the references that provide information about their faunal remains

| Site | Lithic industry | Faunal descriptions |
|------|-----------------|---------------------|
| Bir Tarfawi 14, Egypt | Middle Paleolithic | Gautier (1993) |
| Dar es-Soltan 1, Morocco | Neolithic<br>Aterian | Ruhlmann (1951) |
| Jebel Irhoud, Morocco | Mousterian | Thomas (1981); Amani (1991); Amani and Geraads (1993, 1998) |
| Doukkala II, Morocco | Moustero-Aterian | Michel and Wengler (1993); Geraads (2008) |
| El Harhoura 1 (Zouhrah Cave), Morocco | Neolithic<br>Aterian | Debénath and Sbihi-Alaoui (1979); Aouraghe (2000); Aouraghe and Abbassi (2002); Bailon and Aouraghe (2002); Aouraghe (2004) |
| El Harhoura 2, Morocco | Neolithic<br>Epipaleolithic<br>Aterian | Debénath and Sbihi-Alaoui (1979); Nespoulet et al. (2008) |
| El Mnasra, Morocco | Neolithic<br>Aterian | Nespoulet et al. (2008) |
| Grotte des Contrebandiers (Smugglers' Cave, Témara), Morocco | Neolithic<br>Iberomaurusian<br>Aterian<br>Mousterian | Souville (1973, p. 112); Roche and Texier (1976); Bouzouggar et al. (2002) |
| Haua Fteah (Great Cave), Libya | Historic<br>Neolithic<br>Libyco-Capsian<br>Iberomaurusian<br>Dabban<br>Mousterian<br>"Pre-Aurignacian" | Hey (1967); Higgs (1967); Klein and Scott (1986); MacDonald (1997) |
| Mughâret el 'Aliya (one of the Caves of Hercules), Morocco | Neolithic<br>Aterian | Howe and Movius (1947); Arambourg (1967); Briggs (1967); Wrinn (2001); Wrinn and Rink (2003) |
| Rhafas Cave, Morocco | Neolithic<br>Aterian | Michel (1992) |
| Taforalt (Grotte des Pigeons), Morocco | Iberomaurusian<br>Aterian | Bouzouggar et al. (2007) |

particular taxa fluctuated with the Late Pleistocene glacial cycles. These changes are most clearly visible in the Cape Ecozone (along the southwestern and southern coasts), either because the environmental changes were strongest there or because the faunal record is better documented there. Faunal assemblages in the Cape Ecozone indicate that the abundance of grazers, particularly alcelaphines (wildebeests and hartebeests) and equids, increased during the cooler intervals of the Late Pleistocene (Klein 1980, 1983).

Almost all zooarchaeological studies of Paleolithic assemblages need to consider the impact carnivores may have had on the assemblage before human behavior can be reconstructed. Analysts working in South Africa and elsewhere have studied natural bone occurrences and hyena dens with the same attention that they have given to archaeological assemblages. From these studies, we have learned the characteristics of carnivore and raptor accumulations, such as species and skeletal part abundance, age-at-death of the prey, breakage patterns, and surface damage (Klein 1975b; Avery et al. 1984; Avery 1989; Cruz-Uribe 1991; Cruz-Uribe and Klein 1998, among many others), and are reminded that just because bones and stones are found together, the bones are not all necessarily the result of human activity (Klein et al. 1999a, 2007; Villa and

Soressi 2000; Cruz-Uribe et al. 2003; Villa et al. 2005). This includes large mammals as well as small prey such as hares. These taphonomic analyses, which are necessary to determine the primary accumulator of an assemblage, form the basis of modern zooarchaeological studies.

Klein's (1972, 1975a, 1976, 1977, 1978) seminal descriptions of the mammalian faunas from the now famous MSA sites of Klasies River Main/Mouth and Die Kelders Cave 1 and the nearby LSA fauna from Nelson Bay Cave stimulated great interest in human subsistence during the MSA. Klein's (1976, 1994; Klein and Cruz-Uribe 2000) analysis of these faunas led him to propose that MSA people could not hunt large and dangerous game as readily or consistently as LSA people and that MSA people therefore hunted relatively more docile eland (*Taurotragus oryx*). Klein based his proposal on the high abundance of eland in the Klasies sample, more than would be expected based on the historic abundance of eland near the site. In contrast, the Nelson Bay Cave assemblage accumulated under similar environmental circumstances, but contained many more Cape buffalo (*Syncerus caffer*) and bushpig (*Potamochoerus larvatus*). Buffalo and bushpig respond especially vigorously to predators, and among all the prey species in the assemblages, they were probably the most difficult to capture. Buffalo are known to have killed lions with mobbing attacks (Estes 1992), while bushpigs have been described as being unusually dangerous because of their aggressiveness and "willingness" to use their "razor-sharp" tusks (Estes 1992, p. 214; Skinner and Chimimba 2005). Klein proposed that the abundance of buffalo and bushpigs in the LSA assemblages could reflect the introduction of projectile technology, which would have allowed hunters to strike from a safer distance. MSA people did manage to hunt buffalo at least occasionally, as indicated by a spear tip embedded in the neck vertebrae of an extinct buffalo in the Klasies sample (Milo 1994), and buffalo were probably still difficult for even LSA people to frequently hunt, even with their more sophisticated technology. Mortality profiles illustrating the age-at-death of animals found in archaeological assemblages show that the majority of buffalo in both MSA and LSA assemblages are young and old individuals, the most vulnerable individuals of a herd; in contrast, the majority of eland are prime-aged adults, which are frequently the most difficult age class to hunt (Klein 1978, 1994; Klein and Cruz-Uribe 1996). Prey selection is a result of the combination of available technology and prey behavior.

The MSA of South Africa has also played a central role in studies of skeletal part representation. In his analysis of the Klasies fauna, Klein (1976) identified a pattern in which small bovids are represented by all skeletal parts while larger bovids are represented by fewer proximal limb bones (humeri and femora—considered high-utility parts) and more head and foot bones. This pattern has been identified in many assemblages from many time periods, and discussion of its causes and interpretation has been widespread (Klein 1976, 1989; Binford 1984; Turner 1989; Marean and Frey 1997; Marean 1998; Bartram and Marean 1999; Klein et al. 1999b; Outram 2001). It has been suggested that this pattern has to do either with MSA people (1) hunting small bovids and scavenging large bovids, or (2) hunting all bovids but bringing small bovid carcasses back to the shelter complete while transporting back only certain parts of the large bovids, or with (3) differential preservation in which small bovid parts are more likely to remain complete and identifiable while large bovid parts are more likely to become broken, eliminated, or less identifiable during butchery or by post-depositional processes such as leeching, compaction, and carnivore ravaging. Finally, some have argued that the pattern is a by-product of the tendency for zooarchaeologists to record only more readily identifiable articular ends and to exclude the more difficult to identify long-bone shaft fragments or because the long-bone shaft fragments were not kept by the excavators, as was the case with the Singer and Wymer excavations at Klasies. More recent analyses of the MSA Layers 10 and 11 from new excavations at Die Kelders Cave 1, where all bones were

kept during excavation, suggest that the high-utility parts of large bovids were commonly transported back to the cave (Marean et al. 2000), indicating that these parts were from animals that were hunted, not scavenged, and highlighting the importance of unbiased faunal assemblages. The analyses of the Die Kelders fauna also indicated that the smallest bovids may have been accumulated by raptors while the larger ones were accumulated by MSA people (Klein and Cruz-Uribe 2000; Marean et al. 2000), indicating the complex history of many of the assemblages and the care needed during analyses.

When further investigating the taphonomic history of an assemblage, the abundance of cut-marks and burning has frequently been quantified to gauge the degree of human involvement with accumulating the bones. Green breaks and percussion marks that result from marrow processing have been noted (Milo 1998), but remain unquantified for many MSA assemblages. One exception is the study of Layers 10 and 11 from the recent excavations at Die Kelders, where abundant fresh breakage and hammer-stone percussion marks (Marean et al. 2000) indicate marrow consumption. The exploitation of marrow and bone grease during the MSA (and LSA) needs more detailed research, and fortunately, these data are now being consistently recorded (Dewar et al. 2006; Thompson 2008).

In addition to large mammals, marine birds, such as penguins (*Spheniscus demersus*), cormorants (*Phalacrocorax* sp.), and gannets (*Morus* sp.), are common in coastal MSA and LSA sites (Avery 1990). LSA sites contain more birds overall, and a higher proportion of them are flying birds, relative to penguins. Ostrich (*Struthio camelus*) eggshell (OES) is common in many MSA sites, and usually we assume that they are the remains of people's meals. OES is also found in LSA assemblages, but the remains are usually artifactual, having been made into beads, pendants, and water canteens. However, hyenas occasionally hide or cache the eggs for later consumption (Kandel 2004), so we must investigate if the abundance of OES more closely relates to carnivore involvement than to human accumulation. We can do this by examining if there is a positive correlation between the abundance of carnivores and OES and if there is characteristic carnivore tooth damage on the OES (Kandel 2004). Other small game may include Cape dune mole rats (*Bathyergus suillus*), hares (*Lepus* sp.), and hyraxes (*Procavia capensis*) (Henshilwood 1997; Cruz-Uribe and Klein 1998; Klein and Cruz-Uribe 2000). Only the mole rats commonly occur in high abundances, but they may have entered the deposits either through raptor pellets or carnivore scats, or as natural deaths, not as the result of MSA human subsistence.

The coast of South Africa contains numerous shell middens, most of which are LSA, but nine MSA sites are known as well (Parkington 2003; Klein et al. 2004; Avery et al. 2008; Steele and Klein 2008). Mussels (*Choromytilus meridionalis* and *Perna perna*) always dominate MSA samples, and limpets (*Patella* sp.) provide most of the balance. In contrast, mollusc species diversity is higher in LSA middens, and taxa such as whelks are occasionally dominant. In addition to more marine birds, LSA middens also contain abundant large fish bones and some rock lobsters (*Jasus lalandii*), both of which are nearly absent in MSA middens. Overall, the marine component in MSA diets was much less diverse than in LSA diets (Parkington 2003; Klein et al. 2004; Steele and Klein 2008). In addition to fewer mollusc species, the molluscs that are found in MSA middens are larger than those found in LSA sites (Klein 1979; Parkington 2003; Steele and Klein 2005/2006; Avery et al. 2008). Limpets are slow but continuously growing animals, and under heavy predation, the typical size of limpets in the population will decrease. These data indicate that MSA people preyed less heavily on marine molluscs than LSA people, probably because they were living at lower population densities.

Tortoises (*Testudo* [*Chersina*] *angulata*) are abundant in many sites as well, and the abundance of proximal limbs (humeri and femora) and shells indicates that they were consumed by humans, not raptors (Sampson 2000). Like limpets, tortoises grow slowly and continuously, and are sensitive to over-predation. In general, MSA tortoises are larger than LSA ones (Klein and Cruz-Uribe 1983; Steele and Klein 2005/2006), further supporting the conclusions reached with the limpets that MSA people lived at low population densities.

# Review of the North African Faunal Evidence

## Jebel Irhoud, Morocco

Jebel Irhoud, Morocco, has provided the largest and best studied faunal collection for the late Middle Pleistocene of North Africa. The site was discovered in 1960 and subsequently excavated by Ennouchi. In 1967 and 1969, Tixier and de Bayle des Hermens continued work at the site. The site has yielded numerous human fossil remains, including parts of two crania, two mandibles, a humerus, and a pelvis, which has made it one of the most prominent sites in North Africa (Ennouchi 1963, 1968, 1969; Hublin et al. 1987; Amani and Geraads 1993; Tixier et al. 2001). However, the provenience within the site is known for only one of the fossils, a juvenile humerus found by Tixier and de Bayle des Hermens (Hublin et al. 1987). Early electron spin resonance (ESR) dates indicated that the deposits just above this fossil accumulated between 90 and 190 ka (Grün and Stringer 1991). More recently, combined Uranium-series/ESR dating of a tooth fragment from the juvenile mandible placed the material at $160 \pm 16$ ka (Smith et al. 2007). Work at the site was renewed in 2004 (Ben-Ncer, Hublin, and McPherron), and one of the primary goals is to refine the chronology of the deposits.

Amani and Geraads (Amani 1991; Amani and Geraads 1993, 1998) studied the large mammal remains from Ennouchi's excavation to reconstruct biochronology and paleoenvironment, while Thomas (1981) described the material from the Tixier and de Bayle des Hermens excavations (Table 8.2). Thomas identified approximately 10 species (Number of Identified Specimens [NISP] > 200). Gazelles dominate the assemblage (70–75% of the identified specimens), but only *Gazella cuvieri* was identified. Other animals included hares, jackals, leopards, equids, rhinoceroses, as well as a few additional species of bovids. For the most part, the species indicate a dry, open, perhaps steppe, environment and are quite different from those found in the region today. The represented equid and rhinoceroses are now extinct, and the identified eland, oryx, and addax are no longer present in the area. Thomas uses the lack of Eurasian taxa, such as cervids and boars, to argue that the assemblage accumulated before the Late Pleistocene. Amani and Geraads studied a much larger assemblage than was available to Thomas (Minimum Number of Individuals [MNI] = ca. 99; Amani 1991), and they identified a higher diversity of species (about 24). Notably, more species of gazelles and carnivores were present, and a different gazelle species was dominant. However, Amani and Geraads still did not find cervids or boars. They did identify an Alcelaphine with a simple occlusal pattern and an extinct large gerbil, and together these species support a late Middle Pleistocene date for the assemblage. Their analysis also suggests a dry, open environment with some shrubby cover (Amani and Geraads 1998).

One limitation to both analyses is that all the material is unprovenienced. The differences between the two Irhoud samples, especially in the diversity of gazelles, other bovids, and carnivores, could be the result of multiple factors, which are not mutually exclusive: (1) sample size: the higher species diversity present in the Ennouchi sample is simply because the sample is larger, or (2) stratigraphic or spatial variation in the site as a result of (a) changes in environment through the sequence or (b) changes in site utilization by carnivores and humans through the sequence or across the site. Some variant of the second explanation is more likely than the first because the dominant species of gazelle differs between the two assemblages, indicating that Tixier's smaller sample is not simply a subset of the larger sample. Tixier's excavation may have sampled material more closely associated with the human fossils, so stratigraphic variation is most likely influenced by the composition of the samples (see also Geraads 2012).

Thomas and Amani both conducted limited zooarchaeological analyses of the fossil material. Thomas (1981) divided the gazelle tooth rows into seven ages classes, three juvenile and four adult, based on tooth eruption and wear. Based on left mandibles (MNI = 10), he concluded that three juveniles and seven adults were present. The horn cores also show that at least one very young individual was included in the assemblage. However, among the five adult horn cores, only males were identified. Thomas notes the presence of cut-marked bones, especially two phalanges, and burnt pieces, but no quantification is provided. Therefore, humans must have played a role in accumulating at least part of the assemblage. Thomas found little evidence of carnivore involvement in the assemblage, but a subsequent analysis of a human fossil pelvis revealed intriguing marks on its surface, which were interpreted as carnivore damage (Tixier et al. 2001). This result highlights the intricacies of separating human and carnivore components of this assemblage.

Amani (1991) conducted a taphonomic study of the processes that impacted the faunal assemblage before burial, primarily to evaluate if the Ennouchi Irhoud sample was collected by humans or other carnivores. He considered species abundance and skeletal part presentation, ages-at-death, and bone breakage of gazelles. The high percentage of carnivores, including known bone accumulators such as leopards and hyenas, indicated that they might have played a role in collecting the assemblage, while abundant green breakage, consistent with marrow exploitation, shows that humans also had a role. Gazelles are represented by a higher diversity of skeletal parts than medium (alcelaphines, hippotragines, and tragelaphines) or large bovids (bovines), and in general, higher density parts are better represented. Amani concluded that whole gazelles, often juveniles, were brought to the cave, while only select parts of larger animals were brought back to the site. He also constructed a mortality profile for the gazelles (MNI = 34) using dp4 and m3 crown heights, and young adults are best represented in the sample. Amani (1991) compared the gazelle mortality profile to the Klasies River *Pelorovis* MSA sample, in which juvenile and old individuals are abundant, and to the Elandsfontein *Pelorovis* natural death assemblage, in which there are fewer juveniles (Klein 1981, 1982a, b), but the Irhoud sample did not closely resemble either.

Unfortunately, in both samples calcretions on the bones limited investigation of cut- and chew-marks, which are the best indicators of the accumulating agent. In addition, neither study gave full consideration to the effects of pre- and post-depositional destruction that might bias an assemblage towards higher density skeletal parts, including removing juvenile bones and teeth, because these methods were just being developed at the time. Finally, the authors do not discuss the possibility of excavator bias in the assemblage where the excavators kept only the "identifiable" pieces and disposed of the "unidentifiable" material. We do not know if the sediments were passed through screens and if the small finds were kept and analyzed. These limitations combined with the lack of spatial data mean that the roles of carnivores versus ancient humans, of large versus small game, and of chronological variation remain unclear. It is likely that humans and carnivores alternated in their use of the cave with signs of human activity, including stone artifacts and burnt patches, being more abundant in the lower levels.

Fortunately, new excavations at Irhoud are currently underway. From a faunal perspective, the main goal of this new work is to study a stratigraphically documented and complete faunal sample. In addition, many new zooarchaeological and taphonomic studies have been conducted in the last 20 years, which has greatly contributed to the methods available to further our understanding and interpretation of faunal assemblages. This new research will help separate the roles of humans and carnivores in accumulating the assemblage, allowing a better understanding of how the site was formed and ultimately of ancient human behavior.

## El Harhoura 1 And 2, Morocco

After Irhoud, the faunal assemblage that has been studied in the most detail is the Aterian site of El Harhoura 1, Témara, Morocco, which Debénath and Sbihi-Alaoui excavated as a salvage operation in 1977 (Debénath and Sbihi-Alaoui 1979). Aouraghe and colleagues have studied the faunal remains (Aouraghe 2000, 2004; Aouraghe and Abbassi 2002; Bailon and Aouraghe 2002). This large assemblage (NISP > 21,000) contains 36 species of mammals, plus a variety of birds, reptiles, amphibians, and molluscs (Table 8.2). The species present indicate that the environment when the site was utilized was semi-arid and open, similar to the local environment today but perhaps with more moisture and temporary water ponds. Eurasian wild boar is present in the assemblage, indicating that the site is younger than Irhoud and belongs within the Late Pleistocene. Thermoluminescence dates place the uppermost level between 41 and 25 ka (as cited in Gallois 1980; Aouraghe and Abbassi 2002), but new samples and methods may show that these dates are too young.

As with other assemblages, the relative role of carnivores and humans in accumulating the assemblage was an issue at El Harhoura 1. The 5 m thick sequence was excavated in five layers, with the lowest two levels (Niveau 0 and Sous niveau 1) being very sparse, and the next level (Niveau 1) containing a large assemblage of bovids, especially

**Table 8.2** Species lists for the primary sites discussed in the text and for those assemblages published since Ferring (1975, p. 120)

| Species | Common name | Irhoud (Tixier) Thomas (1981) NISP > 200 | Irhoud (Ennouchi) Amani and Geraads (1998) MNI = 99 | El Harhoura 1 Aouraghe (2004) NISP > 21,000 | El Harhoura 2 Campmas et al. (2008) NISP = 215 | El Mnasra Nespoulet et al. (2008) | Doukkala II Michel & Wengler (1993) NISP > 2,000 | Rhafas Michel (1992) | el 'Aliya Wrinn (2001) NISP = 2,970 | Haua Fteah Klein & Scott (1986) NISP = 466 |
|---|---|---|---|---|---|---|---|---|---|---|
| Elephantidae | Elephant | | | X | | | X | | X Arambourg (1967) | X |
| Rhinocerotidae | Rhinoceros | X | X | X | X | X | X | X | X | X |
| Equus mauritanicus | Zebra | X as cf. | X | X | X | | X | X | X | X Higgs (1967) |
| Equus algericus | Horse | | | X | | | | | | |
| Equus asinus | Ass/donkey | | ? | ? | | | X | X | ? | |
| Equus sp. | | | | | X | X | | X | | |
| Cervus sp. | Deer | | | | | | | X | | |
| Sus scrofa | Wild boar | | | X | X | X | X | | X | |
| Phacochoerus africanus | Warthog | | | X | | X | | X | X | |
| Bos/Pelorovis | Bovini | | | | X | | | | | |
| Bos primigenius | Aurochs | X | X | X | | X | X | X | X | X |
| Pelorovis antiquus | Giant buffalo | | | X | | | X | | X | |
| Alcelaphus buselaphus | Hartebeest | | | X | | X | X | X | X Arambourg (1967) | X |
| Connochaetes taurinus | Blue wildebeest | | X | X | | X | X | | X Arambourg (1967) | |
| Alcelaphus/Connochaetes | Alcelaphini | | | | X | | | | X | |
| Rabaticeras arambourgi | | | ? | | | | | | | |
| Damaliscus sp. | Topi/tiang/tsessebe | | ? | | | | | | | |
| Taurotragus sp. | Eland | ? | cf. *oryx* | X | | | ? *oryx* | | | |
| Tragelaphus sp. | Kudu | | | X | | | | | | |
| Tragelaphus/Taurotragus | Tragelaphine | | | | | | | | X | |
| Oryx sp. | Gemsbok/oryx | *dammah* | cf. *gazella* | X | | ? | X | X | | |
| Hippotragus sp. | Roan/sable antelope | | | X | | | X | | | |
| Addax nasomaculatus | Addax | X | | | | | | | | |
| Redunca sp. | Reedbuck | | | | | | X | | | |
| Gazella spp. | Gazelle | X | X | X | X | X | X | X | X | X |
| Gazella atlantica | | | X | X | | | X | | X | |
| Gazella cuvieri | Cuvier's gazelle | X | X | X | | | X | | X | |

(continued)

**Table 8.2** (continued)

| Species | Common name | Irhoud (Tixier) Thomas (1981) NISP > 200 | Irhoud (Ennouchi) Amani and Geraads (1998) MNI = 99 | El Harhoura 1 Aouraghe (2004) NISP > 21,000 | El Harhoura 2 Campmas et al. (2008) NISP = 215 | El Mnasra Nespoulet et al. (2008) | Doukkala II Michel & Wengler (1993) NISP > 2,000 | Rhafas Michel (1992) | el 'Aliya Wrinn (2001) NISP = 2,970 | Haua Fteah Klein & Scott (1986) NISP = 466 |
|---|---|---|---|---|---|---|---|---|---|---|
| *Gazella tingitana* | | | X | | | | | | X Arambourg (1967) | |
| *Gazella rufina* | Red gazelle | | X | | | | | | X Arambourg (1967) | |
| *Gazella dorcas* | Dorcas gazelle | | | | | | X | | X | |
| *Gazella dracula* | | | | | | | X | | | |
| *Ammotragus lervia* | Barbary sheep/aoudad | | X | | | | *Capra* sp. | X | X | X |
| *Hippopotamus amphibius* | Hippo | | | | | X | | | X | |
| *Ursus arctos* | Bear | | | | | | | | X | |
| *Canis aureus* | Golden jackal | X | X | X | X | | X | ? | X | |
| *Canis* sp. nov. | Large canid | | X | X | | | | ? | | |
| *Lycaon* sp. | Hunting dog | | | | | | X Geraads (2008) | | | |
| *Vulpes vulpes* | Red fox | | X | X | X | X | X | | X | X |
| *Crocuta crocuta* | Spotted hyena | | | X | | | X | | X | |
| *Hyaena hyaena* | Striped hyena | | X | X | | | X | | X | |
| Hyaenidae | Hyena | | | | | | | X | | X |
| *Panthera* sp. | Large felid | | | | | | | | | |
| *Panthera leo* | Lion | | X | X | | | X | | X | |
| *Panthera pardus* | Leopard | X | X | X | | | X | | X Arambourg (1967) | |
| *Caracal caracal* | Caracal | | | X | | | | | | |
| *Felis libyca* | Wildcat | | | X | | | | | | |
| *Felis margarita* | Sand cat | | | X | | | | | | |
| *Herpestes ichneumon* | Egyptian mongoose | | | X | | | | | | |
| *Mustela putorius* | European polecat | | | X | | | | | | |
| *Ictonyx (Poecilictis) libyca* | Striped polecat, zorilla | | | X | | | | | | |
| *Mellivora capensis* | Ratel, honey badger | | | | | X | | | | |
| Small mustelid | | | | | | | | | | X |
| Leporidae | Hares and rabbits | | | | X | | | | | X |

(continued)

**Table 8.2** (continued)

| Species | Common name | Irhoud (Tixier) Thomas (1981) NISP > 200 | Irhoud (Ennouchi) Amani and Geraads (1998) MNI = 99 | El Harhoura 1 Aouraghe (2004) NISP > 21,000 | El Harhoura 2 Campmas et al. (2008) NISP = 215 | El Mnasra Nespoulet et al. (2008) | Doukkala II Michel & Wengler (1993) NISP > 2,000 | Rhafas Michel (1992) | el 'Aliya Wrinn (2001) NISP = 2,970 | Haua Fteah Klein & Scott (1986) NISP = 466 |
|---|---|---|---|---|---|---|---|---|---|---|
| *Lepus* sp. | Hare | *capensis* | X | *capensis* | | X | *capensis* | | *capensis* | |
| *Oryctolagus* sp. | European rabbit | | | X | | | *cuniculus* | ? | *cuniculus* | |
| *Erinaceus* sp. | Hedgehog | | X | | | | X | | *europaeus* | |
| *Hystrix cristata* | Crested porcupine | X | X | X | X | X | X | X | X | X |
| *Struthio* sp. | Ostrich | nov. sp. | X | *camelus* | X | X | | X | | |
| Other birds | | X | | X | X | | | X | X | X |
| *Testudo graeca* | Spur-thighed tortoise | | X | X | X | X | X | X | X | X |
| Other reptiles and amphibians | | | | X | X | | X | X | X | |
| Fish | | | | | | | | | X | |
| Marine mollusks | | | | X | ? | | | | X | X |

Information is provided for only the Mousterian and Aterian levels for each site, as available. For Mugharet el 'Aliya, these are Layers 5 and 6; for Haua Fteah, these are the Mousterian and "Pre-Aurignacian" gazelles, and many carnivores. The level above that (Niveau 2) has fewer herbivores and more carnivores, suggesting that carnivores may have played a larger role in accumulating this material. The highest level (Niveau 3) contains a Neolithic cemetery. El Harhoura 1 contains a diversity of carnivores, comprising 16% of the large mammal assemblage (Aouraghe 2000). The majority of them are jackals and foxes, but spotted and striped hyenas (and their coprolites), lions, and leopards are also present. Despite this, Aouraghe (2004) concluded that humans accumulated the majority of bones in the site because of the presence of numerous cut-marks, burnt bones, intentional breakage for marrow extraction, their association with human remains and stone artifacts, and the presence of living floors. However, in the current publication he does not present frequencies of cut- versus chew-marks and their distribution across species and within the stratigraphic sequence of the site. He did examine skeletal part representation by calculating the ratio of fore- to hind-limbs and of axial elements to both. This analysis suggested to him that complete gazelle carcasses were brought into the site while only the most nutritious pieces of larger animals, mainly aurochs and equids, were brought back. He argued that carnivores would have completely destroyed the small bones of these small antelopes if they were responsible for bringing the carcasses back to the site, and therefore the overwhelming abundance of small bovids indicates that humans were the primary accumulator. However, these results need to be interpreted carefully because only 14 equid bones were included in the analysis of their skeletal part representation.

Aouraghe (Aouraghe and Debénath 1999; Aouraghe 2004) further investigated the El Harhoura 1 assemblage by constructing mortality profiles for gazelles (n = 77, MNI = 51) and equids (n = 5) based on their dental eruption and wear. Adult animals dominate the gazelle sample, and Aouraghe concluded that the Aterian people were selectively hunting adult gazelles and that the site was a gazelle hunting camp. However, all ages of animals are almost equally represented in the small equid sample, and Aouraghe suggested that these animals were occasionally and opportunistically taken and only the choicest pieces were brought to the site.

New excavations at El Harhoura 2 are producing well-stratified faunal samples from the Neolithic (couche 1), Iberomaurusian (couche 2), and Aterian (couches 3–7) that are suitable to zooarchaeological analyses. Although the currently published sample is still small (NISP = 1147; Table 8.2), the Aterian and Neolithic samples have been analyzed from a modern zooarchaeological perspective (Campmas et al. 2008). The Aterian fauna (NISP = 679; MNI = 24) is dominated by gazelles (74%), and alcelaphines, bovines, suids, and equids are also present.

Carnivores are rare in the Aterian, and cut-marks, green breaks, and percussion flakes indicate that humans were the primary accumulator of the bones. Cut-marks are most abundant on long-bone shafts and therefore appear to be the result of butchery. Gazelles are represented by all age classes (based on dental eruption and wear; MNI = 7) and most skeletal parts are present, indicating non-selective hunting and transport. Skeletal part representation is not significantly related to bone density, food utility, or marrow availability. Continued excavations at this site promise to produce assemblages that will greatly inform on Late Pleistocene subsistence.

## Mugharet el 'Aliya, Morocco

The cave of Mugharet el 'Aliya, Tangier, Morocco, was excavated by multiple teams between 1939 and 1947 (Howe and Movius 1947; Coon 1957; Howe 1967). The site preserved Neolithic and Aterian levels, and the Aterian occupations have been dated to 35–60 ka using ESR on ungulate tooth enamel samples (Wrinn and Rink 2003). Allen (in Howe and Movius 1947) and Arambourg (1967) provided paleontological descriptions of the fauna, which is consistent with the other Late Pleistocene assemblages discussed here. Gazelles were most abundant and were diverse. Eurasian elements, such as deer and boar, are present, supporting a Late Pleistocene age for the deposits. One notable feature of Arambourg's (1967) study is the identification of monk seal (*Monachus monachus*) in a lower level of the site. To the best of my knowledge, this is the only published determination of a marine mammal in a Moroccan Late Pleistocene assemblage (Steele and Álvarez-Fernández, 2011). Unfortunately, these pieces were not located during Wrinn's subsequent study, and they must have been lost in the intervening years (Wrinn 2001; Wrinn 2007, personal communication). The assemblage has suffered from excavator bias, lack of screening, and numerous relocations; often pieces considered "unidentifiable," such as limb bone shaft fragments and small pieces, were not kept (Wrinn 2001).

Wrinn (2001) conducted a detailed zooarchaeological analysis of the available Mugharet el 'Aliya fauna from three primary layers (NISP = 3859, MNI = 276) (Table 8.2). He began by calculating the changes in species relative abundance and evenness through the sequence. More carnivores and small mammals are present in the stratigraphically older sample (Layer 9), and chew-marks are also most abundant here, indicating that this layer contains a large non-humanly accumulated component (and therefore is not listed in Table 8.2). However, juvenile hyenas, a characteristic of hyena dens, are abundant in all the studied levels, and a few coprolites were found in both the upper (Layers 5 and 6) and lower levels. Cut-marks are very rare throughout the assemblage, but they were also difficult to identify because of heavy calcium carbonate encrustations on the bones. Prime adult gazelles dominate the upper layers. The presence of deciduous hyena teeth but not deciduous gazelle teeth could indicate that the lack of juvenile gazelles is not due to depositional factors and instead reflects the human hunting of these age groups. However, it could also reflect preservation and collection biases. Wrinn's study provides a solid zooarchaeological analysis, but the unknown history of the assemblage limits our interpretation of the data.

## Haua Fteah, Libya

Outside of Morocco, Haua Fteah in coastal, Cyrenaican Libya has the largest and best-documented faunal assemblage spanning the Late Pleistocene and Holocene (Higgs 1967; Klein and Scott 1986). From 1947 to 1955, McBurney excavated this site and three nearby smaller ones, Hagfet ed Dabba, Hagfet et Tera, and Sidi el Hajj Creiem (Wadi Derna) (McBurney 1967). The sequence at Haua Fteah spans more than 13 m and perhaps from 130 ka to historic times, and at least seven culture-stratigraphic groups are represented, allowing for interesting comparisons between samples that share similar depositional and analytical histories. However, the majority of the sample is from the three richest layers, the Upper Paleolithic (Iberomaurusian and Libyco-Capsian) and the Neolithic; there are small samples associated with an Upper Paleolithic Dabban industry and with Mousterian artifacts, and the Aterian is absent. Approximately 21 large mammal taxa are found in the sequence (NISP = 8003), plus birds, tortoises, snakes, and fish (Klein and Scott 1986; Table 8.2). The material in the site varied in density, and it is likely that the cave went unoccupied by humans or other bone accumulators for some periods. The deposits contain numerous stone artifacts and hearths, which attest to the presence of humans. Carnivores, while present in the deposits, are rare, and jackals, often an indicator of hyena accumulations in southern Africa (Cruz-Uribe 1991), are absent. Surprisingly, nowhere in the sequence are cut- or chew-marks present in significant numbers, although Higgs (1967) noted abundant green breakage consistent with marrow extraction, and Klein and Scott (1986) concluded that the archaeological context of the bones provides the best indicator that ancient humans accumulated the faunal material.

As described by Higgs (1967) and Klein and Scott (1986), Barbary sheep, or aoudads, dominate all of the Late

Pleistocene samples from Haua Fteah, although gazelles and aurochs are better represented in the base of the sequence than they are in the rest of the site and aurochs again increase in abundance in the youngest levels. Barbary sheep are very drought tolerant and their increase in abundance could reflect drying during the last glaciation (MIS 4); tortoise abundance also increases with Barbary sheep abundance. Aurochs need more water and their increased abundance in the lower and upper levels is consistent with proposed moister conditions of MIS 1 and MIS 5. Marine molluscs are more common when aurochs are also more common in these lower and upper levels. Because marine molluscs used for subsistence are rarely transported in large numbers more than 10 km, and often they are consumed much closer to the coastline (Buchanan 1988), the lack of marine molluscs in the Dabban and Mousterian layers may reflect lowered sea-levels during this time; the magnitude of displacement of the sea needs to be investigated with detailed bathymetry studies (for example, Marean et al. 2007; Avery et al. 2008). The mammal remains are also sparse in the Dabban and Mousterian levels and so it may also be that human occupation intensities of the site, and their population densities on the landscape, were lower during this time. A similar pattern is seen in South African coastal sites (Klein 2009).

Klein and Scott (1986) compared skeletal part representation and mortality profiles between the samples to learn more about their depositional histories. Examination of the skeletal parts present for Barbary/domestic sheep and aurochs/cattle shows an underrepresentation of less dense parts, such as the proximal humerus, proximal tibia, and sacrum. Hard, dense parts, such as the teeth and distal humerus, tend to be common. Smaller bovids are represented by more body parts, while the large bovids are primarily represented by dentitions and foot bones, the same pattern as is found in Klasies River and El Harhoura 1. As discussed above, this pattern characterizes most fossil assemblages to varying degrees and likely reflects pre- and post-depositional processes that differentially remove the softer parts. The effect of these processes is further indicated by the increasing abundance of the softer sheep parts as the samples become younger in the sequence, reflecting the decreased amount of time these bones have been subjected to post-depositional processes such as leaching and compaction. Klein and Scott (1986) concluded that the skeletal part representation at Haua Fteah reflects both human behavior and depositional history.

Using traditional methods, mortality profiles could only be constructed for the three largest assemblages. Klein and Scott (1986) measured the tooth crown heights on Barbary and domestic sheep dp4 s and m3 s, and by using Quadratic Crown Heights equations provided in Klein and Cruz-Uribe (1984), they placed each tooth in one of 10 age classes, each encompassing 10% of life span, to construct mortality profiles. Juveniles, represented by deciduous teeth, appear to be underrepresented in the samples, and Klein and Scott suggest that this is a result of post-depositional processes, which differentially remove these relatively soft remains. This conclusion is supported by the fragmented nature of many of the other dental remains. An additional limitation is that sheep shed their dp4 s before their m3 s erupt, so there is a gap between these two age groups because individuals without dp4 s or m3 s will not be included in the analysis. Keeping these limitations in mind, Klein and Scott (1986) interpret the resulting mortality profiles as most closely resembling what would be expected if the cave inhabitants were hunting sheep of different ages in frequencies equal to their natural abundance. More recently, Wall-Scheffler (2007) digitally analyzed dental cementum annuli luminance to investigate the Barbary sheep's age and season of death at Haua Fteah. She found that the majority of animals were prime-aged and that in each of the prehistoric cultural groups, the majority of Barbary sheep were taken during the growth months (summer), although winters kills are present, too.

## Doukkala II, Morocco

In addition to the sites above, only a few other assemblages have been subjected to limited analyses. One of these is Doukkala II, Morocco (Michel and Wengler 1993), a karstic hole, or *aven*, that accumulated many bones (NISP > 2,000) and a few stone tools (n = 25) during the Middle and Late Pleistocene (Table 8.2). Michel and Wengler (1993) argued that the association of the bones and stone artifacts is not accidental and that the stone artifacts did not wash into the hole from elsewhere but were left there by ancient humans. They did not find any cut-marks on the bones, but fresh breaks and burnt bones were present. Michel and Wengler (1993) concluded that while carnivores contributed a significant portion of bones to the site, the site also acted as a place where humans were specialized scavengers, facilitated by nearby wooded areas. According to these authors, humans brought bones into this protected place that were (1) scavenged elsewhere, either from lion kills left under the trees or leopard kills stashed in the trees, (2) scavenged off the carcasses brought into this protected place by other carnivores, or (3) from carcasses of animals the fell into the hole and died naturally.

When Doukkala II fauna was published, there was strong debate about whether or not Middle Stone and Middle Paleolithic people were primarily hunters or scavengers (Klein 1976; Binford 1984, 1985, 1988; Grayson and Delpech 1994; Stiner 1994). The current consensus is that while ancient humans probably occasionally scavenged, they primarily hunted large

game. Recent detailed investigations of sites where stone tools are associated with animal bones in natural setting reminds researchers that we need to be cautious when interpreting these ephemeral stone artifact and bone associations. These associations do not necessarily mean that humans played a significant role in accumulating the bones, either by hunting or by scavenging (Klein et al. 1999a, 2007; Villa and Soressi 2000; Cruz-Uribe et al. 2003; Villa et al. 2005).

## Northeastern Africa

Although paleontologists have investigated faunal assemblages from northeastern Africa, much less is known about Late Pleistocene human subsistence there than in northwestern Africa. Zooarchaeological analyses in the region have been hampered by the fact that most sites are open-air sites, where bone is usually poorly preserved. Species lists are often possible, but quantifications have limited meaning. One exception is the fauna from Bir-Tarfawi 14 Main Excavation (BT-14) in the Western Desert of Egypt, where preservation and sample size were sufficient for preliminary investigations (Gautier 1993). Animals and stone artifacts were found together, but surface preservation was poor enough that it was not possible to confidently identify cut- or chew-marks. As in northwestern Africa, gazelles dominate the BT-14 assemblage and the remains of all skeletal parts are present. However, the same is true of large animals such as rhinoceros and giraffe, so Gautier concluded that whole animals died there. The faunal sample probably represents dry-season deaths around a small pool, and Gautier was rightly cautious in concluding that he could not say if the animals' deaths were natural or the result of hunting or scavenging by humans based only on this evidence. However, he did think that the absence of clusters of animal bones not associated with stone tools indicated that humans were primarily responsible for accumulating the animal remains, by targeting gazelles regularly and opportunistically taking the larger animals. Further investigations of Late Pleistocene human subsistence and paleoecology in this region must wait for the discovery of additional suitable samples.

## Small Game Exploitation

Of the limited amount of work that has been conducted on North African faunas, most has focused on the human exploitation of large mammals and the role of large carnivores. Any attention paid to small animals has focused on microfauna for environmental reconstructions, as seen in

Geraads's work (2012), at El Harhoura 1 (Aouraghe and Abbassi 2002; Bailon and Aouraghe 2002), Doukkala II (Michel and Wengler 1993), and Irhoud (Amani and Geraads 1998), or for assessing chronology (Amani and Geraads 1993). While this work is invaluable, as seen in work in South Africa, Europe, and the Near East, much can also be learned about human subsistence from studying the smaller animals found in archaeological assemblages. Taxa of interest include hares, birds, tortoises and other reptiles, as well as fish, molluscs, and other aquatic resources.

The bird assemblage from Haua Fteah is the best described for the region and times considered here (MacDonald 1997). Around 65 taxa are represented (NISP = 703) throughout the long sequence. Birds were most abundant in the Neolithic by a wide margin, and quite rare in the Late Pleistocene samples. MacDonald attributed the difference to accumulator, where most of the birds from the higher levels were human prey and those from the lower layers were probably natural deaths or raptor meals. However, a few bones of the most common taxa in the Mousterian, dove (*Columba* sp.) and partridge (*Alectoris barbara*), were burnt (2 of NISP = 35), so it is possible that ancient people were occasionally consuming some birds. Burnt bird bones do not reappear until the Neolithic. MacDonald also used the bird remains for environmental reconstruction. During the Holocene, birds are abundant and indicate freshwater, woodland, and marine habitats. Birds are less numerous and less diverse during the Iberomaurusian in the Last Glacial Maximum (MIS 2), which MacDonald found indicative of the aridity during this time. He further considered the lack of birds in the even older time periods to be the result of the combination of lack of human predation and aridity during the last glacial, including the sea being far from the site during this time.

In addition to Haua Fteah, birds are mentioned only for Irhoud, El Harhoura 1 and 2, and Mugharet el 'Aliya. Only a few species were identified from Thomas' (1981) study of the Irhoud material, including quail (*Coturnix coturnix*), probable magpies (*Pica pica*), and a large raptor, perhaps a vulture. The El Harhoura 1 bird material is currently being studied and no analyses are available yet (Aouraghe 2004). The presence of bird remains is mentioned for other sites, but details, even species lists, are rarely provided. However, the presence of one species is frequently noted, the ostrich, which is commonly identified from egg shells in the sites, a feature the northern Africa shares with southern Africa. Irhoud, Haua Fteah, El Harhoura 1 and 2, Rhafas Cave, and BT-14 all preserved ostrich bones or egg shells. These large and sturdy shells were probably a good source of food, and they might also have been used for water containers. However, hyenas may also have brought ostrich eggshell into sites (Kandel 2004). During the Upper Paleolithic, as in the Later Stone Age, OES were used as raw material for

beads and ornaments, such as at Dar es-Soltan 1 (Ruhlmann 1951) and Haua Fteah (McBurney 1967).

Another resource that northern and southern Africa has in common is tortoises. Considering the attention given to South African tortoise exploitation (Klein and Cruz-Uribe 1983; Steele and Klein 2005/2006), and now in other regions of the Mediterranean Basin (Stiner et al. 1999, 2000; Speth and Tchernov 2002), there has been little discussion of tortoise utilization in North African sites, although they are commonly present. The typical species is the spur-thighed tortoise (*Testudo graeca*), whose range extends around the Mediterranean Sea. However, others (*Geochelone* sp.) have been identified in the Western Desert of Egypt (Gautier 1993). Because samples have been quantified for only a few sites (El Harhoura provides an exception; Bailon and Aouraghe 2002), we do not know the relative abundance or taphonomic history of tortoises in most assemblages. Also, as in southern Africa, hares are typically present in faunal assemblages but apparently never abundant. In South Africa, there has been some research into how to distinguish human-collected hares from raptor-collected hares (Cruz-Uribe and Klein 1998), and it should be possible to apply these criteria to North African samples to investigate the human exploitation of these animals. As with all other taxa, more taphonomic work needs to be done.

The coastal setting of many of northwest Africa's Late Pleistocene archaeological sites means that during many periods, aquatic resources would have been within ancient people's reach (Steele and Álvarez-Fernández, 2011). However, only Arambourg (1967) mentions seals, and I have found no other references to marine mammal exploitation. A number of site reports mention the presence of marine molluscs, but none provide any quantification or detailed discussion. Oxygen-isotope ratios found in top shells (*Trochus* [*Osilinus*] *turbinatus)* and limpets (*Patella coerula* [*caerulea*]) from Haua Fteah were used to reconstruct past ocean temperatures (McBurney 1967). Although the samples came from throughout the sequence, a listing of abundance by level is not provided. However, McBurney (1967, p. 59) states that the shells were "collected from dense masses of food debris forming in many cases virtual 'kitchen middens'." This statement implies that marine molluscs were a regular food resource to the inhabitants at the site, at least when the coastline was suitably close, as was also the case in South African coastal sites. Roche and Texier (1976) identified abundant limpets in the upper Aterian levels at Grotte des Contrebandiers, and Bouzouggar (Bouzouggar 1997; Bouzouggar et al. 2002) provided a species list for Contrebandiers that included various limpets (*Patella* sp.) along with mussels (*Mytilus* sp.). Limpets and mussels were also found in the Aterian samples from Dar es-Soltan 1 (Ruhlmann 1951). The Aterian layers at El Harhoura 1 contained only rare shells, but they have been identified to species; the assemblage yielded a higher diversity of species than at Contrebandiers (Debénath and Sbihi-Alaoui 1979; Aouraghe 2004). Mugharet el 'Aliya (Briggs 1967) and El Harhoura 2 (Debénath and Sbihi-Alaoui 1979; Campmas et al. 2008) had marine molluscs, but no analyses are provided. Limpets were found during testing in an Aterian level at the site of Kebibat, near Rabat (Souville 1973) and further away, marine molluscs were mentioned as occurring in the Aterian levels of sites along the coast of western Algeria (Roubet 1969). Previously, researchers were only interested in the species of molluscs for climatic reconstruction, a sentiment that is explicitly expressed in Briggs' (1967, p. 187) brief statement on the shells from Mugharet el 'Aliya where he states that "nothing of value" came out of submitting the shells to a specialist. Fortunately, current investigations will provide more details about the use of marine resources during the Late Pleistocene of northern Africa.

North Africa does have one food resource that is not frequently seen in southern Africa—land snails. Many of these snails are large enough to be profitable as food resources, and they have been used as such in recent times. Haua Fteah contained an abundance of land snails, especially in the Neolithic, Libyco-Capsian, and Iberomaurusian (Hey 1967), and this pattern is repeated many times over in northern Africa and around the Mediterranean (Lubell 2004). However, the evidence for intense exploitation of land snails comes primarily from Upper Paleolithic sites, and this phenomenon has not been described for Mousterian or Aterian sites.

Five crab pincers (chelipeds) were identified in the Neolithic and Libyco-Capsian layers at Haua Fteah (Klein and Scott 1986), and crabs and fish were found in the Neolithic at El Harhoura 2 (Campmas et al. 2008). Crabs are on the species list for El Harhoura 1, but their exact provenience is not specified (Aouraghe 2004). Therefore, no crabs have been identified in Aterian assemblages. Crustaceans are rarely found in South African MSA samples, although they are quite numerous in LSA sites (Jerardino et al. 2001; Klein et al. 2004), so this is a feature shared by the northern and southern coast of Africa. Arambourg (1967) identified a few fish remains for Aterian levels at Mugharet el 'Aliya. While no quantification is provided, the brief discussion indicates that they were sparse. The same is true at Dar es-Soltan 1 (Ruhlmann 1951). Otherwise, there is no mention of marine or freshwater fish bones from Mousterian or Aterian samples, consistent with the lack of fish in MSA samples in South Africa. Sites in eastern Africa provide limited evidence for the exploitation of large catfish (Brooks et al. 1995; Yellen et al. 2005).

## Future Research and New Work

While research into the North African environmental and paleontological past has been active, investigations into human subsistence and paleoecology have been limited. Previous researchers acknowledged that both humans and carnivores were contributing to the accumulation of faunal assemblages, but few studies investigated these interactions in detail. Much of this work was conducted before zooarchaeology developed into a discipline independent of paleontology, and these early researchers did not have the benefit of the ecological theory and comparative data, on both modern and fossil assemblages, which are available to today's researchers. This additional data allows for much more detailed investigations into the meaning of ancient faunal assemblages.

Fortunately, as numerous chapters in this volume indicate, a number of sites in Morocco are currently being re-excavated and analyzed using the most up-to-date techniques possible. This includes not only keeping *all* faunal remains through the use of fine screens, but also recording the 3-dimensional coordinates for larger and identifiable pieces. This will allow researchers to study in detail the spatial relationships between faunal and lithic remains, carnivore and herbivore remains, etc., to reconstruct the taphonomic history of the site. The zooarchaeologists working on these projects now have the benefit of decades of previous research by colleagues focused on how to gain the most information possible from these fossil assemblages. As seen in some of the analyses discussed here, future zooarchaeological research should begin with unbiased samples and include basic quantification of the numbers of identified specimens (NISP), minimum numbers of individuals (MNI) and minimum numbers of elements (MNE) for each taxa and each stratigraphically excavated level, not just species presence/absence lists. Surface damage data should also be quantified by level, including incidences of cut-marks, percussion marks, chew-marks, and breakage. All studies need to consider the potential impact of destruction of the softest skeletal parts, either through mechanical (trampling, compaction, carnivore ravaging) or chemical (acidic soils) processes. This removal of elements from the archaeological record will bias analyses of skeletal part representation and mortality profiles. Analyses of small game exploitation, which was all but ignored in earlier studies of North African faunas, will now be possible because of systematic screening.

Before examining human subsistence, zooarchaeologists, paleontologists, and paleoecologists will reconstruct the ancient environments around their sites. While previous researchers have investigated this with both macro- and micro-fauna, studies were hampered because the lack of chronological controls meant that it was difficult to fit the data into global reconstructions. As the chronology of the assemblages is resolved by the new work highlighted in this volume, I anticipate that we will be able to more directly relate fluctuations in the relative abundances of species with the glacial cycles of the Late Pleistocene. However, how these global glacial cycles manifest themselves locally will vary from region to region (Smith 2012). For coastal northwestern Africa, Geraads (2012) has identified a pattern where alcelaphines, gazelles, and white rhinoceroses, which are characteristic of more open environments, are more abundant during cold and therefore dry times; the taxa present during warmer and wetter times are more variable, characteristic of more closed environments, and include aurochs, buffalo, reedbuck, and hippopotamuses. This pattern is very similar to that identified in southwestern Africa where grazers (alcelaphines and equids) are more abundant during the cooler periods of the Late Pleistocene (Klein 1980). The most humid period of the Late Pleistocene appears to be the earlier part, corresponding to MIS 5 and especially MIS 5e, with marked aridification and Saharan expansion occurring during MIS 4 (Geraads 2012; Larrasoaña 2012; Smith 2012) and more humidity during subsequent interstadials (Moreno 2012).

The new zooarchaeological studies are likely to continue to demonstrate the importance of gazelle and Barbary sheep especially, but also aurochs, hartebeest, and wildebeest, in Late Pleistocene human diets, and to show how the abundance of these taxa varies relative to local environments (for example, Higgs 1967). More detailed studies will provide information on the age and sex structures of the prey. I anticipate that these analyses will indicate that gazelle were hunted opportunistically or even that adult animals were targeted. It will be very interesting to see if mortality profiles for larger aurochs samples will suggest capturing of only the youngest and oldest animals, similar to the MSA sample at Klasies, or of opportunistic hunting or driving, similar to Middle Paleolithic aurochs and bison samples from Europe (Gaudzinski 1996). More detailed taphonomic studies will likely reveal that both humans and carnivores contributed to faunal assemblages, alternating the use of cave resources as is documented in both southern Africa and Europe. The human components will almost certainly demonstrate cut-marks from butchery and disarticulation and percussion marks and breakage patterns consistent with marrow consumption. I anticipate that even once the analysis of small game is consistently and fully incorporated into future zooarchaeological analyses, small terrestrial game, particularly fast game such as rabbits, hares, and birds, will still only be a minor component of Mousterian and Aterian diets, again consistent with what is seen in both southern Africa and Europe. If this pattern holds, the slow,

easy to catch tortoises should be larger than those found in subsequent Iberomaurusian and Neolithic assemblages. Similar to southern Africa and Europe, these results would suggest that the North African Middle Paleolithic people lived at lower population densities than subsequent populations and that they did not have the technology to regularly and efficiently capture these small, fast prey. As in southern Africa and Mediterranean Europe, molluscs should be an important component in the diets of the Mousterian and Aterian (and Iberomaurusian) people when they inhabit sites within easy walking distance of the coast, and mollusc diversity in these sites should be low and the individual molluscs should be large, on average. Current analyses of Mediterranean coastal assemblages (Stiner 1994; Barton 2000; Stringer et al. 2008), however, still do not produce nearly the density of marine molluscs or marine mammals that the South African samples provide (Avery et al. 2008; Klein and Steele 2008), and human behavioral, ecological, and environmental explanations will need to be explored (Steele and Álvarez-Fernández, 2011).

One of the most interesting questions to be addressed with new zooarchaeological analyses will be the investigation of subsistence differences between Mousterian-associated and Aterian-associated assemblages—are these chronological, spatial, functional, stylistic and/or environmental variants? Before zooarchaeologists can fully investigate this question, however, lithic analysts need to provide more detailed investigations of the assemblages. The future of our work in North Africa looks very bright as these various lines of evidence come together. Soon, we will know much more about Aterian and Mousterian subsistence and paleoecology in North Africa, which will help illuminate the relationship these groups had with the modern humans who left Africa 50,000 years ago.

## Conclusions

There is much to be learned from North African faunal samples. The major limitations to date have been the lack of stratified analyses, the lack of quantification, and the lack of consideration of small game. Previous analyses that did attempt to consider zooarchaeological questions lacked the comparative datasets that are now available to assist with interpretations. However, given all the new work outlined in this volume, I am very optimistic about the future of zooarchaeological studies in North Africa. This region holds a great deal of information on the relationship between technology, the environment, human subsistence, human morphology, and human demography, all of which combined will help us understand modern human origins.

**Acknowledgments** I am grateful to A. Ben-Ncer, J.-J. Hublin, and S. McPherron for giving me the opportunity to participate in the new work at Jebel Irhoud; to A. Bouzouggar and J.-J. Hublin for allowing me to study the Rhafas Cave material from their current work at the site; to H. Dibble, M.A. El Hajraoui, and U. Schurmans for providing me with the opportunity to analyze the Contrebandiers molluscs from their recent excavations; and to J.-J. Hublin and S. McPherron for inviting me to participate in the Max Planck Institute for Evolutionary Anthropology's 2007 conference on "Modern Human Origins: A North African Perspective." C. Egeland, D. Geraads, T.D. Weaver and an anonymous reviewer offered helpful comments on earlier drafts of this manuscript. R.G. Klein, L. Niven, K. Reed, and E. Turner contributed useful discussions, and C. Letourneux and M. Soressi kindly helped with some French translations. My research into Moroccan faunas is supported by the Max Planck Society and the University of California, Davis.

# References

Amani, F. (1991). *La Faune de la Grotte à Hominidé du Jebel Irhoud (Maroc)*. Ph.D. Dissertation, University of Rabat, Morocco.

Amani, F., & Geraads, D. (1993). Le gisement Moustérien du Djebel Irhoud, Maroc: Précisions sur la faune et la biochronologie, et description d'un nouveau reste humain. *Compte Rendu de l'Académies des Sciences de Paris, 316*, 847–852.

Amani, F., & Geraads, D. (1998). Le gisement Moustérien du Djebel Irhoud, Maroc: Précisions sur la faune et la paléoecologie. *Bulletin d'Archéologie Marocaine, 18*, 11–17.

Aouadi-Abdeljaouad, N., & Belhouchet, L. (2012). Middle Stone Age in Tunisia: Present status of knowledge and recent advances. In J.-J. Hublin & S. McPherron (Eds.), *Modern origins: A North African perspective*. Dordrecht: Springer.

Aouraghe, H. (2000). Les carnivores fossiles d'El Harhoura 1, Temara, Maroc. *L'Anthropologie, 104*, 147–171.

Aouraghe, H. (2004). Les populations de mammifères atériens d'El Harhoura 1 (Témara, Maroc). *Bulletin d'Archéologie Marocaine, 20*, 83–104.

Aouraghe, H., & Abbassi, M. (2002). Les rongeurs du site Atérien d'El Harhoura 1 (Témara, Maroc). *Quaternaire, 13*, 125–136.

Aouraghe, H., & Debénath, A. (1999). Les équidés du Pléistocène supérieur de la Grotte Zouhrah á El Harhoura, Maroc. *Quaternaire, 10*, 283–292.

Arambourg, C. (1967). Appendix A. Observations sur la faune des Grottes d'Hercule près de Tanger, Maroc. In B. Howe (Ed.), *The Palaeolithic of Tangier, Morocco: Excavations at Cape Ashakar, 1939–1947* (Vol. 22, pp. 181–186). Cambridge, MA: The Peabody Museum.

Avery, G. (1989). Some features distinguishing various types of occurrences at Elandsfontein, Cape Province, South Africa. *Paleoecology of Africa, 19*, 213–219.

Avery, G. (1990). *Avian fauna, palaeoenvironments and palaeoecology in the Late Quaternary of the western and southern Cape, South Africa*. Ph.D. Dissertation, University of Cape Town.

Avery, G., Avery, D. M., Braine, S., & Loutit, R. (1984). Bone accumulation by hyaenas and jackals: A taphonomic study. *South African Journal of Science, 80*, 186–187.

Avery, G., Halkett, D., Orton, J., Steele, T. E., Tusenius, M., & Klein, R. G. (2008). The Ysterfontein 1 Middle Stone Age Rockshelter and the evolution of coastal foraging. *South African Archaeological Society Goodwin Series, 10*, 66–89.

Bailon, S., & Aouraghe, H. (2002). Amphibiens, chéloniens et squamates du Pléistocène supérieur d'El Harhoura 1 (Témara, Maroc). *Geodiversitas, 24*, 821–830.

Barton, R. N. E. (2000). Mousterian hearths and shellfish: Late Neanderthal activities in Gibraltar. In C. B. Stringer, R. N. E. Barton, & C. Finlayson (Eds.), *Neanderthals on the edge: 150th anniversary conference of the Forbes' Quarry discovery, Gibraltar* (pp. 211–220). Oxford: Oxbow Books.

Bartram, L. E., & Marean, C. W. (1999). Explaining the "Klasies Pattern": Kua ethnoarchaeology, the Die Kelders Middle Stone Age archaeofauna, long bone fragmentation and carnivore ravaging. *Journal of Archaeological Science, 26*, 9–20.

Binford, L. R. (1984). *Faunal remains from Klasies River Mouth*. Orlando, FL: Academic Press.

Binford, L. R. (1985). Human ancestors: Changing views of their behavior. *Journal of Anthropological Archaeology, 4*, 292–327.

Binford, L. R. (1988). Etude taphonomique des restes fauniques de la grotte Vaufrey. In J.-P. Rigaud (Ed.), *La Grotte Vaufrey à Cénac et Saint-Julien (Dordogne): Paléoenvironnements, chronologie et activités humaines* (pp. 535–564). Mémoires de la Société Préhistorique Française 19.

Bouzouggar, A. (1997). *Matières premières, processus de fabrication et de gestion des supports d'outils dans la séquence atérienne de la grotte des Contrebandiers à Témara (Maroc)*. Ph.D. Dissertation, Université Bordeaux I.

Bouzouggar, A., & Barton, R. N. E. (2012). The identity and timing of the Aterian in Morocco. In J.-J. Hublin & S. McPherron (Eds.), *Modern origins: A North African perspective*. Dordrecht: Springer.

Bouzouggar, A., Kozlowski, J. K., & Otte, M. (2002). Étude des ensembles lithiques atériens de la grotte d'El Aliya à Tanger (Maroc). *L'Anthropologie, 106*, 207–248.

Bouzouggar, A., Barton, R. N. E., Vanhaeren, M., d'Errico, F., Collcutt, S. N., Higham, T. F. G., et al. (2007). 82,000-year-old shell beads from North Africa and implications for the origins of modern human behavior. *Proceedings of the National Academy of Sciences of the USA, 104*, 9964–9969.

Bräuer, G. (1989). The evolution of modern humans: A comparison of the African and non-African evidence. In P. Mellars & C. B. Stringer (Eds.), *The human revolution: Behavioural and biological perspectives on the origins of modern humans* (pp. 123–154). Edinburgh: Edinburgh University Press.

Briggs, L. C. (1967). Appendix B. The mollusks. In B. Howe (Ed.), *The Palaeolithic of Tangier, Morocco: Excavations at Cape Ashakar, 1939–1947. Vol. 22 (pp. 187)*. Cambridge, MA: The Peabody Museum Cambridge.

Brooks, A. S., Helgren, D. M., Cramer, J. S., Franklin, A., Hornyak, W., Keating, J. M., et al. (1995). Dating and context of three Middle Stone Age sites with bone points in the Upper Semliki Valley, Zaire. *Science, 268*, 548–553.

Buchanan, W. F. (1988). Shellfish in prehistoric diet: Elands Bay, S. W. Cape Coast, South Africa. (Vol. 455). Oxford: British Archaeological Reports International Series.

Campmas, E., Michel, P., Amani, F., Cochard, D., Costamagno, S., Nespoulet, R., et al. (2008). Comportements de subsistence a l'Atérien et au Néolithique au Maroc Atlantique : Premiers résultats de l'étude taphonomique et archéozoologique des faunes d'El Harhoura 2 (Région de Témara, maroc). *Actes RQM4, Oujda*, 236–254.

Coon, C. S. (1957). *The seven caves: Archaeological explorations in the Middle East*. New York: Alfred A. Knopf.

Cruz-Uribe, K. (1991). Distinguishing hyena from hominid bone accumulations. *Journal of Field Archaeology, 18*, 467–486.

Cruz-Uribe, K., & Klein, R. G. (1998). Hyrax and hare bones from modern South African eagle roots and the detection of eagle involvement in fossil bone assemblages. *Journal of Archaeological Science, 25*, 135–147.

Cruz-Uribe, K., Klein, R. G., Avery, G., Avery, M., Halkett, D., Hart, T., et al. (2003). Excavation of buried Late Acheulean

(Mid-Quaternary) land surfaces at Duinefontein 2, Western Cape Province, South Africa. *Journal of Archaeological Science, 30*, 559–575.

Debénath, A., & Sbihi-Alaoui, F. (1979). Découverte de deux nouveaux gisements préhistoriques prés de Rabat (Maroc). *Bulletin de la Société Préhistorique Française, 76*, 11–12.

Dewar, G., Halkett, D., Hart, T., Orton, J., & Sealy, J. (2006). Implications of a mass kill site of springbok (*Antidorcas marsupialis*) in South Africa: Hunting practices, gender relations, and sharing in the Later Stone Age. *Journal of Archaeological Science, 33*, 1266–1275.

Ennouchi, E. (1963). Les Néanderthaliens du Jebel Irhoud (Maroc). *Comptes Rendus de l'Académie des Sciences, Paris, 256*, 2459–2460.

Ennouchi, E. (1968). Le deuxième crâne de l'homme d'Irhoud. *Annales de Paléontologie, 55*, 117–128.

Ennouchi, E. (1969). Présence d'un enfant néanderthalien au Jebel Irhoud (Maroc). *Annales de Paléontologie, 55*, 251–265.

Estes, R. D. (1992). The behavior guide to African mammals: Including hoofed mammals, carnivores, primates. Berkeley and Los Angeles: The University of California Press.

Ferring, C. R. (1975). The Aterian in North African prehistory. In F. Wendorf & A.E. Marks (Eds.), *Problems in prehistory: North Africa and the Levant* (pp. 113–126). Dallas: Southern Methodist University Press.

Gallois, B. (1980). *Thermoluminescence et interaction de couplage dans des cristaux dopés avec des ions terres rares. Application à la mise au point d'une nouelle méthode en chronologie absolue. La gamma thermoluminescence.* Ph.D. Dissertation, Université de Bordeaux I. "Garcea, E. A. A. (2012)". Modern human desert adaptations: A Libyan perspective on the Aterian Complex. In J.-J. Hublin & S. McPherron (Eds.), *Modern origins: A North African perspective*. Dordrecht: Springer.

Gaudzinski, S. (1996). On bovid assemblages and their consequences for the knowledge of subsistence patterns in the Middle Palaeolithic. *Proceedings of the Prehistoric Society, 62*, 19–39.

Gautier, A. (1993). The Middle Paleolithic archaeofaunas from Bir Tarfawi (Western Desert, Egypt). In F. Wendorf, R. Schild, & A. E. Close (Eds.), *Egypt during the Last Interglacial: The Middle Paleolithic of Bir Tarfawi and Bir Sahara East* (pp. 121–143). New York: Plenum Press.

Geraads, D. (2008). Plio-Pleistocene Carnivora of northwestern Africa: A short review. *Comptes Rendus Palevol, 7*, 591–599.

Geraads, D. (2012). The faunal context of human evolution in the late Middle/Late Pleistocene of northwestern Africa. In J.-J. Hublin & S. McPherron (Eds.), *Modern origins: A North African perspective*. Dordrecht: Springer.

Grayson, D. K., & Delpech, F. (1994). The evidence for Middle Palaeolithic scavenging from Couche VIII, Grotte Vaufrey (Dordogne, France). *Journal of Archaeological Science, 21*, 359–375.

Grün, R., & Stringer, C. B. (1991). Electron spin resonance dating and the evolution of modern humans. *Archaeometry, 33*, 153–199.

Harvati, K., & Hublin, J.-J. (2012). Morphological continuity of the face in the late Middle and Upper Pleistocene hominins from northwestern Africa—a 3-D geometric morphometric analysis. In J.-J. Hublin & S. McPherron (Eds.), *Modern origins: A North African perspective*. Dordrecht: Springer.

Hawkins, A. L. (2012). The Aterian of the oases of the Western Desert of Egypt: Adaptation to changing climatic conditions? In J.-J. Hublin & S. McPherron (Eds.), *Modern origins: A North African perspective*. Dordrecht: Springer.

Henshilwood, C. S. (1997). Identifying the collector: Evidence for human processing of the Cape dune mole-rat, *Bathyergus suillus*, from Blombos Cave, southern Cape, South Africa. *Journal of Archaeological Science, 24*, 659–662.

Henshilwood, C. S., & Marean, C. W. (2003). The origin of modern behavior: Critique of the models and their test implications. *Current Anthropology, 44*, 627–651.

Hey, R. W. (1967). Land-snails. In C. B. M. McBurney (Ed.), *The Haua Fteah (Cyrenaica) and the Stone Age of the South-East Mediterranean (pp. 358)*. Cambridge: Cambridge University Press.

Higgs, E. S. (1967). Environment and chronology—the evidence from mammalian fauna. In C. B. M. McBurney (Ed.), *The Haua Fteah (Cyrenaica) and the Stone Age of the South-East Mediterranean* (pp. 149–164). Cambridge: Cambridge University Press.

Howe, B. (1967). The Palaeolithic of Tangier, Morocco: Excavations at Cape Ashakar, 1939–1947. *Bulletin of the American School of Prehistoric Research, 22*, 1–200.

Howe, B., & Movius, H.L. (1947). A stone age cave site in Tangier: Preliminary report on the excavations at the Magharet el 'Aliya, or High Cave, in Tangier. Papers of the Peabody Museum of American Archaeology and Ethnology, Harvard University, 28, 1–32.

Hublin, J.-J. (2001). Northwestern African Middle Pleistocene hominids and their bearing on the emergence of *Homo sapiens*. In L. S. Barham & K. A. Robson-Brown (Eds.), *Human roots: Africa and Asia in the Middle Pleistocene* (pp. 99–121). Bristol, England: Western Academic & Specialist Press Limited.

Hublin, J.-J., Tillier, A.-M., & Tixier, J. (1987). L'humérus d'enfant moustérien (Homo 4) du Jebel Irhoud (Maroc) dans son contexte archéologique. *Bulletin et Mémoires de la Société d'Anthropologie de Paris 4, série, 14*, 115–142.

Jerardino, A., Navarro, R., & Nilssen, P. (2001). Cape rock lobster (*Jasus islandii*) exploitation in the past: Estimating carapace length from mandible sizes. *South African Journal of Science, 97*, 59–62.

Kandel, A. W. (2004). Modification of ostrich eggs by carnivores and its bearing on the interpretation of archaeological and paleontological finds. *Journal of Archaeological Science, 31*, 377–391.

Klein, R. G. (1972). The late Quaternary mammalian fauna of Nelson Bay Cave (Cape Province, South Africa): Its implications for megafaunal extinctions and for cultural and environmental change. *Quaternary Research, 2*, 135–142.

Klein, R. G. (1975a). Middle Stone Age man-animal relationships in southern Africa: Evidence from Die Kelders and Klasies River Mouth. *Science, 190*, 265–267.

Klein, R. G. (1975b). Paleoanthropological implications of the non-archeological bone assemblage from Swartklip 1, south-western Cape Province, South Africa. *Quaternary Research, 5*, 275–288.

Klein, R. G. (1976). The mammalian fauna of the Klasies River Mouth sites, southern Cape Province, South Africa. *South African Archaeological Bulletin, 31*, 75–96.

Klein, R. G. (1977). The ecology of early man in southern Africa. *Science, 197*, 115–126.

Klein, R. G. (1978). Stone Age predation on large African bovids. *Journal of Archaeological Science, 5*, 195–217.

Klein, R. G. (1979). Stone Age exploitation of animals in southern Africa. *American Scientist, 67*, 151–160.

Klein, R. G. (1980). Environmental and ecological implications of large mammals from Upper Pleistocene and Holocene sites in southern Africa. *Annals of the South African Museum, 81*, 223–283.

Klein, R. G. (1981). Later Stone Age subsistence at Byeneskranskop Cave, South Africa. In R. S. O. Harding & G. Teleki (Eds.), *Omnivorous primates: Gathering and hunting in human evolution* (pp. 166–190). New York: Columbia University Press.

Klein, R. G. (1982a). Age (mortality) profiles as a means of distinguishing hunted species from scavenged ones in Stone Age archeological sites. *Paleobiology, 8*, 151–158.

Klein, R.G. (1982b). Patterns of ungulate mortality and ungulate mortality profiles from Langebaanweg (Early Pliocene) and Elandsfontein (Middle Pleistocene), south-western Cape Province, South Africa. *Annals of the South African Museum, 20*, 49–94.

Klein, R.G. (1983). Palaeoenvironmental implications of Quaternary large mammals in the Fynbos Biome. In H.J. Deacon, Q. B. Hendey & J. J. N. Lambrechts (Eds.), *Fynbos palaeoecology: A preliminary synthesis*, Vol. 75 (pp. 116–138). South African National Scientific Programmes Reports.

Klein, R. G. (1989). Why does skeletal part representation differ between smaller and larger bovids at Klasies River Mouth and other archeological sites? *Journal of Archaeological Science, 6*, 363–381.

Klein, R. G. (1994). Southern Africa before the Iron Age. In R. S. Corruccini & R. L. Ciochon (Eds.), *Integrative paths to the past: Paleoanthropological advances in honor of F. Clark Howell* (pp. 471–519). Englewood Cliffs: Prentice Hall.

Klein, R. G. (2009). *The human career: Human biological and cultural origins* (3rd ed.). Chicago: University of Chicago Press.

Klein, R. G., & Cruz-Uribe, K. (1983). Stone Age population numbers and average tortoise size at Byneskranskop Cave 1 and Die Kelders Cave 1, southern Cape Province, South Africa. *South African Archaeological Bulletin, 38*, 26–30.

Klein, R. G., & Cruz-Uribe, K. (1984). *The analysis of animal bones from archeological sites*. Chicago: University of Chicago Press.

Klein, R. G., & Cruz-Uribe, K. (1996). Exploitation of large bovids and seals at Middle and Later Stone Age sites in South Africa. *Journal of Human Evolution, 31*, 315–334.

Klein, R. G., & Cruz-Uribe, K. (2000). Middle and Later Stone Age large mammal and tortoise remains from Die Kelders Cave 1, western Cape Province, South Africa. *Journal of Human Evolution, 38*, 169–195.

Klein, R. G., & Scott, K. (1986). Re-analysis of faunal assemblages from the Haua Fteah and other Late Quaternary archaeological sites in Cyrenaican Libya. *Journal of Archaeological Science, 13*, 515–542.

Klein, R. G., & Steele, T. E. (2008). Gibraltar data are too sparse to inform on Neanderthal exploitation of coastal resources. *Proceedings of the National Academy of Sciences of the USA, 105*, E115.

Klein, R. G., Avery, G., Cruz-Uribe, K., Halkett, D., Hart, T., Milo, R. G., et al. (1999a). Duinefontein 2: An Acheulean site in the western Cape Province of South Africa. *Journal of Human Evolution, 37*, 153–190.

Klein, R.G., Cruz-Uribe, K., & Milo, R.G. (1999b). Skeletal part representation in archaeofaunas: Comments on "Explaining the 'Klasies Pattern': Kua ethnoarchaeology, the Die Kelders Middle Stone Age archaeofauna, long bone fragmentation and carnivore ravaging" by Bartram and Marean. *Journal of Archaeological Science, 26*, 1225–1234.

Klein, R. G., Avery, G., Cruz-Uribe, K., Halkett, D., Parkington, J. E., Steele, T. E., et al. (2004). The Ysterfontein 1 Middle Stone Age site, South Africa, and early human exploitation of coastal resources. *Proceedings of the National Academy of Sciences of the USA, 101*, 5708–5715.

Klein, R. G., Avery, G., Cruz-Uribe, K., & Steele, T. E. (2007). The mammalian fauna associated with an archaic hominin skullcap and later Acheulean artifacts at Elandsfontein, western Cape Province, South Africa. *Journal of Human Evolution, 52*, 164–186.

Larrasoaña, J. C. (2012). A Northeast Saharan perspective on environmental variability in North Africa and its implications for modern human origins. In J.-J. Hublin & S.P. McPherron (Eds.), *Modern origins: A North African perspective*. Dordrecht: Springer.

Lubell, D. (2004). Prehistoric edible land snails in the circum-Mediterranean: The archaeological evidence. In J.-P. Brugal & J. Desse (Eds.), *Petits animaux sociétés humaines: du complément alimentaire aux ressources utilitaires* (pp. 77–98). Antibes: Éditions APDCA.

MacDonald, K. C. (1997). The avifauna of the Haua Fteah (Libya). *ArchaeoZoologia, 9*, 83–102.

Marean, C. W. (1998). A critique of the evidence for scavenging by Neandertals and early modern humans: New data from Kobeh Cave (Zagros Mountains, Iran) and Die Kelders Cave 1 Layer 10 (South Africa). *Journal of Human Evolution, 35*, 111–136.

Marean, C. W., & Frey, C. J. (1997). Animal bones from caves to cities: Reverse utility curves as methodological artifacts. *American Antiquity, 62*, 698–711.

Marean, C. W., Abe, Y., Frey, C. J., & Randall, R. C. (2000). Zooarchaeological and taphonomic analysis of the Die Kelders Cave 1 layers 10 and 11 Middle Stone Age larger mammal fauna. *Journal of Human Evolution, 38*, 197–233.

Marean, C. W., Bar-Matthews, M., Bernatchez, J., Fisher, E., Goldberg, P., Herries, A. I. R., et al. (2007). Early human use of marine resources and pigment in South Africa during the Middle Pleistocene. *Nature, 448*, 905–907.

McBrearty, S., & Brooks, A. S. (2000). The revolution that wasn't: A new interpretation of the origins of modern human behavior. *Journal of Human Evolution, 39*, 453–563.

McBurney, C. B. M. (1967). *The Haua Fteah (Cyrenaica) and the Stone Age of the South-East Mediterranean*. Cambridge: Cambridge University Press.

Mellars, P. (2006). Why did modern human populations disperse from Africa ca. 60,000 years ago? A new model. *Proceedings of the National Academy of Sciences of the USA, 103*, 9381–9386.

Michel, P. (1992). Pour une meilleure connaissance du Quaternaire Continental Marocain: les vertébrés fossiles du Maroc Atlantique, Central et Oriental. *L'Anthropologie, 96*, 643–656.

Michel, P., & Wengler, L. (1993). Le site paléontologique et archéologique de Doukkala II (Maroc, Pléistocène moyen et supérieur): Premier jalon en Afrique du Nord d'un comportement humain assimilable à "charognage contrôlé et actif". *Comptes Rendus de l'Académie des Sciences, Paris, 317, Série II*, 557–562.

Milo, R. G. (1994). *Human-animal interactions in southern African prehistory: A microscopic study of bone damage signatures*. Ph.D. Dissertation, University of Chicago.

Milo, R. G. (1998). Evidence for hominid predation at Klasies River Mouth, South Africa, and its implications for the behaviour of early modern humans. *Journal of Archaeological Science, 25*, 99–133.

Moreno, A. (2012). A multiproxy paleoclimate reconstruction over the last 250 kyr from marine sediments: The Northwest African margin and the western Mediterranean Sea. In J.-J. Hublin & S.P. McPherron (Eds.), *Modern origins: A North African perspective*. Dordrecht: Springer.

Nespoulet, R., El Hajraoui, M. A., Amani, F., Ben Ncer, A., Debénath, A., El Idrissi, A., et al. (2008). Palaeolithic and Neolithic occupations in the Témara region (Rabat, Morocco): Recent data on hominin contexts and behavior. *African Archaeological Review, 25*, 21–39.

Outram, A. K. (2001). The scapula representation could be the key: A further contribution to the "Klasies Pattern" debate. *Journal of Archaeological Science, 28*, 1259–1263.

Parkington, J. E. (2003). Middens and moderns: Shellfishing and the Middle Stone Age of the western Cape, South Africa. *South African Journal of Science, 99*, 243–247.

Rightmire, P. G. (1989). Middle Stone Age humans from eastern and southern Africa. In P. Mellars & C. B. Stringer (Eds.), *The human revolution: Behavioural and biological perspectives on the origins of modern humans* (pp. 109–122). Edinburgh: Edinburgh University Press.

Roche, J., & Texier, J.-P. (1976). Découverte de restes humains dans un niveau atérien supérieur de la grotte des Contrebandiers, à Temara (Maroc). *Comptes Rendus de l'Académie des Sciences de Paris, Série D, 282*, 45–47.

Roubet, F.-E. (1969). Le niveau Atérien dans la stratigraphie côtière a l'ouest d'Alger. *Palaeoecology of African and of the Surrounding Islands and Antarctica, 4*, 124–129.

Ruhlmann, A. (1951). *La Grotte préhistorique de Dar es-Soltan*. Paris: Larose.

Sampson, C. G. (2000). Taphonomy of tortoises deposited by birds and Bushmen. *Journal of Archaeological Science, 27*, 779–788.

Skinner, J. D., & Chimimba, C. T. (2005). *The mammals of the southern African subregion* (3rd ed.). Cambridge: Cambridge University Press.

Smith, J.R. (2012). Spatial and temporal variation in the nature of Pleistocene pluvial phase environments across North Africa. In J.-J. Hublin & S.P. McPherron (Eds.), *Modern origins: A North African perspective*. Dordrecht: Springer.

Smith, T. M., Tafforeau, P., Reid, D. J., Grün, R., Eggins, S., Boutakiout, M., et al. (2007). Earliest evidence of modern human life history in North African early *Homo sapiens*. *Proceedings of the National Academy of Sciences of the USA, 104*, 6128–6133.

Souville, G. (1973). *Atlan préhistorique du Maroc: Le Maroc Atlantique*. Paris: Centre National de la Recherche Scientifique.

Speth, J. D., & Tchernov, E. (2002). Middle Paleolithic tortoise use at Kebara Cave (Israel). *Journal of Archaeological Science, 29*, 471–483.

Steele, T. E., & Álvarez-Fernández, E. (2011). Initial investigations into the exploitation of coastal resources in North Africa during the Late Pleistocene at Grotte des Contrebandiers, Morocco. In N. Bicho, J. Haws & L.G. Davis (Eds.), *Trekking the Shore: Changing Coastlines and the Antiquity of Coastal Settlement* (pp. 383–403). New York: Springer.

Steele, T. E., & Klein, R. G. (2005/2006). Mollusk and tortoise size as proxies for stone age population density in South Africa: Implications for the evolution of human cultural capacity. *Munibe (Antropologia -Arkeologia), 57*, 221–237.

Steele, T. E., & Klein, R. G. (2008). Intertidal shellfish use during the Middle and Later Stone Age of South Africa. *Archaeofauna, 17*, 63–76.

Steele, T. E., & Klein, R. G. (2009). Late Pleistocene subsistence strategies and resource intensification in Africa. In J.-J. Hublin & M. P. Richards (Eds.), *The evolution of hominid diets: Integrating approaches to the study of palaeolithic subsistence* (pp. 111–124). New York: Springer Science.

Stiner, M. C. (1994). *Honor among thieves: A zooarchaeological study of Neandertal ecology*. Princeton: Princeton University Press.

Stiner, M. C., Munro, N. D., Surovell, T. A., Tchernov, E., & Bar-Yosef, O. (1999). Paleolithic population growth pulses evidenced by small animal exploitation. *Science, 283*, 190–194.

Stiner, M. C., Munro, N. D., & Surovell, T. A. (2000). The tortoise and the hare: Small-game use, the Broad-Spectrum Revolution, and Paleolithic demography. *Current Anthropology, 41*, 39–73.

Stringer, C. B., Finlayson, J. C., Barton, R. N. E., Fernández-Jalvo, Y., Cáceres, I., Sabin, R. C., et al. (2008). Neanderthal exploitation of marine mammals in Gibraltar. *Proceedings of the National Academy of Sciences of the USA, 105*, 14319–14324.

Thomas, H. (1981). La faune de la Grotte à Néandertaliens du Jebel Irhoud (Maroc). *Quaternaria, 23*, 191–217.

Thompson, J. C. (2008). *Zooarchaeological tests for modern human behavior at Blombos Cave and Pinnacle Point Cave 13B, southwestern Cape, South Africa*. Ph.D. Dissertation, Arizona State University.

Tixier, J., Brugal, J.-P., Tillier, A.-M., Bruzek, J., & Hublin, J.-J. (2001). Irhoud 5, un fragment d'os coxal non adulte des niveaux moustériens marocains. *Actes des 1ères Journées Nationales d'Archéologie et du Patrimoine Volume 1 : Préhistoire*, 149–153.

Turner, A. (1989). Sample selection, schlepp effects and scavenging: The implications of partial recovery for interpretations of the terrestrial mammal assemblage from Klasies River Mouth. *Journal of Archaeological Science, 16*, 1–11.

Villa, P., & Soressi, M. (2000). Stone tools in carnivore sites: The case of Bois Roche. *Journal of Anthropological Research, 56*, 187–215.

Villa, P., Soto, E., Santonja, M., Pérez-González, A., Mora, R., Parcerisas, J., et al. (2005). New data from Ambrona: Closing the hunting versus scavenging debate. *Quaternary International, 126–128*, 223–250.

Wall-Scheffler, C. M. (2007). Digital cementum luminance analysis and the Haua Fteah hominins: How seasonality and season of use changed through time. *Archaeometry, 49*, 815–826.

Wrinn, P. J. (2001). Reanalysis of the Pleistocene archaeofauna from Mugharet el 'Aliya, Tangier, Morocco: Implications for the Aterian UISPP. Paper presented at the XIVth UISPP Congress; September 2–8, 2001, Liège, Belgium.

Wrinn, P. J., & Rink, W. J. (2003). ESR Dating of tooth enamel from Aterian levels at Mugharet el 'Aliya (Tangier, Morocco). *Journal of Archaeological Science, 30*, 123–133.

Yellen, J., Brooks, A. S., Helgren, D. M., Tappen, M., Ambrose, S. H., Bonnefille, R., et al. (2005). The archaeology of Aduma Middle Stone Age Sites in the Awash Valley, Ethiopia. *PaleoAnthropology, 10*, 25–100.

# Chapter 9
# Modern Human Desert Adaptations: A Libyan Perspective on the Aterian Complex

E. A. A. Garcea

**Abstract** The present Libyan territory extends over a large area that goes from the Mediterranean coast to the Saharan Desert. Aterian sites were found in the central Saharan mountain range of the Tadrart Acacus, the eastern Saharan massif of the Jebel Uweinat, the Maghrebi extension of the Jebel Gharbi, as well as the lowlands of Lake Shati in the central part of the country. Recent research in the Tadrart Acacus and the Jebel Gharbi has provided radiometric dates, geoarchaeological stratigraphic sequences, and lithic assemblages that call for a revision of the chronological, environmental, and functional interpretation of the Aterian Industrial Complex. In Libya, two distinct Aterian variants, one to the northwest (Jebel Gharbi), the other to the southwest (Tadrart Acacus and Messak Settafet) of the country, display specific chronological developments and site settings related to different paleoenvironmental conditions that allow one to trace geographic boundaries, with different latitudinal and altitudinal adaptational patterns. Their differences concur to show that Aterian groups developed different skills and tools to adapt to different dry environments that inevitably conditioned their behavior and settlement systems. This paper reviews the recent evidence from the Libyan Aterian sites and those that immediately preceded and followed, discusses both the general perspective and the regional variants within the Aterian, and addresses the question of the spread of anatomically modern humans in North Africa.

**Keywords** Aterian • Desert adaptation • Jebel Gharbi • Libyan Sahara • Messak Settafet • Modern humans • Tadrart Acacus

E. A. A. Garcea (✉)
Dipartimento di Lettere e Filosofia, Università di Cassino,
Via Zamosch 43, 03043 Cassino, FR, Italy
e-mail: egarcea@fastwebnet.it

## The Sahara: An Empty Land?

*Sahara* is an Arabic word meaning "desert" and is related to another Arabic word, *ashar*, which refers to its reddish, vegetationless sands. These terms are most appropriate as the Sahara is presently the largest desert in the world, covering 8.6 million km$^2$. Past climatic conditions were alternately moister and drier than today, changing between semi-arid, sub-arid, and arid. According to Wilson et al. (2000), in the Late Pleistocene, the desert practically never disappeared between latitude 20° and 30° north, which is the belt that runs precisely across present-day Libya (Fig. 9.1).

Lake Shati was relatively small in size during the mid-OIS 5 (100–110 ka), although Lake Megafezzan, which incorporated Lake Shati, reached its largest extension only in the earliest Holocene, when the next humid phase after the Last Interglacial one occurred (Armitage et al. 2007).

Therefore, human populations of all times have had to cope with unstable conditions with patchy resources, sandy soils, strong winds, and large temperature fluctuations (e.g., Yellen 1977; Clark 1980; Giraudi 2005). Nevertheless, in spite of their challenging conditions, deserts can facilitate human dispersal and circulation of cultural traditions by groups in search of other resources when those in their biome are diminishing or exhausted. To put it in Smith et al.'s (2005, p. 2) words, "Deserts have a special role in human evolution and adaptation. They appear to be the major terrestrial habitat that channeled early human dispersal, representing barriers at some times, corridors at others." This description accurately depicts the Sahara and the circumstances in which Aterian hunter-gatherers lived and alternately moved through favorable corridors, or remained isolated.

Several authors (among others, Alimen et al. 1966; Debénath et al. 1986; Tillet 1995) had inaccurately associated the Aterian with a humid climate, whereas others disagreed with this interpretation. Among the latter, Clark

J.-J. Hublin and S. P. McPherron (eds.), *Modern Origins: A North African Perspective*,
Vertebrate Paleobiology and Paleoanthropology, DOI: 10.1007/978-94-007-2929-2_9,
© Springer Science+Business Media B.V. 2012

**Fig. 9.1** Map showing the relatively northern position of Libya

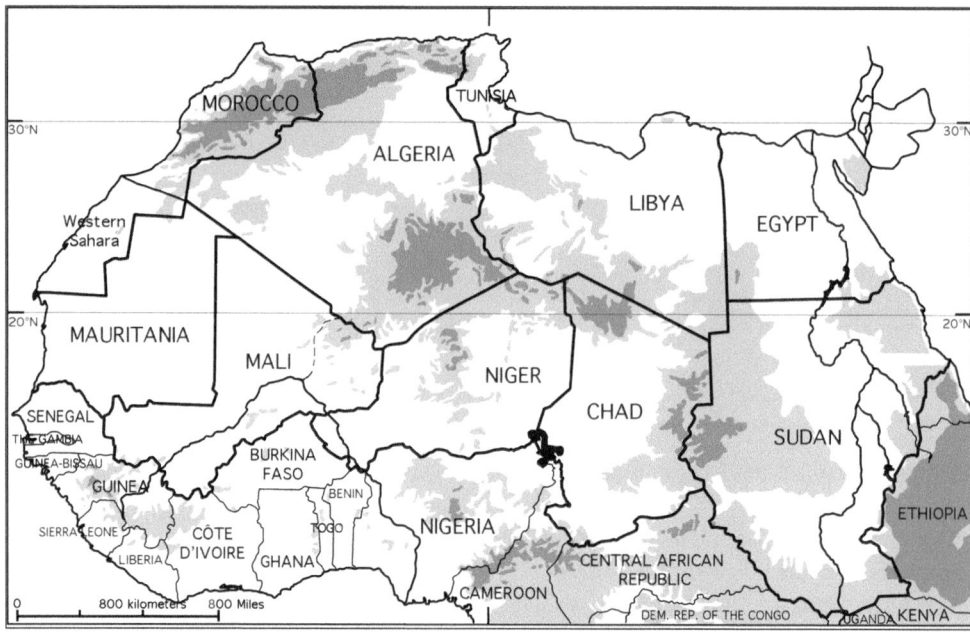

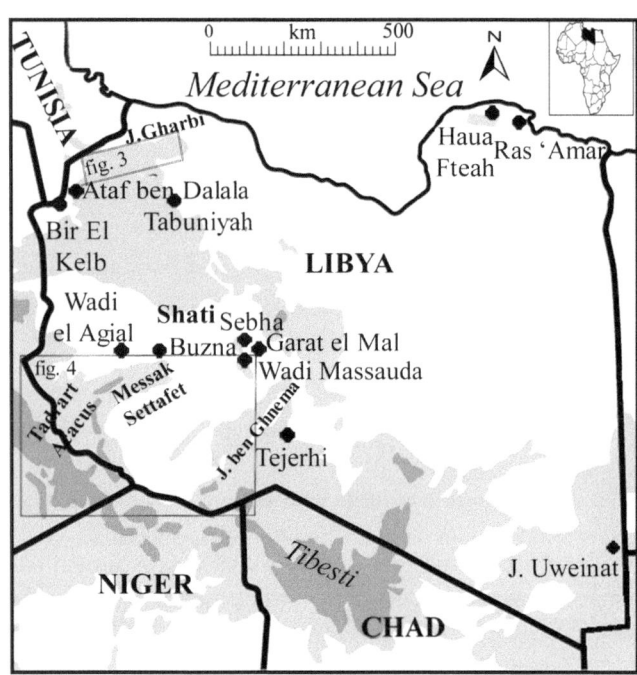

**Fig. 9.2** Map of Aterian sites in Libya (except northwestern sites, in Fig. 9.3, and southwestern sites, in Fig. 9.4)

although it is very often cited, the available data go back to McBurney's excavations made in the 1950s (McBurney 1967) and are waiting to be confirmed or revised by the new excavations resumed in 2007 by Barker (Barker et al. 2007, 2008). In the meantime, research in north and southwestern Libya has provided geoarchaeological and stratigraphic sequences, radiometric dates, and technological evidence of lithic manufacture that call for a revision of the chronological, environmental, and functional interpretation of the Aterian Industrial Complex. In northwestern Libya, specific research on Late Pleistocene human cultures has been, and still is being, conducted in the Jebel Gharbi (Barich et al. 1996, 2003a, b, 2006; Garcea 2004, 2006a, 2009, 2010a; Barich and Giraudi 2005; Garcea and Giraudi 2006; Barich and Garcea 2008); in southwestern Libya, the study area includes the Tadrart Acacus mountain range, the Messak Settafet plateau, and the Edeyen of Murzuq (Garcea 1997, 2001a, b, 2004, 2010a; Cremaschi 1998; Cremaschi and di Lernia 1998; Cremaschi et al. 1998; Martini et al. 1998; di Lernia 1999b) (Fig. 9.2).

(1980, p. 547) was one of the first to recognize that the Aterian was "a desert-orientated adaptation." As a matter of fact, the majority of Aterian sites are spread throughout the present Sahara Desert; only a few are located along the Atlantic Coast and even less are on the Mediterranean Coast (cf. Garcea 2001b; Gifford-Gonzalez et al., in preparation).

With regards to the prehistory of Libya, the Haua Fteah Cave, at the foot of Jebel Akhdar in Cyrenaica (Fig. 9.2), is probably still the first site that comes to mind. However,

## African Versus European Nomenclature

Archaeological research in North Africa was started during colonial times by European scholars trained in European prehistory, who considered North Africa as an extension of Europe and therefore employed the terms they used in Europe to name the cultural units they found in Africa. European terms like "Middle Paleolithic" and "Mousterian" have been carelessly used to describe North African

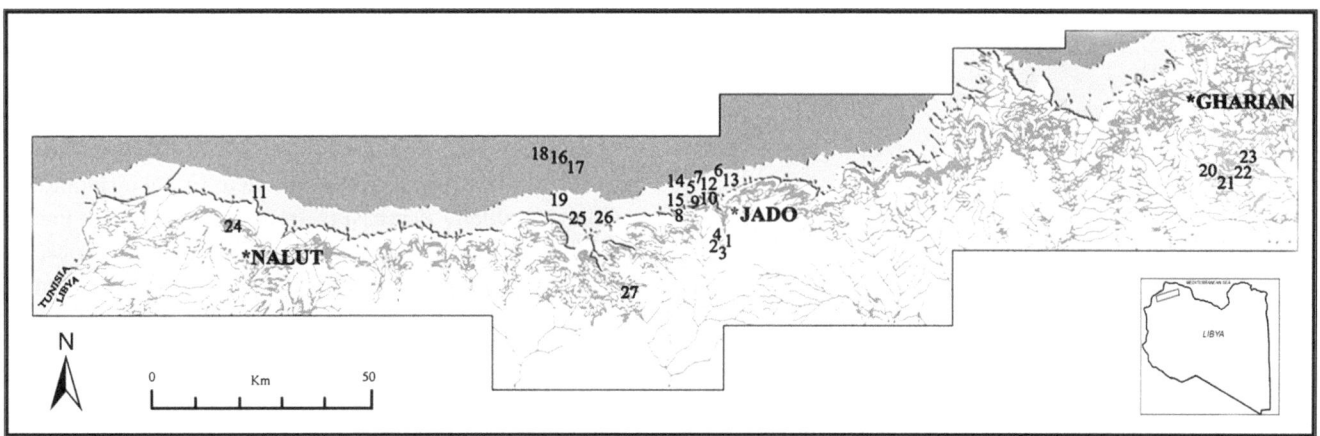

**Fig. 9.3** Map of northwestern Libya. Early MSA site: *11* Wadi Nalut, SJ-00-60. Aterian sites: *1* Ras el Wadi, SJ-90-12; *2* Ras el Wadi, SJ-98-27; *3* Ras el Wadi, SJ-98-27A; *4* Ras el Wadi, SJ-98-28; *5* Shakshuk West, SJ-00-55/Test 2; *6* Shakshuk East, SJ-00-55 East; *7* Shakshuk West, SJ-00-56/Ext. 2; *8* Mahatta Frid, SJ-00-57; *9* Shakshuk West, SJ-00-58; *10* Shakshuk West, SJ-00-58A; *13* Aïn Soda, SJ-02-67; *14* Wadi Sel, SJ-02-68; *15* 3 km W Shakshuk, SJ-02-69; *16* Wadi Ali, SJ-02-70; *17* Wadi Ali, SJ-03-71; *18* Wadi Ali, SJ-03-77; *19* East Badarna, SJ-03-78; *20* Wadi Ghan, SG-99-40; *21* Wadi Ghan, SG-99-41; *22* Wadi Ghan, SG-99-46; *23* Wadi Ghan, SG-00-61; *24* Nalut, SJ-06-89; *25* Josh, SJ-06-88; *26* Josh, SJ-06-87; *27* Josh, SJ-06-86. LSA site: *12* Shakshuk East, SJ-02-66

industries, even though more correct terms have been created for African prehistory (Early, Middle, Later Stone Age). Unfortunately, the use of the European nomenclature for African industries has implied confusing assumptions and erroneous understandings. One of the clearest examples is the past supposition that the so-called "Mousterian/ Middle Paleolithic" industries from Jebel Irhoud and Haua Fteah had to be made by Neandertal peoples, as was the case in Europe. Although we now know that Neandertals never lived in Africa, some cultural techno-complexes are still called "Mousterian" or "Middle Paleolithic." This is hardly acceptable for a variety of reasons. From a physical anthropological point of view, there is a remarkable difference between the makers of Middle Paleolithic industries of Eurasia and those of Middle Stone Age (MSA) industries of Africa. The former were Neandertals, apart from rare exceptions such as Skhul and Qafzeh in the Near East (Vandermeersch 1989), whereas the latter were anatomically modern humans. Moreover, unlike in Eurasia, anatomically modern types did not succeed Neandertal forms in Africa. Consequently, the Middle/Upper Paleolithic transition in Eurasia is usually based on comparative analysis between Neandertal and *sapiens* morphological types, whereas in Africa physical displacement or coexistence of the two types did not occur.

Also, technologically, there is a clear distinction between the Middle Paleolithic of Eurasia and the Middle Stone Age of Africa. In their review of the MSA projectile technologies, Brooks et al. (2006) argue that projectile armatures typify the MSA throughout Africa, whereas they are a common archaeological marker of the Upper Paleolithic in Eurasia. This is a logical discrepancy, considering that tool types are the material products of human beings, not cultural units, and the human beings in question are modern humans in both the Upper Paleolithic of Eurasia and the MSA of Africa. Why then should the Middle Paleolithic of Eurasia be equated with the MSA of Africa?

Furthermore, potential connections emerge between sub-Saharan Africa, East Africa in particular, and North Africa on both environmental and technological bases (Caton-Thompson 1946; Clark 1993; Kleindienst 1998; Van Peer 1998; Garcea 2001b, 2004; Gifford-Gonzalez et al. 2004, in preparation; Clark et al. 2008). It would then be paradoxical to use the same names (Middle Paleolithic) for two unrelated cultural units, one in Eurasia, the other in North Africa, and to use two different names (Middle Paleolithic and MSA) for two related cultural units, one in North Africa, the other in East Africa. As a matter of fact, several Africanist archaeologists (e.g., McBrearty and Brooks 2000; O'Connor and Reid 2003; Marean 2005; Stahl 2005) have disapproved of this artificial separation, arguing that it "obscures the uniqueness of the MSA" (McBrearty and Brooks 2000, p. 487). Last, but certainly not least, the trans-Saharan historical divide, which has been created by Western scholars, has been reasonably defined as a racialist conception (MacEachern 2007). Such a perspective is not simply a terminological issue, but has biased the scientific debate of Western and non-Western archaeologists who assign greater or lesser relevance to the cultural relations between North Africa, the Sahara, and sub-Saharan Africa.

For all the above reasons, I maintain the use of the African nomenclature to describe African cultural complexes.

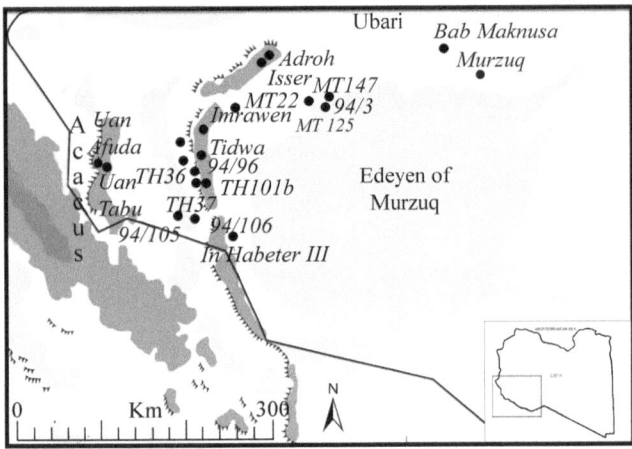

**Fig. 9.4** Map of Aterian sites in southwestern Libya

## Altitudinal/Latitudinal Location of the Libyan Aterian Sites

Fifty-six Aterian sites could be located in the Libyan country. Twenty-five of them were found in the Jebel Gharbi, in northwestern Libya (Fig. 9.3) (Barich et al. 1996, 2003a, b, 2006; Barich and Giraudi 2005; Giraudi 2005; Garcea and Giraudi 2006; Barich and Garcea 2008).

Nineteen were identified in the Tadrart Acacus and in the Messak Settafet, in the central Sahara (Fig. 9.4) (Garcea 1997; 2001a, b; Cremaschi and di Lernia 1998; di Lernia 1999b; Van Peer 2001).

The Jebel Gharbi lies between the Sahara to the south and the Jefara plain to the north and is 60 km south of the Mediterranean Coast at its closest distance. Surveys have tested the archaeological potential of the geomorphological features, which were first identified on satellite images and then located in the field (Fig. 9.5) (Barich et al. 2006). Assessments of the archaeological potential refer to the availability of environmental resources, although they can also depend on archaeological visibility. Valley heads can have high archaeological potentials, and confirmed high concentrations of sites, including Ras el Wadi, which is the valley head of the Wadi Ain Zargha. The next favorable landscape unit is the alluvial fan belt at the foot of the jebel, where the areas of Shakshuk and Wadi Basina are located (Table 9.1). No archaeological sites could be detected either in the areas between the plateau and the alluvial fan belt, or in the Jefara plain.

Site altitudes are uniform in the two main landscape units: sites are located at heights just over 200 m above sea level in the alluvial fan belt, and at around 600 m on the plateau (Fig. 9.6).

To the south, central Saharan sites from the Tadrart Acacus, Messak Settafet, Edeyen of Murzuq, and Erg Uan Kasa are at much higher elevations, with all sites in the Tadrart Acacus over 900 m above sea level (Fig. 9.7).

If all remaining sites are considered, it can be observed that site altitudes are strictly related to their latitudinal position, gradually decreasing from south to north; that is, sites in the lower latitudes, such as the Tadrart Acacus, are at higher elevations, whereas those at higher latitudes, such as the sites in Cyrenaica, including Haua Fteah and Ras 'Amar, are as low as 60 and 20 m above sea level, respectively (Fig. 9.8).

Moreover, site altitudes indicate that Aterian foragers followed an altitudinal gradient in association with the latitudinal belts that offered the most advantageous environmental conditions. In the central Sahara, drier conditions pushed human groups on higher elevations to search for more humid environments, whereas at higher latitudes they could live at lower altitudes.

## Climate, Chronology, and Stratigraphic Sequences

The Jebel Gharbi is currently a semi-arid region, but has been affected by arid spells during various expansions of the Sahara. Geostratigraphic records show that the deposits with a generalized Early MSA industry are separated from the upper layers with Aterian artifacts by a thick accumulation of aeolian sands, which correspond to a northerly expansion of the Sahara dated between 70 and 58 ka. After that major arid phase, the following dry episode only occurred around 20 ka (Giraudi 2005). Therefore, according to the geoarchaeological sequence, the Early MSA is much older than the Aterian. OSL dating gave age estimates of sand samples as early as $146.8 \pm 11$ ka and it is likely that the Early MSA appeared before that date (Garcea et al., in preparation). The Aterian has been dated by several methods; AMS radiocarbon dates gave ages of 43–44 ka, and U/Th, which was employed to date calcretes below and on top of colluvial silts with Aterian artifacts, gave dates of $64 \pm 21$ ka and <60 ka, respectively (Table 9.2). A systematic program for OSL dating is presently in progress (carried out by Schwenninger of the University of Oxford), in order to confirm the other dating methods. In fact, both AMS and U/Th can be problematic considering that in the first case, it is close to the technical limits of radiocarbon dating, and in the second case, calcretes can contain old detrital calcite or can later recrystallize, producing an average of the ages of different events (Giraudi 2004).

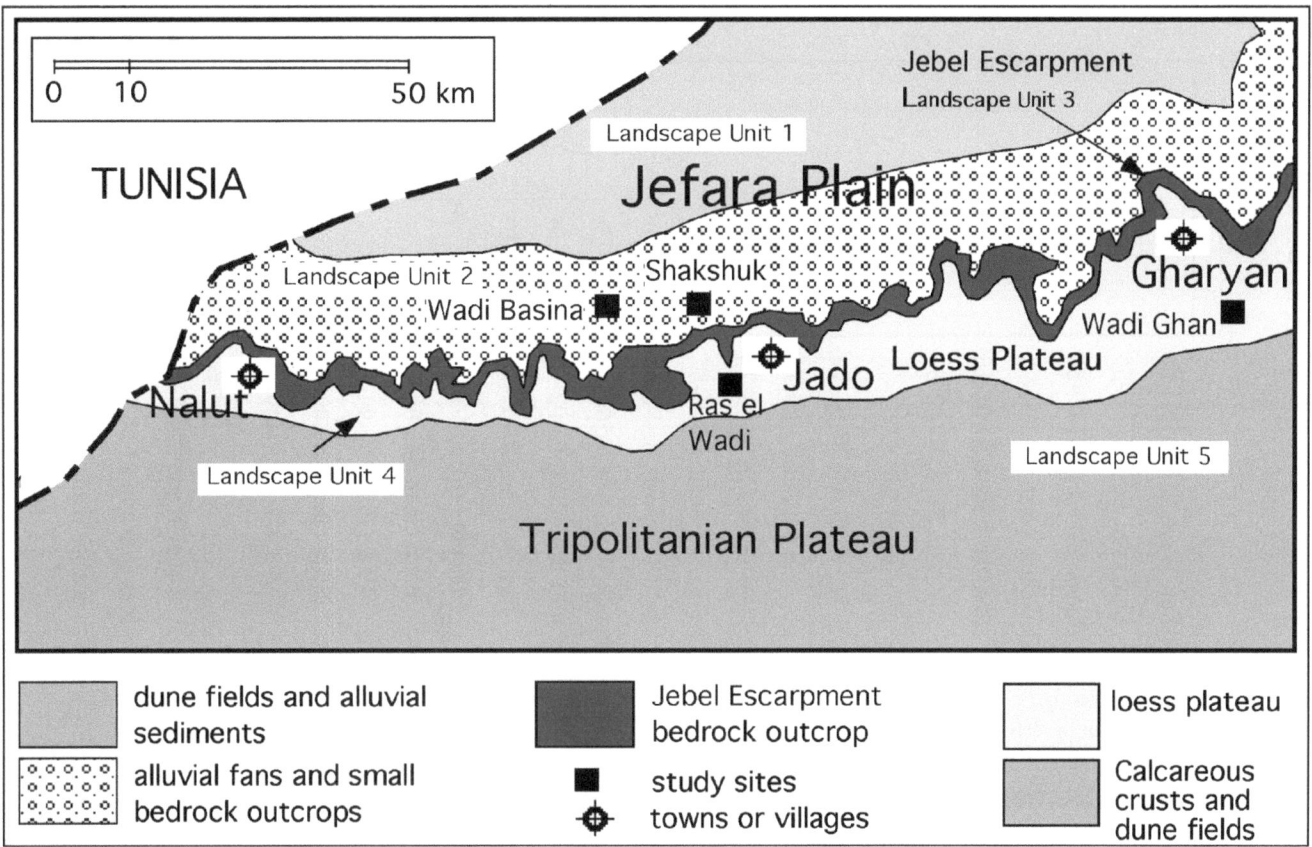

**Fig. 9.5** Landscape Units (LU) of the Jebel Gharbi (modified from Garcea and Giraudi 2006, Fig. 1)

**Table 9.1** Landscape units (LU) and related archaeological potential, according to Barich et al. 2006

| LU No. | Name | Archaeological potential |
|---|---|---|
| 1 | Jefara plain | Low |
| 2 | Alluvial fan belt | Very high |
| 3 | Escarpment | Low |
| 4 | Loess plateau | Low |
| 5 | Calcrete and sandsheet | Low |
| 6 | Valley heads | Low to high |

Further south, in the Tadrart Acacus, the Uan Tabu rock-shelter is the only stratified site with a consistent Aterian techno-complex in a sequence below an unconform erosional surface underlying an Early Holocene deposit (Garcea 2001a). This unit was formed by aeolian sand accumulated under desert conditions and was related to a large dune located outside the rockshelter (Cremaschi and Trombino 2001). The upper Aterian deposit was dated by the OSL method and provided the first absolute age of the Aterian in the Sahara of 61 ± 10 ka (Cremaschi et al. 1998; Martini et al. 1998). This date was confirmed by the TL and OSL dates from Uan Afuda, another site in the Tadrart Acacus. A further OSL date of 90 ± 10 ka was obtained from the sands below the archaeological deposit (Table 9.3) (Cremaschi et al. 1998; Martini et al. 1998; di Lernia 1999a).

These dates confirm precursory hints that the age of the Aterian could go beyond or very close to the limits of conventional radiocarbon dating, as suggested by the earliest finite date at Haua Fteah, which gave a radiocarbon age of 47 ± 3.2 ka (GrN-2023) (McBurney 1967).

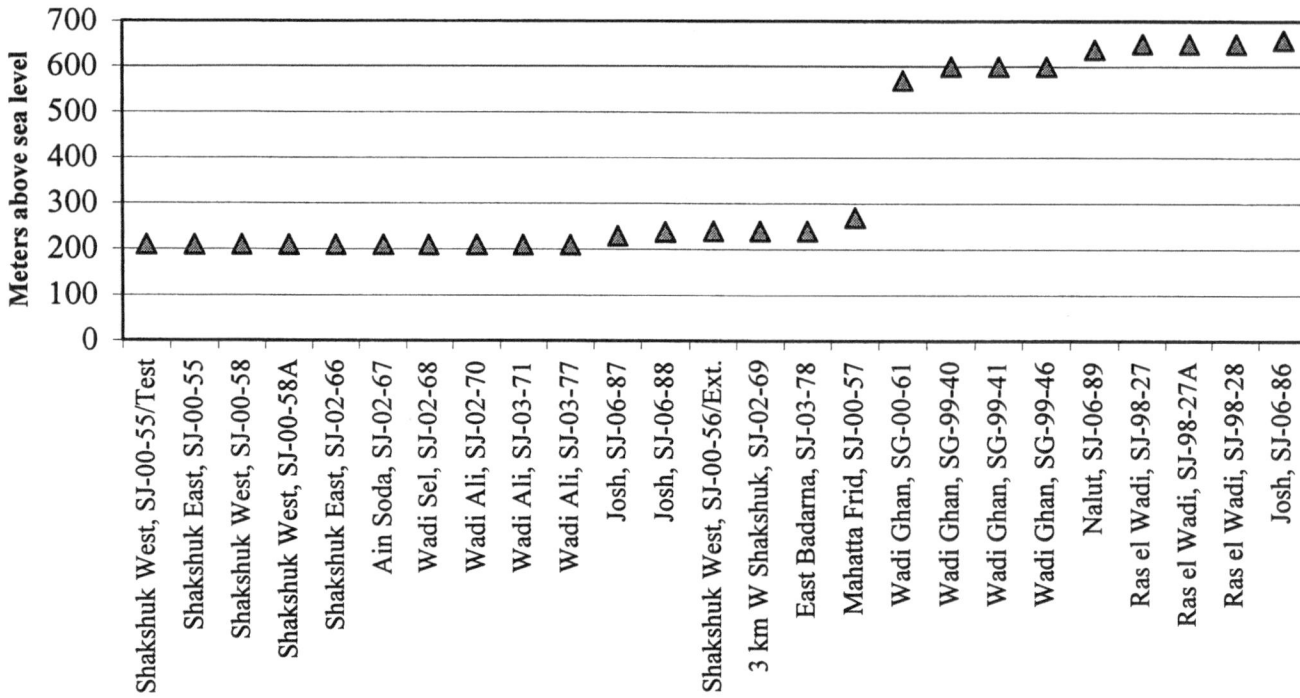

**Fig. 9.6** Site altitudes in the Jebel Gharbi

## The Desert Adaptation of Aterian Foragers

The geographic location of Aterian sites suggests that Aterian foragers developed a strategy to settle by the few available water sources as a form of adaptation to arid environments. Furthermore, the Libyan record provides evidence on the various strategies of adaptation to dry and drying environments by Aterian groups across a north–south section extending from the Mediterranean Coast to the central Sahara.

In the Jebel Gharbi, human populations concentrated near springs formed by outlets of underground aquifers that flew through fissures created by tectonic faults, which were produced by high magnitude earthquakes starting from the Late Pleistocene (Garcea and Giraudi 2006). On the mountain range at Ras el Wadi, a still active aquifer has regularly fed the outlet of the spring, which has provided constant water. In the lowlands, a series of springs between Wadi Sel and Aïn Soda, in the Shakshuk area, indicates that numerous intersections of the fault system guaranteed the outlet of underground water even during arid periods. Here, an excavation at Site SJ-02-68 brought to light a fault in an archaeological sequence including: Layer 1 with charcoal and Iberomaurusian (Upper LSA) artifacts, Layer 2 with a few LSA artifacts, Layer 3 with charcoal, ashes, and a Levallois flake, and Layer 4 with several Aterian artifacts. Layer 3 was radiocarbon dated to 44.6 ± 2.43 ka (Fig. 9.9). The artifacts in both Layers 3 and 4 can be attributed to the

Aterian. They correspond to two phases of occupation of the site and suggest that the date of Layer 3 is a minimal age for the Aterian occupation, which preceded the faulting and tectonic activities.

The presence of underground springs in the Jebel Gharbi should be seen as the main reason for the survival of Aterian groups until around 40 ka, a time when no evidence exists south in the Sahara, where the age of 61 ± 10 ka from Uan Tabu represents the latest date for this cultural unit in the Tadrart Acacus, and no other occupation followed here until the Early Holocene. In fact, although Saharan sites were located along wadis, their subsistence depended on the very ephemeral availability of seasonal water courses.

## Technology and Function of Tanged Tools

Tangs are normally the basal modification of hafted projectile points of thrusting spears (cf. Farmer 1994). Although there are also other specific technological features, tanged tools have often been regarded as the typical cultural marker of the Aterian Industrial Complex (Balout 1955; Tixier 1958–1959, 1967; Bordes 1961). The production of tangs was thought to be an effective innovation that Aterian groups devised for hafting projectile or spear points. Measurements of the tip cross-sectional area (TCSA) of a sample of Aterian points showed than none of them was a projectile point, but they were the points of

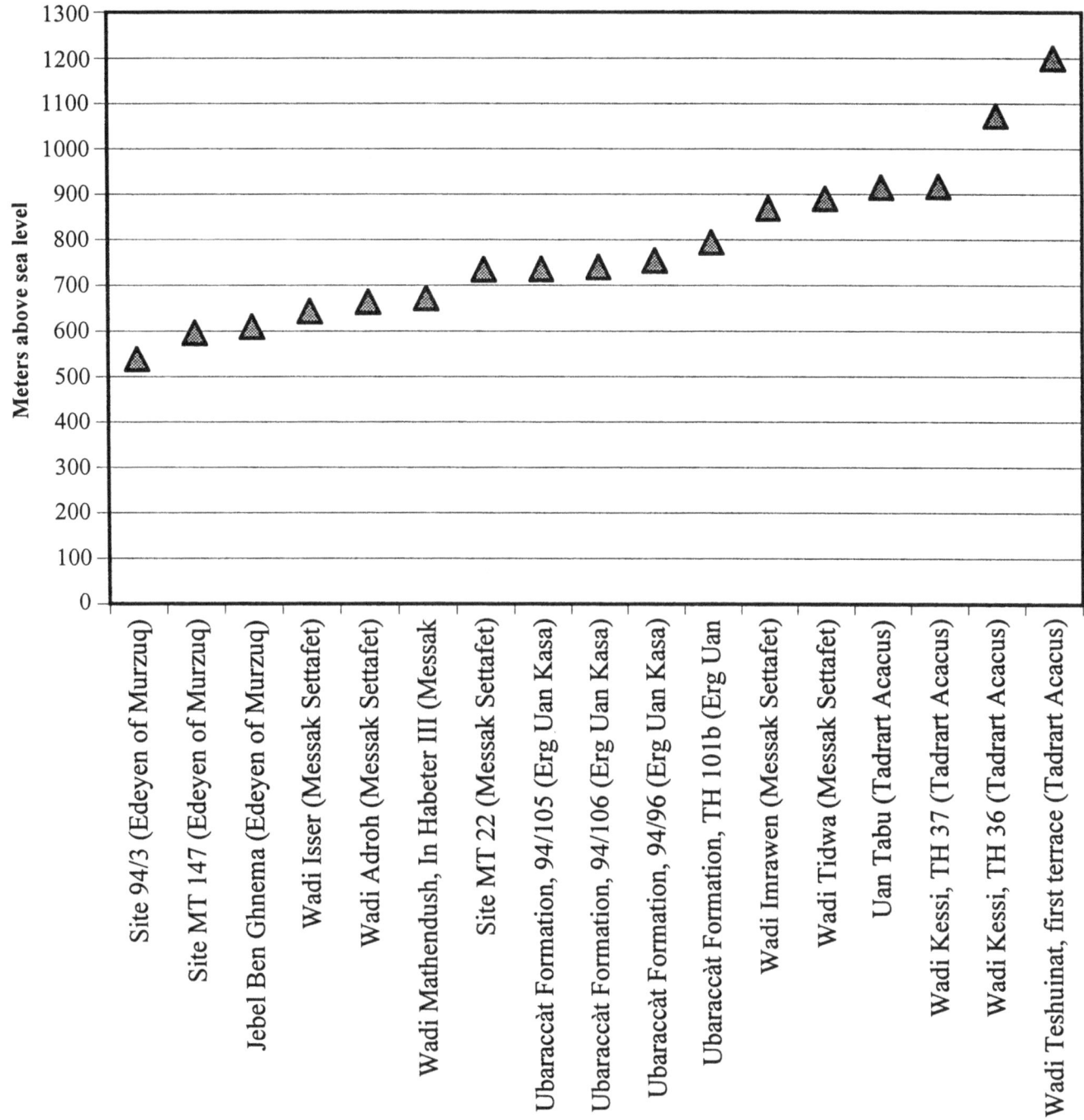

**Fig. 9.7** Site altitudes in southwestern Libya

hand-cast or thrusting spears (Shea 2006). Technological and functional analyses have suggested other explanations for the use of tangs, in addition to hafting. First of all, not all Aterian tanged tools are points; many of them have blunt edges, and some are simple unretouched flakes, the only finishing being on the tang. Second, Aterian tangs are not associated with specific classes of tools, but appear on many of them, such as side scrapers, end scrapers, denticulates, etc. Third, tangs could have been the working parts of hand-held tools (Clark et al. 2008). Fourth, tanged tools could have also been used as digging sticks or for processing tubers (O'Connell et al. 1999; Gifford-Gonzalez et al. 2004, in preparation). Fifth, tangs are often the thickest part of the tool, whereas the thickness at the base of projectile points is usually smaller than at the midsection in order to set the points in the shaft (Knecht 1997).

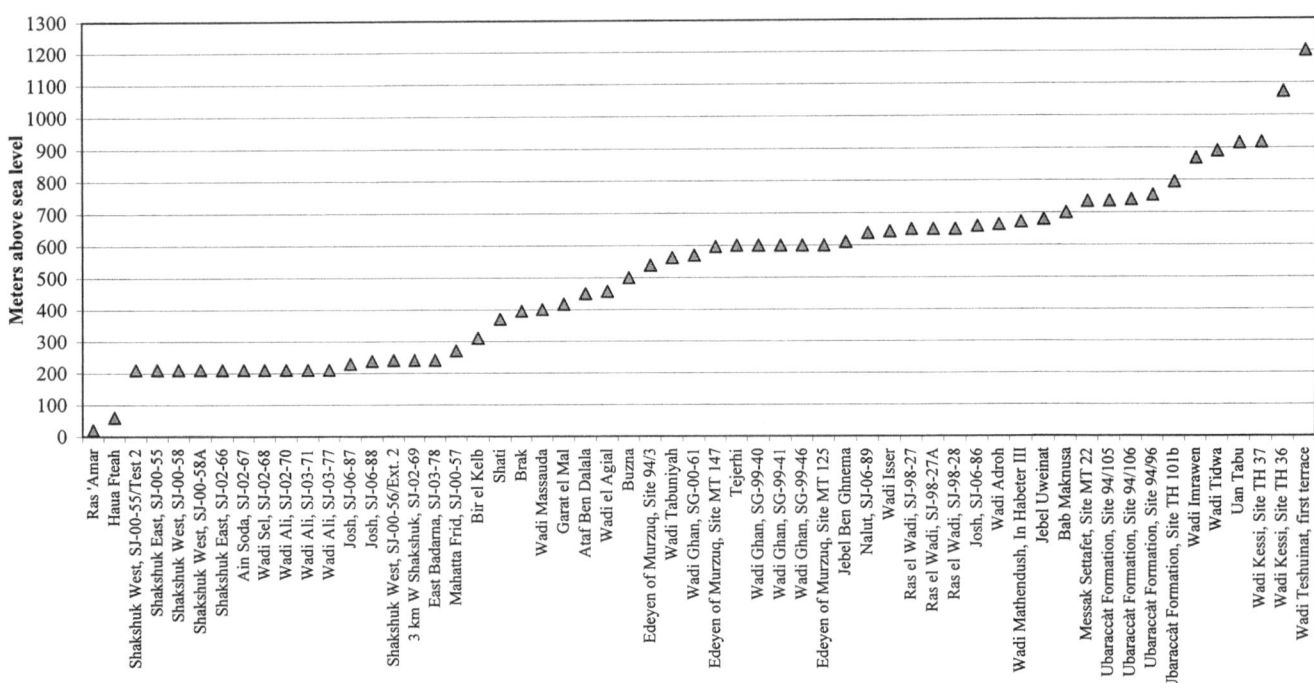

**Fig. 9.8** Site altitudes of all Aterian sites in Libya

**Table 9.2** Absolute dates of Early MSA and Aterian sites in the Jebel Gharbi

| Cultural unit | Locality | Site | Years BP | Method | Dated material | Sample number |
|---|---|---|---|---|---|---|
| Aterian | Shakshuk | Geological profile | 43,530 ± 2,110 | AMS | Charred material | Beta-167098 |
| | Wadi Sel | SJ-02-68 | 44,600 ± 2,430 | AMS | Organic sediment | Beta-167097 |
| | Ain Zargha | 27G N-trench | <60,000 | U/Th | Calcrete | |
| | Ain Zargha | 27E N-trench | 64,000 ± 21,000 | U/Th | Calcrete | |
| Early MSA | Shakshuk | C2000-10 | 49,200 ± 3,500 | OSL | Sand | X1511 |
| | Ain Zargha | N-trench G83 | 114,700 ± 7,400 | OSL | Sand | X1512 |
| | Ain Zargha | 98-27B-G78 | 146,800 ± 11,000 | OSL | Sand | X1514 |

**Table 9.3** Absolute dates of sites in the Tadrart Acacus

| Cultural unit | Site | Years BP | Method | Dated material |
|---|---|---|---|---|
| Aterian | Uan Tabu | 61,000 ± 10,000 | OSL | Sand |
| | Uan Afuda | 69,000 ± 7,000 | OSL | Sand |
| | Uan Afuda | 70,500 ± 9,500 | TL | Sand |
| | Uan Afuda | 73,000 ± 10,000 | TL | Sand |
| Early MSA | Uan Afuda | 90,000 ± 10,000 | OSL | Sand |

Preliminary techno-morphological, experimental, and functional analyses have been carried out on samples from five sites in the Jebel Gharbi (SJ-00-56, SJ-00-57, SJ-00-58, SJ-00-60, SG-00-61), including a total of 477 pieces and 11 tanged tools (Massussi and Lemorini 2004–2005; for other information on the technical features of these lithic assemblages, see Garcea 2006a). The blanks of the studied tanged tools are mostly flakes with a modified proximal or distal end (Fig. 9.10). Two different techniques for producing tangs have been determined and experimentally tested for the two major raw materials, flint and quartzite. For flint, the notches that form the tang are made by pressure flaking with either a small hammerstone or a soft hammer, such as antler. In contrast, quartzite tangs are

**Fig. 9.9** Cross-section with fault at Site SJ-02-68, at Wadi Sel, Jebel Gharbi

produced with direct hard percussion of two bilateral basal notches that are finished by perpendicular abrasion to dull the edges.

Experiments on the ballistic performance of hafted projectile points have shown that a considerable part (up to one-third) of the tool is usually inserted in the shaft in order to be efficient and securely attached (Massussi and Lemorini 2004–2005). With regards to the tanged tools in the analyzed sample, the length of all of the stems is very short, seemingly too small to form an effective hafting device for projectile points.

Moreover, use-wear analysis on the same sample showed remains of macro-traces of edge-damage on some of the tools. Traces of scraping were found on several notches that formed some "tangs." They could be assigned to scraping activities, and not to hafting, as they showed distinct striations located on a larger portion of the tool, which can be distinguished from those, usually on a limited part of the tool, made by friction or abrasion of the proximal part of a stone tool in a haft (Rots 2003; Lombard 2005). Therefore, use-wear corroborated the morpho-functional evidence that these "tangs" were not hafted, but were the working edges of the tools. Considering manufacturing techniques, morphology, use-wear, and edge-damage of the analyzed tanged pieces, Massussi and Lemorini (2004–2005) suggested that the double notches of these Aterian "tangs" are the active functional areas of the tools. Although no use-wear traces were observed on the tangs proper, macro- and micro-wear analysis of a sample of tanged pieces from Rhafas Cave and Pigeons Cave at Taforalt, Morocco, indicated that only very few of them were used as projectiles. They mostly showed distal margins with traces of longitudinal movements and transversal movements on dry skin, hard animal materials, and soft materials (Bouzouggar et al. 2004–2005).

These preliminary results confirm that technological and functional analyses, which had never been made on Aterian Industrial Complexes, are very helpful and instructive (among others, Geneste and Plisson 1990; Plisson and Schmider 1990; Lemorini 2000). They can reveal that our knowledge is often derived from past preconceptions and suggest that other interpretations should be envisaged. This does not imply that none of the Aterian tanged points were used for hafting. Aterian people undoubtedly made some of their tanged tools to make hafted projectile points, but these first functional analyses show that they were also aware that a stem on a tool could be an effective device for performing other activities. Several of these activities, which are attested to in other past and contemporary contexts, need to be further identified in the Aterian assemblages.

## Dispersal of Modern Humans Across Libya

Human fossil remains from Libya are too scant to provide direct evidence on the dispersal of modern humans across the country. Only Haua Fteah yielded two mandibular fragments (McBurney et al. 1953; Dart 1954; Tobias 1967), which could be compared to the early modern humans of Jebel Irhoud in Morocco (Tobias 1967; Hublin 2002) and those of Skhul/Qafzeh in Israel (Howell 1999). Both

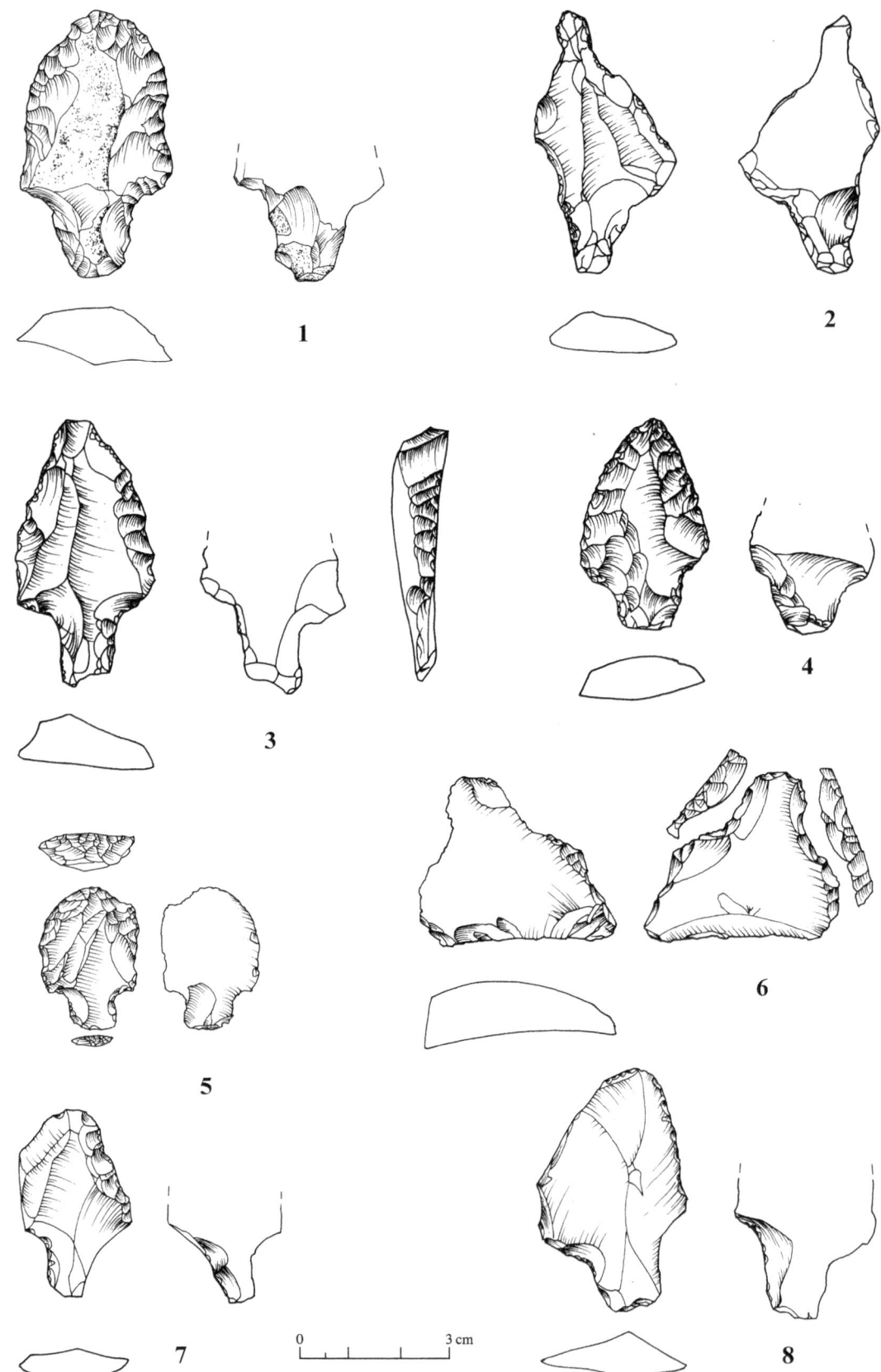

**Fig. 9.10** Variability of "tanged" tools from different sites in the Jebel Gharbi. *1–3* SJ-98-28; *4, 7–8* SJ-98-27A; *5* SJ-99-41; *6* SJ-00-58A

specimens from Haua Fteah came from Layer XXXIII and were attributed to two separate individuals. As for the archaeological artifacts from this layer, McBurney (1967) considered them to be more archaic than those in the underlying layers (XXXIV and XXXV), which he assigned to a developed stage of the "Middle Paleolithic." Current research will hopefully confirm or revise such opinion (cf. Barker et al. 2008). For the time being, no dates are available for these layers, but they may be assigned to at least the Last Interglacial (125–75 ka) or, as the morphology of the human remains suggests, MIS 6 (186–125 ka) or even earlier (Hublin 2000).

The discussion on the dispersal of modern humans across Libya is then left to other sets of evidence. Given the revised chronology of the Aterian, the makers of this industry are the most likely candidates for the last and final out of Africa movement. But from where they came and through where they went is still open to discussion. Passageways between Africa and Eurasia are very limited. The Straight of Gibraltar has been considered a possible corridor due to the narrow distance between the Maghreb and the Iberian Peninsula. Caton-Thompson (1946) was among the first to suggest that a "current of Aterianism" affected the Gravettian and produced the Solutrean in southwestern Europe, supposing similarities between Solutrean foliate points and Aterian tanged points. However, she did recognize higher technical skills in the Solutrean and an incongruity between the blade industries in southwestern Europe and the flake industries in North Africa. Affinities between the Aterian and the Solutrean were also assumed on chronological grounds. However, as the original chronology of a 20–40 ka Aterian is no longer acceptable and the age of 40 ka indicates the end, not the beginning, of the Aterian, this industrial complex is much earlier and cannot be compared with the Solutrean. Furthermore, the natural conditions of the Strait of Gibraltar, with strong currents, must not have encouraged Aterian desert-adapted people to embark on seafaring adventures, as Erlandson (2001) has clearly stated. Therefore, I (Garcea 2004, 2010a) fully concur with those (among others, de Sonneville-Bordes 1966; Hublin 1993; Souville 1998; Kleindienst 2000; Straus 2001; Derricourt 2005) who rule out the Strait of Gibraltar as a potential passageway for Aterian people.

Northeastern Africa and the Levant seem to be a most realistic corridor for the out-of-Africa dispersal of anatomically modern humans. As Aterian hunter-gatherers adapted to live in dry lands, it is not surprising that they did not settle along the Nile Valley, although they had to cross the Nile Delta. In the Levant, they would have found the arid environment they were familiar with in North Africa (Bar-Yosef 2000; Derricourt 2005). As Aterians did not habitually occupy the Nile Valley, the river must not have been a plausible corridor for them, whereas it is likely that they moved into the Levant from the Sahara and/or from the Mediterranean Coast.

The other unanswered question is where Aterian peoples and their technology came from. Increasing evidence indicates that East Africa is a likely region to consider for the origin of the Aterian. In fact, the Aterian bifacial technology shows some affinities with the Lupemban of East and central Africa (Clark 1993; Kleindienst 1998; Gifford-Gonzalez et al. 2004, in preparation; Clark et al. 2008). Furthermore, Aterian bifaces have been linked to the Nubian Middle Paleolithic II, which derived from the Nubian Middle Paleolithic I and the Sangoan (Arkell 1964; Guichard and Guichard 1965; Van Peer 1986, 2001; Van Peer et al. 2003). The groups that split from their East African stock and moved north and northwest later developed their adaptation skills to live in arid lands.

## The Aftermath of the Aterian

Towards the end of MIS 4, when the climate became too harsh to sustain human life in the Sahara, Aterian groups moved out, probably towards the north, whereas northern Libya continued to offer some favorable areas throughout MIS 3 (58–39 ka). In Cyrenaica, the Lower LSA is locally known as Dabban, after the type-site of Hagfet ed Dabba, in the Jebel Akhdar (McBurney and Hey 1955). Layers (from XXV to XVI) with Dabban artifacts were also found at Haua Fteah, on top of the "Middle Paleolithic" and below the "Eastern Oranian" (Iberomaurusian/Upper LSA) units (McBurney 1967). McBurney defined the Dabban as a blade industry and divided it into two phases, Early and Late Dabban, the former dating between about 40 and 30 ka and the latter between about 30 and 20 ka (Table 9.4) (McBurney and Hey 1955; McBurney, 1967). He noted a typological contrast with the underlying "Middle Paleolithic" industries and related the Dabban to the Emiran of the Levant, suggesting a cultural replacement by groups skilled in the laminar technology who supposedly came from the East.

Marks (1975) also observed that in both the Levant and Cyrenaica, blade technology followed the Middle Paleolithic within a short time. However, he pointed out specific technological differences, namely the persistence of Levallois-derived traditions, the prevalence of backed blades and opposed platform cores in Cyrenaica, and the difference of Dabban end scrapers from the finely made pieces of the Levantine Aurignacian. He remarked that the most significant affinities between the Early Dabban and the Levantine Upper Paleolithic were only burins and chamfered blades which, however, occur in coastal Lebanon, but

**Table 9.4** Absolute dates of Lower LSA sites in Libya

| Cultural unit | Locality | Site | Years BP (uncal.) | Method | Dated material | Sample number |
|---|---|---|---|---|---|---|
| Lower LSA/ | Shakshuk East | Aïn Soda area | 24,620 ± 400 | AMS | Charcoal | Beta-167094 |
| Late Dabban | Shakshuk West | SJ-00-55 West (Test 2, −106 cm) | 24,740 ± 140 | AMS | Charred material | Beta-157687 |
| | Shakshuk | SJ-00-56 Extension 2 | 25,410 ± 150 | AMS | Organic sediment | Beta-185497 |
| | Shakshuk West | SJ-00-55 West (Test 2, −17 cm) | 25,500 ± 400 | AMS | Charred material | Beta-167099 |
| | Wadi Basina | Lacustrine series | 26,330 ± 80 | AMS | Organic sediment | Beta-154555 |
| | Jado | SJ-98-12 | 27,310 ± 320 | Standard $^{14}$C | Carbonate sediment | Beta-154576 |
| | Shakshuk | SJ-00-56 Extension 2 (bottom) | 27,800 ± 430 | Standard $^{14}$C | Charcoal | GdA-196 (KIA-17720) |
| | Jebel Akhdar | Haua Fteah | 28,500 ± 80 | Standard $^{14}$C | Charcoal | W-86 |
| | Ain Zargha | 27B1 | 30,000 ± 9,000 | U/Th | Calcrete | |
| | Shakshuk West | SJ-00-55 West (Test 2, −64 cm) | 30,870 ± 200 | AMS | Organic sediment | Beta-157688 |
| Lower LSA/ | Jebel Akhdar | Haua Fteah | 33,100 ± 400 | Standard $^{14}$C | Charcoal | GrN-2550 |
| Early Dabban | Jebel Akhdar | Hagfet ed Dabba | 40,500 ± 1,600 | Standard $^{14}$C | Charcoal | |

not in Israel, which is geographically closer to Africa. In conclusion, with regards to North Africa, as the Dabban stratigraphically and chronologically follows the "Middle Paleolithic" at Haua Fteah, direct evolution, rather than replacement, seems to be the most plausible interpretation (Garcea 2006b). On the other hand, connections between North Africa and the Levant's Emiran have been reconsidered, suggesting that the Emiran is an indigenous innovation made by *sapiens* populations who spread to the Levant from Africa (Shea 2006).

As the Dabban is an industry peculiar to Cyrenaica, I have suggested to use the general term "Lower LSA" for the other techno-complexes succeeding the Aterian and preceding the Iberomaurusian that occur in other parts of Libya (Garcea 2006b). In the Jebel Gharbi, some Lower LSA artifacts are scattered at different sites (Barich et al. 1996, 2003a, b; Garcea 2004, 2006b, 2009, 2010b; Barich and Giraudi 2005; Giraudi 2005; Garcea and Giraudi 2006). At Ras el Wadi, near the same permanent spring used in the Aterian, a paleosol containing Lower LSA artifacts laid between two discontinuous layers of calcrete. The lower calcrete was dated to 27.31 ± 0.320 ka by conventional radiocarbon and 30 ± 9 ka by the U/Th method, and the upper calcrete was radiocarbon dated to 18.02 ± 0.190 ka (21,610 to 20,090 cal. BP) (Giraudi, 2004). Furthermore, at Ain Shakshuk, a test excavation at Site SJ-00-55 West including Lower LSA artifacts gave AMS dates between 30.87 ± 0.20 ka and 24.74 ± 0.140 ka. These dates indicate that the Lower LSA occupation took place between about 30 and 18 ka (Table 9.4), suggesting a hiatus between the latest Aterian and the Lower LSA techno-complexes in the Jebel Gharbi, which are contemporary with the Late Dabban in Cyrenaica. Therefore, the Jebel Akhdar appears

to be the only region that was continually inhabited, whereas the Jebel Gharbi seems to have remained uninhabited for several millennia, between about 40 and 30 ka.

## Conclusions

Aterian sites are widespread throughout the Sahara, and occasionally occur in the Atlantic and Mediterranean Coasts, but never in the Nile Valley, suggesting that the makers of the Aterian Industrial Complex had developed successful forms of adaptation to dry environments. During the Late Pleistocene, climatic conditions shifted between arid and semi-arid in the periphery of the Sahara, but remained basically dry in its central belt, where the mountain ranges are located. Research in north- and southwestern Libya has provided geoarchaeological stratigraphic sequences, radiometric dates, and technological data from two study areas, one on the periphery of the desert, and another in the central Sahara, which show different adaptational strategies, chronology, and cultural developments. In northwestern Libya, only two major dry episodes occurred, one between 70 and 58 ka (MIS 4), and the other around 20 ka (Last Glacial Maximum), with a relatively mild climate during MIS 3. Furthermore, a network of underground aquifers was identified in the Jebel Gharbi, providing constant water access to Aterian groups, who were mostly settled near permanent springs, and assuring water and food resources under any climatic condition. Consequently, in the Jebel Gharbi, Aterian foragers were able to live through the first part of MIS 3, until about 40 ka. Conversely, in the South, a very dry desert developed during MIS 4, as indicated by the deposit at Uan Tabu, and no human occupation occurred after the last Aterian occupation at ca. 60 ka, until the Early Holocene.

Aterian groups were able to live at any altitude, spanning from almost 1,000 m above sea level to the coastal level of the Mediterranean Sea. The geographic distribution of their sites shows that they had a well-organized settlement system structured on an inversely proportional relation between altitude and latitude that recurred from the central Sahara to the Mediterranean Coast. That is, at lower latitudes, such as on the Tadrart Acacus mountain, they were settled at higher elevations, whereas at higher latitudes, such as near the coast, they occupied sites at lower elevations. Such a strict altitudinal/latitudinal ratio should be related to the local environmental conditions, which were drier in the desert to the south, but moister in the mountain range habitats.

Technological, experimental, and functional analyses on a preliminary sample of tanged tools from the Jebel Gharbi have suggested other uses, in addition to hafting. In fact, the stems of the studied sample were found to be the active part of the tools, showing scraping use-wear and edge damage. These first results do not mean to rule out that some Aterian tanged tools were actually hafted on projectile points, but suggest that some other tools could be hand-held and that other functions should also be taken into consideration.

Aterian peoples are the most likely candidates for the last out-of-Africa dispersal of anatomically modern humans into Eurasia. The Nile Valley did not play a major role in this movement, confirming that Aterian foragers were adapted to live in dry environments. Therefore, the corridor(s) leading out of the continent must have been across the Sahara and/or the Mediterranean Coast, or even the peridesertic regions between the Mediterranean and the Sahara, such as the Jebel Gharbi in Tripolitania and the Jebel Akhdar in Cyrenaica. Obviously there is no need to envisage a population exodus from Africa to southwestern Asia, but Aterian groups certainly became more mobile when they had to move out of the Sahara. It is probably during this time that their radius of expansion reached Greater Africa, which includes the Levant.

After the end of the latest Aterian occupation in northern Libya, human settlements became very dispersed and patchy. The Jebel Akhdar appears to be the only region that was continually occupied, whereas the Jebel Gharbi exhibits a chronological gap between around 40 and 30 ka, when groups with a Lower LSA industry resettled the area. The Lower LSA of North Africa, including the Dabban, shows very few similarities with the Upper Paleolithic of the Levant and it seems very unlikely that groups from the two regions were culturally similar to each other. The LSA of North Africa suggests a considerable population decrease and technological stagnation. This is the exact opposite of

what occurred in Eurasia during the Upper Paleolithic, which underwent an explosion of different and complex cultural units.

To sum up, dry regions can stimulate interesting responses in modern human behavior. As environmental conditions change, natural habitats develop, some species become extinct, and others move to less dry environments, implying a biological shift after the environmental one. The occupation of refuge areas takes place during these periods. It is at this time that humans can create new forms of adaptation as their survival is at stake and may require transformations in their diet, technology, and social relations. Adaptive strategies involve risk and stress management with a high degree of organizational and technological flexibility that is successfully performed by anatomically modern humans.

**Acknowledgments** I wish to sincerely thank Jean-Jacques Hublin and Shannon McPherron for inviting me to contribute to this volume. I am grateful to them for bringing North Africa into the latest debate of modern human origins. I would also like to express my appreciation to the Department of Human Evolution of the Max Planck Institute for Evolutionary Anthropology for supporting the organization of such a fruitful and productive meeting. Finally, I wish to thank the three anonymous reviewers for their comments and suggestions, which greatly contributed to improve this paper.My research in northwestern Libya is part of the Italian-Libyan Joint Mission in the Jebel Gharbi, co-directed by Barbara E. Barich of the University of Rome "La Sapienza" and myself. My research in southwestern Libya was part of the Joint Italo-Libyan Mission for Prehistoric Research in the Sahara, formerly directed by Fabrizio Mori and presently directed by Savino di Lernia, both from the University of Rome "La Sapienza," Italy.

# References

Alimen, M. H., Beucher, F., & Conrad, G. (1966). Chronologie du dernier cycle pluvial au Sahara Nord-occidental. *Comptes Rendus de l'Académie des Sciences de Paris, 263*, 5–8.

Arkell, A. J. (1964). *Wanyanga and an archaeological reconnaissance of the South-West Libyan Desert. The British Ennedi expedition, 1957*. London: Oxford University Press.

Armitage, S. J., Drake, N. A., Stokes, S., El-Hawat, A., Salem, M. J., White, K., et al. (2007). Multiple phases of North Africa humidity recorded in lacustrine sediments from the Fazzan Basin, Libyan Sahara. *Quaternary Geochronology, 2*, 181–186.

Balout, L. (1955). *Préhistoire de l'Afrique du Nord: essai de chronologie*. Paris: Arts et Métiers Graphiques.

Barich, B. E., & Garcea, E. A. A. (2008). Ecological patterns in the Upper Pleistocene and Holocene in the Jebel Gharbi, northern Libya: Chronology, climate and human occupation. *African Archaeological Review, 25*, 87–97.

Barich, B.E., & Giraudi, C. (2005). The late hunting societies of Jebel Gharbi, Libya—settlement and landscape. In B.E. Barich, T. Tillet & K.H. Striedter (Eds.), Hunters vs. pastoralists in the Sahara: Material culture and symbolic aspects (pp. 2–10). Oxford: Acts of the XIVth UISPP Congress, University of Liège, Belgium. BAR International Series 1338.

Barich, B. E., Conati Barbaro, C., & Giraudi, C. (1996). The archaeology of Jebel Gharbi (Northwest Libya) and the Libyan sequence. In L. Krzyzaniak, K. Kroeper, & M. Kobusiewicz (Eds.), *Interregional contacts in the later prehistory of northeastern Africa* (pp. 37–49). Poznan: Poznan Archaeological Museum.

Barich, B. E., Bodrato, G., Garcea, E. A. A., Conati Barbaro, C., & Giraudi, C. (2003a). Northern Libya in the final Pleistocene. The late hunting societies of Jebel Gharbi. *Quaderni di Archeologia della Libya, 18*, 259–265.

Barich, B.E., Garcea, E. A. A., Conati Barbaro, C., & Giraudi, C. (2003b). The Ras El Wadi sequence in the Jebel Gharbi and the Late Pleistocene cultures of northern Libya. In L. Krzyzaniak, K. Kroeper & M. Kobusiewicz (Eds.), *Cultural markers in the later prehistory of northeastern Africa and recent research* (pp. 11–20). Poznan: Poznan Archaeological Museum.

Barich, B. E., Garcea, E. A. A., & Giraudi, C. (2006). Between the Mediterranean and the Sahara: The geoarchaeological reconnaissance in the Jebel Gharbi, Libya. *Antiquity, 80*, 567–582.

Barker, G., Hunt, C., & Reynolds, T. (2007). The Haua Fteah, Cyrenaica (Northeast Libya): Renewed investigations of the cave and its landscape, 2007. *Libyan Studies, 38*, 1–22.

Barker, G., Basell, L., Brooks, I., Burn, L., Cartwright, C., Cole, F., et al. (2008). The Cyrenaican Prehistory Project 2008: The second season of investigations of the Haua Fteah cave and its landscape, and further results from the initial (2007) fieldwork. *Libyan Studies, 39*, 175–221.

Bar-Yosef, O. (2000). *The Middle and Early Upper Paleolithic in southwest Asia and neighboring regions. In O. Bar-Yosef & D. Pilbeam (Eds.), The geography of Neandertals and modern humans in Europe and the Greater Mediterranean (pp. 107–156).* Cambridge: Peabody Museum Bulletin 8.

Bordes, F. (1961). *Typologie du Paléolithique Ancien et Moyen.* Bordeaux: Publications de l'Institut de Préhistoire de l'Université de Bordeaux.

Bouzouggar, A., Barton, R. N. E., & De Araújo Igreja, M. (2004–2005). A brief overview of recent research into the Aterian and Upper Palaeolithic of northern and eastern Morocco. *Scienze dell'Antichità, 12*, 473–488.

Brooks, A. S., Yellen, J. E., Nevell, L., & Hartman, G. (2006). Projectile technologies of the African MSA: Implications for modern human origins. In E. Hoovers & S. Kuhn (Eds.), *Transitions before the transition: Evolution and stability in the Middle Paleolithic and Middle Stone Age* (pp. 233–256). Berlin: Springer.

Caton-Thompson, G. (1946). *The Aterian Industry: Its place and significance in the palaeolithic world. Huxley Memorial Lecture for 1946.* London: Royal Anthropological Institute of Great Britain and Ireland.

Clark, J. D. (1980). Human populations and cultural adaptations in the Sahara and Nile during prehistoric times. In M. A. J. Williams & H. Faure (Eds.), *The Sahara and the Nile* (pp. 527–582). Rotterdam: Balkema.

Clark, J. D. (1993). The Aterian of the Central Sahara. In L. Krzyzaniak, M. Kobusiewicz, & J. Alexander (Eds.), *Environmental change and human culture in the Nile Basin and northern Africa until the second millennium B.C* (pp. 49–67). Poznan: Poznan Archaeological Museum.

Clark, J. D., Schultz, U., Kroll, E. M., Freedman, E. E., Galloway, A., Batkin, J., et al. (2008). The Aterian of Adrar Bous and the Central Sahara. In J. J. Clark & D. Gifford-Gonzalez (Eds.), *Adrar Bous: Archaeology of a Central Saharan granitic ring complex in Niger* (pp. 91–162). Tervuren: Royal Museum for Central Africa.

Cremaschi, M. (1998). Late Quaternary geological evidence for environmental changes in south-western Fezzan (Libyan Sahara). In M. Cremaschi & S. di Lernia (Eds.), *Wadi Teshuinat. Palaeoenvironment and prehistory in south-western Fezzan (Libyan Sahara)* (pp. 13–48). Firenze: All'Insegna del Giglio.

Cremaschi, M., & di Lernia, S. (1998). The geoarchaeological survey in central Acacus and surroundings (Libyan Sahara). Environment and cultures. In M. Cremaschi & S. di Lernia (Eds.), *Wadi Teshuinat. Palaeoenvironment and prehistory in south-western Fezzan (Libyan Sahara)* (pp. 243–296). Firenze: All'Insegna del Giglio.

Cremaschi, M., & Trombino, L. (2001). The formation processes of the stratigraphic sequence of the site and their palaeoenvironmental implications. In E. A. A. Garcea (Ed.), *Uan Tabu in the settlement history of the Libyan Sahara* (pp. 15–23). Firenze: All'Insegna del Giglio.

Cremaschi, M., di Lernia, S., & Garcea, E. A. A. (1998). Some insights on the Aterian in the Libyan Sahara: Chronology, environment, and archaeology. *African Archaeological Review, 15*, 261–286.

Dart, R. A. (1954). The phylogenetic implications of African and Palestinian mandible profiles. *American Journal of Physical Anthropology, 12*, 487–502.

Debénath, A., Raynal, J.-P., Roche, J., Texier, J.-P., & Ferembach, D. (1986). Stratigraphie, habitat, typologie et devenir de l'Atérien marocain: données récentes. *L'Anthropologie, 90*, 233–246.

Derricourt, R. (2005). Getting "out of Africa": Sea crossings, land crossings and culture in the hominin migrations. *Journal of World Prehistory, 19*, 119–132.

di Lernia, S. (1999a). The cultural sequence. In S. di Lernia (Ed.), *The Uan Afuda Cave: Hunter-gatherer societies of Central Sahara* (pp. 57–130). Firenze: All'Insegna del Giglio.

di Lernia, S. (Ed.). (1999b). *The Uan Afuda Cave: Hunter-gatherer societies of Central Sahara.* Firenze: All'Insegna del Giglio.

Erlandson, J. M. (2001). The archaeology of aquatic adaptations: Paradigms for a new millennium. *Journal of Archaeological Research, 9*, 287–350.

Farmer, M. F. (1994). The origins of weapon systems. *Current Anthropology, 35*, 679–681.

Garcea, E. A. A. (1997). Prehistoric surveys in the Libyan Sahara. *Complutum, 8*, 33–38.

Garcea, E. A. A. (2001a). The Pleistocene and Holocene archaeological sequences. In E. A. A. Garcea (Ed.), *Uan Tabu in the settlement history of the Libyan Sahara* (pp. 1–14). Firenze: All'Insegna del Giglio.

Garcea, E. A. A. (2001b). A reconsideration of the Middle Palaeolithic/Middle Stone Age in Northern Africa after the evidence from the Libyan Sahara. In E. A. A. Garcea (Ed.), *Uan Tabu in the settlement history of the Libyan Sahara* (pp. 25–49). Firenze: All'Insegna del Giglio.

Garcea, E. A. A. (2004). Crossing deserts and avoiding seas: Considerations on the theory of Aterian North Africa-European relations. *Journal of Anthropological Research, 60*, 27–53.

Garcea, E. A. A. (2006a). Aterians in Libya. In Le Secrétariat du Congrès (Ed.), *Acts of the XIVth UISPP Congress, Section 15: African prehistory (pp. 41–48).* Oxford: BAR International Series 1522.

Garcea, E. A. A. (2006b). The "Upper Palaeolithic" seen from northern Libya. In J. L. Sanchindrián Torti, A. M. Márquez Alcántara & J. M. Fullola i Pericot (Eds.), IV Simposio de Prehistoria, Cueva de Nerja. La Cuenca Mediterránea durante el Paleolítico Superior (38.000-10.000 años). Reunión de la VIII Comisión U.I.S.P.P. del Paleolítico Superior (pp. 152–160). Nerja: Fundación Cueva de Nerja.

Garcea, E. A. A. (2009). The evolutions and revolutions of the Late Middle Stone Age and Lower Later Stone Age in north-west Africa. In M. Camps & C. Szmidt (Eds.), *The Mediterranean from 50 000 to 25 000 BP: Turning points and new directions* (pp. 51–66). Oxford: Oxbow Books.

Garcea, E. A. A. (2010a). The spread of Aterian peoples in North Africa. In E. A. A. Garcea (Ed.), *South-eastern Mediterranean peoples between 130,000 and 10,000 years ago* (pp. 37–53). Oxford: Oxbow.

Garcea, E. A. A. (2010b). The Lower and Upper Later Stone Age of North Africa. In E.A.A. Garcea (Ed.), *South-eastern Mediterranean peoples between 130,000 and 10,000 years ago* (pp. 54–65). Oxford: Oxbow.

Garcea, E. A. A., & Giraudi, C. (2006). Late Quaternary human settlement patterning in the Jebel Gharbi, Northwestern Libya. *Journal of Human Evolution, 51*, 411–421.

Garcea, E. A. A., Giraudi, C., & Schwenninger, J.-L. (in preparation). The Sahara expansion in the Jebel Gharbi (Northwest Libya) during MIS 6 and its influence on the Middle Stone Age peopling of North Africa.

Geneste, J.-M., & Plisson, H. (1990). Technologie fonctionnelle des pointes à cran solutréennes: l'apport des nouvelles données de la grotte de Combe-Saunière (Dordogne). In M. Otte (Ed.), *Fleurs de pierre. Les industries à pointes foliacées du Paléolithique supérieur européen* (pp. 293–320). Liège: ERAUL.

Gifford-Gonzalez, D., Garcea, E. A. A., & Clark, J. D. (2004). The Aterian seen from Adrar Bous, Niger: Southern affinities and further questions. *Nyame Akuma, 61*, 59.

Gifford-Gonzalez, D., Garcea, E. A. A., & Clark, J. D. (in preparation). Regionalism and parochialism in the Aterian: Plant resources as the ultimate limiting factor of a desert-adapted population.

Giraudi, C. (2004). The Upper Pleistocene to Holocene sediments on the Mediterranean island of Lampedusa (Italy). *Journal of Quaternary Science, 19*, 537–545.

Giraudi, C. (2005). Eolian sand in the peridesert northwestern Libya and implications for Late Pleistocene and Holocene Sahara expansion. *Palaeogeography, Palaeoclimatology, Palaeoecology, 218*, 161–173.

Guichard, J., & Guichard, G. (1965). The Early and Middle Paleolithic of Nubia. In F. Wendorf (Ed.), *Contributions to the prehistory of Nubia* (pp. 57–116). Dallas: Southern Methodist University Press.

Howell, F. C. (1999). Paleo-demes, species clades, and extinctions in the Pleistocene hominin record. *Journal of Anthropological Research, 55*, 191–243.

Hublin, J.-J. (1993). Recent human evolution in Northwest Africa. In M. J. Aitken, C. B. Stringer, & P. A. Mellars (Eds.), *The origin of modern humans and the impact of chronometric dating* (pp. 118–131). Princeton: Princeton University Press.

Hublin, J.-J. (2000). Modern-nonmodern hominid interactions: A Mediterranean perspective. In O. Bar-Yosef & D. Pilbeam (Eds.), The geography of Neandertals and modern humans in Europe and the Greater Mediterranean (pp. 157–182). Cambridge: Peabody Museum Bulletin 8.

Hublin, J.-J. (2002). Northwestern African Middle Pleistocene hominids and their bearing on the emergence of *Homo sapiens*. In L. Barham & K. Robson-Brown (Eds.), *Human roots: Africa and Asia in the Middle Pleistocene* (pp. 99–121). Bristol: Western Academic & Specialist Press.

Kleindienst, M. R. (1998). What is the Aterian? The view from Dakhleh Oasis and the Western Desert, Egypt. In M. Marlow & A. J. Mills (Eds.), *The Oasis Paper: Proceedings of the First International Symposium of the Dakhleh Oasis Project* (pp. 1–14). Oxford: Oxbow.

Kleindienst, M. R. (2000). On the Nile Corridor and the Out-of-Africa model. *Current Anthropology, 41*, 107–109.

Knecht, H. (1997). Projectile points of bone, antler, and stone: Experimental explorations of manufacture and use. In H. Knecht (Ed.), *Projectile technology* (pp. 191–212). New York: Plenum Press.

Lemorini, C. (2000). *Reconaître des tactiques d'éxploitation du milieu au Paléolithique moyen. La contribution de l'analyse fonctionelle à l'étude des industries lithiques de Grotta Breuil (Latium, Italie) et de La Combette (Bonnieux, Vaucluse, France)*. Oxford: BAR International Series 858.

Lombard, M. (2005). Evidence of hunting and hafting during the Middle Stone Age at Sibidu Cave, KwaZulu-Natal, South Africa: A multianalytical approach. *Journal of Human Evolution, 48*, 279–300.

MacEachern, S. (2007). Where in Africa does Africa start? Identity, genetics and African studies from the Sahara to Darfur. *Journal of Social Archaeology, 7*, 393–412.

Marean, C. W. (2005). From the tropics to the colder climates: Contrasting faunal exploitation adaptations of modern humans and Neanderthals. In F. d'Errico & L. Backwell (Eds.), *From tools to symbols: From early hominids to modern humans* (pp. 333–371). Johannesburg: Witwaterstrand University Press.

Marks, A. E. (1975). The current status of Upper Paleolithic studies from the Maghreb to the Northern Levant. In F. Wendorf & A. E. Marks (Eds.), *Problems in prehistory: North Africa and the Levant* (pp. 439–459). Dallas: Southern Methodist University Press.

Martini, M., Sibilia, E., Zelaschi, C., Troja, S. O., Forzese, R., Gueli, A. M., et al. (1998). TL and OSL dating of fossil dune sand in the Uan Afuda and Uan Tabu rockshelters, Tadrart Acacus (Libyan Sahara). In M. Cremaschi & S. di Lernia (Eds.), *Wadi Teshuinat. Palaeoenvironment and prehistory in south-western Fezzan (Libyan Sahara)* (pp. 67–72). Firenze: All'Insegna del Giglio.

Massussi, M., & Lemorini, C. (2004–2005). I siti Ateriani del Jebel Gharbi: Caratterizzazione delle catene di produzione e definizione tecno-funzionale dei peduncolati. *Scienze dell'Antichità, 12*, 19–28.

McBrearty, S., & Brooks, A. S. (2000). The revolution that wasn't: A new interpretation of the origin of modern human behavior. *Journal of Human Evolution, 39*, 453–563.

McBurney, C. B. M. (1967). *The Haua Fteah (Cyrenaica) and the Stone Age in the South-East Mediterranean*. Cambridge: Cambridge University Press.

McBurney, C. B. M., & Hey, R. W. (1955). *Prehistory and Pleistocene geology of Cyrenaican Libya*. Cambridge: Cambridge University Press.

McBurney, C. B. M., Trevor, J. C., & Wells, L. H. (1953). The Haua Fteah fossil jaw. *Journal of the Royal Anthropological Institute of Great Britain and Ireland, 83*, 71–85.

O'Connell, J. F., Hawkes, K., & Blurton Jones, N. (1999). Grandmothering and the evolution of *Homo erectus. Journal of Human Evolution, 36*, 461–485.

O'Connor, D., & Reid, A. (Eds.). (2003). *Ancient Egypt in Africa*. Philadelphia: University of Pennsylvania Press.

Plisson, H., & Schmider, B. (1990). Etude préliminaire d'une série de pointes de Châtelperron de la Grotte du Renne à Arcy-sur-Cure: Approche morphometrique, technologique et tracéologique. In C. Farizy (Ed.), *Paléolithique Moyen Récent et Paléolithique Ancien en Europe* (pp. 313–318). Nemours: Mémoires du Musée de Préhistoire d'Ile de France.

Rots, V. (2003). Towards an understanding of hafting: The macro- and microscopic evidence. *Antiquity, 77*, 805–815.

Shea, J. J. (2006). The origins of lithic projectile point technology: Evidence from Africa, the Levant, and Europe. *Journal of Archaeological Science, 33*, 823–846.

Smith, M., Veth, P., Hiscock, P., & Wallis, L. A. (2005). Global deserts in perspective. In P. Veth, M. Smith, & P. Hiscock (Eds.), *Desert peoples: Archaeological perspectives* (pp. 1–13). Oxford: Blackwell.

de Sonneville-Bordes, S. (1966). L'évolution du Paléolithique supérieur en Europe Occidentale et sa signification. *Bulletin de la Société Préhistorique Française, 63*, 3–34.

Souville, G. (1998). Contacts et échanges entre la péninsule ibérique et le nord-ouest de l'Afrique durant les temps préhistoriques et protohistoriques. *Comptes Rendus de l'Académie des Inscriptions & Belles Lettres, Janvier-Mars*, 163–177.

Stahl, A. B. (2005). Introduction: Changing perspectives on Africa's pasts. In A. B. Stahl (Ed.), *African archaeology: A critical introduction* (pp. 1–23). Oxford: Blackwell.

Straus, L. G. (2001). Africa and Iberia in the Pleistocene. *Quaternary International, 75*, 91–102.

Tillet, T. (1995). Recherches sur l'Atérien du Sahara méridional (bassins Tchadien et de Taoudenni): Position chrono-stratigraphique, définition et étude comparative. In R. Chenorkian (Ed.), *L'homme méditerranéen. Mélanges offerts à Gabriel Camps* (pp. 29–56). Aix-en-Provence: Publications de l'Université de Provence.

Tixier, J. (1958–1959). Les pièces pédonculées de l'Atérien. *Libyca, 6-7*, 127–158.

Tixier, J. (1967). Procédés d'analyse et questions de terminologie concernant l'étude des ensembles industriels du Paléolithique récent et de l'Epipaléolithique dans l'Afrique du Nord-Ouest. In N. W. Bishop & J. D. Clark (Eds.), *Background to evolution in Africa* (pp. 771–820). Chicago: University of Chicago Press.

Tobias, P. V. (1967). The hominid skeletal remains of Haua Fteah. In C. B. M. McBurney (Ed.), *The Haua Fteah (Cyrenaica) and the Stone Age of the South-East Mediterranean* (pp. 338–352). Cambridge: Cambridge University Press.

Van Peer, P. (1986). Présence de la technique Nubienne dans l'Atérien. *L'Anthropologie, 90*, 321–324.

Van Peer, P. (1998). The Nile Corridor and the Out-of-Africa model: An examination of the archeological records. *Current Anthropology, Supplement, 39*, S115–S140.

Van Peer, P. (2001). Observations on the Palaeolithic of the southwestern Fezzan and thoughts on the origin of the Aterian. In E. A. A. Garcea (Ed.), *Uan Tabu in the settlement history of the Libyan Sahara* (pp. 51–62). Firenze: All'Insegna del Giglio.

Van Peer, P., Fullagar, R., Stokes, S., Bailey, R. M., Moeyersons, J., Steenhoudt, F., et al. (2003). The Early to Middle Stone Age transition and the emergence of modern human behaviour at site 8-B-11, Sai Island, Sudan. *Journal of Human Evolution, 45*, 187–193.

Vandermeersch, B. (1989). The evolution of modern humans: Recent evidence from southwest Asia. In P. A. Mellars & C. B. Stringer (Eds.), *Behavioural and biological perspectives in the origin of modern humans* (pp. 155–164). Edinburgh: Edinburgh University Press.

Wilson, R. C. L., Drury, S. A., & Chapman, J. L. (2000). *The great Ice Age: Climate change and life*. London: Routledge.

Yellen, J. E. (1977). Long term hunter-gatherer adaptation to desert environments: A biogeographical perspective. *World Archaeology, 8*, 262–274.

# Chapter 10
# Middle Stone Age in Tunisia: Present Status of Knowledge and Recent Advances

N. Aouadi-Abdeljaouad and L. Belhouchet

**Abstract** During the twentieth century, five famous archaeological sites attributed to the Middle Paleolithic period were discovered in Tunisia. All of them are open-air sites and have yielded mixed material from Mousterian and/or Aterian cultures. Recent field prospecting and surveys in the Meknassy Basin (central Tunisia) have revealed many prehistoric sites, some of which belong to the Middle Stone Age (MSA). The excavation of the Aïn El-Guettar Mousterian open-air site, which began in 2005, has yielded a faunal assemblage dominated by bovids and equids. The stratigraphic sequence contains charcoal-rich occupation layers with faunal and lithic finds. One human tooth was found in situ. A level with an Aterian industry was found beneath the Mousterian layer. The excavations at the open-air site of Aïn Oum Henda 2 (Jebel Maloussi) provided many Middle Paleolithic lithic artifacts, including some with tangs characteristic of the Aterian. Unfortunately, there were no faunal remains. In this chapter, we propose a framework for the succession of Paleolithic cultures in Tunisia and for the relationships between humans, fauna, and paleoenvironmental conditions.

**Keywords** Aterian • Human remains • Lithic industry • Mammalian fauna • Meknassy • Mousterian • Tunisia

## Middle Paleolithic Sites in Tunisia

All Middle Paleolithic archaeological discoveries in Tunisia have been from open-air sites. They are in direct contact with the artesian springs. First, we will describe some famous Middle Paleolithic sites from Tunisia.

### Aïn Métherchem

This site is located 40 km north of Feriana. It was excavated by Vaufrey in 1933 and was published in 1955 by the same author (Vaufrey 1955). Two Mousterian levels were found in a sandy tuff, in a ground rich in faunal remains. Few data are available for this site. The fauna is composed of *Rhinoceros mercki, Equus mauritanicus, Bos primigenius, Bubalis boselaphus, Alcelaphus bubalis* (Vaufrey 1955).

Bordes (1976–1977) analyzed the lithic material from the site. He distinguished a "Proto-Aterian" from an Aïn Métherchem soil and Mousterian with Aterian tendencies from an Aïn Métherchem tuff.

### Sidi Zin

Located 11 km south of the city of Kef, this site is particularly known for its Acheulian levels. A tufa containing Mousterian tools covers the Acheulian horizons. It was excavated and published by Gobert in 1950. Artifacts from the Mousterian level include thick points, small bifaces, sidescrapers, and several abruptly retouched pieces of flint (Clark 1982).

N. Aouadi-Abdeljaouad (✉)
Institut National du Patrimoine, Musée National de Raqqada, 3100 Kairouan, Tunisia
e-mail: aouadi73@yahoo.fr

L. Belhouchet
Institut National du Patrimoine, 4 Place du Château, 1008 Tunis, Tunisia
e-mail: lotfi_belhouchet@yahoo.fr

J.-J. Hublin and S. P. McPherron (eds.), *Modern Origins: A North African Perspective*, Vertebrate Paleobiology and Paleoanthropology, DOI: 10.1007/978-94-007-2929-2_10, © Springer Science+Business Media B.V. 2012

## Wadi Akarit

This site is located 30 km north of Gabes (southern Tunisia), on the right bank of Wadi El Akarit. The site is at the bottom of a spring that is still active today. It was first excavated by Gobert and Howe in 1951 (Gobert and Howe 1955), then by Page in 1967 and 1968 (Page 1972). New excavation work by Roset (IRD, France) and Harbi-Riahi (INP, Tunisia) began in 1991. A human occupation structure was discovered. Thermoluminescence dating of burnt lithics that came from an area containing a small proportion of tanged artifacts attributed to a Proto-Aterian industry has resulted in an age for this site of at least 90 ka (Roset 1996, 2005; Roset and Harbi-Riahi 2007). Pollen analysis indicates a steppic vegetation with graminean, crassulescent, and Chamaephytes (Brun et al. 1988). Faunal remains include *Ceratotherium mauritanicum* Pomel, 1895, *Equus mauritanicus* Pomel, 1897, *Pelorovis antiquus* Duvernoy, 1851, *Bos primigenius* Bojanus, 1827, and *Hippotragus cf. equinus* Desmarest, 1804 (Guérin and Faure 2007). Fauna indicate a savannah biotope with small woody regions.

## El Guettar

This site is located approximately 15 km southeast of Gafsa. El Guettar constituted a prototype for Mousterian culture in the Maghreb. The discovery, excavation, and publication of the site were done by Gruet in 1954. This site is located at a spring deposit with a rich and varied fauna and is associated with a lithic industry with Levallois technology (Gruet 1950, 1954, 1955, 1958–1959). Faunal remains include *Rhinoceros mercki, Equus mauritanicus, Bos primigenius, Bubalis boselaphus, Gazella cuvieri, Ammotragus lervia, Hyaena hyaena, Camelus dromedarius* (Vaufrey 1955). Even though some tanged tools were discovered here, Gruet considered the site to be a Mousterian culture.

In contrast, Clark (1982) attributed El Guettar to the Aterian culture. A cairn of stone balls and stone tools (with one pedunculate tool), including bones and ochre, was also found. It was interpreted as having a spiritual significance. Pollen analysis indicates a predominance of grasses and compositae, in addition to cedar, deciduous oaks, and cypress (Leroi-Gourhan 1958). Faunal remains and pollen analysis indicate an arid environment.

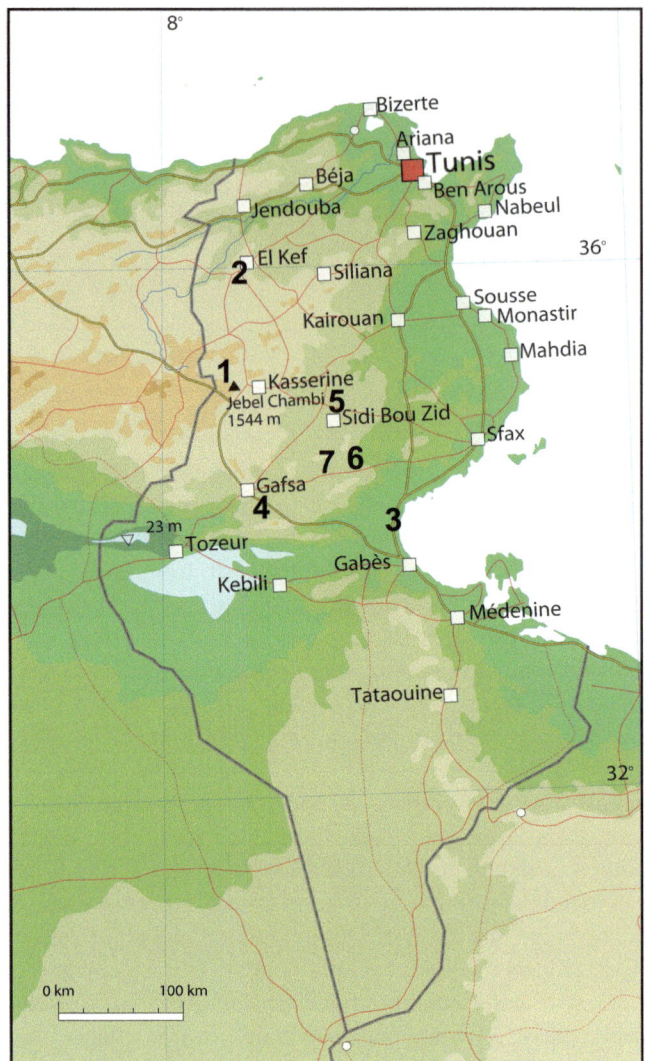

**Fig. 10.1** Map of the most important Mousterian sites in Tunisia

## Aïn Mghotta

This site is located southeast of Nasrallah city (governorship of Kairouan). It is south of an artesian spring that is located at the bottom of a slope with an east/west orientation, which opens into Jebel Cherahil. Harson discovered the site while collecting the Aïn Mghotta waters. The site was announced for the first time by Gobert and Harson in 1958. A survey was undertaken by Harson in the southern part of the terrace of the Capsian shell midden. Only Mousterian artifacts (traditional type) were found in a conglomerate level. This site was not excavated and has produced no reliable data (Fig. 10.1).

# Recent Discoveries: The Meknassy Basin

## Introduction

Located at the southern end of central Tunisia, the Meknassy region looks like an almost closed basin that is bounded by a series of uplands, including Jebel El Maloussi, Jebel Majoura, and Jebel El Goussa in the north and west, Jebel Bou Hedma and Jebel Bou Douaou in the south, and Jebel Mehiri-Zebbeus, Kef Abdallah, Jebel Dribika El Hamra, and Jebel Nadhour in the east. The only access to the basin is via the Strait of El Bkakrya in the north, Wadi Leben in the east, and the Snad region in the south. After systematically prospecting in this region, especially around the wadis and water points, we discovered several sites that we excavated at in order to evaluate their archaeological potential. Furthermore, we tried to understand paleoenvironmental and landscape change in relation to human settlement during the various prehistoric phases. Our overall objective is to understand the nature of the prehistoric settlement, and therefore migrations, and the related exchanges with other populations of the Gafsa region in the southwest, the alluvial plain of the Sahel to the east, and more open areas in the north.

### Aïn El-Guettar

*Mousterian level.* The site is at approximately 130 m to the southeast of the spring called Aïn El-Guettar and on the right bank of Wadi Leben. A first test, S1 (1 × 1 m), was carried out in the sandy dune. The greenish clay-like level yielded a few archaeological artifacts. A second test, S2 (1 × 2 m), oriented north/south, was carried out further upstream from the dune (Fig. 10.2). Finally, a 28 m² surface area was excavated stratigraphically with all finds piece provenienced.

The lithology of the layers, from top to bottom, is as follows (Fig. 10.3):
- Unit 1: a compact current sandy layer, with chalk intercalations but without any archaeological finds
- Unit 2: lenticular black clay
- Unit 3: a bed of greenish clay, with bones of bovids and equids and some flints
- **Unit 4: a bed of dark clay, with abundant bones and stone artifacts (Mousterian occupation)**
- Unit 5: a bed of greenish clay with some bones
- Unit 6: ferruginous lenticular
- Unit 7: greenish clay
- Unit 8: ferruginous lenticular
- Unit 9: greenish clay with some equid bones

- Unit 10: lenticular black clay
- Unit 11: greenish clay with some bones
- Unit 12: greenish clay
- Unit 13: ferruginous lenticular
- Unit 14: greenish clay
- Unit 15: greenish clay with ferruginous lenticular
- **Unit 16: sandy limestone (Aterian level)**
- **Unit 17: sandy level (Aterian horizon)**
- Unit 18: greenish clay with no archaeological remains

We collected 20 lithic pieces from the surface. Following the lithic typology of Bordes (1961), these included three Levallois cores, two Levallois flakes, nine side scrapers, and six Mousterian points. All lithic materials from tests S1 and S2 are shown in Table 10.1.

The lithic series clearly shows Levallois technology. The analysis of objects from S2 seems to indicate that the production of flakes and their transformation into tools was done on site. The groups of tools seem to indicate that two types of activities occurred: hunting (the Mousterian points) and domestic tasks (skin treatment, sidescrapers; butchery, knives). Analysis of use-wear from tools showed that butchering activities, in particular, occurred at the site. Faunal cranial remains (mandibles and teeth) are numerous. The postcranial bones are very fragile and crumble quickly once exposed. Bones of equids and bovids are dominant.

In March 2006, we began to systematically excavate the site in order to understand the spatial distribution of the archaeological objects and to establish the different stratigraphic layers. A thin bed of small fluvial stones that are rolled and much rounded covers the black clay bed. The archaeological material collected from the different layers included a total of 504 faunal remains and lithic objects. We analyzed 360 lithic objects. The results presented here are limited to the complete flakes, cores, and tools (Table 10.2).

## Lithic Technology

Despite the low number of lithic objects, we attempted to analyze these techno-typologically. This approach indicated diverse production systems. Indeed, at least two distinct operational schemes coexist within the assemblage.

### Diagrams of Production of Flakes

*Unipolar recurrent debitage.* The only core attributable to this technology is on brown translucent flint. At discard, this object shows traces of at least three predetermined

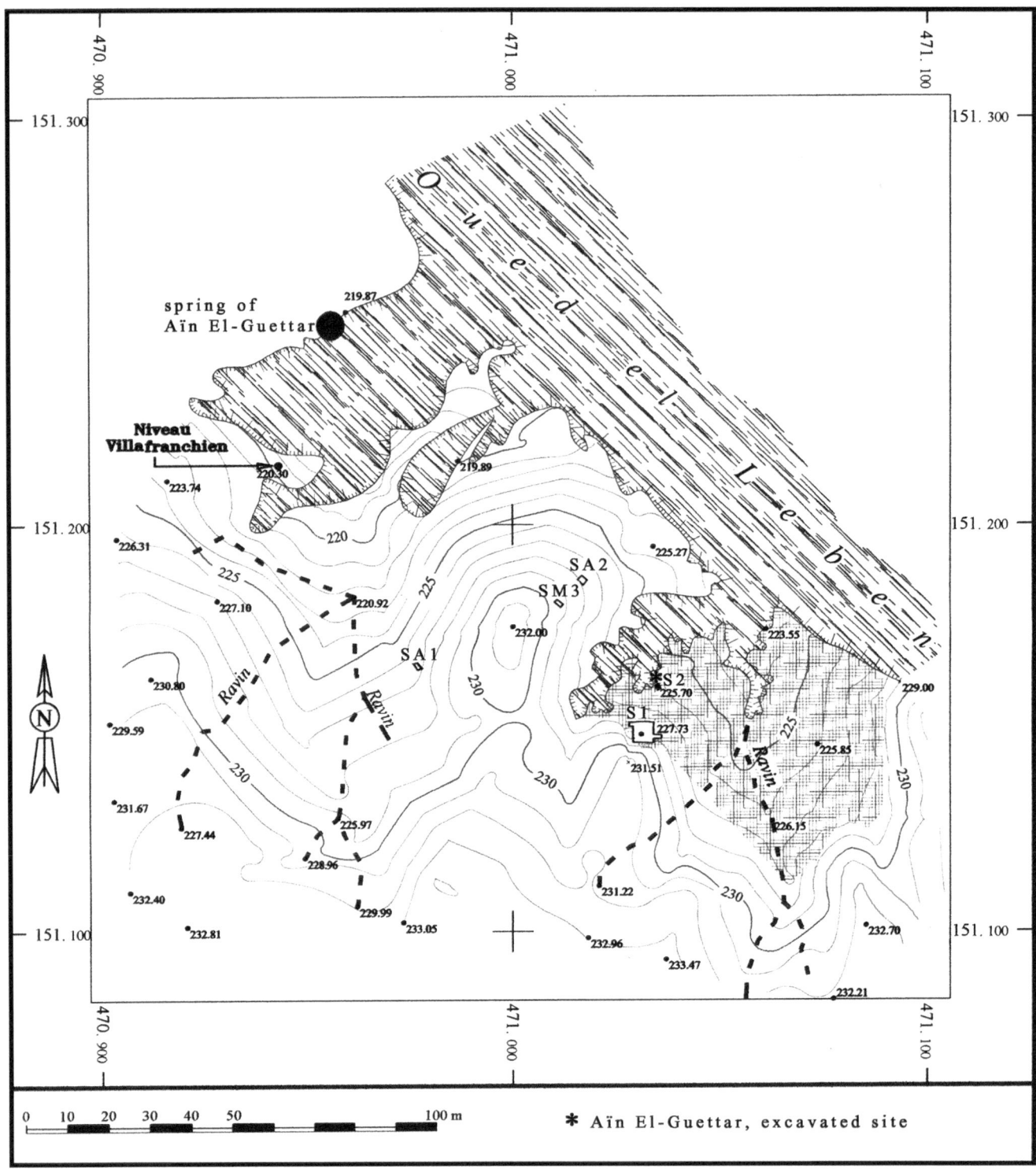

**Fig. 10.2** Topographic drawing of the Aïn El-Guettar site

removals. The products of this type of debitage with only one direction of removal are generally long and have a triangular form. Lateral removals were done in order to recover the surface. These maintenance flakes were generally transformed into points and knives.

*Centripetal recurrent debitage.* The lithic series includes only one Levallois core in which the phase of full debitage is characterized by a centripedal series of invasive removals. Larger in size (6.3 × 6 × 2.7 cm), this object still preserves a broad cortical zone on its

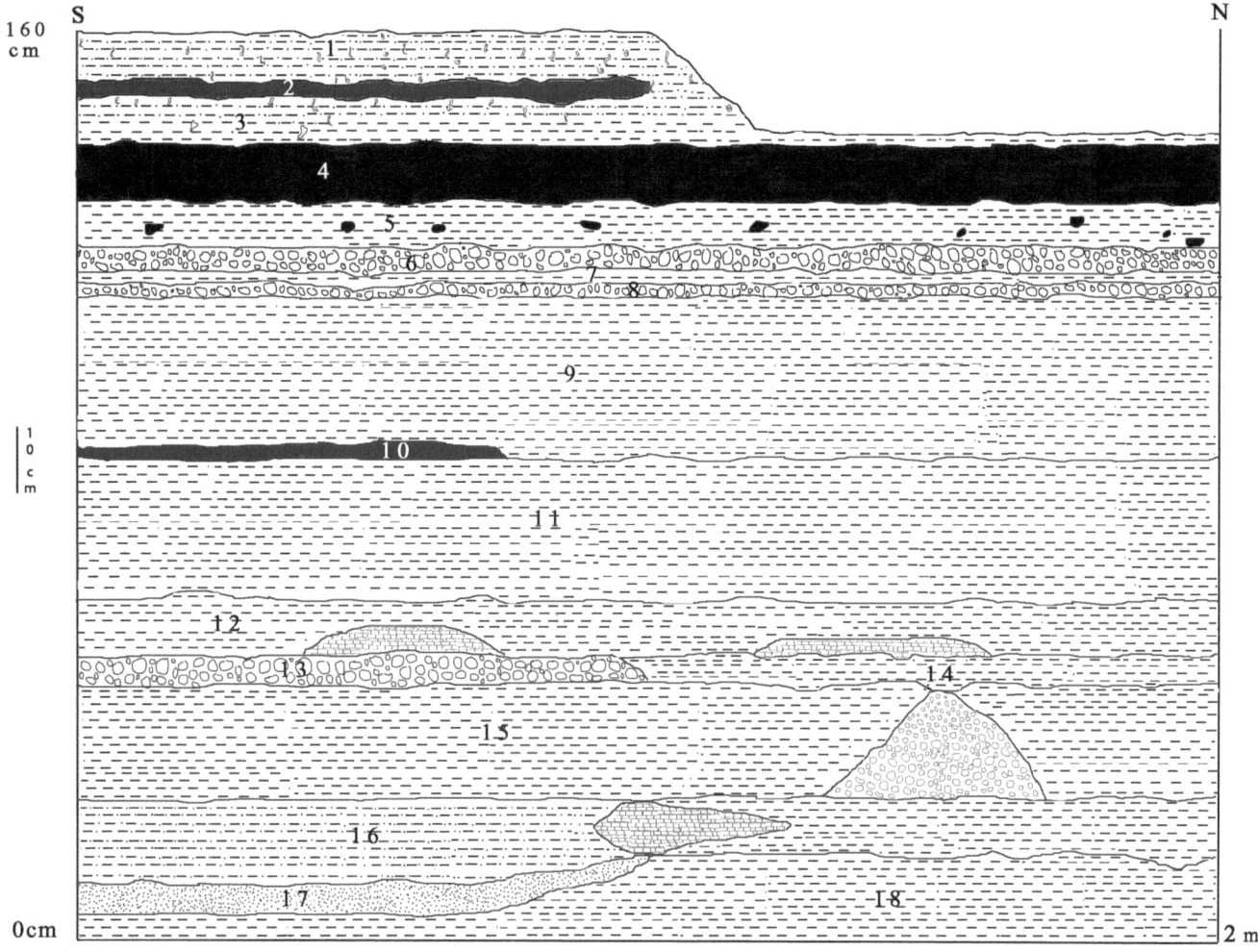

**Fig. 10.3** Stratigraphic section of the Aïn El-Guettar site

**Table 10.1** Lithic material from test units S1 and S2

| Test | Cores | Flakes | Sidescrapers | Mousterian points | Debris |
|------|-------|--------|--------------|-------------------|--------|
| S1 | 0 | 21 | 1 | 0 | 14 |
| S2 | 0 | 23 | 4 | 2 | 61 |

**Table 10.2** Percentage of different lithic categories

| Categories | N | % |
|------------|---|---|
| Cores | 5 | 4.5 |
| Cortical flake (>1 cm) | 14 | 12.7 |
| Non-cortical (>1 cm) | 53 | 48.2 |
| Tools | 38 | 34.5 |
| Retouching flakes (<1 cm) | 250 | |

lower surface. Three other cores are on flakes; two are still in a preparation phase and the third shows traces of two perpendicular invasive removals. During the shaping phase, the flakes resulting from this method of flake production were transformed especially into notches. Other flakes were used unretouched or were very slightly worked.

In conclusion, the most significant characteristics are as follows:

• The study of the entire series revealed the existence of only one mode of acquisition: average-sized flint nodules probably coming from the area of Gafsa.
• The studied series shows the use of only one production sequence: Levallois flaking with two alternatives (unipolar and centripetal).
• The resulting flakes were used in particular to make sidescrapers and points, characteristic Middle Paleolithic tools (Fig. 10.4).

**Fig. 10.4** Two sidescrapers from the Mousterian level of Aïn El-Guettar

## Faunal Remains

Five taxonomic groups make up the faunal remains (Table 10.3).

The faunal assemblage is composed of 19.5% cranial and 80.5% postcranial remains (NR = 364). Of the total number of bones for which age determination was possible (N = 192), 94.3% came from adult individuals. Young-adult individuals make up 4.7%, while young individuals make up only 1% of the total. Some bones have cut-marks.

### Aterian Level

We collected several lithic artifacts that resemble Aterian tools around the excavated Mousterian site. As they were primarily found in the layers eroding below the Mousterian layer, we decided to excavate deeper in square M5, starting from the Mousterian layer. At a depth of approximately 227 cm, in a green clay level with a high percentage of white gypsum below the Mousterian level, we found fragments of ostrich eggshell and some lithic artifacts with white patina, of Aterian aspect, similar to those collected on the eroded slopes.

In March 2007, we excavated a surface of 10 m² under the Mousterian level. The lithic artifacts (N = 50) are all patinaed and from microlithic Levallois flaking (Plate 3). The spatial distribution shows that the artifacts have a low density, and become increasingly dense towards the hill side of the

unit. We carried out two other test excavations, SM3 and SA2. The two excavations are separated by 8.5 m and the slope is steep towards the north. The archaeological level in unit SM3 is located in dark clay and belongs to the Mousterian layer. In the sandy levels of unit SA2, mixed with chalk concretions, we found some tanged tools (Fig. 10.5).

### Lithic Technology

*Technology of flake production.* The recent discovery of an Aterian level in Aïn El-Guettar allowed for a techno-typological analysis of some lithic tools. This analysis revealed the coexistence of distinct reduction sequences within the assemblage, underlining a diverse range of production systems. Local flint (grey or beige) makes up the raw material of artifacts.

The study of almost all of the lithic assemblage revealed that only one mode of acquisition existed: small or average flint blocks coming from terraces of the wadis. Within this lithic industry, which shows an obvious microlithic aspect (average length of flakes <3 cm), we identify two types of Levallois cores:

1) Levallois cores with only one platform: these artifacts were carefully formed by the preparation of convexities on the side and distal parts. The phase of full flaking generates only one, very invasive Levallois flake. Finally, the abandonment of these cores is generally preceded by a reworking of surfaces.
2) Centripetal Levallois cores: the phase of full flaking is characterized by a series of invasive removals. Six objects also belong to this technology: a tanged point, two sidescrapers, tanged end scrapers, and two notches.

Thus, flakes were used to make sidescrapers and points, which are characteristic Middle Paleolithic tools, as well as end scrapers, which are typical Upper Paleolithic elements. The technique used to produce a tang leaves no doubt that this assemblage should be attributed to the Aterian.

*Human remains:* one upper molar from a robust hominin was found in the greenish clay layer attributed to the Mousterian (Fig. 10.6).

### Aïn Oum Henda 2

*Localization.* The southern slope of Jebel El Maloussi (stratotype of the Lower Cretaceous Maloussi formation) is full of prehistoric sites, especially around springs. We chose

**Table 10.3** Percentage of each taxon (NISP = 172)

| Bovids (%) | Equids (%) | Canids (%) | Birds (%) | Rodents (%) |
| --- | --- | --- | --- | --- |
| 50 | 41.3 | 7.5 | 0.6 | 0.6 |

**Fig. 10.5** Tanged flint from the Aterian level of Aïn El-Guettar

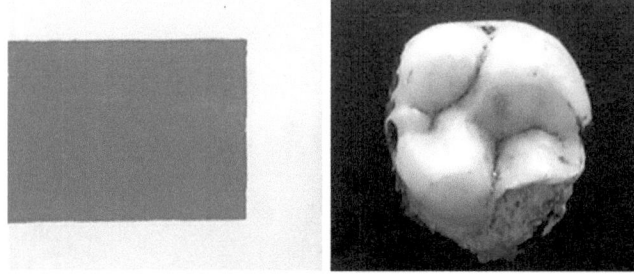

**Fig. 10.6** Human upper tooth from the Mousterian level of the Aïn El-Guettar site, scale units in cm

to survey and test one Middle Paleolithic site that is located in the immediate vicinity of the Aïn Oum Henda spring (Fig. 10.7).

Upstream from and a few meters to the east of the spring, we inventoried an interesting Middle Paleolithic site where, due to water and wind erosion, flints are scattered on the surface. The S1 test excavation (1 × 1.5 m) was carried out on the right bank of a small ravine, not far from a recent rectangular basin built to capture rainwater. The direction of the test unit was northwest/southeast and opening to the east. At the top, the level contains a yellowish, compact, sterile crust. Just below this is an archaeological deposit, formed by green sandy clay and containing flint (Fig. 10.8).

On the surface, we collected 93 lithic pieces that indicate an Aterian origin, including 8 Levallois cores, 40 flakes, 31 sidescrapers, 11 Mousterian points, 1 end scraper, and 2 tanged tools. The S1 test unit produced 71 additional artifacts (Table 10.4).

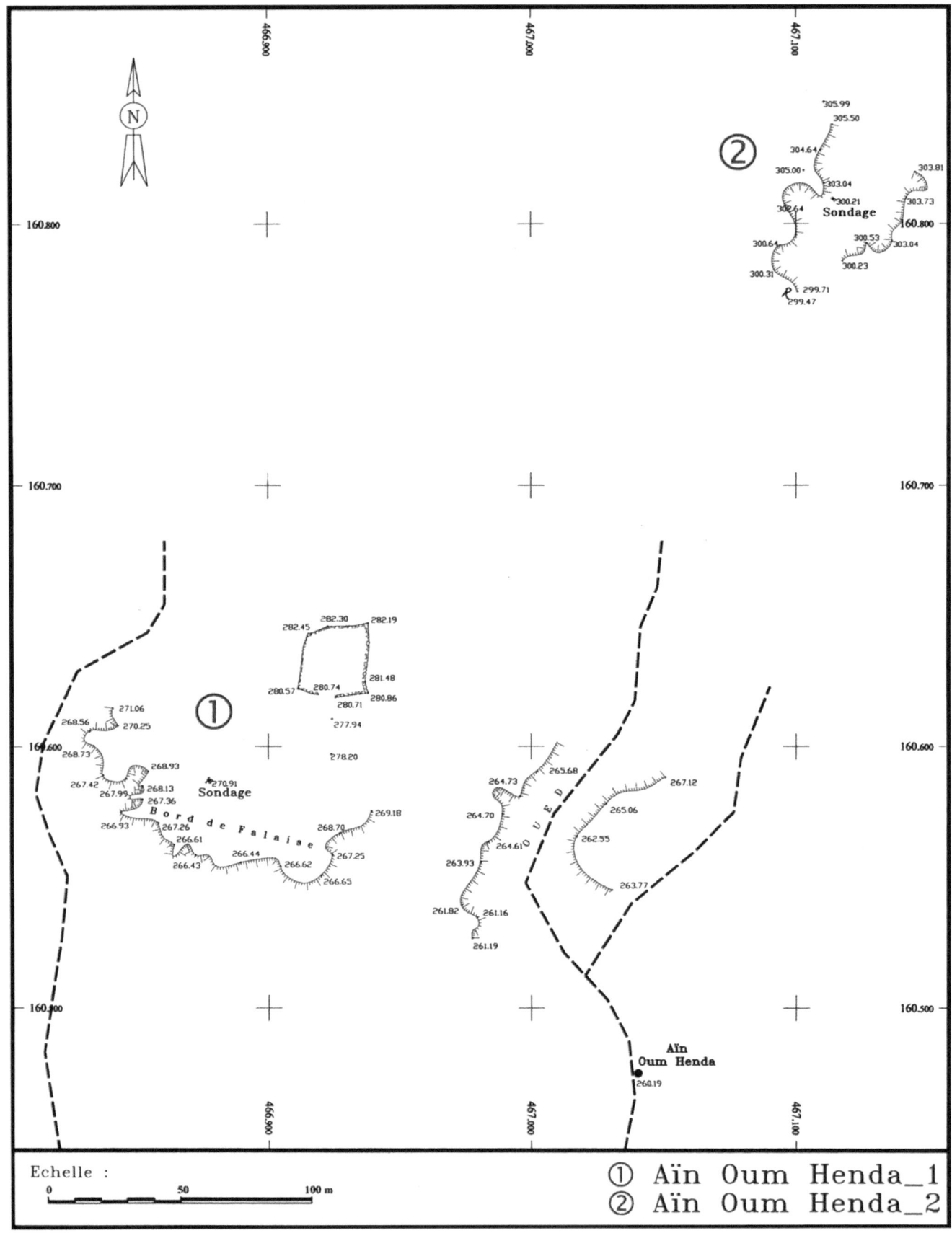

**Fig. 10.7** Topographic drawing of sites Aïn Oum Henda 1 (Capsian) and Aïn Oum Henda 2 (Aterian)

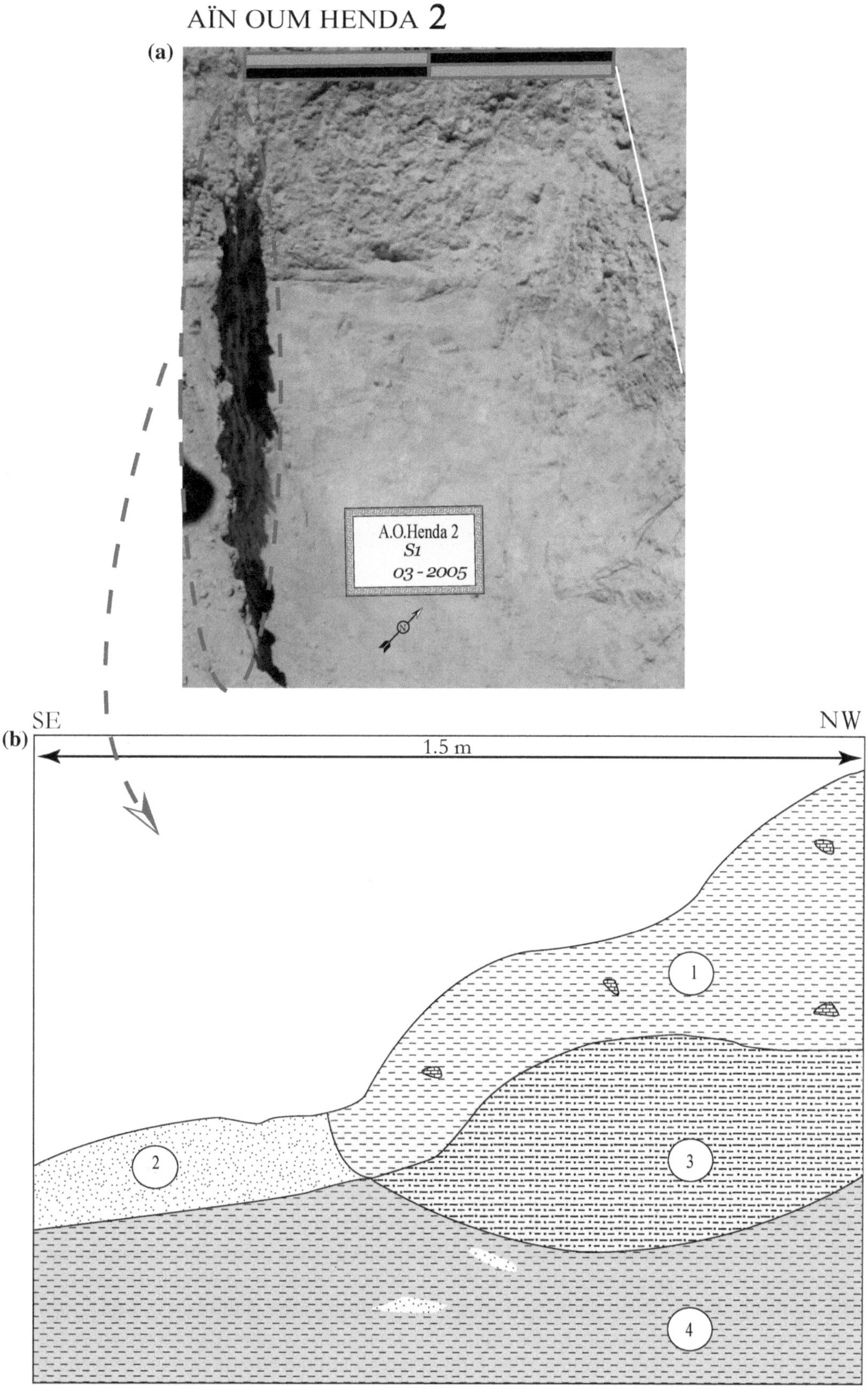

AÏN OUM HENDA 2

**Fig. 10.8  a** Aïn Oum Henda 2: test S1. **b** Stratigraphic section from test S1: *1 Beige clay* with limestone; *2 Yellow sand*; *3 Greenish silt*; *4 Greenish clay* with some *yellow sand* intercalations

**Table 10.4** Lithic material from the two levels of S1

| Levels | Cores | Flakes | Sidescrapers | Debris |
|---|---|---|---|---|
| Level I | 1 | 9 | 3 | 19 |
| Level II | 0 | 18 | 0 | 21 |

**Fig. 10.9** Flint nodule in the Snad region used in prehistoric sites of Meknassy Basin

## Flint Origin

One of our research interests in the Meknassy Basin is to locate prehistoric flint sources. We surveyed Wadi Ad Drag and Jebel Zagoufta in the area of Snad (Gafsa region). The flint used in the prehistoric sites of the Meknassy Basin probably comes from these two locations. Indeed, the Senonian flint nodules are in the Upper Cretaceous limestone deposits of Jebel Zagoufta. The flint source is far away, about 30 km west of the Aïn El-Guettar site (Fig. 10.9).

## Discussion

The Aterian is a proper prehistoric culture from North Africa. However, we know very little about its origin, chronology, and geographic extension. According to Tixier (1967), Aterian is a Mousterian variant (with *Levallois technology*, often with blades). End scrapers are numerous, but only 25% of lithic objects are tanged. The tang could be placed onto the base of almost all types of tools, including sidescrapers, end scrapers, *percoirs*, denticulates, and burins. The majority of tools are pedunculate (Tixier 1958–1959);

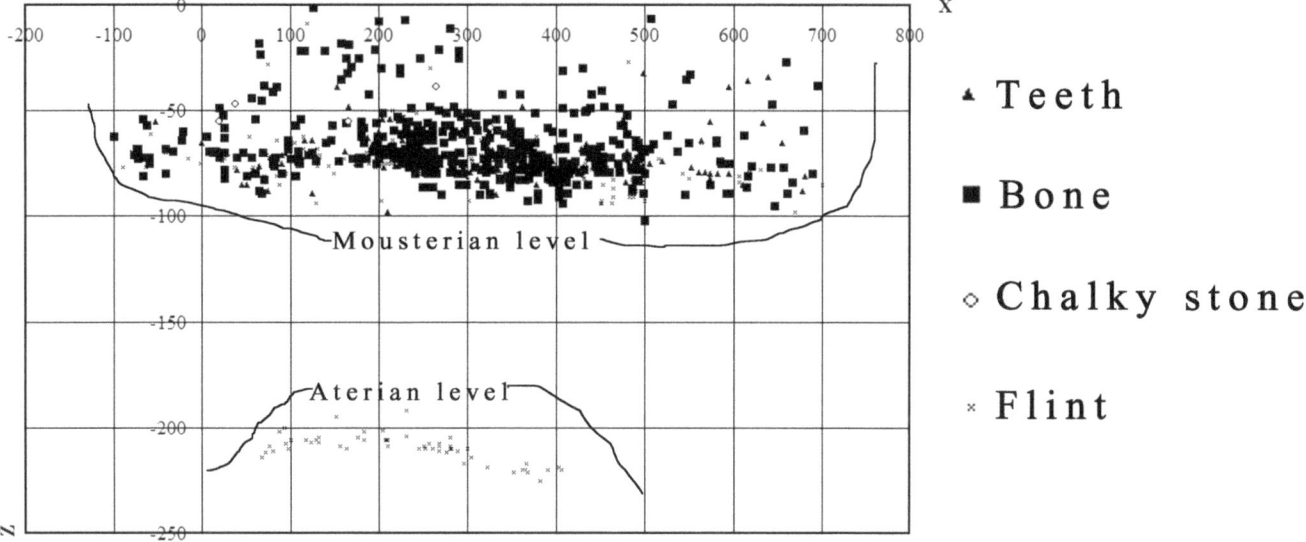

**Fig. 10.10**  Spatial distribution of artifacts (according to depth) within different cultural horizons of the Aïn El-Guettar site

however, a bifacial foliate point or a pedunculate tool does not make an Aterian culture (Bordes 1976–1977).

According to Van Peer et al. (2003), at around 200 ka, during the Middle Stone Age, modern human behavior emerged and the Sangoan culture spread throughout Africa. The presence of *Homo sapiens (sensu stricto)* in North Africa between 190 and 130 ka has been established (Smith et al. 2007). It seems that *Homo sapiens* are responsible for the Aterian culture in the Maghreb (Ferembach 2001). Therefore, the idea of an Aterian between 40 and 20 ka (Debénath et al. 1986; Debénath 1992) has been revised, and the Aterian may be much older in these new sites, such as in Egypt (Van Peer and Vermeersch 2000), in Libya (Cremaschi et al. 1998), and in Tunisia. In the Sahara Desert, the age of the Aterian seems to be more older than was suggested (Vernet 2004).

In Libya, two sites in *Tadrart Acacus* (central Sahara) have yielded Aterian levels.

The site of Uan Tabu was dated by OSL to 61 ± 10 ka (Garcea 2001) and the site of Uan Afuda was dated by TL to 71 ± 10 ka and 73 ± 10 ka (OSL dates vary between 69 ± 7 ka and 90 ± 10 ka; Di Lernia 1999). It seems that, in this region, the first Aterian occupation occurred between 60 and 90 ka, when the climate was dry (Cremaschi et al. 1998). In Jebel Gharbi, the Aterian is between 60 and 70 ka. The Aterian sequence corresponds to a hyperarid period (OIS 4) (Barich et al. 2006; Garcea and Giraudi 2006).

In Morocco, the Aterian level in Mugharet el 'Aliya Cave was dated between 39 and 51 ka (Wrinn and Rink 2003). In Rhafas Cave, the transition from the Mousterian to the Aterian was dated to between 80 and 70 ka (Mercier et al. 2007). In Algeria, one of the cavities of Sidi Sa (Tipaza) contains a microlithic Mousterian level which surmounts an Aterian level in place (Betrouni 1997, 2001). The Aterian level was dated by radiocarbon to 38.130 ± 1.320 ka.

As discussed above, several Tunisian archaeological sites have been attributed to the Middle Paleolithic. Recently, sites such as Aïn Métherchem, El Guettar, and Wadi el Akarit have been recognized as "Proto-typical Aterian" (name given by Bordes 1976–1977, to the lithic industry of Aïn Métherchem and Taforalt F) because of the presence of tanged tools. We think that at least two levels attributable to Aterian and Mousterian cultures were present at the first two sites. New excavations are needed to check the stratigraphic sequences. The recently excavated site of Wadi El Akarit yielded an archaeological level attributed to the "Proto-Aterian." The occupation soil is dated by thermoluminescence to 90 ka (Roset and Harbi-Riahi 2007). The Meknassy Basin is one of the richest areas of previously unexploited prehistoric sites in Tunisia. With its wadis and springs, the area has attracted animal and human groups for a long time. At Aïn El-Guettar, two archaeological levels have been identified. The top level is Mousterian and the bottom level is Aterian (Fig. 10.10), with no transition from the Aterian level to the Mousterian.

The two levels are separated by at least 1.4 m of sterile deposit. The Mousterian level has yielded remains of large bovids and equids. These herbivores are characteristic of open environments. The inhabitants of the Meknassy Basin practiced selective hunting of bovids and equids. Whole carcasses were brought back to camp to be butchered. The presence of the other small herbivores is probably a sign that they occasionally hunted these kinds of animals. The presence of an Aterian lithic industry under the Mousterian indicates that what is called "Proto-Aterian" or "Old Aterian" is contemporaneous with the Mousterian in other regions.

**Acknowledgments** We thank the *Institut National du Patrimoine* for providing financial and other valuable resources for this project. Our gratitude is extended to Pr. Hédi Ben Ouezdou, Mr. Abderrazac Graguebr, Noura Rahmani, and Sophie Acheche for their help during the excavation. Warm thanks to our friend and colleague Lamine Bouazizi for his usual technical support. We also thank anonymous reviewers for helping us to improve this manuscript.

# References

Barich, B. E., Garcea, E. A. A., & Giraudi, C. (2006). Between the Mediterranean and the Sahara: Geoarchaeological reconnaissance in the Jebel Gharbi, Libya. *Antiquity, 80*, 567–582.

Betrouni, M. (1997). Le paléokarst de Sidi Sa (Tipasa, Algérie) et la question du Paléolithique supérieur maghrébin. In J. M. Fullola & N. Soler (Eds.), *El Món Mediterrani Després del Pleniglacial (18.000–12.000 BP): Sèrie Monografica*, 17 (pp. 57–68). Girone: Centre d'Investigacions Arqueologiques.

Betrouni, M. (2001). Le paleokarst de Sidi-Said. Aspects chrono-culturels. In C. N. R. PH (Ed.), *L'homme Maghrébin et son environnement depuis 100 000 ans* (pp. 101–112). Alger: Centre national de recherches préhistoriques, anthropologiques et historiques.

Bordes, F. (1961). *Typologie du Paléolithique Ancien et Moyen. 2 vols.* Bordeaux: Delmas.

Bordes, F. (1976–1977). Moustérien et Atérien. *Quaternaria, XIX*, 19–34.

Brun, A., Guérin, C., Levy, A., Riser, J., & Rognon, P. (1988). Steppic environments at the end of the Upper Pleistocene in Southern Tunisia (oued El Akarit). *Journal of African Earth Science, 7*, 969–980.

Clark, J. D. (1982). The cultures of the Middle Palaeolithic/Middle Stone Age. In J. D. Clark (Ed.), The Cambridge history of Africa, volume I: From earliest times to c. 500 B.C. (pp. 348–341). Cambridge: Cambridge University Press.

Cremaschi, M., Di Lernia, S., & Garcea, E. A. A. (1998). Some insights on the Aterian in the Libyan Sahara: Chronology, environment, and archaeology. *African Archaeological Review, 15*, 261–286.

Debénath, A. (1992). Hommes et cultures matérielles de l'Atérien marocain. *L'Anthropologie, 96*, 711–720.

Debénath, A., Raynal, J. P., Roche, J., Texier, J.-P., & Ferembach, D. (1986). Stratigraphie, habitat, typologie, et devenir de l'Atérien marocain: Données récentes. *L'Anthropologie, 90*, 233–246.

Di Lernia, S. (1999). Uan Afuda Cave. Hunter-gatherer societies of Central Sahara Rome: Arid Zone Archaeology, Monographs 1.

Ferembach, D. (2001). Evolution du peuplement du Maghreb des origines au Néolithique. In C. N. R. PH (Ed.), *L'homme Maghrébin et son environnement depuis 100 000 ans* (pp. 123–129). Alger: Centre national de recherches préhistoriques, anthropologiques et historiques.

Garcea, E. A. A. (2001). *Uan Tabu. In the settlement history of the Lybian Sahara*. Rome: Arid Zone Archaeology, Monographs 2.

Garcea, E. A. A., & Giraudi, C. (2006). Late Quaternary human settlement patterning in the Jebel Gharbi. *Journal of Human Evolution, 51*, 411–421.

Gobert, E. G. (1950). Le gisement paléolithique de Sidi Zin. *Karthago, 1*, 3–51.

Gobert, E. G., & Harson, L. (1958). Recherches de préhistoire tunisienne. *Karthago, 9*, 3–43.

Gobert, E. G., & Howe, B. (1955). L'Ibéromaurusien de l'oued El Akarit (Tunisie). *Actes du IIᵉ Congrès Panafricain de Préhistoire*, Alger, 575–594.

Gruet, M. (1950). Note préliminaire sur le gisement moustérien d'El Guettar. *Bulletin de la Société Préhistorique Française, VLVII*, 232–241.

Gruet, M. (1954). Le gisement Mousterien d'el Guettar. *Karthago, V*, 3–79.

Gruet, M. (1955). Amoncellement pyramidal de sphères calcaires dans une source fossile moustérienne à El-Guettar (Sud Tunisien). In L. Balout (Ed.), *Actes du congrès panafricain de préhistoire, IIe session, Alger* (pp. 449–460). Paris: Arts et Métiers Graphiques.

Gruet, M. (1958–1959). Le gisement d'El Guettar et sa flore. *Libyca, VI-VII*, 79–126.

Guérin, C., & Faure, M. (2007). Etude paléontologique des mammifères du Pléistocène supérieur de l'oued El Akarit. In J.-P. Roset & M. Harbi-Riahi (Eds.), El Akarit un site archéologique du Paléolithique Moyen dans le Sud de la Tunisie. Recherches sur les Civilisations (ERC) (pp. 365–390). Paris: Culturesfrance (ex ADPF et AFAA).

Leroi-Gourhan, A. (1958). Résultats de l'analyse pollinique du gisement d'El Guettar, Tunisie. *Bulletin de la Société Préhistorique Française, LV*, 546–551.

Mercier, N., Wengler, L., Valladas, H., Joron, J.-L., Froget, L., & Reyss, J.-L. (2007). The Rhafas Cave (Morocco): Chronology of the mousterian and aterian archaeological occupations and their implications for Quaternary geochronology based on luminescence (TL/OSL) age determinations. *Quaternary Geochronology, 2*, 309–313.

Page, W. D. (1972). *The geological setting of the archaeological site at Oued el Akarit and the palaeoclimatic significance of gypsum soils, Southern Tunisia*. Ph.D. Dissertation, University of Colorado.

Roset, J.-P. (1996). Nouvelles recherches sans l'Oued el Akarit, sud de la Tunisie: état actuel de la question. *XIIIe Congrès UISPP, Italie, Forli, 1*, 464–465.

Roset, J.-P. (2005). *El-Akarit et le Paléolithique moyen en Tunisie. Archéologies, 20 ans de recherches françaises dans le monde* (pp. 225–226). Ministères des Affaires étrangères: Edit. ADPF.

Roset, J.-P., & Harbi-Riahi, M. (2007). *El Akarit : Un site archéologique du Paléolithique moyen dans le sud de la Tunisie*. Recherche sur les Civilisations (ERC). Paris: Cultures France (ex ADPF et AFAA).

Smith, T. M., Tafforeau, P., Reid, D. J., Grün, R., Eggins, S., Boutakiout, M., et al. (2007). Earliest evidence of modern human life history in North African early *Homo sapiens*. *Proceedings of the National Academy of Sciences of the USA*, 104, 6128–6133.

Tixier, J. (1958–1959). Les pièces pédonculées de l'Atérien. *Libyca, VI/VII*, 127–158.

Tixier, J. (1967). *Pièces pédonculées Atériennes du Maghreb et du Sahara; Types 1–30*. Paris: AMG/Muséum National d'Histoire Naturelle.

Van Peer, P., & Vermeersch, P. M. (2000). The nubian complex and the dispersal of modern humans in North Africa. *Studies in African Archaeology, 7*, 47–59.

Van Peer, P., Fullagar, R., Stokes, S., Bailey, R. M., Moeyersons, J., Steenhoudt, F., et al. (2003). The Early to Middle Stone Age transition and the emergence of modern human behaviour at site 8-

B-11, Sai Island, Sudan. *Journal of Human Evolution, 45*, 187–193.

Vaufrey, R. (1955). *Préhistoire de l'Afrique, tome premier: Maghreb*. Paris: Masson.

Vernet, R. (2004). Le Sahara préhistorique entre Afrique du Nord et Sahara: état des connaissances et perspectives. In A. Bazzana & H. Bocoum (Eds.), *Du Nord au Sud du Sahara. Cinquante ans d'Archéologie française en Afrique de l'Ouest et au Maghreb* (pp. 89–100). Saint Maur: Bilan et Perspectives, édition Sépia.

Wrinn, P. J., & Rink, W. J. (2003). New ESR dating results for Aterian levels at Mugharet el Aliya, Tanger, Marocco. *Journal of Archaeological Science, 30*, 113–133.

# Chapter 11
# The Aterian of the Oases of the Western Desert of Egypt: Adaptation to Changing Climatic Conditions?

A. L. Hawkins

**Abstract** The Aterian is well-represented in arid eastern North Africa, particularly in the Egyptian oases and other formerly watered areas. In this region, study of the Middle Stone Age (MSA), including the Aterian, has been hindered by the rarity of buried sites. However, work by a number of teams suggests that the Levallois-based industries associated with significantly higher moisture during Marine Isotope Stage 5 are not Aterian. The artifact inventory of Aterian differs from that of the earlier MSA industries, as does the distribution of sites on the landscape. Taking a technological viewpoint, I suggest that the Aterian represents an elaboration of earlier industries arising in response to changing climatic regimes.

**Keywords** Adaptation • Aterian • Climatic change • Dakhleh oasis • Kharga oasis • Lithic technology • Middle Stone Age

## Introduction

The presence of Middle Stone Age (MSA)[1] material in the Western Desert of Egypt[2] has been attested since the early twentieth century (Winlock 1936). Caton-Thompson and Gardner's seminal study of the stone age material from Kharga Oasis (Caton-Thompson 1952) has proven to be an invaluable baseline for further work in the area by Simmons and Mandel (1986; Mandel and Simmons 2001), the Combined Prehistoric Expedition (Wendorf and Schild 1980; Wendorf et al. 1993a), the Dakhleh Oasis Project (DOP) (Churcher and Mills 1999; Churcher et al. 1999; Kleindienst 2003), the Kharga Oasis Prehistory Project (KOPP) (Smith et al. 2004, 2007), and the Czech Institute of Egyptology (Bárta et al. 2002; Svoboda 2004). Despite the considerable amount of research that has been undertaken in the Western Desert, a number of fundamental long-standing questions about the MSA, including the Aterian, of the Western Desert remain. Kleindienst (2001) has articulated some of these: can lithic assemblages be identified as Aterian if they lack tanged tools by using the presence or absence of other types or technological traits?; can temporal trends within the Aterian be identified?; did the Aterian evolve from an earlier MSA in North Africa or does it represent a population influx, perhaps from the south or from the Nile Valley?; what is the relation of the Aterian of the Western Desert of Egypt to that of the central Sahara and further west?; and, does the presence of tanged tools

---

[1] In this chapter I use the term Middle Stone Age (MSA) to refer to Pleistocene-aged Levallois-based units. The Aterian, using predominantly Levallois reduction, is therefore included in the MSA. See below for a discussion of the various units described for the MSA of the Eastern Sahara.

[2] In Egypt, the Nile Valley serves as a dividing line between the "Eastern Desert" and the "Western Desert." Archaeologists also employ the term "Eastern Sahara" to refer to the area that broadly overlaps with the "Western Desert," but is not confined to the borders of present-day Egypt.

A. L. Hawkins (✉)
Laurentian University, 935 Ramsey Lake Road,
Sudbury, ON P3E 2C6, Canada
e-mail: ahawkins@laurentian.ca

J.-J. Hublin and S. P. McPherron (eds.), *Modern Origins: A North African Perspective*,
Vertebrate Paleobiology and Paleoanthropology, DOI: 10.1007/978-94-007-2929-2_11,
© Springer Science+Business Media B.V. 2012

and bifacial foliates in the Aterian represent a fundamental change in behavior? To this we can also add: to what extent were the humans who produced earlier MSA and/or Aterian artifacts modern in their behavior?

In this chapter, I argue that, while the archaeological record from the Western Desert differs from that of the Maghreb, we can gain insights into the question of behavioral change by taking a regional approach and by comparing the Aterian with earlier MSA units. There are no stratified MSA sites in the Western Desert where the Aterian occurs in situ with other MSA material. In most cases, it is impossible to obtain chronometric dates on MSA archaeological materials. However, a number of recent publications help to clarify the chronology of changing climatic conditions of the Western Desert (Wendorf et al. 1993a; Churcher et al. 1999; Smith et al. 2004, 2007; Kieniewicz 2007; Kleindienst et al. 2008). By comparing the Aterian with earlier MSA units, we can move beyond description and chronology to address issues of behavioral change. Models for human behavior that consider different technological strategies and the desired design characteristics of tools may contribute to an understanding of the reasons for the lithic technology changes that are evident in the Aterian.

In this chapter, I focus on the archaeology of two of the important Western Desert oases, Dakhleh and Kharga, and the Bir Tarfawi region. I describe the oasis setting, including the geology, the chronology of climatic variation and MSA occupations, and the available fauna. The organization of technology approach that is employed to compare the different units is explained and is followed by a comparison of certain attributes of the Aterian with those of the earlier MSA. I conclude by considering the differences between the complexes and evidence for emergence of behavioral modernity.

The interpretation proposed in this chapter is, admittedly, based on a dataset with many limitations. Among the most important of these is the lack of secure dating for much of the material discussed. Further, I have chosen to focus on the MSA of Kharga and Dakhleh Oases and the Bir Tarfawi/Bir Sahara area because these are the materials with which I am most familiar. However, I argue that it is through proposing interpretations such as the one outlined here that we may be better able to focus future research on questions related to the behavior.

## Oasis Database

The Western Desert Pleistocene archaeological record differs significantly from that of the Maghreb in that the majority of archaeological material is found in surface context (Kleindienst 1999). This is particularly true for the Aterian; the only known locations with definite sub-surface Aterian artifacts are Mound-Spring KO6E at Kharga Oasis and silts at Bulaq Pass, which contain redeposited material (Caton-Thompson 1952; Hawkins et al. 2001). A positive aspect of this limitation is that archaeological sites are highly visible. Distributions are easily identified and samples can be collected in relatively short periods of time. It is also reasonably straightforward to determine the size of concentrations and the density of material, although it is necessary to consider the processes acting differentially upon surface assemblages. Raw material outcrops and geomorphic features associated with sites are also easily identified. Archaeological concentrations, however, should be considered time-averaged. Artifacts that have laid on the desert surface for over 40,000 years are frequently abraded and desert varnished. The degree of abrasion and varnish depends on the location of deposition in that objects lying on high elevation surfaces are subject to less abrasion.

Because this chapter presents an interpretation based in large part on material recovered from surface locations, a few cautionary notes are in order.

The comparative obtrusiveness of the different archaeological units must be considered. In some parts of the oases, there are large areas of gravel pavement composed of naturally fragmented chert and limestone, intermingled with artifacts. Surveyors decide to designate locations as "localities" based on considerations such as artifact density and the presence of diagnostic artifacts. Following Schiffer et al. (1978), obtrusiveness refers to the properties of artifacts, such as their colors, shapes, and sizes, while visibility refers to the relationship between an artifact and its background. While we have not tested this empirically, it is likely that Aterian sites, which include distinctive tanged tools and bifacial foliates, are more visible to surveyors than earlier MSA ones. Additionally, earlier MSA units are more difficult to assign to specific units because they lack distinctive tools types (cf. Caton-Thompson 1952).

Second, the cultural affiliation of some site types may be more easily identified than that of others; for example, tanged tools at former occupation areas make it easy to identify the sites as Aterian. However, the cultural affiliation of workshop localities is more difficult to ascertain and, in the absence of chronometric dates, any such attribution must be considered tentative. Isolated tanged tools can be identified and mapped, but the same is rarely true for isolated earlier MSA tools.

The possibility that surface sites may be "cumulative palimpsests" must be acknowledged. These are accumulations of material that represent activities on a number of occasions; most strictly defined, each occupation adds to the previous material and the patterning that once existed becomes increasingly blurred as a result of this addition and activities such as trampling (Bailey 2007). There is also the

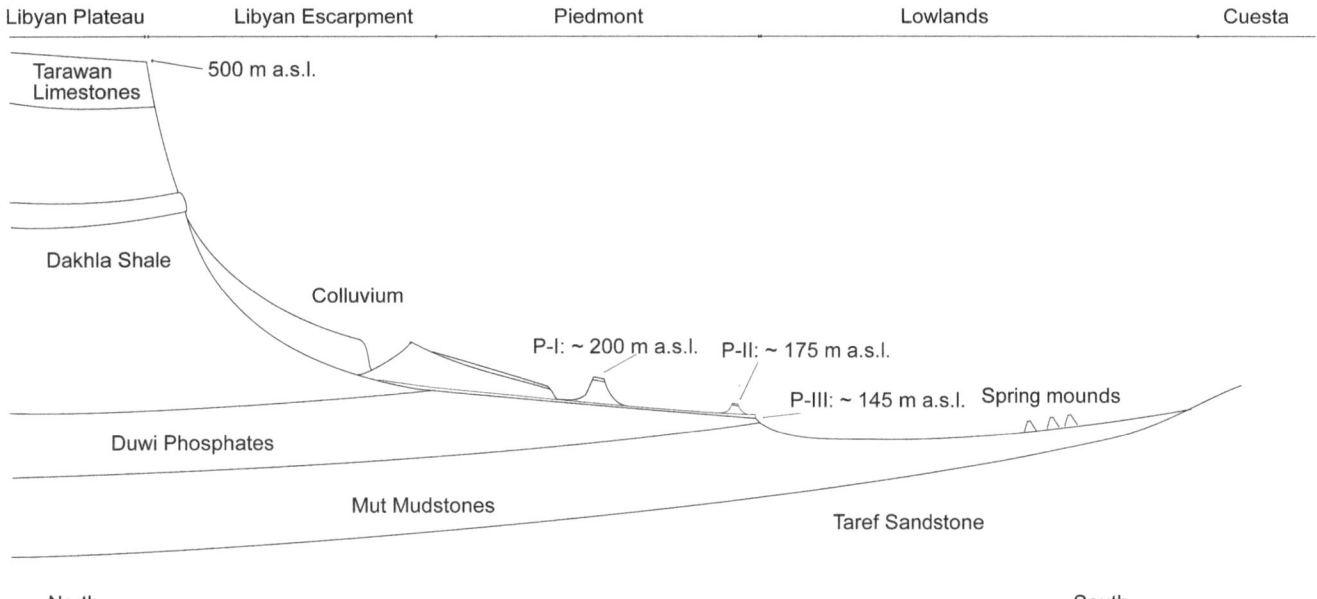

**Fig. 11.1**  Schematic section of Dakhleh Oasis showing the location of geomorphic zones described in the text. The P-I, -II, and -III surfaces represent three generations of pediment terraces mantled with gravels. After Kleindienst et al. (1999)

possibility that material from earlier uses of a site have been removed or reused. A relatively dense accumulation found close to a known resource location may represent a similar type of activity through time (i.e., lithic reduction or resource extraction).

The comparability of sites assigned to different times may be questioned. Surface sites dating to Marine Isotope Stage (MIS) 7 have undergone surface processes for a longer period of time than those assigned to MIS 3–4 (Kleindienst 2003).

Abrasion of artifacts in surface context precludes any form of use-wear analysis, and also makes retouch difficult to identify in some cases. Examination of the degree of reworking through use of edge angles is problematic, both because of the error inherent in measuring these on abraded artifacts, but also because the degree of abrasion varies between and within locations (Hawkins 2001; Kleindienst 2003). As such, tool functions are suggested based on form but these remain unverified hypotheses.

There is such a high density of material in surface context that surveyors sometimes selectively collect representative objects, such as cores, tools, and "specialized flakes"[3] from concentrations. While this strategy is potentially useful for identifying a site's cultural affiliation, it places limits on the behavioral interpretations that can be based on such material.

---

[3] Kleindienst (2003) uses the term "specialized flake" to refer to flakes, points, and blades produced using Levallois, discoidal, and blade reduction. Similarly, "specialized cores" are Levallois, discoidal, and blade cores.

## Generalized Geological and Climatic Background for the Oases

To understand strategies for using resources and to consider whether these changed, it is necessary to examine the distribution of archaeological artifacts in relation to former sources of lithic raw materials, water, and associated flora and fauna.

### Geology and Geomorphology

The geology and geomorphology of Dakhleh and Kharga oases are not identical but the oases are broadly similar with respect to resources that would have been of interest to prehistoric human occupants. Both are structurally-controlled depressions bordering the Libyan plateau; the depression lies south of the plateau in Dakhleh Oasis, and west of the plateau in Kharga. The Dakhleh Oasis Project (DOP) archaeologists have used a broad geomorphic division of this oasis as a framework for understanding the locations of resources and archaeological localities (Kleindienst et al. 1999) (Fig. 11.1, Table 11.1).

The Libyan Plateau is not part of the DOP study area and has not been researched in depth at Dakhleh. Simmons and Mandel (1986) included the plateau in their research at Kharga Oasis, and interestingly, in their survey, the only Aterian sites that they located occurred on the plateau. The

**Table 11.1** Summary of the geomorphic zones in Dakhleh and Kharga and the associated resources and archaeological technocomplexes

| Geomorphic zone | Resources present | General age of surface or deposits | Associated technocomplexes |
|---|---|---|---|
| Libyan plateau | Wadis and pans | Holocene, Pleistocene | Earlier MSA, Aterian[a] |
| | Chert lag | Indeterminate | |
| Libyan escarpment | Tufa deposits | Pleistocene (see below for dates) | Earlier MSA (MIS 7 and 5)[b] |
| | Chert nodules | Indeterminate | Earlier MSA (MIS 5)[c] |
| Piedmont terraces | Chert/lookouts on P-I terraces | Indeterminate but significantly older than the P-II surface | Earlier MSA (MIS 7 and 5)[d] |
| | Chert/lookouts on P-II terraces | >200,000? | Earlier MSA (MIS 7 and 5) Aterian,[d] |
| | Chert on P-III terraces | <90,000–ca. 20,000 | Aterian, Khargan[d] |
| Lowland | Springs | Pleistocene and Holocene | Earlier MSA, Aterian, Khargan[e] |
| | Lakes | Middle Pleistocene | Earlier MSA (MIS 7)[f] |
| | Lithic raw materials (chert in lag, small chert nodules, chalcedony, quartzite) | Indeterminate | Workshop concentrations associated have not been documented to date |
| Southern Cuesta | Quartzite | Indeterminate | Aterian, Khargan[d] |

[a] Mandel and Simmons (2001)
[b] Caton-Thompson (1952); Smith et al. (2007); Kleindienst et al. (2008)
[c] Smith et al. (2007)
[d] Kleindienst et al. (1999); Kleindienst (1999)
[e] Caton-Thompson (1952)
[f] Churcher et al. (1999)

Tarawan limestone capping the plateau dates from the Paleocene to the Eocene. In both Dakhleh and Kharga, the plateau limestone is chert-bearing in some areas; chert lag is reported on the surface north of Dakhleh and east of Kharga (Kleindienst et al. 1999). The plateau at Kharga has wadis and pans that contained water in times of higher precipitation (Caton-Thompson 1952); the only well-documented basin with lacustrine sediments on the plateau in Dakhleh dates to the Holocene (Kleindienst et al. 1999), but this area in Dakhleh has not been well-explored.

In Dakhleh, the Libyan Escarpment runs approximately east–west on the northern edge and north–south on the eastern edge of the oasis (Fig. 11.2). In Kharga, the escarpment runs north–south. There is a drop of approximately 300 m from the plateau to the piedmont zone (Kleindienst et al. 1999). The morphology of the escarpment crest varies locally, and in some areas, steep limestone cliffs make access to the plateau impossible. In Kharga, the escarpment is mantled in tufa deposits (Caton-Thompson 1952), but in Dakhleh there is less evidence of tufa. Kleindienst et al. (1999) argue that, in Dakhleh, the escarpment reached its present form before 350 ka. At both Kharga and Dakhleh, the escarpment would be a place of interest to humans for several reasons: (1) at various times in the Pleistocene, springs issued from it; (2) the limestone capping the plateau is chert-bearing in places and chert nodules occur on the escarpment; (3) certain places on the

escarpment would have served as conduits for animal and human migration to eastern and northern locations; and (4) the escarpment provides a view of the surrounding landscape, including locations of springs and fauna.

The piedmont zone is composed of a series of complex alluvial fans that stretch from the escarpment into the lowland area (Kleindienst et al. 1999). It is argued that there were several periods of pediment formation and erosion and the resulting gravel terraces are differentially preserved at different elevations (Brooks 1986; Kleindienst et al. 1999). In Dakhleh, the earliest formed terraces (referred to as P–I) are the highest and are found mainly near the escarpment. Successive lower terraces (P-II and P-III) are more extensive, and extend well into the lowland oasis. The alluvial gravels of the piedmont zone were a significant source of lithic raw material for humans living in Dakhleh in the Pleistocene (Hawkins 2001; Kleindienst 2003). The presence of chert in the Tarawan limestone and in the piedmont gravels varies. In some areas, the remnants of the piedmont gravels stand high above the surrounding area, providing ancient occupants of the oases with good vantage points for observing locations of fauna and other resources.

Present-day springs and wells are found in the lowland oasis, where there is also ancient evidence for water. Mounds mark the locations of former springs in Dakhleh and Kharga, and in the oases, archaeological artifacts within the mound deposits attest to human use of the springs.

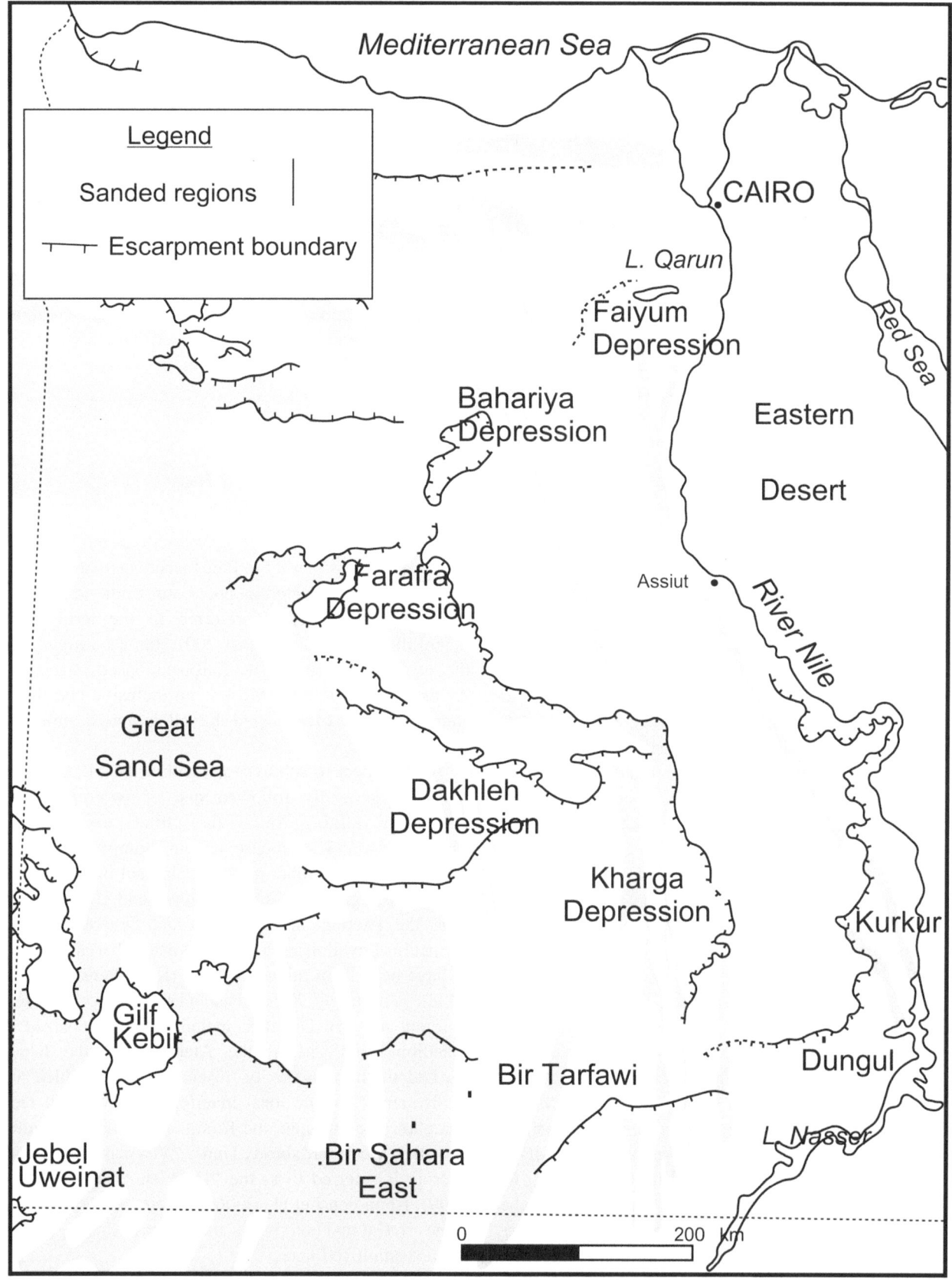

**Fig. 11.2** The location of the Dakhleh and Kharga depressions and Bir Sahara and Bir Tarfawi. After Kleindienst

Spring mounds have been investigated in Kharga, where Caton-Thompson (1952) found Earlier Stone Age (ESA), MSA (including Aterian), and Neolithic artifacts in the deposits. At Dakhleh, Wendorf and Schild (1980) excavated a mound containing Terminal ESA artifacts and the DOP researchers investigated a mound with MSA artifacts that may be attributed to the Aterian (Hawkins 2001). At least three types of lithic raw materials occur in the lowland at Dakhleh: nodular chert that is found in nodules less than 10 cm in diameter, chalcedony, and ferruginous quartzite. While the presence of artifacts made from these raw materials attests to their use, no quarries or workshops of these materials are known. In addition, Tarawan chert fragments occur on the stone-paved surface or reg.

Lake sediments in the southern and eastern parts of the Dakhleh lowland overlay Nubian sandstone (Churcher et al. 1999; Kleindienst et al. 1999; Kieniewicz 2007). These are interbedded calcareous (CSS—Calcareous Silty Sediments) and ferruginous (FSS—Ferruginous Silty Sediments) deposits. Churcher et al. (1999) posit the existence of three large paleolakes that possibly date to the Middle Pleistocene. Earlier MSA concentrations occur in association with these lakes, and later MSA Khargan technocomplex material is found on the surface of CSS deposits (Kleindienst 1999).

The southern edge of Dakhleh Oasis is bordered in some areas by a sandstone ridge. There are large deflation basins in the sandstone, some with Pleistocene laminated sediments, but no association between artifacts and these sediments has yet been discovered (Churcher et al. 1999). Raw material available in this area includes quartzite. Although several expeditions have surveyed south of the oases, the extent of investigations by the DOP has been limited to the southern margins of present-day Dakhleh Oasis.

## Chronology

Two aspects of the Pleistocene chronology of the Western Desert that should be considered are evidence for climatic changes and evidence for human occupation. Although there is an obvious connection between the two, it is necessary to consider them separately. Archaeological artifacts occur in association with tufa or water-laid deposits in Kharga (Caton-Thompson 1952), Bir Sahara and Bir Tarfawi (Wendorf et al. 1993a), and Dakhleh (Churcher et al. 1999). While some archaeological units are clearly associated with periods of relatively high humidity, this is not the case for all archaeological units.

Figure 11.3 summarizes some of the Middle and Late Pleistocene U-series dates that have been obtained by KOPP

and DOP researchers from tufas and a calcareous deposit at Kharga and Dakhleh (Kleindienst et al. 1999, 2008; Smith et al. 2004, 2007).[4] Wendorf et al. (1993b), when summarizing the chronology of the lakes at Bir Sahara and Bir Tarfawi, assert that most of the lithic and bone accumulations at Bir Tarfawi can be assigned to 130 ka and later. It is noteworthy that between ca. 50 and 90 ka, there are very few chronometric dates on sediments indicative of higher humidity (Smith et al. 2007). There is one date for a calcareous deposit from Dakhleh at 40 ± 10 ka (Kleindienst et al. 1999), but this age is considered "rough." At Kharga, a U-series determination of 49.8 ± 0.1 ka on tufa from Mata'na falls into MIS 3 and roughly correlates with a few other determinations from the Western Desert (Szabo et al. 1989; Sultan et al. 1997; Smith et al. 2004).

## Archaeological Terminology

Archaeologists working in the Western Desert have used a plethora of terms to refer to Pleistocene-aged archaeological material. In this chapter, as noted above, the term "Middle Stone Age" (MSA) is preferred to the term "Middle Paleolithic" (see Kleindienst 2001, for a consideration of the use of African versus European-based terminology). As used here, the term MSA is an inclusive one that subsumes all Pleistocene-aged Levallois based units under discussion.

With respect to subdivisions within the MSA, researchers have generally followed one of two approaches—lumping or splitting. Among the splitters, one finds Caton-Thompson and Kleindienst. Caton-Thompson (1952) recognized a large number of archaeological units of different ages at Kharga, including the Lower and Upper Levalloisian, the Aterian, and the Khargan. Several other units (Acheulio-Levalloisian and Levalloisio-Khargan) are likely to have been from mixed deposits (Kleindienst et al. 2006, 2008). Similarly, Kleindienst (1999) describes several different units in Dakhleh, including a "large-sized" and "medium-sized" MSA, the Aterian, and the Khargan. To add to the complexity, researchers at Dakhleh assign "unit names" to the local manifestations of each technocomplex. For example, the Khargan at Dakhleh is referred to as the "Sheikh Mabrouk Unit" (Wiseman 1999), and the Aterian is referred to as the "Dakhleh Unit" (Kleindienst 1999). Kleindienst et al. (2006) also suggest new names for some of Caton-Thompson's units from Kharga, but these are not employed here.

---

[4] Infinite dates of >350 and >400 ka have been excluded.

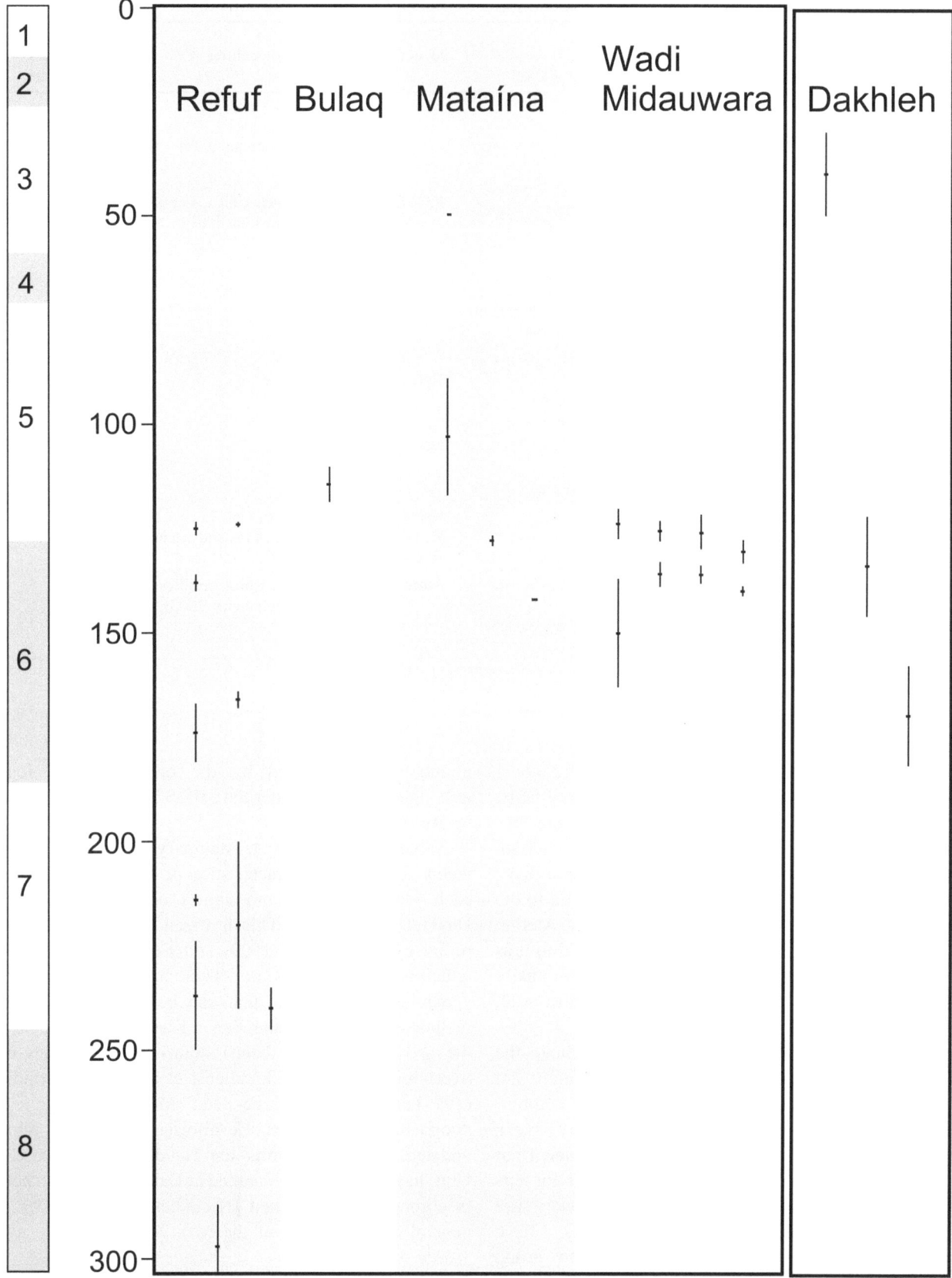

**Fig. 11.3** U-series determinations on tufa and a calcareous deposit from Kharga and Dakhleh, based on published data in Smith et al. (2004, 2007) and Kleindienst et al. (1999, 2008)

**Table 11.2** Framework for Pleistocene archaeological units from Dakhleh and Kharga Oasis

| Estimated Age, Ka BP | Technocomplex | Characteristics (Caton-Thompson 1952; Kleindienst 2003) | Kharga Oasis localities | Dakhleh Oasis localities | Other Western Desert locations |
|---|---|---|---|---|---|
| 30–50 | Khargan Technocomplex | Cores reduction <br>   Small-sized Levallois <br><br> Tools <br>   Scrapers | Bulaq solution pans | Sheik Mabruk Unit Localities (Wiseman 1999) | |
| MIS 3: <br>   > 50 < 100 | Aterian Technocomplex | Cores reduction <br>   Levallois flake cores <br>   Nubian point cores <br><br> Tools <br>   Range of scrapers <br>   Tanged points & other Tools <br>   Basally thinned tools <br>   Bifacial foliates | **KO6E** | Dakhleh Unit Localities (Hawkins 2001) | |
| MIS 5: ca. 120 | Medium-sized MSA | Core reduction <br>   Levallois flake cores <br>   Nubian point cores | **Refuf VII** | | **Bir Tarfawi** |
| | Upper Levalloisian | Tools <br>   Borers <br>   Basally thinned flakes | **Mata'na G** | | |
| MIS 7: ca. 220 | Large-sized MSA | Core reduction <br>   Levallois flake cores <br>   Nubian point cores | **Refuf IV** <br> **Refuf VIII** <br> **Refuf VI** | Teneida Unit Localities (Hawkins and Kleindienst 2002, 2003) | |
| | Lower Levalloisian | Tools <br>   Endscrapers <br>   Little evidence of retouch <br>   Gifata points | **Mata'na F** | Gifata Unit Localities (Kleindienst 2003) | |

Localities in bold are in situ within deposits. After Churcher et al. (1999) and Kleindienst (2003)

Other researchers (e.g., Hester and Hobler 1969; Mandel and Simmons 2001) recognize only a small number of units, and it is noteworthy that frequently no divisions are recognized in Levallois-based units that predate the Aterian. Members of the Combined Prehistoric Expedition have varied in their use of terms; they sometimes referred to the Levallois-based materials from Bir Tarfawi as Aterian (Schild and Wendorf 1975), and sometimes referred to it as Denticulate Aterian (Wendorf et al. 1987), but eventually they eschewed the use of labels in general (Wendorf et al. 1993a).

It is not my intention in this chapter to evaluate the validity of any of these divisions or terms, although I am conscious that the choices archaeologists make in combining or dividing archaeological entities ultimately have a bearing on behavioral interpretations. In this chapter, I put forward interpretations regarding adaptation, and for this reason I have chosen, as much as possible, to sidestep issues of terminology. In order to structure this analysis, I have chosen to use very broad climatic periods to group archaeological units. Furthermore, I have accepted the assignment of units to time periods as published by the original authors.[5] Finally, where I do use unit names, I have

maintained those used by the original archaeologists, although I do not consider the MIS 5 MSA at Bir Tarfawi to be Aterian.

Table 11.2 provides a summary of the chronology, technocomplexes, characteristics of the stone tools from each technocomplex, and names of the localities from Dakhleh and Kharga. This is based on a framework proposed by Kleindienst (1999) and Churcher et al. (1999), which is a revision of Caton-Thompson's (1952) scheme.

Archaeological units that can be attributed to MIS 7 include the Lower Levalloisian at Kharga (Refuf Locus IV), the age of which has been confirmed by U-series dates overlying this locus (Kleindienst et al. 2008). Kleindienst (1999) considers the "large-sized" MSA at Dakhleh to be a "cognate to the Lower Levalloisian," but this remains undated. Two MIS 7 units, the Tenida Unit and the Gifata Unit, have recently been named at Dakhleh although neither is chronometrically dated (Kleindienst et al. 2006). The general characteristics of the MIS 7 MSA include use of

---

[5] In some cases, the age of these units has been verified by chronometric dating (Wendorf et al. 1993a; Smith et al. 2007; Kleindienst et al. 2008), but in other cases they have not.

Levallois flake cores and Nubian point cores, production of end scrapers but few other retouched tools, and, according to Kleindienst (2003), production of an elongated biface referred to as a "Gifata point."

Materials that are ascribed to MIS 5 include the Upper Levalloisan at Kharga, which is dated based on U-series determinations on tufa associated with Mata'na G and Refuf Locus VII (Smith et al. 2007; Kleindienst et al. 2008). The "medium-sized" MSA at Dakhleh is thought to be a "cognate of the Upper Levalloisian" (Kleindienst 1999). Most of the sub-surface archaeological remains from Bir Tarfawi are dated to MIS 5 (Wendorf et al. 1993b). The MIS 5 MSA is characterized by the continued use of Levallois flake cores and Nubian point cores and the addition of borers and basally thinned flakes to the tool repertoire.

The Aterian technocomplex is represented at both Dakhleh and Kharga, but remains undated. It is ascribed to MIS 3–4 based on its position on relatively younger geomorphic surfaces than the MSA units above, and based on the position of finds at Bir Tarfawi. The clearly pedunculated pieces from Bir Tarfawi are surface finds affected by aeolian processes. Wendorf and Schild (1993) attribute them to dry phases of the Green Lake. Wendorf et al. (1993b) suggest that the Green Lake phase of higher humidity, which is not well dated, ended before 60 ka. The MIS 3–4 Aterian assemblages are characterized by an expansion of formal tool types to include tanged points and other tools and finely produced bifacial foliates. One finds more examples of basally thinned tools and numerous scrapers. Cores include both Levallois flake cores and Nubian point cores.

The Khargan technocomplex may date to between 30 and 50 ka, based on its location on younger surfaces at Dakhleh Oasis (Churcher et al. 1999). Therefore, it is also ascribed to MIS 3. Although the Khargan does not form part of this analysis, its place and chronology is noted because Aterian technocomplex sites do not occur on the younger surface on which Khargan sites are found. It is therefore suggested that the Aterian in the Western Desert may predate 30–50 ka.

## Fauna

Pleistocene-aged faunal assemblages from the Western Desert are few in number; Churcher et al. (1999) report on materials recovered from Dakhleh Oasis, and Gautier (1993), Van Neer (1993), and Kowalski (1993) describe the material from Bir Tarfawi and Bir Sahara. Churcher et al. (1999) place the Dakhleh material between 200 ka and greater than 300 ka, based on the "elevations and field relationships" of

the deposit in which the fauna are found. The fauna from Bir Tarfawi were recovered from several strata associated with different phases of the lake (Gautier 1993). Wendorf et al. (1993b) place them at ca. 130 ka and later, with the ages of the more recent phases not well-established.

The Middle Pleistocene fauna from Dakhleh suggests a "riverine gallery forest backed by savanna grasslands" to Churcher et al. (1999, p. 310). They draw a comparison between East Africa and Middle Pleistocene Dakhleh, indicating that a diverse array of resources would have been available in this environment. The presence of freshwater snails, in particular, is indicative of flowing freshwater (Churcher et al. 1999). Churcher et al. (1999) do not report any evidence of traces on the bone that could have resulted from hunting, scavenging, or butchering, and they do not elaborate on the association between clusters of lithic artifacts and faunal remains. This suggests that the faunal remains from Dakhleh should be treated, at this time, as paleontological rather than archaeological specimens. The reported hippopotamus, buffalo, antelope, gazelle, and zebra were at least available to humans using the oasis approximately 200 ka.

The fauna from Bir Tarfawi is interpreted as indicative of a Sudano-Sahelian dry savannah, with some variation through time (Gautier 1993). The presence of small gazelles and the decrease in the representation of large animals in the upper parts of the sequence suggests increasingly dry conditions closer to the present (Gautier 1993). During the Grey Lake 1 phase, a number of large mammals would have been present in the region, including rhinoceros, giraffe, buffalo, and camel. Gautier (1993) is equivocal about how they were obtained or used by humans, if they were at all, but he does indicate that they likely died at Bir Tarfawi and were not carried there by people. In contrast, he does assert that gazelles were hunted seasonally (Gautier 1993).

There is no clear association between any faunal assemblage in the Western Desert and the Aterian, nor is there any faunal assemblage that can be ascribed to the period subsequent to the Tarfawi Lakes and before the Holocene. A single equid metacarpal shaft found in the spring deposits at Location 80 is attributed to the Aterian by Churcher et al. (1999), but no unequivocally Aterian artifacts were discovered in this location (Hawkins 2001).

## An "Organization of Technology" Framework for Examining the Western Desert MSA

Study of the Aterian in much of northern Africa was initiated and has been subsequently carried out largely within a theoretical framework developed for use on European materials (Hawkins 2001; Kleindienst 2001). While there has been discussion of the relationship between the Aterian and the Mousterian (e.g., Bordes 1975–1976; Wengler 1994), and subdivisions within the Aterian (e.g., Caton-Thompson 1946; Ferring 1975), studies of the Aterian have generally not applied theoretical frameworks developed mainly by North American lithic analysts to try to investigate human behavior and behavioral changes, particularly among mobile foragers (e.g., Bamforth 1986; Bleed 1986; Shott 1986; Dibble 1987). To contribute to our understanding of the origins of modern behavior in northern Africa, it is important that research is framed in terms of understanding adaptation and changes to it, rather than by seeking the presence or absence of markers of modernity such as those suggested by Mellars (1989) and elaborated on in many subsequent publications (see McBrearty and Brooks 2000). Indicators of modernity, such as improved hunting ability, production of bone and antler tools, use of ochre, fishing, and production of items used for personal adornment, are impossible to apply in a region with poor archaeological preservation, and where most archaeological material is found in surface context. Other traits that may indicate modern behavior, such as production of blades and standardization of tools, are best considered contextually in terms of adaptation, mobility, raw material availability, and other requirements.

Nelson (1991, p. 57) describes the organization of technology approach as one focused on "the selection and integration of strategies for making, using, transporting and discarding tools and materials needed for their manufacture." Several key behavioral variables have been the focus of analysis, although Nelson cautions that to understand the dynamics of the past, it is best to employ several levels of analysis. At a general level, the concept of technological strategy can be considered a set of adaptive behaviors that people use to cope with variation in the availability of resources and human constraints on accessing them.

Resources differ in their availability in space and time, varying along a continuum of distribution from even to clumped. Exploitation of resources may involve careful scheduling and may require interception at specific locations. Technological strategies are methods or behaviors that address challenges such as these (Nelson 1991). Different technological strategies result in a range of tool design characteristics. Variable characteristics include reliability, maintainability, and transportability (Nelson 1991). Technological strategies also have outcomes with respect to the distribution of different types of sites on the landscape (Nelson 1991). While the distribution of Aterian sites at Dakhleh is reasonably well-known (Hawkins 2001), representative comparative databases for the earlier MSA are not available. Therefore, this comparison will focus on examining design characteristics of stone tools and the technological strategies of curation and expediency.

## Technological Strategies: Curation and Expediency

People employ technological strategies to extract resources from the environment, in particular, to balance competing requirements of the timing and location of resource-gathering activities. Three strategies, which are not mutually exclusive, are curation, expediency, and opportunism (Nelson 1991).

Curation involves advance energy investment in tools and toolkits, which may be manifested in, for example, fashioning, reworking, storing, or transporting of toolkits or parts thereof. Curation is a strategy that enables people to cope with the occurrence of raw materials and resources to be exploited in different locations. It helps overcome problems related to short-term availability of resources.

The concept of curation has been the subject of much debate, owing to a lack of clarity in its introduction into the archaeological literature (Binford 1973). Shott (1996, p. 267) argues that it should be defined as "the degree or utility extracted, expressed as a relationship between how much utility a tool starts with—its maximum utility—and how much of that utility is realized before discard." In this chapter, I follow Nelson's (1991, p. 62) broader concept of curation as "a strategy of caring for tools and toolkits than can include advanced manufacture, transport, reshaping, and caching or storage… a critical variable differentiating curation from expediency is preparation of raw materials in anticipation of inadequate conditions… for preparation at the time and place of use."

Expedient strategies are employed when raw materials are available at the location of need, which may be used on multiple occasions, and there is no time stress involved in preparing them for use. According to Nelson (1991), use of expedient strategies depends upon resource extraction where raw materials are present, either naturally or through caching, lack of time stress to prepare tools for use, and use of a resource extraction site on multiple occasions. Curated and expedient strategies both involve anticipation of raw material needs in the future. While people who employ curation strategies anticipate future need within a narrow time frame, those who use expedient strategies expect that both time and raw materials for tool preparation will be available at the location of resource exploitation. Use of different technological strategies will result in different assemblage composition and artifact forms (Nelson 1991).

In contrast to curation and expediency, opportunistic strategies are employed in unanticipated conditions. When a hunter, collector, or scavenger is confronted with an unanticipated opportunity for resource exploitation and finds a technological solution that allows for exploitation of that resource, such a strategy is considered opportunistic (Nelson 1991). A crucial difference between opportunism and expediency, as defined by Nelson, is that expediency is a planned strategy, while opportunism is not.

Kuhn (1992) addressed the issue of whether it is possible to apply frameworks that imply planning to premodern humans. He points out that other animals engage in behaviors that could be considered "planned" (Kuhn 1992), but that humans and premodern humans relied upon tools and, therefore, by necessity engaged in planning behavior. Rather than considering whether or not there is evidence for planning, Kuhn argues that it is more instructive to consider the conditions for planning and how these may have differed over time, between groups, and according to other variables. He distinguishes broadly between provisioning of places and provisioning of people. The latter entails the concept of "personal gear," described by Binford (1979) as the tools individuals always carry with them in general anticipation of need. These tools are curated, in that they are prepared and cared for in advance, and that their design characteristics include portability and maintainability (Kuhn 1992). People provision places by carrying raw materials, tools, and partially prepared materials to locations where they anticipate a future need. The transport of materials to a location in anticipation of future use is a type of curation, however, the later of use of these at the location is likely to be expedient.

In the following sections, I will consider how the differences and similarities in the nature of recovered material of different ages may relate to technological organization. I will first examine concentrations that are found near raw material outcrops and I will follow this by comparing concentrations located near former sources of water and further from raw materials. Finally, I will consider what can be learned from the characteristics of the tools themselves.

## Technological Strategies Near Raw Material Outcrops

Although there are several types of lithic raw material available at Dakhleh and Kharga, the preferred material is chert, specifically chert derived from Tarawan formation limestone (Hawkins and Kleindienst 2002). A number of localities in the escarpment and piedmont zones at Dakhleh and Kharga record where people took advantage of raw material outcrops. The MIS 7 MSA is represented by published descriptions of material from the Kharga Lower Levalloisian sites (Caton-Thompson 1952) and the Dakhleh Gifata Unit (Kleindienst 2003). The MIS 5 MSA is represented by the Kharga Upper Levalloisian material (Caton-Thompson 1952) and possibly surface material from a workshop at Wadi Midauwara (Smith et al. 2007). Aterian material comes from several P-II surfaces in Dakhleh (Hawkins 2001). While the Aterian localities are undated, tanged tools, basally thinned flakes, and/or bifacial foliates

of the same degree of abrasion and patina as flaking debris were recovered from these locations (Hawkins 2001).

Springs issued from the scarp during MIS 5 and 7, and based on plant casts in tufa, it was vegetated and would have been attractive to fauna (Smith et al. 2007). Thus, the escarpment localities may have served as *both* resource extraction and lithic raw material procurement areas. To interpret the nature of the activities carried out on the scarp, it is worthwhile to consider what type of remains would be expected from the different technological strategies.

Resource extraction may have been opportunistic (sensu Nelson 1991), in which case we would expect to find nearly all of the remains of complete reduction sequences in small scatters. Resource extraction in this area may also have been based on an expedient strategy in which humans knew that both raw materials and resources were present. In this case, at raw material outcrops, we may anticipate finding little more than a few cortical flakes that may have been removed to examine the quality of the nodule. At resource extraction locations, we would expect to find most of the elements of complete or nearly complete reduction sequences, similar to the remains resulting from opportunistic behavior. However, if people consciously or unconsciously provisioned places for future anticipated use, the concentrations of artifacts would be higher than in locations of chance, opportunistic resource extraction. Further, as locations are reused, the materials found there may be re-sharpened or reworked, so cores may be smaller, more non-cortical flakes will be present, and larger numbers of retouched tools will be found. Finally, it is possible that resource extraction on the scarp took place within a system in which tools were curated (i.e., tools were brought to the site as personal gear). In this case, one would expect to find raw material outcrops with evidence of flake preparation for use as tools and possibly objects that are non-local in origin having been exchanged at the knapping site. In a technological strategy focusing on provisioning of people as opposed to places, resource extraction sites would have less evidence for on-site knapping.

Table 11.3 shows the types of materials that were recovered from different MSA localities near lithic raw material outcrops. The samples from the in situ sites at Kharga are all very small and it is impossible to know whether the numbers of artifacts published by Caton-Thompson (1952) represent the total number of pieces at these aggregates. Recent observations of sections at localities at Refuf and Mata'na suggest that these scatters are relatively small and can be contrasted with the dense concentrations of knapping material found on the P-II remnants in Dakhleh and at Midauwara 10.

The samples of MIS 7 MSA (Lower Levalloisian) materials are the smallest. Despite the small size, the lack of

**Table 11.3** Types of lithic material recovered from lithic raw material exploitation sites on or near raw material outcrops

| | | Primary and secondary flakes | Initial or failed cores | Struck cores | Core trimming flakes | Failed Levallois flakes | Edge modified tools | Levallois flakes | Exhausted tools parts |
|---|---|---|---|---|---|---|---|---|---|
| MIS 7—Lower Levalloisian | Refuf IVa | X | X | X | ? | | | | |
| | Refuf IVb | X | X | X | ? | | X | | |
| | Refuf VIII | | X | X | X | | X | | |
| | Refuf VI | | X | X | | | X | X | |
| MIS 5—Upper Levalloisian | Mata'na G | X | X | X | X | X | X | X | X |
| | Midauwara 10 | X | X | X | X | X | X | X | |
| MIS 3?—Aterian | L. 216 TP | X | X | X | X | X | X | X | X |
| | L. 342 | X | X | X | X | X | X | X | X |
| | L. 216 BC | X | X | X | X | X | X | X | X |

Compiled from data in Caton-Thompson (1952), Hawkins (2001) and fieldnotes

complete reduction sequences suggests that these do not represent opportunistic use of resources. In some cases, primary and secondary flakes are lacking, suggesting that debris represents material removed from partially prepared cores. At other sites, Levallois flakes are lacking, suggesting that they were transported elsewhere. Edge modified tools are present at most MIS 7 MSA sites, possibly indicating that these locations were used for resource extraction or maintenance activities.

The MIS 7 MSA from Dakhleh is represented by the Gifata Unit, which is found dispersed on a P-II surface near the base of the escarpment. According to Kleindienst (2003, p. 10), "the aggregates appear to represent highly dispersed clusters," but she also notes that a conjoining flake and core were found on this surface less than 2 m apart. The high proportion of primary cortical flakes suggests to Kleindienst that these were workshop locations. The density of material is low, ranging from 0.02 to 0.12 artifacts per m$^2$. While the different site situations make it difficult to compare the Kharga and Dakhleh materials, it is noteworthy that the artifact concentrations are small in both locations. The Gifata materials may be most similar to the Refuf IVa and IVb concentrations in which cores and core preparations materials are present, suggesting workshops and preparation of flakes for use as personal gear.

The MIS 5 MSA (Upper Levalloisian) material from Mata'na G is found in a similar context to the clusters of MIS 7 MSA Refuf material but the sample is larger and a wider range of lithic material is represented. Some of the increased diversity may be explained by the greater sample size. Table 11.4 shows that, in addition to flaking debris, retouched tools were recovered. These may have been made for use at this location, but thinning of the base suggests hafting, and potential exchange of tool parts at this location.

Thus, the Mata'na G materials may be interpreted as evidence that both provisioning of place and provisioning of people were part of the technological strategies of MIS 5 MSA humans here.

Midauwara 10 is located on a tufa sheet on the escarpment at Kharga where chert outcrops naturally (Smith et al. 2007). The archaeological site consists of clusters of flaking debris, including initial and struck cores, primary flakes, failed Levallois flakes, and some edge retouched tools. The only good example of a formed tool can be attributed to more recent manufacture, based on abrasion and patina (Smith et al. 2007). The tufa surface on which the artifacts lie has been dated to 124.8 ± 4.0 ka, representing a minimum age for this outcrop. The lack of formed tool parts suggests that this location was not used for retooling. The number of edge retouched tools is low, and the cores indicate production of flakes and points, which would have been produced for transport elsewhere.

The Aterian locations differ from all of the others in that they would not have been located near water; lying on isolated remnants of the P-II surface, they would, however, have provided good vantage points for viewing the surrounding oasis. These localities show the range of knapped material, including many examples of both failed flakes and points, and of exhausted tools. This suggests that one of the main knapping activities here was production of flakes and points for transport away from the knapping location.

There seems to be evidence for the use of special locations to produce flakes and possibly cores throughout the MSA, but in the MIS 7 MSA, these locations appear to be small and may represent use on only a few occasions. Later in time, such locations appear to have been used repeatedly. During MIS 5 and 7, the escarpment likely served as a location for procuring both stone and other

**Table 11.4** Comparison of the distribution of cores, flakes and flaking debris and tools from localities near lake deposits (L. 211 and BT 14 A) and on the southern oasis margin (L. 130)

|  | MIS 7 MSA | | MIS 5 MSA | | Aterian | |
|---|---|---|---|---|---|---|
|  | Dakhleh L. 211 | | Bir Tarfawi 14 A | | Dakhleh L. 130 | |
|  | N | % | N | % | N | % |
| Cores and core fragments | 32 | 15 | 48 | 2 | 45–57 | 5–6 |
| Flakes, broken flakes and debris | 158 | 74 | 1535 | 80 | 783–819 | 91–86 |
| Retouched tools | 24 | 11 | 340 | 18 | 37–71 | 4–7 |

Collections from all three locations were systematically carried out (Kleindienst 2003; Wendorf and Schild 1993; Hawkins 2001)
The totals from L. 130 are given as a range with the first number indicating the number of objects collected in a grid collection and the second number indicating the total number of objects collected. Selective collection of tools and cores occurred before the grid collection, depressing the numbers of tools and cores in the grid. The selective collection, however, extended beyond the area of the grid

resources; the Aterian is much better known from locations within the oasis and has not been found in association with tufa deposits on the scarp. This suggests that during Aterian times, people made less use of the scarp and more use of central oasis locations. Within the central oasis, there appears to be more intensive use of the same places.

## Technological Strategies at Resource Exploitation Locations Near Former Bodies of Water

A second context in which MSA sites are located in the Western Desert is near former water bodies and some distance from raw material outcrops. Again, it is useful to consider how different technological strategies might be reflected in stone tool assemblages. These locations may be places of opportunistic behavior, in which case we would expect to find complete reduction sequences of raw materials that are found in the immediate area. More likely, these places were locations of planned use at some level. If places were provisioned with raw materials, we would expect to find a range of materials, including cores. At least some of these would likely be highly reduced as they are used and reused on subsequent visits. It is also possible that some of the cores may bear cortex or be only partially reduced. In addition, assemblages would include flaking debris and tools used on site for resource extraction or maintenance. If people employed a strategy of provisioning people, one would expect to find two types of sites, remains of camps at which tool maintenance activities were conducted (Binford 1979) and locations of resource extraction. The latter site type is likely to be represented by isolated tools or fragments, or small scatters of tools. The former would likely include some cores and flaking debris, because cores may have formed part of the personal gear. If this were the case, one would anticipate that most of the discarded cores would be highly reduced as they would be removed from the

personal gear because they would no longer be useful. Binford (1979) asserts that people are not likely to embark on resource extraction excursions without preparing personal gear in advance.

The difference between a strategy in which people were provisioned with artifacts and one in which places were provisioned with raw materials may be difficult to observe archaeologically, particularly when there are no accompanying ecofacts that may be used to interpret site function.

In eastern Dakhleh, MIS 7 MSA (Teneida Unit) localities (Churcher et al. 1999; Kleindienst 2003) are found associated with lake deposits.[6] Kleindienst (2003) contends that, although much of this material occurs in surface context, its condition indicates that it has eroded out of deposits recently, and, therefore, that it is representative of MIS 7 MSA material and is not mixed with later material. Locality 211 will be used to exemplify the Teneida Unit.[7] The MIS 5 MSA will be exemplified by the Bir Tarfawi 14 Area A (BT 14 A), the material described by Wendorf and Schild (1993). It is also associated with lake deposits. Given the association with lake environments, it is possible that the Dakhleh L. 211 and Bir Tarfawi locations were used for extraction of resources such as animal carcasses and aquatic plants or that such activities were carried out nearby.

---

[6] Kleindienst (2003, p. 32) notes that "the only chronometric determinations available for the Teneida Palaeobasin are Ar/Ar determinations on lagged Dakhleh Glass ... the isochron age limiting the age of some pre-existing surface of the Lake Teneida Formation onto which the glass was deposited is 122,000 ± 40,000 (Schwarcz et al. 2008)." Churcher et al. (1999, p. 305) assert that the deposit from which the faunal remains derive "dates broadly to the later middle Pleistocene." Some of the earlier MSA material from the Teneida Paleobasin is found in situ in the same deposit, some is found in surface context. Kleindienst (2003, p. 26) indicates that the Teneida unit material is "probably within the older Middle Stone Age time range."

[7] Material from L. 211 was selected for examination because "all artefacts seen were collected" (Kleindienst 2003, p. 32), making the assemblage generally comparable with the grid collection from 130 and the excavated material from BT 14 A.

**Table 11.5** Percentage breakdown of tool types recovered from MSA localities found near former water sources. Compiled from Kleindienst (2003), Wendorf and Schild (1993) and Hawkins (2001)

|  | MIS 7 | MIS 5 | MIS 3–4 |
|---|---|---|---|
| Locality | 211 | BT 14A | 130 |
| N | 17 | 263 | 63 |
| Retouched points | 6 | 8 | 24 |
| Retouched scrapers | 76 | 27 | 13 |
| Burins | 0 | 1 | 0 |
| Bifacial foliates | 0 | 2 | 2 |
| Basally thinned tools | 0 | 0 | 14 |
| Tanged tools | 0 | 3 | 3 |
| Tanged points | 0 | 1 | 3 |
| Edge modified tools (notches and denticulates) | 18 | 58 | 41 |

Finally, Locality 130 is an Aterian concentration in southeastern Dakhleh. The presence of ancient and modern wellheads near it suggests that this site was also located near a former water source, but it is difficult to infer its exact use (Hawkins 2001).

All three of these MSA occurrences are found some distance from outcrops of raw material. The MIS 7 MSA (Teneida Unit) localities are at least 6 km from sources of Tarawan chert (Hawkins and Kleindienst 2002), Tarfawi is approximately 3 km east of sources of quartzitic sandstone used by MIS 5 MSA inhabitants of BT 14 A (Wendorf et al. 1993b), and the Aterian L. 130 is between 5 and 10 km from Tarawan chert sources (Hawkins and Kleindienst 2002).

Table 11.4 compares the proportion of different artifact types from the three locations. At all three places, in which a range of material, including cores, flaking debris, and tools, is represented, we find evidence for knapping of stone that has been carried in from some distance. The question under consideration is whether the stone was brought to the location for use at that specific place or whether it was brought there in the course of normal travel and was further prepared there for use at some other nearby location.

Two pieces of evidence suggest that the three locations likely record some form of habitual behavior. It is unlikely that the amount of material records use on a single occasion, particularly for BT 14A. At the other two locations, a range of raw materials is present, suggesting procurement from different areas (Hawkins and Kleindienst 2002).

The proportion of cores at the Bir Tarfawi location is quite low, despite Wendorf and Schild's (1993, p. 268) assertion that "[a] major activity at almost all of the sites adjacent to lakes was the final preparation and initial exploitation of partially prepared Levallois cores." If cores were only initially exploited, we may infer that they formed part of the personal gear and were transported away from this location.

Data on the sizes of cores is unfortunately lacking for L. 211 but at another MIS 7 MSA (Teneida Unit) locality (L. 374), mean length of 7 "specialized cores" is 54 mm. This does not differ greatly from the 55 mm mean length of 23 Levallois cores from the Aterian L.130. Although the sample sizes are very small, this may suggest that cores were abandoned mainly when they were too small to be of further use. Cores abandoned at Aterian workshop locations are significantly larger (Hawkins 2001). Abandonment of exhausted cores is more likely to occur in a pattern of provisioning of people than of place. Future research should consider the degree to which these concentrations include cortical flakes and the range of sizes of abandoned cores.

The suite of tools may also help to determine the nature of the technological strategies employed at different locations. Table 11.5 shows a summary of the types of tools found at the three localities. Terminology and analysis methods employed by different researchers vary, and summarization doubtlessly simplifies the complexity of the tool assemblages. However, two things are noteworthy. First, scrapers outnumber points at both of the earlier MSA sites, while points outnumber scrapers at the Aterian site. Second, a large number of the tools from BT 14 A are tools such as denticulates and notches.

These differences suggest that a different suite of resources was exploited during the earlier MSA than during the Aterian. It is possible that all of these locations served essentially as occupation locations or camps in which tools were repaired and/or resources that had been brought to the location were used. The difference in the tool assemblages suggests a greater emphasis on tools used as part of mobile personal gear in the Aterian and greater use of tools employed on-site in the earlier MSA.

Finally, although comparison is not possible because published accounts of the earlier MSA do not include tallies

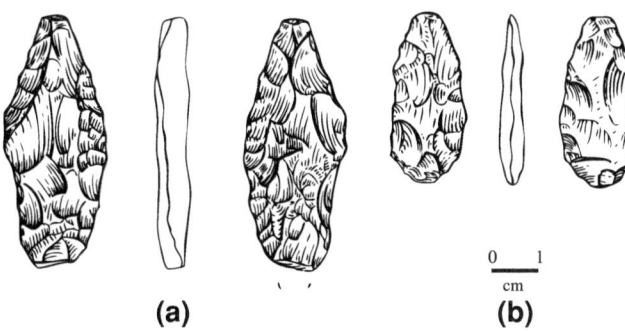

**Fig. 11.4** Bifacial foliates recovered from L. 325, an Aterian surface locality at Dakhleh Oasis

of broken tools, at L. 130, 34% of tools collected were broken, suggesting that a major activity in this area was retooling.

During all three of the periods concerned, people appear to have exploited resources associated with bodies of water. It would appear that they revisited the same locations, bringing stone from several kilometers away with them. The fact that these locations were subject to revisits suggests that they provided humans with something attractive, which may have been water, the resources associated, and/or stone materials brought on previous occasions. Without further analysis, it appears that the differences among the three types of sites lie mainly in the nature of the associated tools, with the Aterian site having greater evidence for tools likely to be used away from the site.

## Tool Design

Technological strategies will also affect design characteristics of tools. Requirements of tools needed for exploiting mainly non-mobile resources available at predictable times will be quite different from the requirements of tools needed for exploiting mobile resources available for only a short time, or at an unpredictable time. Several variables of tool design include reliability, maintainability, flexibility, versatility, and transportability (Nelson 1991).

Reliably designed tools function when they are needed, and are therefore considered appropriate for both encounter hunting of game that occurs in unpredictable locations and for hunting of game available in short time frames (i.e., migratory species) (Bleed 1986). To ensure that the tool functions when needed, it is over-designed (Bleed 1986). Examples of reliable design outcomes are secure fittings, standardized replacement parts, and redundancy (Nelson 1991).

Maintainable designs can work easily under different circumstances. One design strategy that allows for this is

flexibility in that tools can be reformed to meet different needs. A second strategy that also results in maintainability is termed "versatility." In this case, tools are maintained in a generalized form to allow for use in multiple tasks.

The repertoire of formed tools assigned to the MIS 7 MSA is very small. Caton-Thompson (1952, p. 28) describes retouch in the Lower Levalloisian as follows, "retouch, if any, which is exceptional apart from very rare end-scrapers ... is intermittent and nibbling." The list of formed tools assigned to the Gifata Unit includes "side-scrapers, core axes, ventrally-thinned flakes, and a distinctive type of bifacial point" (Kleindienst 2003, p. 25). This point type, which is termed a "Gifata point," is long (ca. 14 cm), narrow (ca. 5 cm), and bifacially worked, frequently with a cortical butt (Kleindienst 2003). These have been discovered in surface context associated with the L. 187 workshop material; the presence of such points on younger surfaces is attributed to redeposition (Kleindienst 2003). Attribution of these tools to the MIS 7 MSA requires confirmation by recovery of similar tools from dated deposits. ·

A somewhat expanded suite of tools is assigned to the MIS 5 MSA. At Kharga, Caton-Thompson (1952) notes that retouch is more invasive and includes basal thinning. The formed tools from Bir Tarfawi include a number of scrapers, Mousterian points (including some with ventral retouch at the tip), and basally-thinned or truncated pieces (Wendorf et al. 1993a). Although they also describe tanged tools, these are either found in surface context (Wendorf and Schild 1993) or they are unlikely to be true tangs, being made on the distal end of flakes (Wendorf and Schild 1993). Similarly, the tools described as bifacial foliates do not closely resemble the thin points typical of the Aterian (compare Wendorf et al. 1993a, with Hawkins 2001, and Caton-Thompson 1952).

Clearly the Aterian is marked by a number of changes in tool technology, including production of small bifacial foliates (Fig. 11.4). It is difficult to imagine that the main function of these was not as points. Tanged tools include points, scrapers, burins, and borers (Fig. 11.5). There is a continuation and perhaps expansion of the use of basal thinning, and both pointed tools and flakes are thinned (Fig. 11.6).

Examined in terms of tool design, it seems clear that there is an increased concern with hafting through time. The Gifata point, if it can be attributed to the MIS 7, was likely hafted, as were the basally thinned pieces of the Upper Levalloisian at Kharga. However, the Aterian bifacial foliates, basally thinned tools, and tanged tools are all designed to be placed within handles. A significant portion of these tools made for hafting are points (Fig. 11.4a), which we can only surmise were used in hunting. I interpret this change as indicative of greater concern with reliability of tools.

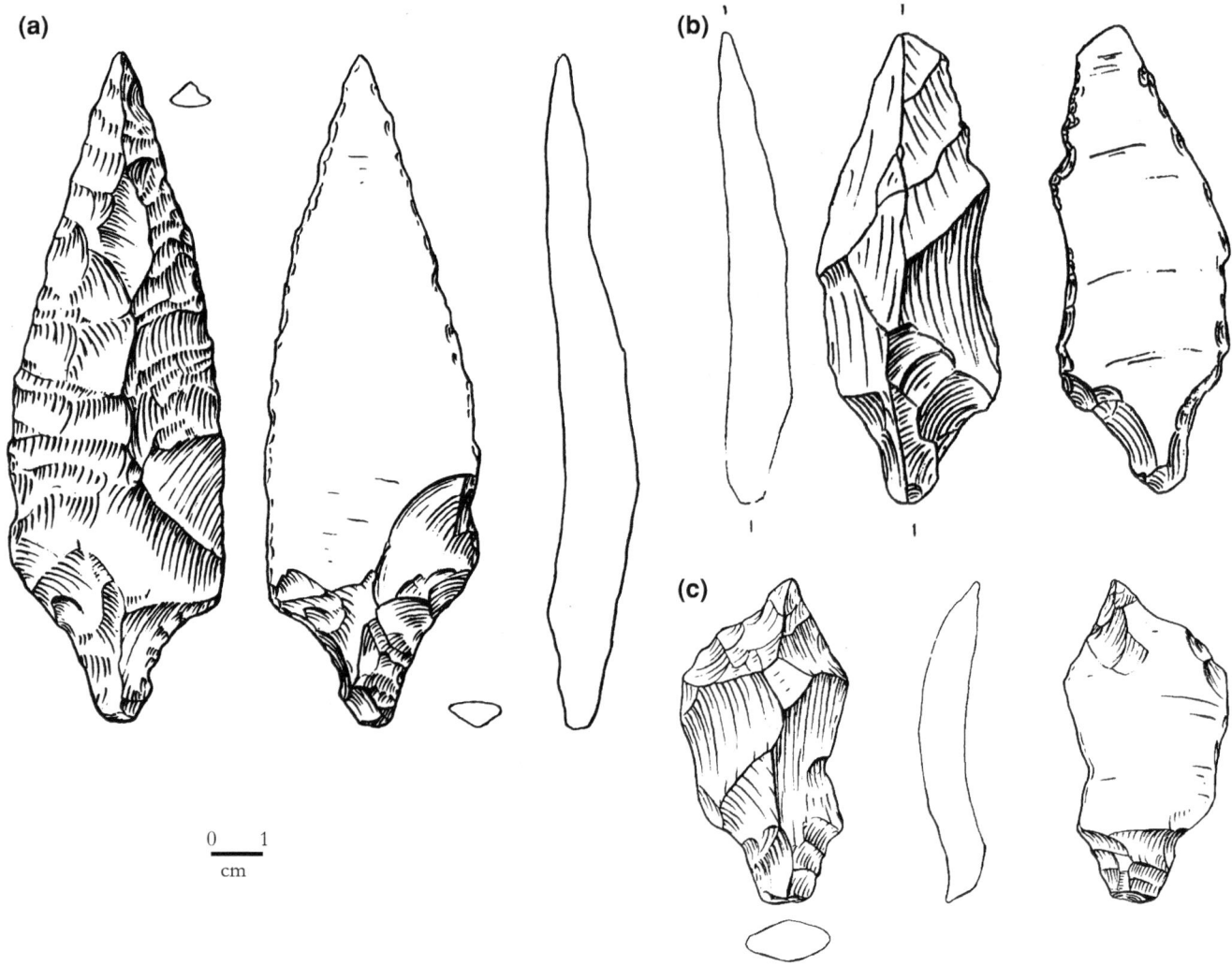

**Fig. 11.5** Tanged tools recovered from Aterian surface localities at Dakhleh Oasis: **a** tanged point from L. 328, **b** tanged flake from L. 283, and **c** tanged borer from L. 130

The wide range of tools that are designed for hafting, both tanged and basally thinned, is also worth considering. These tools are frequently asymmetrical and are often unretouched above the tang or basally thinned area (Fig. 11.4b). At Dakhleh, the tang was designed around a major arête on the dorsal face of the flake; the flint knapper appears to have been concerned with producing a tool with a maximally strong tang, rather than with producing a symmetrical tool. In terms of design, this can be interpreted as concern with reliability of the haft; the hafted tool should not break in the handle when in use. The range and high number of tools that are not retouched above the tang may indicate concern with maintainability, obtained through using a flexible design. Tools may be used for cutting in their unretouched form, but they may also be further modified to produce scrapers, borers, and possibly points.

## Conclusions

The database from the Western Desert demonstrates the following:

1. There were periods of higher precipitation at a number of times in the Pleistocene and these resulted in the presence of lakes in Dakhleh, Bir Tarfawi, and probably also Kharga.
2. Human use of the oases and the Bir Tarfawi area during these periods is well attested.
3. Numerous archaeological localities in the Western Desert have been dated to MIS 5–7 but true tanged tools have not been recovered from any of these; they are not considered to be Aterian.

As discussed above, there are many limitations to the available database for the Western Desert. In this chapter, I propose an explanation for some of the differences that

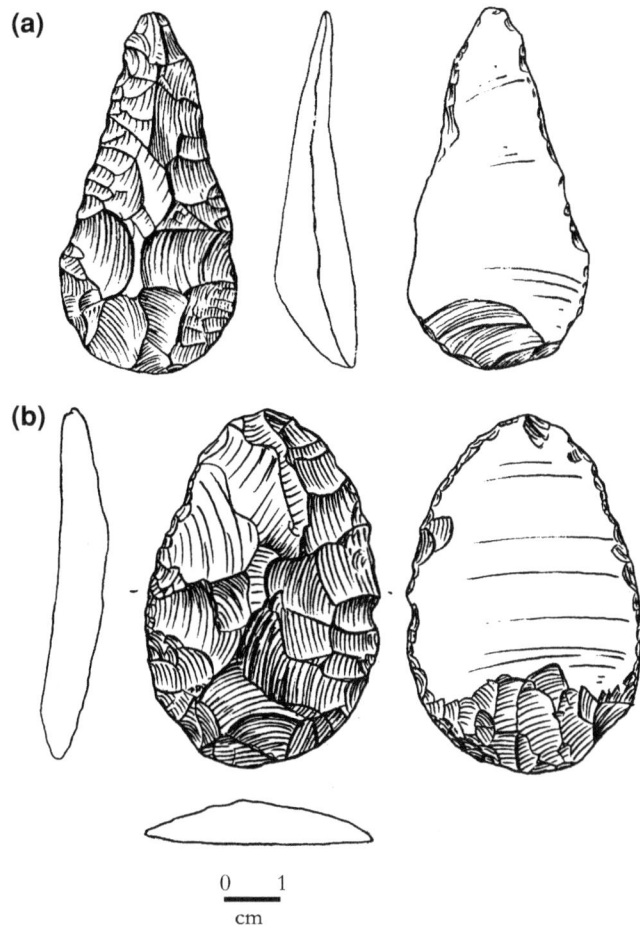

**Fig. 11.6** Basally thinned tools from Aterian surface localities at Dakhleh Oasis: **a** pointed tool from L. 325, **b** basally thinned flake from L. 319

have been observed between earlier MSA units and the Aterian. I have attempted, as much as possible, to use comparable samples for this analysis, however, this has not always been possible and it is my hope that future research will focus on direct comparison of the earlier MSA and the Aterian.

The model that is proposed here suggests that the differences among the MSA units attributed to MIS 7–3 are ones of degree rather than kind. However, the change from the MIS 5 MSA to the Aterian is more pronounced than earlier changes. The differences can be explained reasonably well by a model in which different technological strategies were adopted. These are solutions to problems, or adaptations. Ample evidence points to relatively humid periods during MIS 7 and 5, while the evidence for higher precipitation and spring activity during MIS 3–4 exists but is much more ephemeral. I suggest that, in more humid periods, humans living in the Western Desert would have encountered adequate non-mobile resources in their environment for survival. They may have also pursued game, as

is suggested by both Gautier (1993) and the Gifata points at Dakhleh. However, the archaeological record suggests that this was a comparatively minor contributor to subsistence.

We may anticipate that different resources would have been available in drier periods. Aquatic vegetation and/or carcasses that could be scavenged at lakeshore settings were likely fewer and the escarpment would not have been an attractive location from a subsistence perspective. The exploitation of fauna may have required careful scheduling if, for example, people exploited migratory species such as some species of gazelle. Alternatively, if the hunting strategy was mainly through encounter because locations of game were less predicable, there would also be a need for reliable tools.

Desiccation and the decrease in available resources may have resulted in increased human mobility and focused human activity on a few known locations of resource exploitation. The high number of point cores at the Aterian workshop locations, and the relatively high density of material suggest reuse on many occasions, with the primary focus of lithic reduction being the production of flakes and points for transport, rather than the production of tools for use at the site (Hawkins 2001). Further, higher mobility may have made a focus on personal gear as opposed to provisioned places advantageous.

To what extent were the Aterians who used the Western Desert behaviorally modern? The lithic reduction strategies employed during the Aterian are largely a continuation of those used by earlier groups. The basal thinning of tools to aid in hafting also occurs first in the MIS 5 MSA. If Kleindienst's (2003) attribution of the Gifata points is correct, production of foliates also occurs earlier than the Aterian. Repeated use of workshop locations such as at Midauwara 10 occurs first in MIS 5, but is in greater evidence in the Aterian. Using sites near water as base camps or possible resource extraction locations appears to occur from MIS 7 though MIS 3/4, but the nature of the tools discarded at these locations changes.

Based on the evidence from lithics and the distribution of sites, at this time it is only possible to say that the Aterian represents an elaboration on behavior and technology that occur earlier, and that this elaboration is likely related to changes in subsistence, probably arising in part from climatic changes. At the same time, it must be remembered that this analysis is based largely on surface material, and that many of the indicators of modern behavior—if they existed—would not have survived several tens of thousands of years on the desert surface.

**Acknowledgments** This chapter is based on several seasons of research in the Dakhleh and Kharga Oases, many of which were made possible through National Geographic Foundation Grants. Additional funding came from University of Toronto travel grants and a Halbert Foundation post-doctoral fellowship to Alicia Hawkins, and a Leakey Foundation grant to Jennifer R. Smith. I thank Dakhleh Oasis Project director A.J. Mills and Kharga Oasis Prehistory Project directors M.R.

Kleindienst and M.M.A. McDonald. I thank the Supreme Council of Antiquities in Egypt for granting concessions to allow work in Kharga and Dakhleh. Jonathan O'Carroll illustrated the artifacts pictured in Figs. 11.4, 11.5, 11.6. Numerous individuals collaborated and assisted in different ways. I particularly wish to thank J. R. Smith, M. R. Kleindienst, M.F. Wiseman, and T. Ormerod. I thank Jean-Jacques Hublin and Shannon McPherron for organizing the "Modern Origins: A North African Perspective," and for inviting me to participate. The comments of three anonymous reviewers on an earlier version of this manuscript were very helpful.

# References

Bailey, G. (2007). Time perspectives, palimpsests and the archaeology of time. *Journal of Anthropological Archaeology, 26*, 198–223.

Bamforth, D. B. (1986). Technological efficiency and tool curation. *American Antiquity, 51*, 38–50.

Bárta, M., Brůna, V., Svoboda, J. A., & Verner, M. (2002). El-Hayez, Bahariya Oasis, Egypt. 1st survey report by the Czech Institute of Egyptology. *Přehled výzkumů, 44*, 11–14.

Binford, L. R. (1973). Interassemblage variability: The Mousterian and the "Functional" argument. In C. Renfrew (Ed.), *The explanation of culture change: Models in prehistory* (pp. 227–254). London: Duckworth.

Binford, L. R. (1979). Organization and formation processes: Looking at curated technologies. *Journal of Anthropological Research, 35*, 255–273.

Bleed, P. (1986). The optimal design of hunting weapons: Maintainability or reliability. *American Antiquity, 51*, 737–747.

Bordes, F. (1975–1976). Moustérien et Atérien. *Quaternaria, 21*, 19–34.

Brooks, I. A. (1986). Quaternary geology and geomorphology of the Dakhleh Oasis region and its environs, South-central Egypt: Reconnaissance findings. *York University, Department of Geography, Discussion Paper, 32*, 1–90.

Caton-Thompson, G. (1946). The Aterian industry: Its place and significance in the Palaeolithic world. *Journal of the Royal Anthropological Institute of Great Britain and Ireland, 76*(1), 87–130.

Caton-Thompson, G. (1952). *The Kharga Oasis in prehistory*. London: Athalone Press.

Churcher, C. S., & Mills, A. J. (Eds.). (1999). *Reports from the survey of Dakhleh Oasis, 1977–1987*. Oxford: Oxbow.

Churcher, C. S., Kleindienst, M. R., & Schwarcz, H. P. (1999). Faunal remains from a Middle Pleistocene lacustrine marl in Dakhleh Oasis, Egypt: Palaeoenvironmental reconstructions. *Palaeogeography, Palaeoclimatology, Palaeoecology, 154*, 301–312.

Dibble, H. (1987). The interpretation of Middle Paleolithic scraper morphology. *American Antiquity, 52*, 109–117.

Ferring, C. R. (1975). The Aterian in North African prehistory. In F. Wendorf & A. Marks (Eds.), *Problems in prehistory: North Africa and the Levant* (pp. 113–126). Dallas, TX: SMU Press.

Gautier, A. (1993). The Middle Paleolithic archaeofaunas from Bir Tarfawi (Western Desert, Egypt). In F. Wendorf, R. Schild, & A. E. Close (Eds.), *Egypt during the last Interglacial: The Middle Paleolithic of Bir Tarfawi and Bir Sahara East* (pp. 121–143). New York: Plenum Press.

Hawkins, A.L. (2001). Getting a handle on tangs: Defining the Dakhleh Unit of the Aterian technocomplex—a study in surface archaeology from Dakhleh Oasis, Western Desert, Egypt. Ph.D. Dissertation, University of Toronto.

Hawkins, A. L., & Kleindienst, M. R. (2002). Lithic raw material usages during the Middle Stone Age at Dakhleh Oasis, Egypt. *Geoarchaeology, 17*, 601–624.

Hawkins, A. L., Smith, J. R., Giegengack, R., McDonald, M. M. A., Kleindienst, M. R., Schwarcz, H. P., et al. (2001). New Research

on the prehistory of the escarpment in Kharga Oasis, Egypt. *Nyame Akuma, 55*, 8–14.

Hester, J. J., & Hobler, P. M. (1969). *Prehistoric settlement patterns in the Libyan Desert*. University of Utah Anthropological Papers, 92, University of Utah.

Kieniewicz, J.M. (2007). *Pleistocene pluvial lakes of the Western Desert of Egypt: Paleoclimate, paleohydrology and paleolandscape reconstruction*. Ph.D. Dissertation, Washington University.

Kleindienst, M. R. (1999). Pleistocene archaeology and geoarchaeology of the Dakhleh Oasis: A status report. In C. S. Churcher & A. J. Mills (Eds.), *Reports from the survey of Dakhleh Oasis, Western Desert of Egypt, 1977–1987* (pp. 83–107). Oxford: Oxbow Books.

Kleindienst, M. R. (2001). What is the Aterian? The view from Dakhleh Oasis and the Western Desert, Egypt. In C. A. Marlow & A. J. Mills (Eds.), *The Oasis Papers 1: The proceedings of the first conference of the Dakhleh Oasis Project* (pp. 1–14). Oxford: Oxbow Books.

Kleindienst, M. R. (2003). Strategies for studying Pleistocene archaeology based upon surface evidence: First characterization of an older Middle Stone Age Unit, Dakhleh Oasis, Western Desert, Egypt. In G. Bowen & C. A. Hope (Eds.), *The Oasis Papers 3: Proceedings of the Third International Conference of the Dakhleh Oasis Project* (pp. 1–42). Oxford: Oxbow Books.

Kleindienst, M. R., Churcher, C. S., McDonald, M. M. A., & Schwarcz, H. P. (1999). Geology, geography, geochronology and geoarchaeology of the Dakhleh Oasis region: Interim report. In C. S. Churcher & A. J. Mills (Eds.), *Reports from the survey of Dakhleh Oasis, Western Desert of Egypt, 1977–1987* (pp. 1–53). Oxford: Oxbow Books.

Kleindienst, M. R., McDonald, M. M. A., Wiseman, M. F., Hawkins, A. L. Smith, J. R., Kieniewicz, J. M., et al. (2006). Walking in the footsteps of Gertrude Caton-Thompson and Elinor W. Gardner: Surveys by Kharga Oasis Prehistory Project (KOPP). *Proceedings of the 18th Bienniel Conference of the Society of Africanist Archaeologists* (pp. 24–29)

Kleindienst, M. R., Schwarcz, H. P., Nicoll, K., Churcher, C. S., Frizano, J., Giegengack, R., et al. (2008). Water in the desert: First report on Uranium-series dating of Caton-Thompson's and Gardner's "classic" Pleistocene sequence at Refuf Pass, Kharga Oasis. In M. F. Wiseman (Ed.), *Oasis Papers II: Proceedings of the second Dakhleh Oasis Project Research Seminar, Royal Ontario Museum and University of Toronto, June 1997* (pp. 25–54). Oxford: Oxbow Books.

Kowalski, K. (1993). Remains of small vertebrates from Bir Tarfawi and their paleoecological significance. In F. Wendorf, R. Schild, & A. E. Close (Eds.), *Egypt during the last Interglacial: The Middle Paleolithic of Bir Tarfawi and Bir Sahara East* (pp. 155–204). New York: Plenum Press.

Kuhn, S. L. (1992). On planning and curated technologies in the Middle Paleolithic. *Journal of Anthropological Research, 48*, 185–214.

Mandel, R. D., & Simmons, A. H. (2001). Prehistoric occupation of Late Quaternary landscapes near Kharga Oasis, Western Desert of Egypt. *Geoarchaeology, 16*, 95–117.

McBrearty, S., & Brooks, A. S. (2000). The revolution that wasn't: A new interpretation of the origin of modern human behavior. *Journal of Human Evolution, 39*, 453–563.

Mellars, P. (1989).Technological changes across the Middle-Upper Palaeolithic transition: Economic, social, and cognitive perspectives. In P. Mellars & C. Stringer (Eds.), *The human revolution: Behavioural and biological perspectives on the origins of modern humans* (pp. 338–365). Princeton, NJ: Princeton University Press.

Nelson, M. C. (1991). The study of technological organization. *Archaeological Method and Theory, 3*, 57–100.

Schiffer, M. B., Sullivan, A. P., & Klinger, T. C. (1978). The design of archaeological surveys. *World Archaeology, 10*, 1–28.

Schild, R., & Wendorf, F. (1975). New explorations in the Egyptian Sahara. In F. Wendorf & A. Marks (Eds.), *Problems in prehistory: North Africa and the Levant* (pp. 65–112). Dallas, TX: SMU Press.

Schwarcz, H. P., Szkudlarek, R., Kleindienst, M. R., & Evensen, N. (2008). Fire in the desert: The occurrence of a high-Ca silicate glass near the Dakhleh Oasis, Egypt. In F. Wiseman (Ed.), *The Oasis Papers II: Proceedings of the Second International Conference of the Dakhleh Oasis Project* (pp. 55–71). Oxford: Oxbow Books.

Shott, M. (1986). Technological organization and settlement mobility: An ethnographic examination. *Journal of Anthropological Research, 42*, 15–51.

Shott, M. J. (1996). An exegesis of the curation concept. *Journal of Anthropological Research, 52*, 259–280.

Simmons, A. H., & Mandel, R. D. (1986). *Prehistoric occupation of a marginal environment: An archaeological survey near Kharga Oasis in the Western Desert of Egypt.* Oxford: BAR.

Smith, J. R., Giegengack, R., Schwarcz, H. P., McDonald, M. M. A., Kleindienst, M. R., Hawkins, A. L., et al. (2004). A reconstruction of Quaternary pluvial environments and human occupations using stratigraphy and geochronology of fossil-spring tufas, Kharga Oasis, Egypt. *Geoarchaeology, 19*, 407–439.

Smith, J. R., Hawkins, A. L., Asmerom, Y., Polyak, V., & Giegengack, R. (2007). New age constraints on the Middle Stone Age occupations of Kharga Oasis, Western Desert, Egypt. *Journal of Human Evolution, 52*, 690–701.

Sultan, M., Sturchio, N., Hassan, F. A., Hamdan, M. A. R., Mahmood, A. M., El Alfy, Z., et al. (1997). Precipitation source inferred from stable isotopic composition of Pleistocene groundwater and carbonate deposits in the Western Desert of Egypt. *Quaternary Research, 48*, 29–37.

Svoboda, J. A. (2004). The Middle Palaeolithic of Southern Bahariya Oasis, Western Desert, Egypt. *Anthropologie, 42*, 253–267.

Szabo, B. J., McHugh, W. P., Shaber, G. G., Haynes, C. V., & Breed, C. S. (1989). Uranium-Series dated authigenic carbonates and Acheulian sites in Southern Egypt. *Science, 243*, 1053–1056.

Van Neer, W. (1993). Fish remains from the last Interglacial at Bir Tarfawi (Eastern Sahara, Egypt). In F. Wendorf, R. Schild, & A. E. Close (Eds.), *Egypt during the last Interglacial: The Middle Paleolithic of Bir Tarfawi and Bir Sahara East* (pp. 144–154). New York: Plenum Press.

Wendorf, F., & Schild, R. (1980). *Prehistory of the eastern Sahara.* New York: Academic Press.

Wendorf, F., & Schild, R. (1993). Work at BT-14 during 1974. In F. Wendorf, R. Schild, R. & A. E. Close (Eds.), Egypt during the last Interglacial: The Middle Paleolithic of Bir Tarfawi and Bir Sahara East (pp. 265–287). New York: Plenum Press.

Wendorf, F., Close, A. E., & Schild, R. (1987). Recent work on the Middle Palaeolithic of the Eastern Sahara. *African Archaeological Review, 5*, 49–63.

Wendorf, F., Schild, R., Close, A., Associates. (1993a). Egypt during the last Interglacial: The Middle Palaeolithic of Bir Tarfawi and Bir Sahara East. New York: Plenum Press.

Wendorf, F., Schild, R., & Close, A. (1993b). Summary and conclusions. In F. Wendorf, R. Schild & A. E. Close (Eds.), Egypt during the last Interglacial: The Middle Paleolithic of Bir Tarfawi and Bir Sahara East (pp. 522–573). New York: Plenum Press.

Wengler, L. (1994). La transition du Moustérien à l'Atérien. *L'Anthropologie, 101*, 448–481.

Winlock, H. E. (1936). *Ed Dakhleh Oasis: Journal of a camel trip made in 1908.* New York: Metropolitan Museum of Art.

Wiseman, M. F. (1999). Late Pleistocene prehistory in the Dakhleh Oasis. In C. S. Churcher & A. J. Mills (Eds.), *Reports from the survey of Dakhleh Oasis, Western Desert of Egypt, 1977–1987* (pp. 109–115). Oxford: Oxbow Books.

# Part III
# The Fossil Hominins

# Chapter 12
# Morphological Continuity of the Face in the Late Middle and Late Pleistocene Hominins from Northwestern Africa: A 3D Geometric Morphometric Analysis

K. Harvati and J.-J. Hublin

**Abstract** Facial morphology comprises some of the most distinctive features of early modern humans. The rich fossil record of Morocco allows assessing changes in facial morphology from the late Middle Pleistocene through the Late Pleistocene. Specimens associated with the Aterian industry in Morocco were originally thought to be relatively recent (40–20 ka), but could be much older (35–90 ka). Predating this population are the late Middle Pleistocene specimens of Irhoud. Later in the same geographical area, larger samples are represented by the Iberomaurusian series. We conducted a 3D geometric morphometric analysis of the facial shape of the Aterian specimen Dar es-Soltan II-5, with the aim of deciphering the affinities of this specimen with earlier North African and Levantine fossils, later Upper Paleolithic Eurasian specimens, as well as later North African populations. We used a large comparative sample ($n = 191$) comprising seven geographic populations of recent humans, Iberomaurusians from Afalou and Taforalt ($n = 22$), and Middle and Late Pleistocene Eurasian and African fossils. The 3D coordinates of 19 facial landmarks were collected. Specimen landmark configurations were processed with Generalized Procrustes Analysis. Principal Components, Canonical Variates, and cluster analyses were performed and Procrustes distances and Mahalanobis squared distances were calculated. Both Irhoud 1 and Dar es-Soltan II-5 are similar to the early anatomically modern humans from Qafzeh, and the Iberomaurusian sample is closely connected to the Upper Paleolithic European sample.

**Keywords** Aterian • Facial morphology • Modern human origins • Neandertals • Upper Paleolithic

## Introduction

Northwestern Africa has yielded a rich series of human fossils documenting human evolution throughout the Middle and Late Pleistocene (e.g., Hublin 1985, 1992, 2001). Recent developments in the determination of a secure chronology for the North African fossil record have raised new questions about the role of this area in the emergence of our species and the origins of non-African modern humans. Specifically, the Aterian assemblages that yielded robust modern looking human remains (Ferembach 1976) and which were initially assigned to a period between 40 and 20 ka (e.g., Debénath et al. 1986) are now considered to be much older. The bulk of the Aterian industries is likely to be dated between 90 and 35 ka and could well be rooted further back in time during the MIS 6 (see Bouzouggar and Barton 2012; Raynal and Occhietti 2012; Richter et al. 2012). Among the Aterian sites that have yielded fossil hominins, the cave of Dar es-Soltan II, near Rabat (Morocco), is best known for the rather complete cranial elements (a partial skull and the associated mandible) discovered in the site (Debénath 1976; Ferembach 1976). The specimen generally reported as Dar es-Soltan 5 was discovered under a sandstone plate within the marine sand deposits at the bottom of the stratigraphic sequence of the site (layer 7). For the purpose of clarity, it is designated here as Dar es-Soltan II-5. This archaeologically sterile layer was overlaid by a distinct reddish layer (layer 6) where a hearth and some Aterian elements have been described (Debénath 1976). Racemization ratios in molluscs from layer 7 indicate an age between 85 and 75 ka (Raynal and Occhietti 2012), compatible with other Aterian dates recently established in northern Africa.

K. Harvati (✉)
Senckenberg Center for Human Evolution and Paleoecology, Institut für Ur- und Frühgeschichte und Archäologie des Mittelalters, Eberhard Karls Universität Tübingen, Rümelinstrasse 23, 72070 Tübingen, Germany
e-mail: katerina.harvati@ifu.uni-tuebingen.de

J.-J. Hublin
Department of Human Evolution, Max Planck Institute for Evolutionary Anthropology, Deutscher Platz 6, 04103 Leipzig, Germany
e-mail: hublin@eva.mpg.de

J.-J. Hublin and S. P. McPherron (eds.), *Modern Origins: A North African Perspective*, Vertebrate Paleobiology and Paleoanthropology, DOI: 10.1007/978-94-007-2929-2_12, © Springer Science+Business Media B.V. 2012

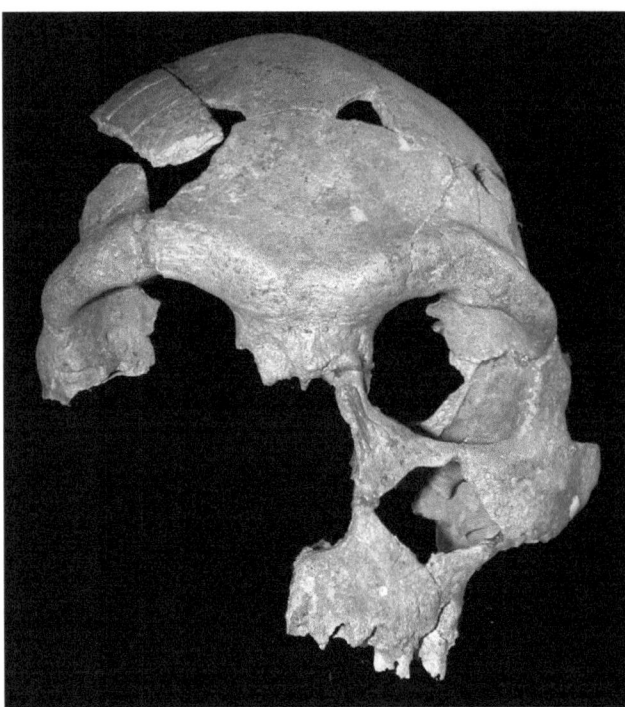

**Fig. 12.1** Dar es-Soltan II-5

**Table 12.1** Recent human comparative sample

| Total | ($n = 213$) |
|---|---|
| Sub-Saharan African (E. & S. Africa) | ($n = 38$) |
| Andaman Islanders | ($n = 29$) |
| Australian | ($n = 26$) |
| Asian (China, Thailand) | ($n = 38$) |
| Inuit (Alaska, Greenland) | ($n = 14$) |
| Near Eastern (Syria) | ($n = 20$) |
| European (Germany, Greece, Czech Republic, Norway) | ($n = 25$) |

**Table 12.2** Fossil and subfossil samples included in the analysis

| **Mid-Pleistocene Europe (MPE)** | ($n = 3$) |
|---|---|
| Arago 21, Petralona, Sima de los Huesos 5 | |
| **Mid-Late Pleistocene Africa (MPA)** | ($n = 3$) |
| Bodo, Kabwe, Irhoud 1 | |
| **Neandertals** | ($n = 6$) |
| Gibraltar 1, Shanidar 5, Guattari 1, La Chapelle, La Ferrassie 1, Shanidar 1 | |
| **Late Pleistocene Near East** | ($n = 2$) |
| Qafzeh 6, Qafzeh 9 | |
| **Late Pleistocene Eurasia** | ($n = 14$) |
| Chancelade, Abri Pataud 1, Cro-Magnon 1 & 2, Mladec 1, Predmost 3 & 4, Grimaldi 4, Upper Cave 101 & 103, Ohalo II | |
| **Late Pleistocene Africa** | ($n = 1$) |
| Wadi Kubbaniya | |
| **Iberomaurusian** | ($n = 22$) |

In light of these developments, the issue of continuity between the Aterian remains and those of the later Ibero-maurusians (dated to 20–10 ka) becomes critical as the chronological distance between the two series of fossil remains increases. Is there evidence of morphological affinities, suggesting population history links and possibly a late chronology, between Dar es-Soltan II-5 and the Iberomaurusians? Or does the morphological evidence point to an evolutionary discontinuity and/or a large chronological gap in the human occupation record? Similarly, the relationship of the Aterians to the earlier specimens from Irhoud, dated to approximately 160 ka (Smith et al. 2007), also becomes relevant.

We conducted a 3D geometric morphometric analysis of the facial morphology of the Dar es-Soltan II-5 individual (Fig. 12.1) in comparison with the Jebel Irhoud and Iberomaurusian specimens, as well as with a large sample of recent humans, Neandertals and early modern humans from the Levant. Our goal was to evaluate the hypothesis of phylogenetic affinities among these fossil samples and, therefore, of continuity or discontinuity in North African human occupation. If there is a strong phylogenetic relationship and continuity among the North African fossil samples, it is expected that they will show greater similarity to older and younger specimens from the same region than to specimens from other parts of the world.

## Materials and Methods

Our comparative sample comprised a total of 213 individuals from seven geographical samples of recent humans, which, together, span a large extent of the modern human geographic range (Table 12.1). It also included several Middle/late Middle Pleistocene fossil humans from Africa and Eurasia and a large Iberomaurusian sample from Afalou and Taforalt (Table 12.2). Nineteen facial osteometric landmarks were digitized using a Microscribe by one observer (KH) (Table 12.3, Fig. 12.2). Data were collected in the form of 3D coordinates, and processed with Generalized Procrustes Analysis. In cases of only minimal bone loss, missing data were reconstructed during data collection using anatomical clues from the preserved surrounding areas. Bilateral landmarks missing on one side were estimated during data processing by mirror-imaging using reflected relabelling (Mardia and Bookstein 2000; Gunz and Harvati 2007), which reflects the paired landmarks without having to specify a mirroring plane.

**Table 12.3** Facial landmarks digitized and their definitions

| 1. Post-toral Sulcus | Minima of concavity on midline post-toral frontal squama |
|---|---|
| 2. Glabella | |
| 3. Nasion | |
| 4. Nasospinale | |
| 5. Prosthion | |
| 6, 7. Mid-orbit Torus Superior | Point on superior aspect of supraorbital torus, approximately at the middle of the orbit |
| 8, 9. Mid-orbit Torus Inferior | Point on inferior margin of supraobrital torus, approximately at the middle of the orbit |
| 10, 11. Dacryon | |
| 12, 13. Zygoorbitale | |
| 14, 15. Frontomalare Orbitale | |
| 16,17. Zygomaxillare | |
| 18, 19. Alare | |

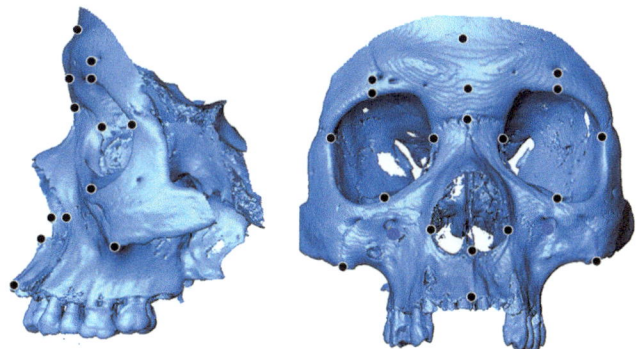

**Fig. 12.2** Facial landmarks digitized for this study

The superimposed coordinates were then analyzed using an array of multivariate statistics: Principal Components Analysis (PCA), Canonical Variates Analysis (CVA), Mahalanobis squared distances ($D^2$, corrected for unequal sample sizes following Marcus (1993)), Procrustes distances (PD), and cluster analysis with the Neighbor Joining Tree method.

In contrast to conventional linear and angular measurements, our coordinate-based approach preserves the geometry of the object studied and allows the intuitive visualization of shape differences between specimens or group averages as landmark displacements (Rohlf and Marcus 1993; O'Higgins 2000; Harvati 2001, 2003a, b). Geometric morphometric methods also provide a way of quantifying shape variability of traits that are difficult to measure with traditional measurement methods, and are therefore usually described qualitatively (Harvati 2001, 2003a; Nicholson and Harvati 2006).

## Results

PCA: The first PC (19.5% of total variance) reflects variation in facial morphology among all groups, with the Inuit samples separating somewhat from other groups on the negative end of this axis (Fig. 12.3). Specimens scoring negatively on this PC show supero-inferiorly tall and medio-laterally narrow faces, with more coronally oriented zygomatics and less subnasal prognathism compared to specimens scoring positively on PC 1. PC 2 (12.4% of total variance), in contrast, separates Neandertals and other "archaic" humans from

modern humans (Fig. 12.3). The Middle Pleistocene specimens from both Africa and Europe occupy an intermediate position: Kabwe, Arago, Sima 5, and Petralona place closer to the Neandertals, while Bodo is closer to modern humans and to the early anatomically modern specimens Qafzeh 6 and 9. Jebel Irhoud 1 is also close to the Qafzeh specimens and well within the recent human range of variation along PC 2, while Dar es-Soltan II-5 falls within the modern human cloud. The Iberomaurusians and Upper Paleolithic specimens from Eurasia and Africa are well within modern human variation on this axis. The shape differences along PC 2 include a heavier browridge, longer face, more projecting midface, more inflated maxilla, and more posteriorly sloping zygomatics on the Neandertal end (Fig. 12.3).

CVA: The CVA was calculated using the first 13 principal components (representing about 85% of the total variance). The first canonical axis accounts for 37.4% and separates archaic from modern humans (Fig. 12.4). Here, all the Middle Pleistocene specimens fall around the Neandertal cloud, with Bodo again occupying the most intermediate position. Qafzeh 6 and 9 fall very close to Bodo and to each other towards the modern human cloud. Irhoud 1 and Dar es-Soltan II-5 are very close to the Qafzeh individuals. They are intermediate to Neandertals/Middle Paleolithic European and African specimens on the one hand and modern humans on the other. As in the PCA, the Iberomaurusians and Upper Paleolithic specimens from Eurasia and Africa are well within the range of modern human variation.

$D^2$: The Mahalanobis squared distances among groups are reported in Table 12.4. They were calculated on the basis of the first 13 principal components. Irhoud 1 was found to be closest to the Upper Cave specimens ($d^2 = 1.91$) and to Qafzeh 9 ($d^2 = 3.4$). Dar es-Soltan II-5 showed much larger overall distances to all groups included in the analysis. It was closest to Petralona ($d^2 = 5.46$), to Qafzeh 6 and 9 ($d^2 = 10.65$ and $12.57$, respectively), and to Wadi Kubbaniya ($d^2 = 12.7$). The Iberomaurusian sample was closest to the European Upper Paleolithic sample

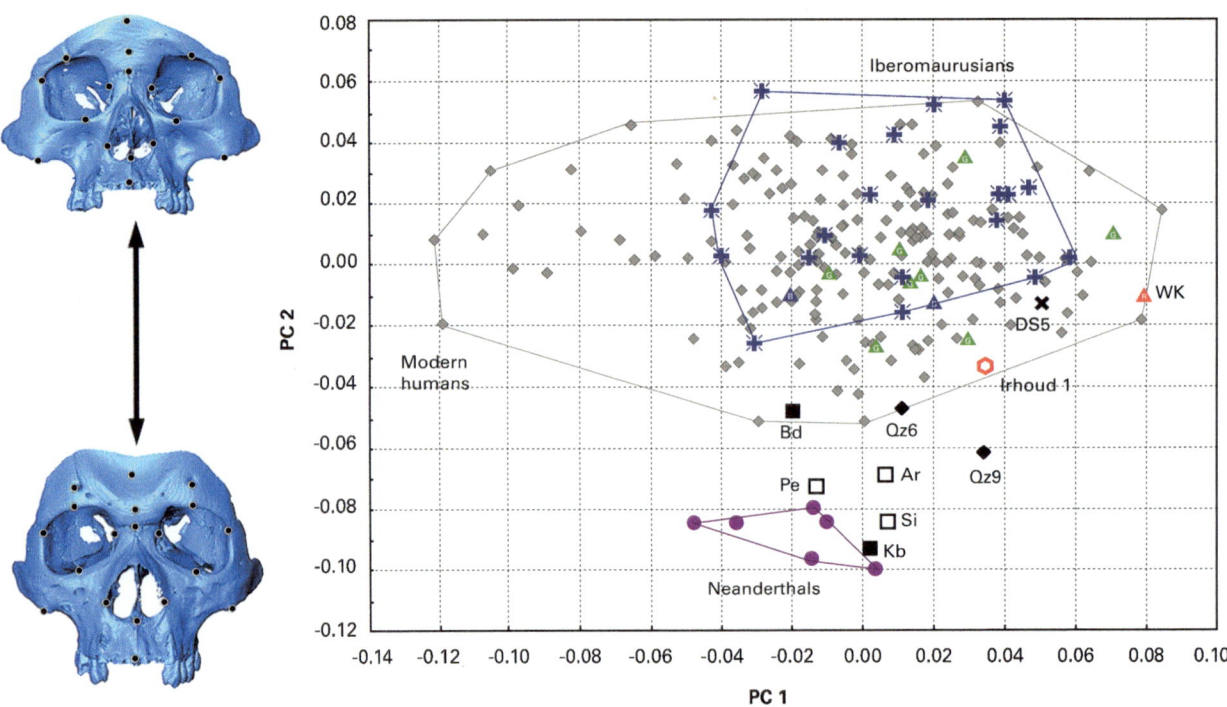

**Fig. 12.3** Principal components analysis. On the *left side* are shown the shape changes along PC 2 in frontal view. Symbols are as follows: *Red star*: Jebel Irhoud 1 (Irhoud1); *Black star*: Dar es-Soltan II-5 (DS5); *Blue stars*: Iberomaurusians; *Purple dots*: Neandertals; *Black squares*: Bodo (Bd), Kabwe (Kb); *Open squares*: Arago (Ar), Sima 5 (Si), Petralona (Pe); *Green triangles*: Up. Paleolithic Europeans; *Blue triangles*: *Upper* Cave 101 and 103; *Red triangle*: Wadi Kubbaniya (WK); *Black diamonds*: Qafzeh 6 & 9 (Qz6, Qz9); *Grey diamonds*: recent humans

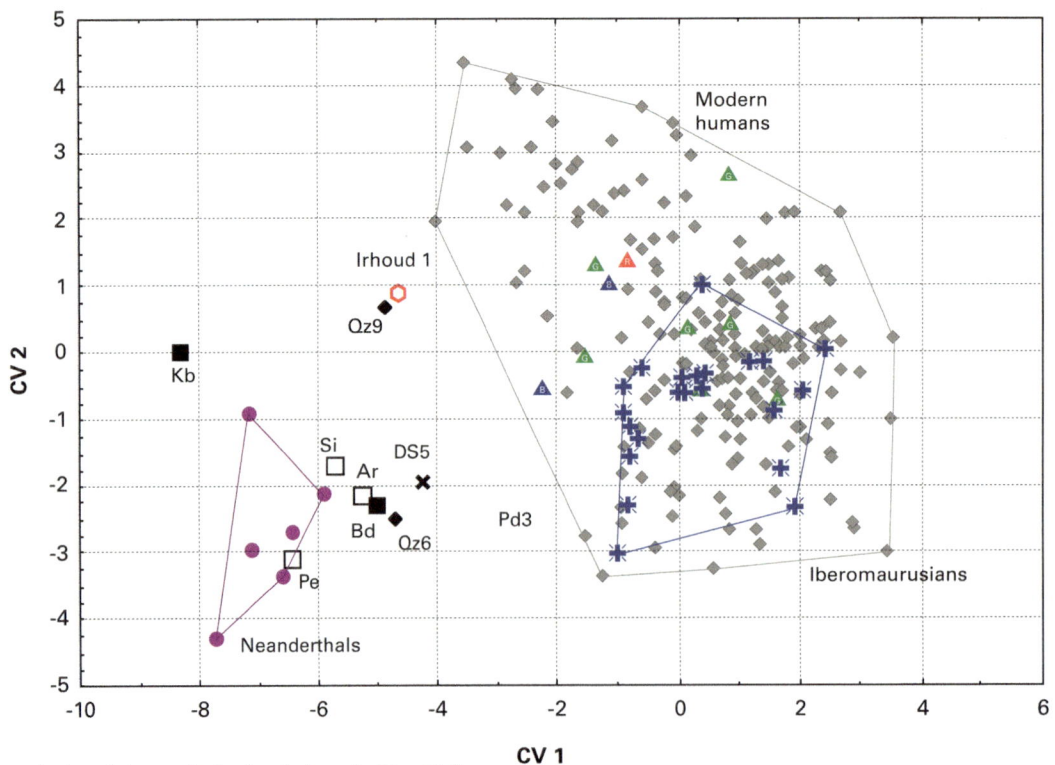

**Fig. 12.4** Canonical variates analysis. Symbols as in Fig. 12.3

**Table 12.4** Mahalanobis squared distances

| | Arago | Bodo | Irhoud1 | Kabwe | Petral | Sima5 | Neandertha | Andaman | African | Asian | Austr |
|---|---|---|---|---|---|---|---|---|---|---|---|
| Arago | 0 | | | | | | | | | | |
| Bodo | 1.53 | 0 | | | | | | | | | |
| Irhoud1 | 17.54 | 9.53 | 0 | | | | | | | | |
| Kabwe | 52.5 | 29.56 | 21.13 | 0 | | | | | | | |
| Petral | 24.3 | 3.36 | 16.98 | −7.27 | 0 | | | | | | |
| Sima5 | 35.93 | 21.53 | 24.99 | 5.77 | −11.5 | 0 | | | | | |
| Neandertal | 27.81 | 5.12 | 14.15 | 13.61 | 3.32 | 20.42 | 0 | | | | |
| Andaman | 68.3 | 51.55 | 34.41 | 90.06 | 70.07 | 62.87 | 74.64 | 0 | | | |
| African | 64.67 | 40.26 | 23.07 | 69.58 | 54.49 | 56.83 | 63.4 | 5.83 | 0 | | |
| Asian | 57.2 | 42.27 | 29.94 | 89.85 | 62.67 | 65.15 | 71.05 | 4.51 | 6.55 | 0 | |
| Austr | 53.07 | 34.84 | 8.44 | 46.88 | 54.24 | 58.67 | 52.2 | 18.25 | 12.78 | 21.45 | 0 |
| DS5 | **39.46** | **20.99** | **18.72** | **26.23** | **5.46** | **24.89** | **37.33** | **54.26** | **39.64** | **49.66** | **44.3** |
| Inuit | 78.95 | 48.26 | 51.45 | 108.3 | 77.35 | 85.36 | 82.45 | 12.75 | 15.86 | 6.91 | 34 |
| Europe | 69.44 | 42.71 | 29.64 | 69.16 | 49.19 | 59.79 | 56.07 | 12.24 | 7.49 | 10.17 | 21.8 |
| Iberom | 39.08 | 29.64 | 27.63 | 84.88 | 54.15 | 63.55 | 63.07 | 12.28 | 9.28 | 7.23 | 24 |
| Qafz6 | 22.07 | 17.7 | 20.67 | 62 | 38.31 | 55.62 | 39.91 | 71.55 | 71.73 | 60.36 | 57.7 |
| Qafz9 | 17.1 | −6.06 | 3.4 | 15.64 | 13.45 | 17.8 | 16.23 | 35.7 | 24.74 | 36.37 | 14 |
| N.East | 57.81 | 37.96 | 32.32 | 72.09 | 50.33 | 59.97 | 50.17 | 10.43 | 11.42 | 9.78 | 23.9 |
| WadiK | 64.54 | 43.63 | 34.26 | 59.44 | 42.57 | 27.79 | 71.94 | 12.11 | 16.65 | 23.63 | 25.2 |
| Up.Cave | 29.13 | 15.83 | 1.91 | 36.88 | 27.37 | 39.44 | 35.02 | 10.99 | 4.78 | 5.9 | 1.34 |
| Ohalo | 23.25 | 29.8 | 19.13 | 52.48 | 36.54 | 58.97 | 62 | 30.46 | 23.29 | 16 | 18 |
| Up.Paleol | 33.32 | 36.54 | 19.67 | 78.26 | 58.31 | 65.14 | 62.11 | 10.89 | 8.03 | 8.12 | 16.1 |

**Table 12.4** (continued)

|  | DS5 | Inuit | Europe | Iberom | Qafz6 | Qafz9 | N.East | WadiK | Up.Cave | Ohalo | Up.Paleol |
|---|---|---|---|---|---|---|---|---|---|---|---|
| Arago |  |  |  |  |  |  |  |  |  |  |  |
| Bodo |  |  |  |  |  |  |  |  |  |  |  |
| Irhoud1 |  |  |  |  |  |  |  |  |  |  |  |
| Kabwe |  |  |  |  |  |  |  |  |  |  |  |
| Petral |  |  |  |  |  |  |  |  |  |  |  |
| Sima5 |  |  |  |  |  |  |  |  |  |  |  |
| Neandertal |  |  |  |  |  |  |  |  |  |  |  |
| Andaman |  |  |  |  |  |  |  |  |  |  |  |
| African |  |  |  |  |  |  |  |  |  |  |  |
| Asian |  |  |  |  |  |  |  |  |  |  |  |
| Austr |  |  |  |  |  |  |  |  |  |  |  |
| DS5 | **0** |  |  |  |  |  |  |  |  |  |  |
| Inuit | **61.3** | 0 |  |  |  |  |  |  |  |  |  |
| Europe | **36** | 13.72 | 0 |  |  |  |  |  |  |  |  |
| Iberom | **30.3** | 16.59 | 10.48 | 0 |  |  |  |  |  |  |  |
| Qafz6 | **10.7** | 71.98 | 67.05 | 48.39 | 0 |  |  |  |  |  |  |
| Qafz9 | **12.6** | 53.83 | 36.85 | 26.15 | 16.08 | 0 |  |  |  |  |  |
| N.East | 44.8 | **12.49** | 5.06 | 14.25 | 61.83 | 34.45 | 0 |  |  |  |  |
| WadiK | 12.7 | **39.52** | 29.23 | 21.6 | 45.02 | 12.48 | 32.39 | 0 |  |  |  |
| Up.Cave | 24.1 | **15.02** | 6.21 | 8.44 | 38.97 | 11.09 | 4.88 | 21.74 | 0 |  |  |
| Ohalo | 23.6 | **33.35** | 22.96 | 12.18 | 37.8 | 26.38 | 24.97 | 31.7 | −3.83 | 0 |  |
| Up.Paleol | 39.2 | **20.59** | 14.09 | 4.63 | 55.06 | 27.22 | 10.6 | 27.88 | 1.67 | 9.53 | 0 |

**Table 12.5** Procrustes distances among sample mean configurations

| | Kabwe | Bodo | Petralona | Arago | Sima5 | Irhoud1 | Neand | African | Asian | Australian | Inuit | European | Iberom | N. Eastern | Up. Pal | Ohalo2 | Up. Cave | Qafzeh9 | DS5 | Qafzeh6 | WadiK. | Andaman |
|---|---|---|---|---|---|---|---|---|---|---|---|---|---|---|---|---|---|---|---|---|---|---|
| Kabwe | 0.000 | | | | | | | | | | | | | | | | | | | | | |
| Bodo | 0.125 | 0.000 | | | | | | | | | | | | | | | | | | | | |
| Petralona | 0.085 | 0.097 | 0.000 | | | | | | | | | | | | | | | | | | | |
| Arago | 0.140 | 0.095 | 0.111 | 0.000 | | | | | | | | | | | | | | | | | | |
| Sima5 | 0.108 | 0.112 | 0.086 | 0.113 | 0.000 | | | | | | | | | | | | | | | | | |
| Irhoud1 | 0.120 | 0.113 | 0.120 | 0.113 | 0.123 | 0.000 | | | | | | | | | | | | | | | | |
| Neand | 0.085 | 0.096 | 0.081 | 0.108 | 0.093 | 0.106 | 0.000 | | | | | | | | | | | | | | | |
| African | 0.132 | 0.111 | 0.120 | 0.131 | 0.126 | 0.088 | 0.120 | 0.000 | | | | | | | | | | | | | | |
| Asian | 0.143 | 0.105 | 0.120 | 0.126 | 0.130 | 0.100 | 0.119 | 0.047 | 0.000 | | | | | | | | | | | | | |
| Australian | 0.117 | 0.115 | 0.121 | 0.127 | 0.129 | 0.078 | 0.118 | 0.053 | 0.073 | 0.000 | | | | | | | | | | | | |
| Inuit | 0.163 | 0.119 | 0.139 | 0.153 | 0.157 | 0.140 | 0.139 | 0.089 | 0.056 | 0.110 | 0.000 | | | | | | | | | | | |
| European | 0.134 | 0.117 | 0.121 | 0.144 | 0.137 | 0.104 | 0.119 | 0.049 | 0.059 | 0.070 | 0.083 | 0.000 | | | | | | | | | | |
| Iberom | 0.148 | 0.108 | 0.123 | 0.120 | 0.135 | 0.092 | 0.126 | 0.049 | 0.056 | 0.073 | 0.092 | 0.056 | 0.000 | | | | | | | | | |
| N.Eastern | 0.123 | 0.106 | 0.111 | 0.126 | 0.122 | 0.102 | 0.099 | 0.052 | 0.051 | 0.071 | 0.081 | 0.051 | 0.071 | 0.000 | | | | | | | | |
| Up.Pal | 0.137 | 0.111 | 0.122 | 0.113 | 0.130 | 0.085 | 0.121 | 0.049 | 0.063 | 0.062 | 0.103 | 0.072 | 0.049 | 0.063 | 0.000 | | | | | | | |
| Ohalo2 | 0.144 | 0.130 | 0.136 | 0.128 | 0.157 | 0.113 | 0.149 | 0.099 | 0.099 | 0.093 | 0.124 | 0.103 | 0.094 | 0.107 | 0.093 | 0.000 | | | | | | |
| Up. Cave | 0.113 | 0.102 | 0.105 | 0.118 | 0.124 | 0.091 | 0.111 | 0.059 | 0.061 | 0.052 | 0.090 | 0.067 | 0.071 | 0.057 | 0.057 | 0.081 | 0.000 | | | | | |
| Qafzeh9 | 0.113 | 0.113 | 0.118 | 0.120 | 0.119 | 0.095 | 0.102 | 0.102 | 0.116 | 0.097 | 0.153 | 0.121 | 0.113 | 0.107 | 0.105 | 0.135 | 0.106 | 0.000 | | | | |
| DS5 | 0.143 | 0.141 | 0.132 | 0.141 | 0.154 | 0.106 | 0.145 | 0.115 | 0.128 | 0.113 | 0.160 | 0.122 | 0.102 | 0.130 | 0.110 | 0.121 | 0.118 | 0.118 | 0.000 | | | |
| Qafzeh6 | 0.161 | 0.134 | 0.149 | 0.127 | 0.157 | 0.112 | 0.139 | 0.145 | 0.134 | 0.137 | 0.160 | 0.151 | 0.132 | 0.143 | 0.135 | 0.139 | 0.134 | 0.115 | 0.117 | 0.000 | | |
| WadiK. | 0.168 | 0.158 | 0.152 | 0.160 | 0.142 | 0.126 | 0.170 | 0.114 | 0.131 | 0.112 | 0.170 | 0.136 | 0.115 | 0.135 | 0.112 | 0.149 | 0.129 | 0.124 | 0.119 | 0.151 | 0.000 | |
| Andaman | 0.143 | 0.114 | 0.128 | 0.130 | 0.124 | 0.093 | 0.122 | 0.042 | 0.048 | 0.061 | 0.091 | 0.066 | 0.058 | 0.057 | 0.058 | 0.111 | 0.071 | 0.104 | 0.123 | 0.140 | 0.104 | 0.000 |

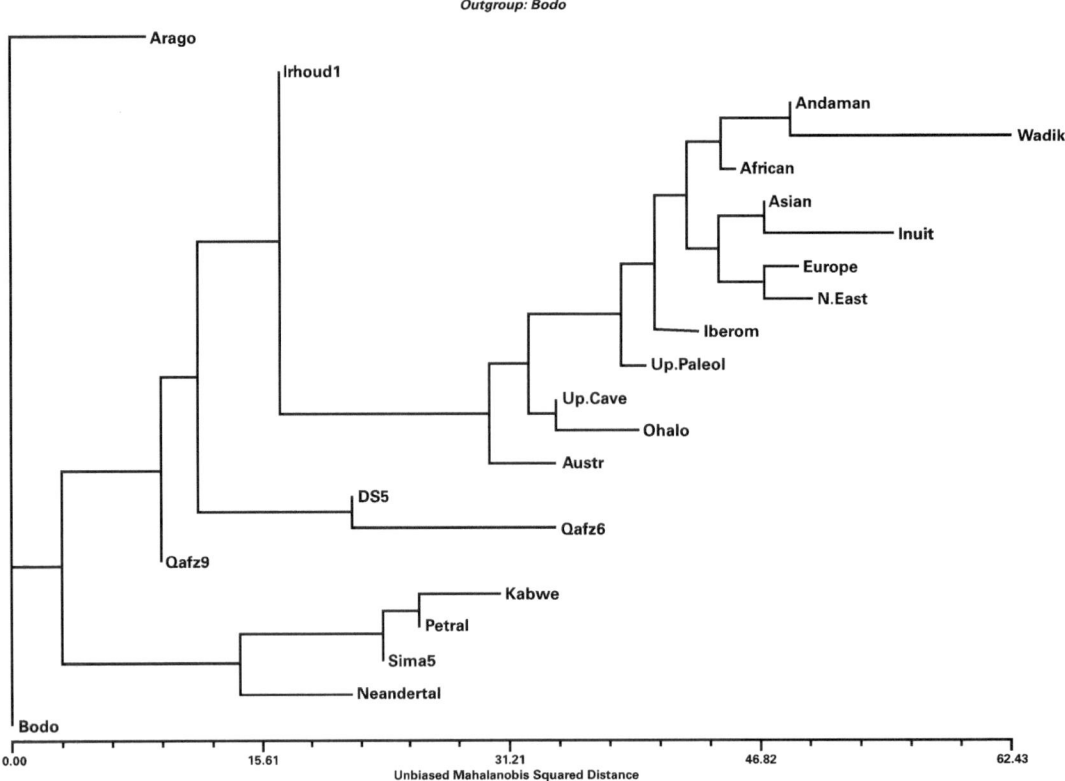

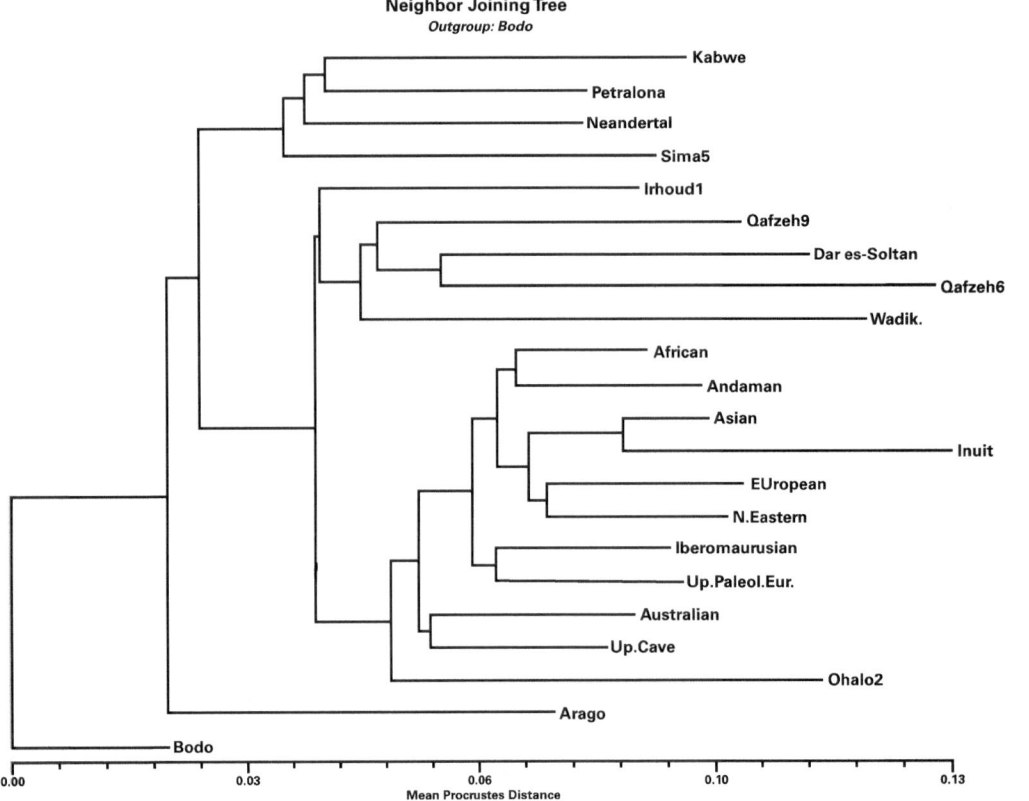

**Fig. 12.5** Neighbor Joining Trees, based on (*top*) Mahalanobis D$^2$ and (*bottom*) Procrustes Distances among population mean configurations. In both cases Bodo was used as an outgroup

$(d^2 = 4.63)$. It was quite distant from all archaic fossil specimens and from Dar es-Soltan II-5 and Irhoud 1.

PD: The Procrustes distances among population mean configurations were calculated and reported in Table 12.5. Irhoud 1 was closest to the mean recent Australian configuration (PD $=$ 0.078) and to the mean Upper Paleolithic European configuration (PD $=$ 0.085). Irhoud 1 showed a rather large distance to Dar es-Soltan II-5 (PD $=$ 0.106). Dar es-Soltan II-5 again showed larger overall distances to all other samples. The smallest distances were to the Iberomaurusian sample (PD $=$ 0.102), and Irhoud 1 and the European Upper Paleolithic specimens (PD $=$ 0.110). The Iberomaurusians were much closer to recent humans, and especially to the mean European Upper Paleolithic configuration (PD $=$ 0.049).

Cluster Analysis: Neighbor Joining trees were built from both the Mahalanobis $D^2$ and the Procrustes mean distances. In both instances, Bodo was used as an outgroup. The two neighbor joining trees are shown in Fig. 12.5. In both cases, Irhoud 1 and Dar es-Soltan II-5 fall, together with the Qafzeh specimens, within the modern human larger branch but as outgroups to the recent human, Upper Paleolithic, and Iberomaurusian samples. The Iberomaurusian specimens, on the other hand, are in both cases placed with the recent human groups and close to the Upper Paleolithic Europeans. The Procrustes distance tree further shows Irhoud 1, DS5, the two Qafzeh specimens, and Wadi Kubbaniya clustering together in one of the two modern human branches—the other branch comprising all recent human groups, all Eurasian Upper Paleolithic specimens, as well as the Iberomaurusian sample.

## Discussion and Conclusions

Our results indicate that the Dar es-Soltan II-5 remains do not show strong morphological similarities to the later Iberomaurusian population. Although this specimen is clearly a modern human, it seems more similar to the early anatomically modern humans from Qafzeh, to the Late Pleistocene North African specimen from Wadi Kubbaniya, and also to the earlier Moroccan specimen Jebel Irhoud 1. Some level of regional continuity is suggested by the Procrustes distance analysis, where Dar es-Soltan II-5 was found to be closest to the Iberomaurusian sample and to Jebel Irhoud 1. However, these distances were very large (approximately double the distances among neighboring recent human populations like the European and near eastern samples), and do not seem to us to imply a close relationship between Dar es-Soltan II-5 and either the Jebel Irhoud individual or the Iberomaurusian sample. Keeping in mind the caveats inherent in the comparative analysis of a single specimen, the overall large Procrustes distances of Dar es-Soltan II-5 to all samples and

its similarities to Late Pleistocene specimens in the PCA, CVA, and in both Mahalanobis and Procrustes distances suggest to us that a characterization as an early modern human is the most appropriate for this specimen.

Our findings are consistent either with an evolutionary discontinuity between the Aterian and the later Ibermaurusian populations, or with a large chronological period between these two samples. As such, they are also consistent with the recent dating results suggesting that the Dar es-Soltan II Aterian human remains could be placed ca. 80 ka (Raynal and Occhietti 2012). The Iberomaurusian remains, on the other hand, show strong morphological similarities with the European Upper Paleolithic sample used here. This result supports previous suggestions that the two groups might have shared a common population history (Ferembach 1962, 1985).

Our analysis supports an earlier chronology and taxonomic placement of Dar es-Soltan II-5 with early anatomically modern humans, as well as affinities between the Upper Paleolithic peoples of Europe and those of the somewhat later Moroccan Iberomaurusians. However, it must be noted that this study did not address the issue of allometry, which has been documented to affect facial morphology in important ways (Rosas and Bastir 2002, 2004; Strand-Viðarsdóttir et al. 2002; Maddux and Franciscus 2008). Such allometric effects may account for some of our unexpected results, such as the small Mahalanobis $d^2$ distance between Dar es-Soltane 5 and Petralona (Table 12.4). Analysis of size-related effects would help clarify these results. Finally it should be kept in mind that the face has previously been argued not to be phylogenetically as informative as other parts of the cranium (e.g., Harvati and Weaver 2006a, b). Analysis of other anatomical regions, such as the cranial vault and base and the dental morphology, will help resolve the issues explored here more conclusively.

**Acknowledgments** We thank all curators for access to fossil and recent human skeletal collections. Special thanks go to the Direction de l'INSAP (Royaume du Maroc), Direction du Patrimoine, Ministère de la Culture (Royaume du Maroc), Monsieur le Conservateur du Musée Archéologique de Rabat, and Prof. Henry de Lumley (Institut de Paleontologie Humaine) for allowing access to the Jebel Irhoud, Dar es-Soltan, and Iberomaurusian material. Comments provided by two anonymous reviewers helped to greatly improve the manuscript. This research was supported by the Max-Planck Society and the EVAN Marie Curie Research Training Network MRTN-CT-019564. NYCEP morphometrics contribution No. 35.

## References

Bastir, M., & Rosas, A. (2004). Facial heights: Evolutionary relevance of postnatal ontogeny for facial orientation and skull morphology in humans and chimpanzees. *Journal of Human Evolution, 47*, 359–381.

Bouzouggar, A., Barton, R. N. E. (2012). The identity and timing of the Aterian in Morocco. In J.-J. Hublin & S. McPherron (Eds.), *Modern origins: A North African perspective*. Dordrecht: Springer.

Debénath, A. (1976). Le site de Dar es-Soltan 2, à Rabat (Maroc). *Bulletins et Mémoires de la Société d'Anthropologie de Paris, 3*, 181–182.

Debénath, A., Raynal, J.-P., Roche, J., Texier, J.-P., & Ferembach, D. (1986). Stratigraphie, habitat, typologie et devenir de l'Atérien Marocain: Données récentes. *L'Anthropologie, 90*, 233–246.

Ferembach, D. (1962). *La nécropole épipaléolithique de Taforalt (Maroc Oriental). Étude des squelettes humains. Avec la collaboration de J. Dastugue et M. J. Poitrat-Targowla.* Casablanca: Edita.

Ferembach, D. (1976). Les restes humains de la grotte de Dar es-Soltan 2 (Maroc), campagne 1975. *Bulletins et Mémoires de la Société d'Anthropologie de Paris, 3*, 183–193.

Ferembach, D. (1985). On the origin of the Iberomaurusians (Upper Paleolithic: North Africa). A new hypothesis. *Journal of Human Evolution, 14*, 393–397.

Gunz, P., & Harvati, K. (2007). The Neanderthal 'chignon': Variation, integration and homology. *Journal of Human Evolution, 52*, 262–274.

Harvati, K. (2001). *The Neanderthal problem: 3D geometric morphometric models of cranial shape variation within and among species.* Ph.D. dissertation, City University of New York.

Harvati, K. (2003a). Quantitative analysis of Neanderthal temporal bone morphology using 3-D geometric morphometrics. *American Journal of Physical Anthropology, 120*, 323–338.

Harvati, K. (2003b). The Neanderthal taxonomic position: Models of intra- and inter-specific craniofacial variation. *Journal of Human Evolution, 44*, 107–132.

Harvati, K., & Weaver, T. D. (2006a). Reliability of cranial morphology in reconstructing Neanderthal phylogeny. In K. Harvati & T. Harrison (Eds.), *Neanderthals revisited: New approaches and perspectives* (pp. 239–254). Dordrecht: Springer.

Harvati, K., & Weaver, T. D. (2006b). Human cranial anatomy and the differential preservation of population history and climate signatures. *Anatomical Record, 288A*, 1225–1233.

Hublin, J.-J. (1985). Human fossils of the North African Middle Pleistocene and the origin of *Homo sapiens*. In E. Delson (Ed.), *Ancestors: The hard evidence* (pp. 283–288). New York: Alan R. Liss.

Hublin, J.-J. (1992). Recent human evolution in northwestern Africa. In M. Aitken, P. Mellars, C. B. Stringer (Eds.), *The origin of modern humans, the impact of science-based dating. Philosophical Transactions of the Royal Society B, 337*, 185–191

Hublin, J.-J. (2001). Northwestern African Middle Pleistocene hominids and their bearing on the emergence of *Homo sapiens*. In L. Barham & K. Robson-Brown (Eds.), *Human roots. Africa and Asia in the Middle Pleistocene* (pp. 99–121). Bristol: CHERUB, Western Academic and Specialist Press Ltd.

Marcus, L. F. (1993). Some aspects of multivariate statistics for morphometrics. In L. F. Marcus, E. Bello, & A. García-Valdecasas (Eds.), *Contributions to morphometrics* (pp. 99–130). Madrid: Monografias Museo Nacional de Ciencias Naturales.

Mardia, K. V., & Bookstein, F. L. (2000). Statistical assessment of bilateral symmetry of shapes. *Biometrika, 87*, 285–300.

Maddux, S. D., & Franciscus, R. G. (2008). Allometric scaling of infraorbital surface topography in *Homo*. *Journal of Human Evolution, 56*, 161–174.

Nicholson, E., & Harvati, K. (2006). Quantitative analysis of human mandibular shape using 3–D geometric morphometrics. *American Journal of Physical Anthropology, 131*, 368–383.

O'Higgins, P. (2000). The study of morphological variation in the hominid fossil record: Biology, landmarks and geometry. *Journal of Anatomy, 197*, 103–120.

Raynal, J.-P., Occhietti, S. (2012). Amino-chronology and an earlier age for the Aterian. In J.-J. Hublin & S. McPherron (Eds.), *Modern origins: A North African perspective.* Dordrecht: Springer.

Richter, D., Moser, J., Nami, M. (2012). New data from the site of Ifri n'Ammar (Morocco) and some remarks on the chronometric status of the Middle Paleolithic in the Maghreb. In J.-J. Hublin & S. McPherron (Eds.), *Modern origins: A North African perspective.* Dordrecht: Springer.

Rohlf, F. J., & Marcus, L. F. (1993). A revolution in morphometrics. *Trends in Ecology & Evolution, 8*, 129–132.

Rosas, A., & Bastir, M. (2002). Thin-plate spline analysis of allometry and sexual dimorphism in the human craniofacial complex. *American Journal of Physical Anthropology, 117*, 236–245.

Rosas, A., & Bastir, M. (2004). Geometric morphometric analysis of allometric variation in the mandibular morphology from the hominids of Atapuerca, Sima de los Huesos Site. *The Anatomical Record Part A, 278A*, 551–560.

Strand-Viðarsdóttir, U., O'Higgins, P., & Stringer, C. (2002). A geometric morphometric study of regional differences in the ontogeny of the modern human facial skeleton. *Journal of Anatomy, 201*, 211–229.

Smith, T. M., Tafforeau, P., Reid, D. J., Grün, R., Eggins, S., Boutaiout, M., et al. (2007). Earliest evidence of modern human life history in North African early *Homo sapiens*. *Proceedings of the National Academy of Sciences of the USA, 104*, 6128–6133.

# Chapter 13
# Dental Evidence from the Aterian Human Populations of Morocco

J.-J. Hublin, C. Verna, S. Bailey, T. Smith, A. Olejniczak, F. Z. Sbihi-Alaoui, and M. Zouak

**Abstract** The Aterian fossil hominins represent one of the most abundant series of human remains associated with Middle Stone Age/Middle Paleolithic assemblages in Africa. Their dates have been revised and they are now mostly assigned to a period between 90 and 35 ka. Although the Aterian human fossil record is exclusively Moroccan, Aterian assemblages are found throughout a vast geographical area extending to the Western Desert of Egypt. Their makers represent populations that were located close to the main gate to Eurasia and that immediately predated the last out-of-Africa exodus. In this chapter, we present an analysis of the Aterian dental remains. The sizes of the Aterian dentitions are particularly spectacular, especially for the post-canine dentition. This massiveness is reminiscent of the Middle Paleolithic modern humans from the Near East, but also of the early *Homo sapiens* in North and East Africa. Morphologically, this megadontia is expressed in the development of mass-additive traits. The Aterian dentition also displays relatively thick enamel. These features help to set some of the traits observed in Neandertals in perspective and highlight their primitive or derived nature. The Aterian morphological pattern is also important to consider when interpreting the dental morphology of the first modern humans in Eurasia.

**Keywords** Aterian • *Homo sapiens* • Middle Stone Age • Modern humans • Morocco • Neandertal • Sahara • Teeth

J.-J. Hublin (✉) · A. Olejniczak
Department of Human Evolution, Max Planck Institute for Evolutionary Anthropology, Deutscher Platz 6, 04103 Leipzig, Germany
e-mail: hublin@eva.mpg.de; anthony.olejniczak@gmail.com

C. Verna
Department of Human Evolution, Max Planck Institute for Evolutionary Anthropology, Deutscher Platz 6, 04103 Leipzig, Germany
and
UPR 2147 CNRS, Paris, France
e-mail: christine.verna@evolhum.cnrs.fr

S. Bailey
Department of Anthropology, New York University, Rufus D. Smith Hall, 25 Waverly Place, New York, NY 10003, USA
and
Department of Human Evolution, Max Planck Institute for Evolutionary Anthropology, Deutscher Platz 6, 04103 Leipzig, Germany
e-mail: sbailey@nyu.edu

T. Smith
Department of Human Evolutionary Biology, Harvard University, Peabody Museum, 11 Divinity Ave, Cambridge, MA 02138, USA
and
Department of Human Evolution, Max Planck Institute for Evolutionary Anthropology, Deutscher Platz 6, 04103 Leipzig, Germany
e-mail: tsmith@fas.harvard.edu

F. Z. Sbihi-Alaoui
Institut National des Sciences de l'Archéologie et du Patrimoine (INSAP), Rabat, Morocco
e-mail: fsbihialaoui@yahoo.fr

M. Zouak
Direction Régionale de la Culture (Tanger Tétouan), Inspection Régionale des Monuments Historiques et Sites, 2 Rue Ben Hsain, 93000 Tétouan, Morocco
e-mail: m_zouak@yahoo.fr

## Introduction

Moroccan sites have yielded the richest Middle and Late Pleistocene human fossil record in North Africa. This material has often been discovered in the course of archaeological excavations and is mostly associated with Acheulean, Mousterian, Aterian, or Iberomaurusian assemblages. The oldest and the youngest parts of this fossil record have attracted the attention of scholars, but the Aterian associated human remains are poorly described and mostly unpublished. However, they represent one of the

J.-J. Hublin and S. P. McPherron (eds.), *Modern Origins: A North African Perspective*,
Vertebrate Paleobiology and Paleoanthropology, DOI: 10.1007/978-94-007-2929-2_13,
© Springer Science+Business Media B.V. 2012

most abundant series of human remains associated with a Middle Stone Age/Middle Paleolithic assemblage in Africa.

In 1939, Howe and Movius discovered the first human remains assigned to an Aterian assemblage in the cave of Mughairet el 'Aliya near Tangier. They were represented by an immature fragmentary maxilla and some isolated teeth (Senyürek 1940). Later, at the "Grotte des Pigeons" (Taforalt), another site in northern Morocco, a fragmentary human parietal was discovered by Roche in 1951 in a "late Aterian" context (Roche 1953). Subsequent discoveries of human fossils associated with Aterian artifacts occurred in sites along the Moroccan coast south of Rabat in the 1960s and 1970s.

The "Grotte des Contrebandiers" at Temara, south of Rabat, was first explored by Roche, who conducted some test excavations between 1955 and 1957. A human mandible yielded by the site was then described by Vallois (Vallois and Roche 1958). The unusual robusticity of the bone and dentition led these authors to assign it to an early stage of human evolution, contemporary to Acheulean assemblages that ultimately were never identified in the archaeology of the site. Subsequently, the excavation by Roche developed mostly from 1967 to 1975 and yielded several other specimens assigned to Aterian contexts.

Another series of discoveries resulted from the work by Debénath in other cave sites along the Atlantic coast. In the cave of Dar es-Soltan II, in the suburb of Rabat (Dar es-Soltan I is a distinct cave located nearby and the eponym site for the Soltanian), a partial skull associated with a hemimandible as well as an immature calvaria and a mandible were found in 1975 at the bottom of the stratigraphy, in a marine sand deposit (layer 7). This sterile layer was overlaid by a distinct reddish layer (layer 6), in which a hearth and some Aterian elements have been described (Debénath 1976).

Finally, another discovery was made in the cave of Zouhrah at El Harhoura, between Dar es-Soltan and Grotte des Contrebandiers. The Zouhrah site is usually called "El Haroura I" in the literature as there is an El Haroura II which, so far, has not yielded Pleistocene hominins. There, another mandible and an isolated canine were unearthed during a salvage excavation in 1977 (Debénath 1980).

In line with the notion that the Aterian was a North African late Middle Paleolithic industry, the Mughairet el 'Aliya material was initially said to display "Neandertal-like" features (Senyürek 1940), as was the case in the mid-twentieth century for virtually any Middle Paleolithic or Middle Stone Age associated human specimen from the Old World. However, these very fragmentary remains looked to many scholars to be "nearly indecipherable" (Piveteau 1957). It was not before the 1970s, when more complete specimens from Dar es-Soltan II and El Haroura I were first described, that some authors started to emphasize the primarily anatomically modern nature of these fossils. By 1976, Ferembach (Ferembach 1976) assigned the newly discovered Dar es-Soltan specimens to "Homo sapiens sapiens," although they displayed remarkably wide facial dimensions and strong supraorbital reliefs. Minugh-Purvis (1993) re-examined the Mughairet el 'Aliya specimen and also rejected the notion that it displayed any Neandertal facial features, supporting instead that few diagnosable features were observable on the specimen and that it could instead be a representative of "Homo sapiens sapiens."

Another issue on which opinions and interpretations have widely varied is the age of these specimens. In northwestern Africa, the Aterian has long been considered a local late Middle Paleolithic assemblage, widely overlapping in time with the first Upper Paleolithic assemblages in Europe between ca. 40 and 20 ka (e.g., Debénath et al. 1986). This chronological assignment was mostly based on $^{14}$C dates, which were later revealed to be mostly infinite dates. However, in the last two decades, the development of new methods based on luminescence has led to the production of a new set of dates that pushed the chronology of the Aterian outside of and later in Morocco much further back in time (for review, see Bouzouggar and Barton 2012; Raynal and Occhietti 2012; Richter et al. 2012). Today, the Aterian assemblages are mostly assigned to a period between 90 and 35 ka and could well be rooted further back in time. This evolution of our perception of the Aterian and their makers occurred in the context of a complete change of view on recent human evolution in Africa, putting an increasing emphasis on the African origins of non-African extant humans (Bräuer 1984; Cann et al. 1987; Stringer and Andrews 1988), and more recently highlighting the early occurrence in the African archaeological record of behavior unique to recent modern humans (McBrearty and Brooks 2000; Henshilwood et al. 2004; Texier et al. 2010). We have gradually shifted from a situation where Aterians were seen as Neandertal-like hominins still producing a delayed Middle Paleolithic at a time when fully anatomically modern humans were producing the Upper Paleolithic in Europe, to a scheme in which they are essentially modern humans producing a Middle Stone Age with features such as the production of personal body ornaments (d'Errico et al. 2009) at a time when Neandertals were still producing Mousterian assemblages in Europe.

In its original form and still in many debates around it, the out-of-Africa model for the origin of non-African modern populations emphasizes a hypothesized sub-Saharan source of an ancestral group. However, it is important to underline that the present day distribution of African populations might be very different from their original location in the late Pleistocene. The episodes of high aridity and extensions of the Sahara during MIS 4 and 2 resulted in the displacement of many human groups further south and their almost complete separation from Maghreb populations. There are also clues to the ancient structure of early modern populations and for more complex scenarios that were once thought to account for the out-of-Africa exodus (Gunz et al. 2009). In this context, elucidating the affinities of North African populations older than 50 ka is of major importance. To date, the Aterian human fossil record is exclusively Moroccan. However, the archaeological record demonstrates that Aterian populations occupied a very large geographical area extending from the Atlantic coast to the Western Desert of Egypt, and from the Mediterranean to south of the modern Sahara (Bouzouggar and Barton 2012; Hawkins 2012). The makers of these assemblages can therefore be seen as (1) a group of *Homo sapiens* predating and/or contemporary to the out-of-Africa exodus of the species, and (2) geographically one of the (if not *the*) closest from the main gate to Eurasia at the northeastern corner of the African continent. Although Moroccan specimens have been discovered far away from this area, they may provide us with one of the best proxies of the African groups that expanded into Eurasia. Comparing them with the European and Near-Eastern human groups that immediately pre- and post-dated this exodus is therefore of crucial importance in order to elucidate the nature of the populations involved in it.

Dental material represents a large portion of the Aterian hominins available for study, either in the form of isolated teeth or more or less complete jaws. Although this material is fragmentary, it is most helpful in elucidating the biological affinities of these populations. Dental development seems to depend little on environmental conditions and mostly on genetic control and several recent studies have demonstrated that when combining metrical and non-metrical approaches, dental morphology is a useful tool to assess taxonomic affinities of Pleistocene hominins (Bailey 2006a; Martinón-Torres et al. 2007; Olejniczak et al. 2008a, b; Bailey et al. 2009; Smith et al. 2009). In this chapter, we present a systematic analysis of the Aterian dental remains. This includes dental metric and morphological analyses, as well as an analysis of the internal dental structures, specifically enamel thickness. Each of these aspects of the dentition provides important information on the affinity of these early North African humans. Ultimately, we compare these variables with those of other Late Pleistocene as well as more recent humans.

## Material

Table 13.1 lists the Aterian dental material included in this study and images of the specimens can be found in Figs. 13.1 and 13.2. At the time that this text was written, this material was housed in either the Musée Archéologique de Rabat or the Institut National des Sciences de Archéologie et du Patrimoine (Rabat). This list includes 51 teeth. In most cases the human remains were found directly in association with Aterian assemblages. However, in the case of the site of Dar es-Soltan II, it should be noted that the human remains were in fact found in a sterile layer immediately underlying an Aterian archaeological layer and are therefore considered to be intrusive from this overlying layer (Debénath 1976). Recent dates in the site (Bouzouggar and Barton 2012; Raynal and Occhietti 2012) as well as geological evidence indicate that this sterile layer is a transgressive horizon corresponding to the MIS 5 interglacial, which therefore represent a *terminus post quem* for the specimens. A similar situation is encountered at the Grotte des Contrebandiers in Temara, where archaeological layers immediately preceding the "genuine" Aterian layers have been identified either as MIS 5 "Ouljian breccias" (Vallois and Roche 1958) or "Mousterian" layers post-dating MIS 5e. However, these uncertainties should also be considered in the broader context of the debate around the age of the initial Aterian and the lack of sharp separation with underlying Mousterian assemblages within MIS 5 (Bouzouggar and Barton 2012; Richter et al. 2012).

A different problem results from the presence of six teeth in the Grotte des Contrebandiers series that Ménard (1998) assigned to the Iberomaurusian (IBM) (i.e., belonging to an epipaleolithic assemblage). Unfortunately, there is no consistency between the numbering and identification provided by Ménard (1998) and the current labeling of the specimens at the Musée Archéologique de Rabat. However, we could identify that an Upper M1 and M2 attached to a maxilla and reported by Ménard as "T4" is today numbered IB19 (Fig. 13.1). Our initial observations suggest that Menard was correct in attributing this specimen to the IBM. The teeth from this individual are smaller and morphologically simpler than the rest of the teeth in our sample. Moreover, one has a well-developed carious lesion. The frequency of caries in samples older than the Holocene is rather low, especially root-type caries of this size. Therefore, this specimen has been excluded from the Aterian sample in this study. For the remaining teeth, the specimen reported by Ménard (1998) as "T6" is today labeled "T1" and is most likely the lower M3 missing on the right side of the Grotte des Contrebandiers 1 Aterian mandible. The specimens reported as "T3" by Ménard do not correspond to the specimens today labeled "T3a" and "T3b," which display a size that clearly matches the rest of the Aterian series and is out of the IBM variation.

For our metrical study, the comparative database encompasses original and bibliographical data for three dental reference samples:

• Neandertals (NEAN) (from MIS 7 to MIS 3) (Maximun number of individuals, i.e., individuals plus unrelated isolated teeth (N = 157)) from the sites of: Abri Agut, Amud, Arcy-sur-Cure, Bau de l'Aubésier, Biache-Saint-Vaast,

**Table 13.1** List of the dental material

| Site | Specimen | Anatomical part | Analysed teeth |
|---|---|---|---|
| Dar es-Soltan II | Mandible H4 | Mandible | Left: C, M1-M3; right: I2-P3, M1-M3 |
| Dar es-Soltan II | Mandible H5 | Mandible fragment (left) | Left: C, P4-M3 |
| Dar es-Soltan II | NN | Isolated tooth | UM1 r |
| Dar es-Soltan II | H5 | Maxilla fragment (right) | Right: M2-M3 |
| Dar es-Soltan II | H6 | Maxilla fragment (left) | Left: P3-M1 |
| Dar es-Soltan II | H9 | Isolated tooth | UM2 l |
| Dar es-Soltan II | H10 | Isolated tooth | UM2 l |
| Contrebandiers | Mandible | Mandible | Left: I1, P4-M3; right: I1-M2 |
| Contrebandiers | T1 | Isolated tooth | LM3 r |
| Contrebandiers | T5 | Isolated tooth | UP4 r |
| Contrebandiers | T2 | Isolated tooth | Udm2 l |
| Contrebandiers | T4 | Isolated tooth | UM3 l |
| Contrebandiers | H7 | Isolated tooth | UM1 r |
| Contrebandiers | T3b | Isolated tooth | UM1 l |
| Contrebandiers | T3a | Isolated tooth | LM1 l |
| Contrebandiers | IB19 | Maxilla fragment (left) | Left : M1, M2 |
| El Haroura | Mandible | Mandible | Left: P4-M3; right M1-M3 |

In Dar es-Soltan, the maxilla fragment H5 likely belongs to the same individual as the mandible H5. The maxilla H6, the mandible H4, and the tooth H9 are also likely from the same individual. In Temara, the LM3 T1 likely belongs to the mandible

Grotta Breuil, Châteauneuf-sur-Charente, Columbeira, Combe Grenal, Dederiyeh, Devil's Tower, Croze del Dua, Le Fate, Fenera, Fossellone, Guattari, Genay, Hortus, Kebara, Krapina, Kůlna, La Chaise–Abri Suard, La Chaise–Bourgeois-Delaunay, La Ferrassie, Le Moustier, La Quina, Macassargues, Monsempron, Montmaurin, Ochoz, Payre, Le Placard, Le Portel, Petit-Puymoyen, Regourdou, les Rivaux, Rochelot, Saccopastore, Shanidar, Sipka, Soulabé-las-Maretas, Spy, Subalyuk, Tabun, Teshik-Tash, Vergisson, Vindija, Grotte Vaufrey, Zafarraya.

- Middle Paleolithic modern humans from Qafzeh and Skhul (MPMH) (N = 13).
- European Upper Paleolithic modern humans (UPMH) from Aurignacian to Solutrean (N = 75): les Abeilles, Les Battut, Bacho-Kiro, Brassempouy, El Castillo, La Crouzade, Dolní Věstonice, La Ferrassie, Kent, Mladeč, Les Rois, Les Vachons, Abri Pataud, Brno, Cro-Magnon, Grotte des Enfants, Lagar Velho, Peştera cu Oase, Paglicci, Pavlov, Předmostí, Isturitz, Lafaye, Cap Blanc, Le Morin, Saint-Germain-la-Rivière, Le Peyrat, Pech de la Boissière, Oberkassel.
- Similar groups were used for the comparison of non-metrical features:
- Neandertals (from MIS 7 to MIS 3; N = 101 specimens, individuals, or isolated teeth): Arcy-sur-Cure, Ciota Ciara, Combe Grenal, Devil's Tower, Ehringsdorf, Guattari, Hortus, Grotta Taddeo, Krapina, Kůlna, La Fate, La Ferrassie, La Quina, Malarnaud, Marillac, Monsempron, Montgaudier, Obi Rakhmat, Ochoz, Petit-Puymoyen,

Pontnewydd, Regourdou, Roc de Marsal, Saccopastore, Spy, St Cesaire, Subalyuk, Taubach,Vindija.
- Middle Paleolithic modern humans from Qafzeh and Skhul (N = 11).
- Upper Paleolithic modern humans from Europe (Aurignacian and Gravettian) (N = 42): Abri Labatut, Abri Pataud, Dolní Věstonice, Derava Skala, Font de Gaume, Fontéchevade, Grotte des Abeilles, Grotte des Rois, Istallosko, Le Ferrassie, Lagar Velho, La Gravette, Lespugue, Les Vachons, Miesslingtal, Mladeč, Peştera cu Oase, Pavlov.

In addition to the European UPMH groups, it was also possible to take measurements on a series of Moroccan IBM specimens from Afalou and Taforalt (N = 106) from the collections of the Institut de Paléontologie Humaine in Paris. Also, Middle Stone Age specimens from Die Kielders, Klasies River Mouth, and Blombos (South Africa) (N = 9) were plotted in our distributions.

Three-dimensional enamel thickness studies require that the entire volume of enamel is unbroken and unworn, and none of the teeth examined here met these criteria. Two-dimensional enamel thickness measurements require that, at most, only minimal wear appears on the mesial cusps, and two of the molar teeth we examined met this condition (H7 and T3b, both maxillary first molars). To assess enamel thickness in these specimens, data describing first maxillary molar enamel thickness in other taxa were taken from the literature on modern humans (Smith et al. 2006a), Neandertals (Olejniczak et al. 2008a), australopiths (Olejniczak et al. 2008b), and chimpanzees (Smith et al. 2005).

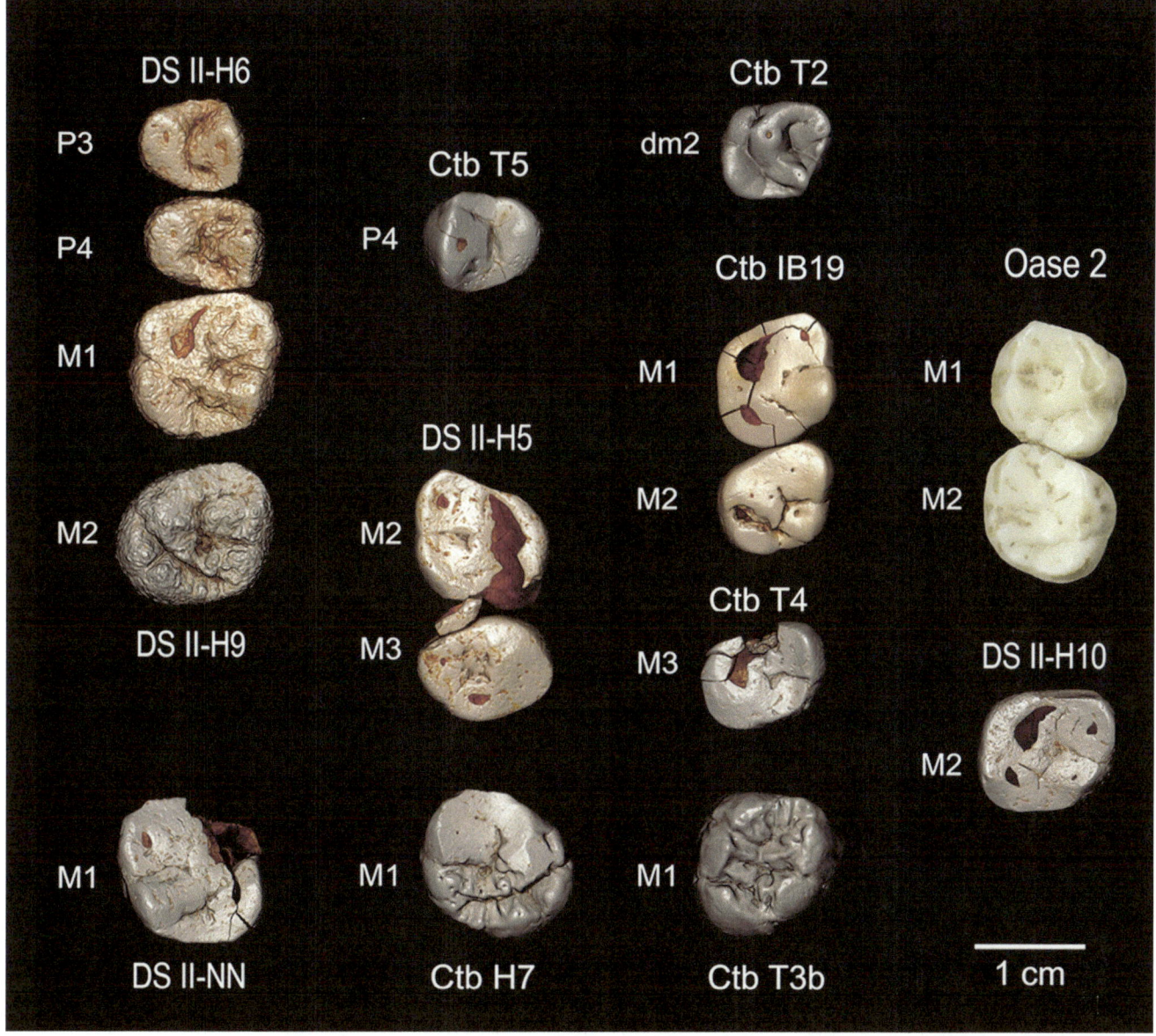

**Fig. 13.1** Upper teeth from the Aterian layers of Dar es-Soltan II (DeS II) and Grotte des Contrebandiers (Ctb). Note that "Temara IB19" is not part of the Aterian sample

## Methods

To assess the variations in size of the teeth, maximum mesio-distal and bucco-lingual crown diameters (M81 and M81(1)) (Braüer, 1988) were measured with calipers. In the case of the Dar es-Soltan "mandible H4" (Dar es-Soltan II-4), these measures were obtained from 3D reconstructions using Amira software 4.1.2. A bivariate assessment of these metrics was made by plotting the crown diameters for each tooth for the different groups; ellipses, including 95% of the population variation, were computed using Statistica 7.0 software for each reference population.

The non-metric analysis is based on 15 traits (4 in the upper dentition and 11 in the lower dentition) that are preserved on the Aterian specimens (see Table 13.4). Many are included in the Arizona State University Dental Anthropology System (ASUDAS) and have been chosen because they are easy to score and can be scored even in worn teeth (Turner et al. 1991). Additional traits found by Bailey (2002, 2006b) to show meaningful differences among fossil hominin groups and which are particularly important to the study of Middle-Late Pleistocene hominins have been added (maxillary premolar accessory ridges or MxPAR: Burnett 1998; lower P4 asymmetry, lower P4

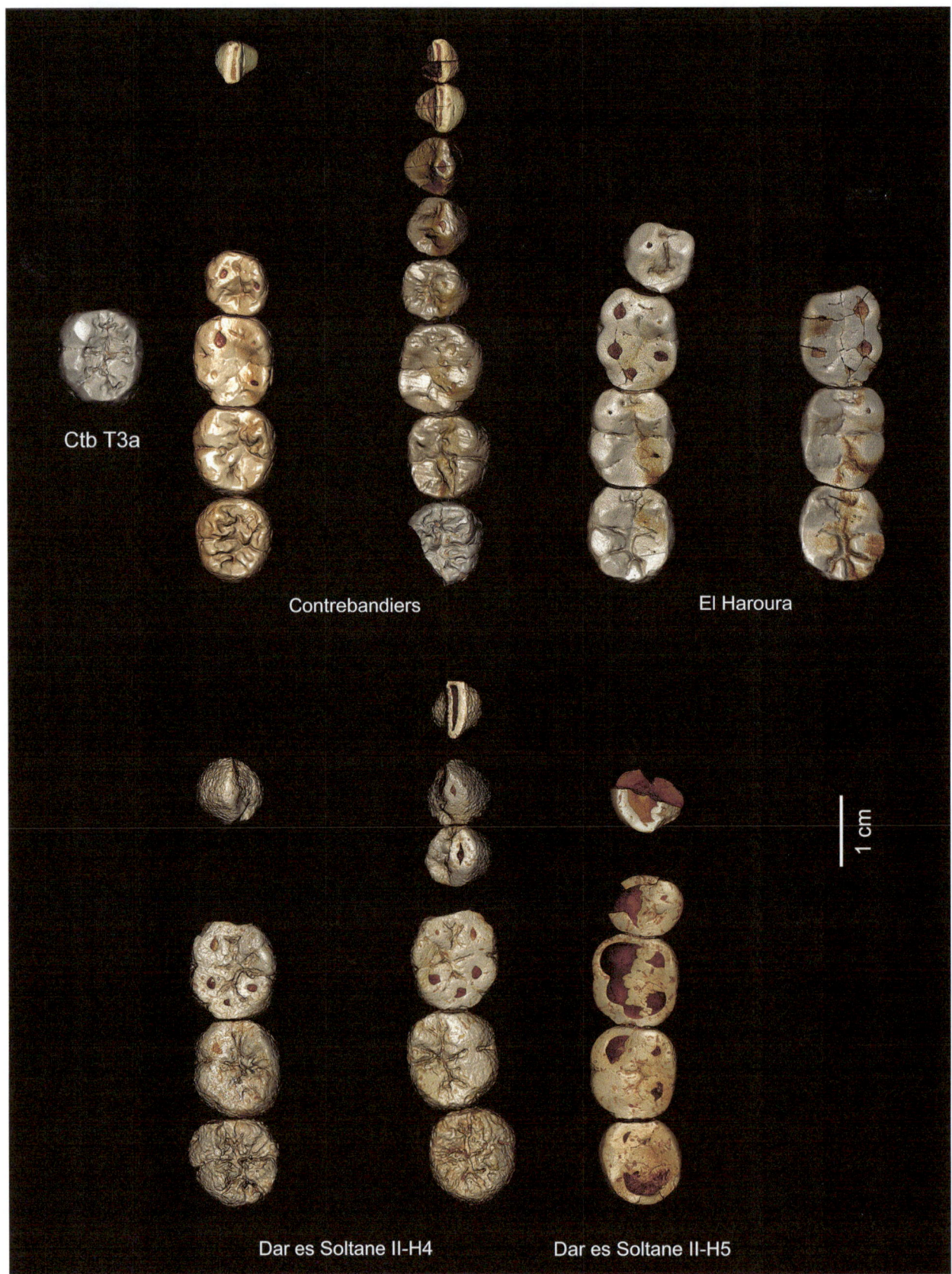

**Fig. 13.2** Lower teeth from the Aterian layers of Dar es-Soltan II, Grotte des Contrebandiers (Ctb), and El Harhoura. Note that the specimen T1 (M3) is positioned behind the lower right M2 of the Temara mandible since it likely belongs with it

transverse crest, and the lower molar mid-trigonid crest). These traits are also easy to score and tend to be preserved even on moderately worn teeth.

In order to analyze enamel thickness, the molars were microCT scanned using a BIR Arctis 225/300 high-resolution industrial CT scanner with X-ray energy at 130 kV, 100 μA, and a 0.25 mm brass filter. Isometric voxel dimensions were kept below 30 microns. Mesial planes of section were produced from the microCT scans using VoxBlast software (Vaytek, Inc.), following techniques described by Olejniczak (2006). We therefore produced virtual planes of section through the mesial cusps to perform traditional, two-dimensional enamel thickness measurements, consistent with those of previous studies (e.g., Martin 1985; Smith et al. 2005). Measurements of mesial sections that were recorded on printed images using a digitizing tablet are defined following Martin (1985) and appear in Fig. 13.3: the area of the enamel cap (mm²), the coronal dentine area (mm²), and the length of the enamel-dentine junction (mm). Average enamel thickness (mm), the average straight-line distance between the enamel-dentine junction and the outer enamel surface, is calculated as enamel area divided by enamel-dentine junction length. Relative enamel thickness, a scale-free measurement facilitating inter-specific comparisons when tooth sizes differ among taxa, is calculated as average enamel thickness divided by the square root of the coronal dentine area.

## Results

### Metrical Analysis

We analyzed 51 teeth from the site of Dar es-Soltan II, Grotte des Contrebandiers, and El Haroura. Of these 51 teeth, 10 are isolated and 41 are attached to 4 mandibles or 3 fragmentary maxilla.

#### Upper Teeth

The UP3 from "Dar es-Soltan H6" (Dar es-Soltan II-6) is mesio-distally wide, falling outside the range of the IBM and close to the limit of the NEAN and UPMH range (see Figs. 13.1, 13.4; Table 13.2). Its bucco-lingual diameter, however, falls well within the overlap area of the comparative samples.

The two UP4s from Dar es-Soltan II and the Grotte des Contrebandiers are robust. They plot well outside the range of variation of IBM and UPMH but also rather far from the

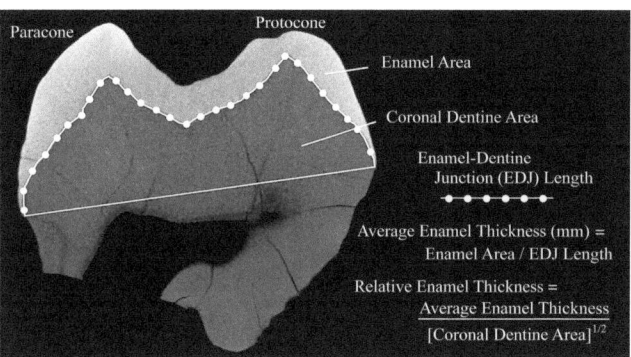

**Fig. 13.3** MicroCT-based virtual mesial cusp cross-section of the Aterian maxillary first molar T3b, demonstrating the enamel thickness measurements and calculations performed in this study

MPMH. Dar es-Soltan II-6 falls close to some specimens from Krapina, and is included in the 95% ellipse of the NEAN. The UP4 Grotte des Contrebandiers T5 has even higher diameters and falls outside the NEAN range.

All the UM1s fall well outside the range of variation of NEAN, UPMH, and IBM. It should be noted that this does not apply to the Grotte des Contrebandiers IB19 specimen, confirming its IBM assignment by Ménard (1998). The latter has lower diameters and falls well within the range of the reference groups.

Although large, UM2 and UM3 from Dar es-Soltan II and Grotte des Contrebandiers 1 are relatively less robust than UM1. When compared to our reference samples, two of them fall within the ranges of variation of NEAN, MPMH, UPMH, and IBM but above the means. Dar es-Soltan II H9 UM2 is very robust, with a BL higher than any other individual included in our analysis. It plots at the upper limit of the NEAN range, and is interestingly close to the UM2 from Peştera cu Oase 2. The UM2 of Grotte des Contrebandiers IB19 is much more gracile, which supports its assignment to the IBM group.

There is little separation between our comparative samples for the Udm2, and the single deciduous tooth, Grotte des Contrebandiers T2, falls well within their range of variation.

#### Lower Teeth

The two lower incisors from the Grotte des Contrebandiers mandible are not particularly large, and fall within the range of IBM and UPMH, with rather low bucco-lingual diameters compared to Neandertals (see Figs. 13.2, 13.5, 13.6; Table 13.3). The lower I2 from Dar es-Soltan, however, falls outside the range of all the reference groups due to its very large MD.

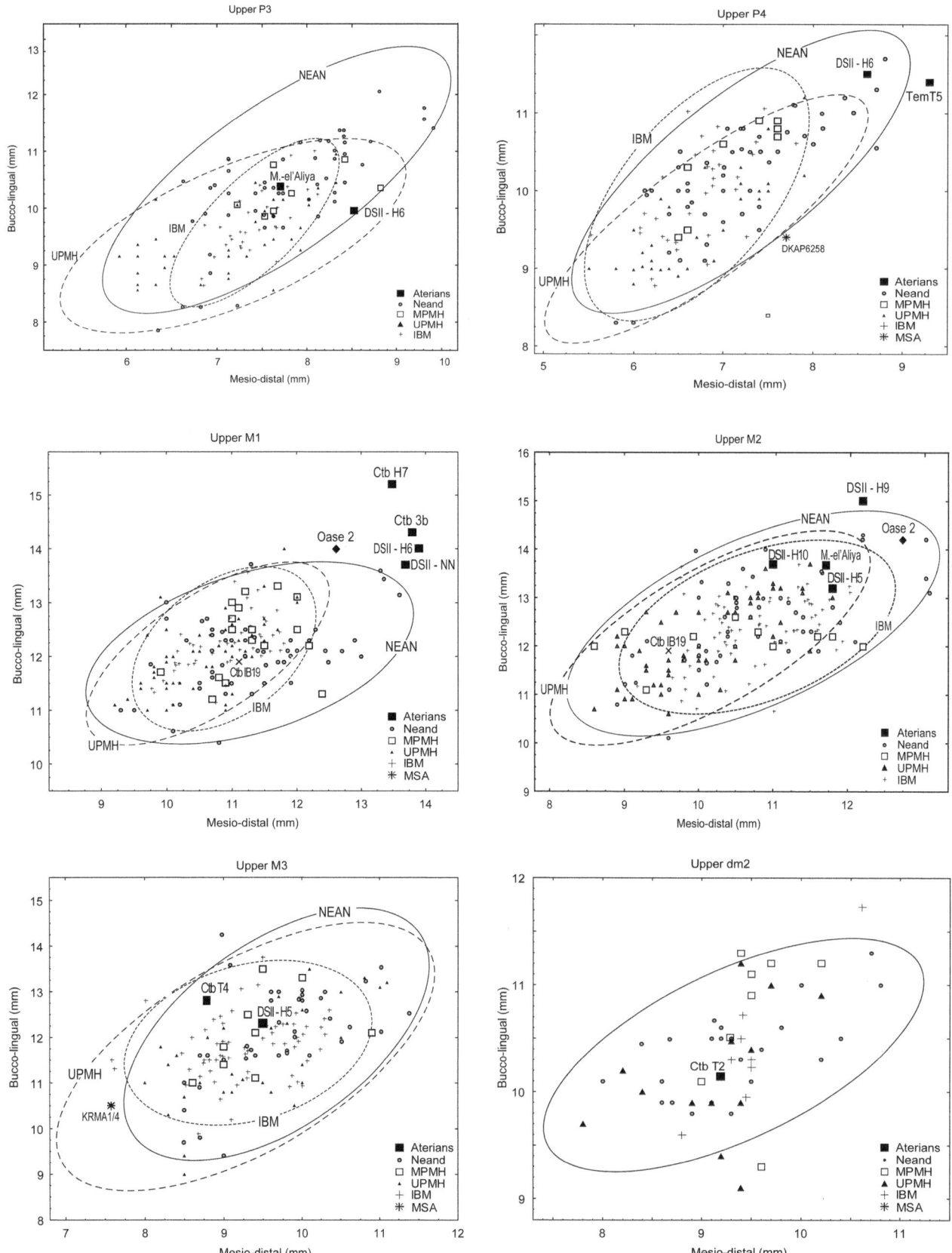

**Fig. 13.4** Bivariate plots of the crown diameters of the upper teeth The graph includes equiprobable ellipses (95%) of Neandertals (Neand), Upper Paleolithic modern humans (UPMH) and Ibero-maurusians (IBM), as well as individual dots for Middle Paleolithic modern humans (MPMH) and South-African Middle Stone Age modern humans (MSA)

**Table 13.2** Crown diameters of the upper teeth

| Site | N° | Side | Udm2 | | UP3 | | UP4 | | UM1 | | UM2 | | UM3 | |
|---|---|---|---|---|---|---|---|---|---|---|---|---|---|---|
| | | | MD | BL | MD | BL | MD | BL | MD | BL | MD | BL | MD | BL |
| Contrebandiers | T2 | L | 8.6 | 10 | | | | | | | | | | |
| Contrebandiers | T3b | L | | | | | | | 14 | 14 | | | | |
| Contrebandiers | T4 | L | | | | | | | | | | | 8.8 | 13 |
| Contrebandiers | T5 | R | | | | | 9.3 | 11 | | | | | | |
| Contrebandiers | T7 | R | | | | | | | 14 | 15 | | | | |
| Contrebandiers | IB19 | L | | | | | | | 11 | 12 | 9.6 | 12 | | |
| Dar es-Soltan II | H5 | R | | | | | | | | | 12 | 13 | 9.5 | 12 |
| Dar es-Soltan II | H6 | L | | | 8.4 | 10 | 8.6 | 12 | 14 | 14 | | | | |
| Dar es-Soltan II | H9 | L | | | | | | | | | 12 | 15 | | |
| Dar es-Soltan II | H10 | L | | | | | | | | | 11 | 14 | | |
| Dar es-Soltan II | NN | R | | | | | | | 14 | 14 | | | | |

The lower canines from Dar es-Soltan are large, with high bucco-lingual diameters falling among the highest values of our comparative samples. Their mesio-distal diameters are even higher and exceed the values of all the individuals included in our analysis. The LC from the Grotte des Contrebandiers mandible is slightly smaller, and falls at the limit of the IBM range and within the range of the NEAN and UPMH. Large canines are also found on the MPMH Qafzeh 9.

Due to their large mesio-distal diameters, the two lower P3s from Dar es-Soltan and Grotte des Contrebandiers fall outside the UPMH and IBM ranges and at the higher limit of the NEAN range. They also plot close to Qafzeh 9. Only three specimens from Krapina show higher mesio-distal diameters.

The lower P4s are very large, falling outside the range of all the comparative samples (Dar es-Soltan H5, El Haroura) or close to their upper limit (Grotte des Contrebandiers mandible).

The lower molars are also very robust, falling either outside or at the limit of the 95% confidence limits of the reference groups. Interestingly, the LM1s stand out the most, followed by the LM2 and the LM3, which are more variable and relatively less robust compared to our reference samples. Once again, the mandible from Grotte des Contrebandiers shows smaller teeth than the Dar es-Soltan and El Haroura specimens. We also found that the LM2 and LM3 from Dar es-Soltan and El Haroura are close to those of the Peştera cu Oase 1 specimen, whereas Nazlet Khater is more gracile and plots closer to the Grotte des Contrebandiers 1 mandible.

None of the specimens preserve the entire dentition but the two mandibles from Grotte des Contrebandiers and Dar es-Soltan H4 allow us to assess dental proportions. We plotted the ratios of the summed I2 to P3 versus P4 to M2

BL diameters (Fig. 13.7a). The results show the known contrast between Neandertals and modern humans for the relative anterior dental size. The mandible from Grotte des Contrebandiers plots within the range of the IBM, UPMH, and MPMH but outside the Neandertal range (Fig. 13.7a), which shows higher relative anterior dentition size. Interestingly, the ratio of the Grotte des Contrebandiers mandible dentition falls among the highest values of the IBM but among the lowest values of the MPMH and UPMH. We also plotted the ratio [I2-C]/[M1-M2] (Fig. 13.7b) which shows a similar trend across our comparative samples. Again, the two North African specimens plot far from the Neandertals for this ratio, and well within the IBM range. The Grotte des Contrebandiers mandible ratio falls within the distribution of the MPMH and at the lowest limit of the UPMH one. The Dar es-Soltan H4 ratio is lower and outside the range of the UPMH and just at the lower limit of the MPMH.

## Non Metrical Traits

In Table 13.4, we present the trait frequencies observed in the Aterians and comparative groups. It is important to point out that the Aterian sample is quite small, ranging from two to five individuals for any given trait.

The single upper P4 shows well-developed accessory ridges, a trait that is frequently observed in Neandertals and other archaic humans (Bailey 2006b). Both permanent upper M1s that preserve the lingual surface show the cusp form of Carabelli's trait (ASUDAS grades 6 and 5). A moderately- to well-developed Carabelli's trait is relatively common in the other fossil hominin permanent M1s, particularly those in the Neandertal group. The upper dm2 (Grotte des Contrebandiers T2) also possesses a Carabelli's

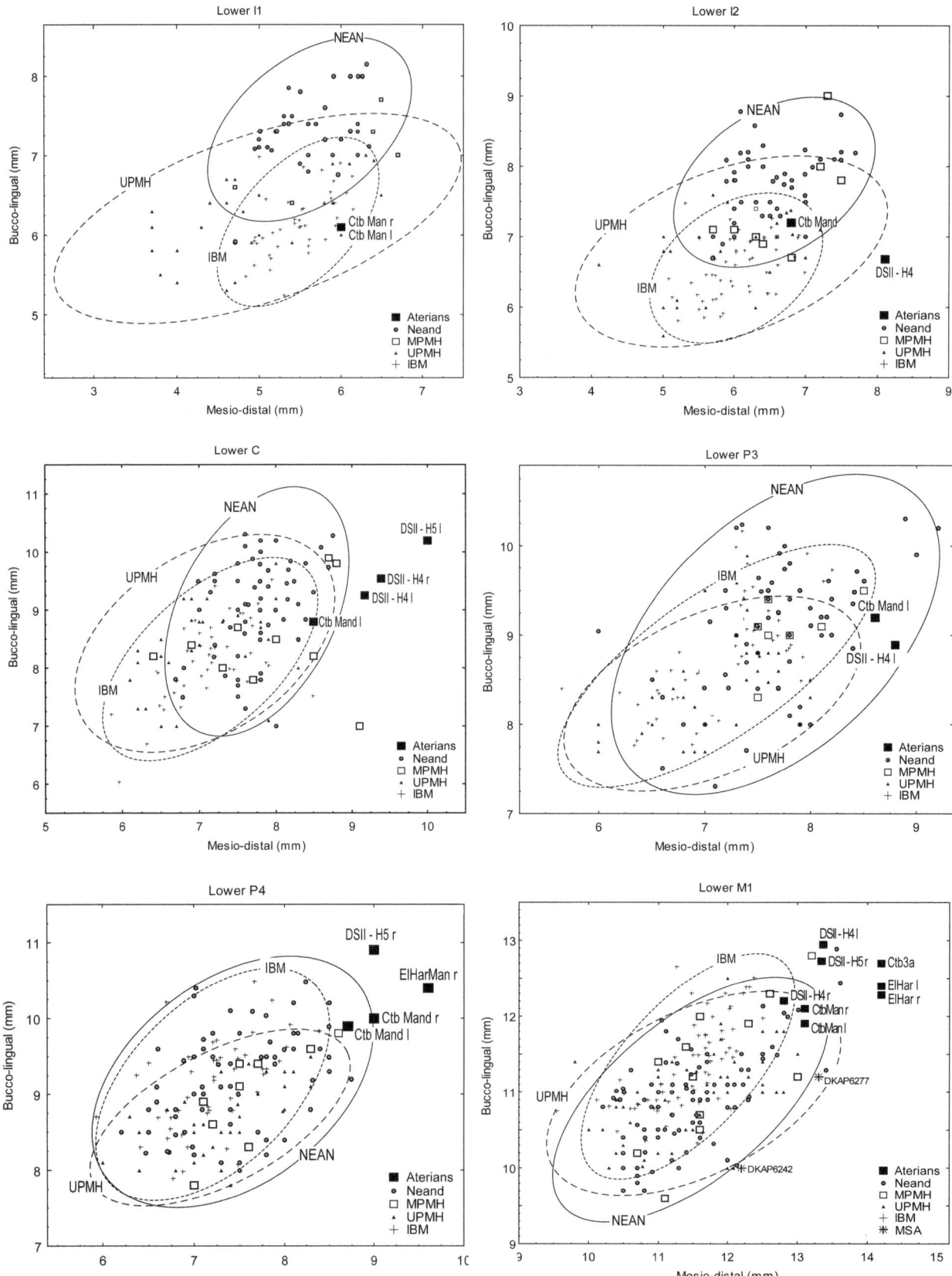

**Fig. 13.5** Bivariate plots of the crown diameters of the lower teeth (I1 to M1)

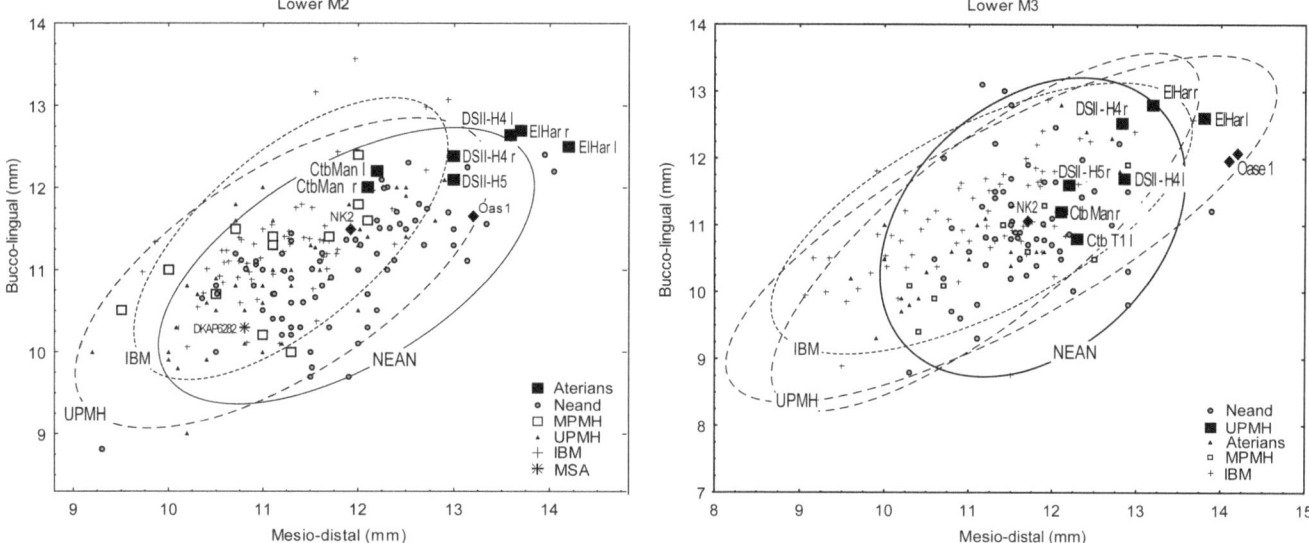

**Fig. 13.6** Bivariate plots of the crown diameters of the lower M2 and M3

cusp (ASUDAS grade 6). In this case, it displaces the protocone in a buccal direction and is continuous with the hypocone. Our comparative morphological sample for deciduous dm2s is small but both fossil hominins (*Homo sapiens* from Die Kelders and Neandertal from Subalyuk) show moderately sized Carabelli's cusps, although slightly smaller than that seen in the Grotte des Contrebandiers tooth. Two of the three M1s also show an accessory distal cusp (Cusp 5 or hypoconule), which is also relatively common in other fossil hominins in our sample, particularly the early Upper Paleolithic and Neandertal groups. Neither of the M2 possesses a reduced hypocone. This is the condition found in most fossil hominins, although the early Upper Paleolithic group shows higher frequencies than the other groups for this trait. In addition, two new variants are observed on the upper molars. The first variant is found in one UM1 from Grotte des Contrebandiers (3b), which possesses a large accessory cusp between the protocone and the paracone. This may be an exceptionally large protoconule or perhaps an entocone, although the latter is usually expressed as a ridge not a cusp. The second variant is seen in two of UM (H7 from Grotte des Contrebandiers and H9 from Dar es-Soltan). Both of these teeth express exceptionally large hypocones that appear to be divided into two or more cusps (Fig. 13.1). This trait is not scored by the ASUDAS and it was not one of the traits studied in Bailey (2002) so comparative data are unavailable.

In the lower dentition, the single P4 that can be scored for the trait possesses multiple lingual cusps. However, neither P4 possesses a transverse crest or marked asymmetry. This is similar to the condition found in nearly all fossil *Homo sapiens* samples, which is strikingly different from the Neandertal condition, where all three traits occur

in high frequency and are often expressed on the same tooth (Bailey 2002). One of the M1s (Dar es-Soltan H4) possesses a Cusp 7 and one individual (El Harhoura) possesses a midtrigonid crest, while Cusp 6 and the deflecting wrinkle are both absent. On the M2, the Y-groove pattern is present on four of the five teeth and one tooth exhibits an asymmetrical expression (Y pattern on the left and X on the right). The Y-groove pattern is the primitive pattern and the most frequently observed condition in the comparative groups with the exception of the early Upper Paleolithic group, in which 50% of the teeth possess either an X or + pattern. One of the five M2s possesses four, instead of five, cusps. This is similar to the frequency found in the comparative groups, except for the Neandertals, who lack this feature. As is the case with the other *Homo sapiens* in our comparative sample, none of the Aterian M2s and M3s possess a midtrigonid crest. This feature appears to be present on these teeth only in the Neandertal sample.

**Enamel Thickness**

The results of enamel thickness measurements in the Aterian maxillary first molars and the comparative sample are presented in Fig. 13.8 and Table 13.5. In terms of relative (size-scaled) enamel thickness, the Aterian molars fall within the range of recent *Homo sapiens* individuals, and do not overlap with the thinner Neandertal molars. Likewise, average enamel thickness (expressed in mm units) shows that the Aterian molars have thicker enamel than the Neandertal sample; specimen H7 falls within the range of recent *Homo sapiens*, and specimen T3b is above the range of recent *Homo sapiens*. Compared to other hominins (all of

**Table 13.3** Crown diameters of the lower teeth

| Site | N° | Side | LI1 MD | LI1 BL | LI2 MD | LI2 BL | LC MD | LC BL | LP3 MD | LP3 BL | LP4 MD | LP4 BL | LM1 MD | LM1 BL | LM2 MD | LM2 BL | LM3 MD | LM3 BL |
|---|---|---|---|---|---|---|---|---|---|---|---|---|---|---|---|---|---|---|
| Contrebandiers | Mand. | L | 6.0 | 6.1 | | | | | | | | | 13.1 | 12.1 | 12.2 | 12.2 | 12.1 | 11.2 |
| | | R | 6.0 | 6.1 | 6.8 | 7.2 | 8.5 | 8.8 | 8.6 | 9.2 | 8.7 | 9.9 | 13.1 | 11.9 | 12.1 | 12.0 | | |
| Contrebandiers | T1 | L | | | | | | | | | | | | | | | 12.3 | 10.8 |
| Dar es-Soltan II | Mand. H5 | L | | | | | 10.0 | 10.2 | | | 9.0 | 10.9 | 12.8 | 12.2 | 13.0 | 12.1 | 12.2 | 11.6 |
| Dar es-Soltan II | Mand. H4 | R | | | 8.3 | 6.8 | 9.3 | 9.3 | 8.8 | 8.9 | | | 13.4 | 12.9 | 13.6 | 12.6 | 12.9 | 11.7 |
| | | L | | | | | 9.5 | 9.5 | | | | | 13.4 | 12.7 | 13.0 | 12.4 | 12.8 | 12.5 |
| Contrebandiers | T3a | R | | | | | | | | | | | 14.2 | 12.7 | | | | |
| El Haroura | Mand. | L | | | | | | | | | 9.6 | 10.4 | 14.2 | 12.3 | 13.7 | 12.7 | 13.2 | 12.8 |
| | | R | | | | | | | | | | | 14.2 | 12.4 | 14.2 | 12.5 | 13.8 | 12.6 |

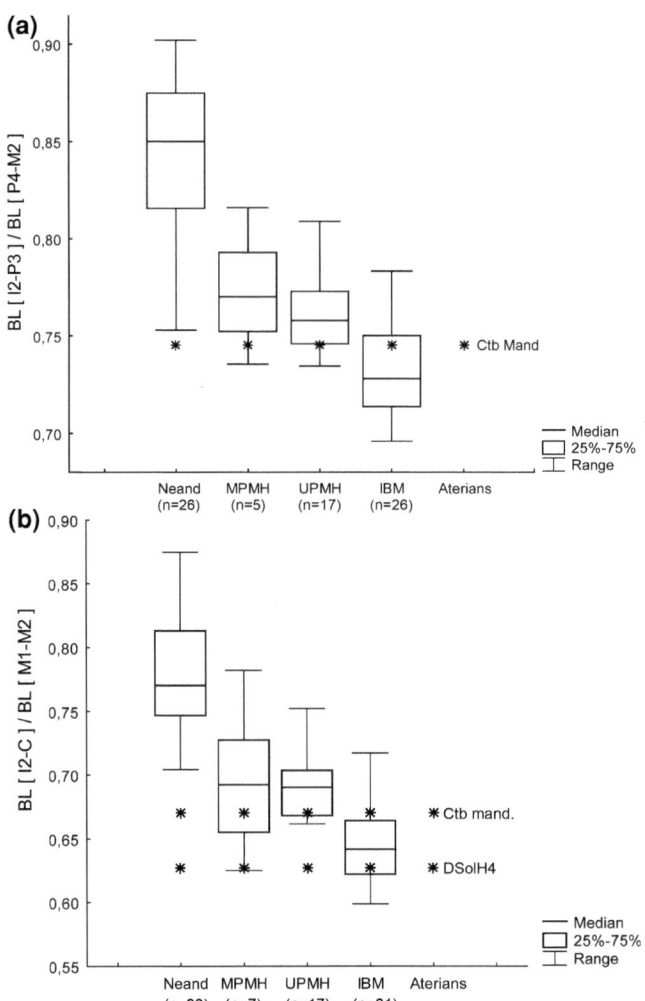

**Fig. 13.7** Box plots showing the ratio of the summed bucco-lingual breadths for our comparative samples. On the left is the ratio of the summed I2 P3 versus P4 M2 bucco-lingual breadths for our comparative samples and the values of the Aterian mandible from Grotte des Contrebandiers. On the right is the ratio of I2 C versus M1 M2 bucco-lingual breadths between our samples and those from both Dar es-Soltan H4 and Grotte des Contrebandiers

which have absolutely and relatively thicker enamel than the *Pan troglodytes* comparative sample), the Aterian molars demonstrate similar relative and absolute enamel thickness values to those of recent *Homo sapiens*, are thinner-enameled than those of *Australopithecus africanus* and *Paranthropus robustus*, and are thicker-enameled than Neandertals.

## Discussion

Overall, the Aterians show large or very large teeth. In several instances, their dimensions lie outside the ranges of our comparatives samples. Most noticeable in this respect

**Table 13.4** Trait frequencies in Aterians and comparative samples. Sample sizes/number with trait presence in parentheses

| | *Homo sapiens* | | | *H. neanderthalensis* |
|---|---|---|---|---|
| | Aterian | Qafzeh/ Skhul | Upper Paleolithic modern | |
| Upper Dentition | | | | |
| P4 MxPAR + = grades 1–2 | 100 (1/1) | 50.0 (2/1) | 0.0 (2/0) | 77.8 (18/14) |
| M1 Carabelli's trait + = grades 3–7 | 100 (2/2) | 40.0 (5/2) | 60.0 (10/6) | 72.0 (25/18) |
| M1 C5 + = grades 1–5 | 66.7 (3/2) | 40.0 (5/2) | 72.7 (11/8) | 63.6 (22/14) |
| M2 Hypocone reduction + = grades 0–2 | 0.0 (1/0) | 0.0 (6/0) | 25.0 (12/3) | 2.9 (34/1) |
| Lower Dentition | | | | |
| P4 asymmetry + = grade 2 | 0.0 (2/1) | 0.0 (4/0) | 0.0 (5/0) | 75.0 (32/24) |
| P4 transverse crest + = grades 1–2 | 0.0 (2/0) | 33.3 (3/1) | 0.0 (5/0) | 90.6 (32/29) |
| P4 lingual cusps + = grades 2–9 | 100 (1/1) | 66.7 (3/2) | 66.7 (6/4) | 93.9 (32/30) |
| M1 Cusp 6 + = grades 1–5 | 0.0 (1/0) | 0.0 (4/0) | 20.0 (15/3) | 36.4 (22/8) |
| M1 Cusp 7 + = grades 2–4 | 25.0 (4/1) | 14.3 (7/1) | 5.3 (19/1) | 16.7 (36/6) |
| M1 Mid-trigonid crest + = grades 1–2 | 25.0 (4/1) | 33.3 (3/1) | 0.0 (15/0) | 93.5 (31/29) |
| M1 Deflecting wrinkle + = grades 2–3 | 0.0 (2/0) | 50.0 (2/1) | 7.7 (13/1) | 3.8 (26/1) |
| M2 Groove pattern + = Y | 80.0 (5/4) | 100 (3/3) | 50.0 (14/7) | 75.0 (33/27) |
| M2 Cusp No. + = 4 cusps | 20.0 (5/1) | 16.7 (6/1) | 36.4 (11/4) | 0.0 (40/0) |
| M2 Mid-trigonid crest + = grades 1–2 | 0.0 (2/0) | 0.0 (2/0) | 0.0 (15/0) | 96.3 (27/26) |
| M3 Mid-trigonid crest + = grades 1–2 | 0.0 (3/0) | 0.0 (2/0) | 0.0 (10/0) | 93.3 (93.8) |

**Table 13.5** Mean 2D average and relative enamel thickness in hominoid upper first molars (mesial sections)

| Taxon | N | Mean AET | Range | Mean RET | Range |
|---|---|---|---|---|---|
| *Pan troglodytes* | 6 | 0.66 | 0.61–0.73 | 10.3 | 8.5–12.2 |
| *Homo neanderthalensis* | 5 | 1.04 | 0.97–1.19 | 15.4 | 14.8–16.9 |
| Modern *Homo sapiens* | 37 | 1.22 | 0.98–1.50 | 18.8 | 14.0–23.9 |
| **Aterian *H. sapiens* (Ctb H7 & T3b)** | **2** | **1.42** | **1.25–1.58** | **19.6** | **18.3–20.9** |

are the UP4 and UM1 and the LC, LP4, and LM2. The LC, UP3, LP3, and LM1 are close in size to some MPMH specimens. In our modern human comparative samples, the only specimen close to the Aterian values is Peştera cu Oase 2. In particular, the size of the very large UM1s and LM3 from El Harhoura are reminiscent of the Peştera cu Oase 1 and 2 values. Within the Neandertal sample, one finds similarly large post-canine dentition only in some very robust MIS 5 Neandertals from Krapina. However, the most complete Aterian specimens (the mandibles Grotte des Contrebandiers and Dar es-Soltan H4) display a pattern of dental proportions clearly different from that of the Neandertals and closer to that of the modern series (which show reduced anterior dentitions relative to the post-canine dentitions). This is confirmed by the lack of bucco-lingual expansion of the anterior dentition that is observed in the Neandertals.

Morphologically, the Aterian teeth possess mass-additive traits including extra crests, distal cuspules, Carabelli's cusp, as well as large, often divided, hypocones on the upper teeth. Mass-additive traits are found in other fossil hominin groups as well, and contribute to the overall primitive morphology of the Aterians. They do not, however, possess in any appreciable frequencies the traits or trait combinations that are diagnostic of Neandertals. For example, the lower fourth premolars are symmetrical and lack a transverse crest. Likewise, with a single exception, the lower molars lack the mid-trigonid crest that is consistently present in Neandertals. More importantly, the M2s and M3s all lack this feature, which occurs nearly ubiquitously in Neandertals. Instead, the dental pattern observed in the Aterians is closer to that of the early *Homo sapiens* from Qafzeh/Skhul and to early Upper Paleolithic modern humans. In particular, the specimens from Peştera cu Oase

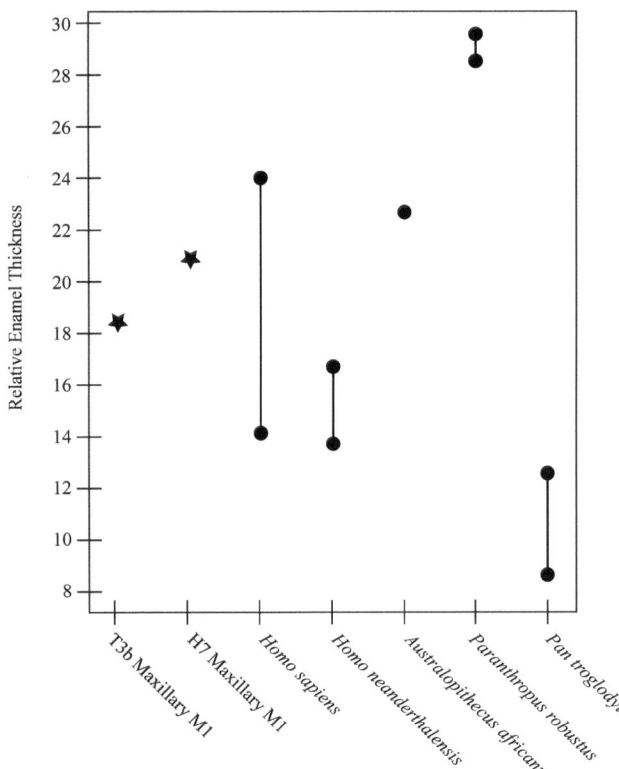

**Fig. 13.8** Ranges of relative enamel thickness in mesial sections of hominin maxillary first molars. The Aterian molars measured here (*stars*) fall within the range of values reported for recent human molars (Smith et al. 2006a, b) and are above the range of values reported for Neandertals (Olejniczak et al. 2008a)

are morphologically similar in several ways. Like the permanent Aterian upper M1s, the upper M1s of Peştera cu Oase 2 possess the large cusp form of Carabelli's trait. In addition, the Peştera cu Oase 2 left UM1 exhibits several distal cusps somewhat contiguous with the hypocone that, while not identical, are similar to the divided hypocones of the Grotte des Contrebandiers UM1 and Dar es-Soltan UM2.

Molar enamel thickness has a long history of study among scholars of hominoid evolution (e.g., Martin 1985; Kono 2004; Smith et al. 2005), and recent studies have demonstrated that hominin taxa exhibit important differences in enamel thickness as well. Specifically, Neandertals have thinner molar enamel than both modern humans (Olejniczak et al. 2008a) and Plio-Pleistocene hominins (Olejniczak et al. 2008b; Smith et al. 2009). A sample of South African Middle Stone Age human molars was previously found to be similar to modern humans in terms of enamel thickness (Smith et al., 2006b) and the authors of that study concluded that enamel thickness in *Homo sapiens* "has remained stable for at least 60,000 years" (p. 516). None of the molars in that study were from the maxillary dentition, and thus were not directly comparable to the data

presented here, but the present study can be taken as additional evidence that enamel thickness has remained relatively constant within *Homo sapiens*, and that enamel thickness in fossil *Homo sapiens* is similar across a wide geography: northern and southern African Middle Paleolithic specimens demonstrate similar enamel thicknesses, which is comparable to recent humans. In light of limited data describing fossil *Homo sapiens*, the relatively thin enamel of Neandertals has been cautiously interpreted as a derived condition relative to other hominins (Olejniczak et al. 2008a). This relatively thin enamel stems from a difference in tooth conformation between *Homo sapiens* and Neandertals, wherein Neandertal molars are characterized by a larger volume of coronal dentine underlying a similar volume of enamel, resulting in absolutely thinner enamel. It is unclear whether there is an adaptive significance to this unique molar tissue conformation, and perhaps it is a trait that has been fixed by neutral evolution (e.g., Weaver et al. 2008). Nonetheless, differences between Neandertals and contemporaneous *Homo sapiens* such as the Aterian molars measured here are of taxonomic utility in the assessment of isolated dental remains. These differences in dental conformation also underscore the importance of sampling earlier *Homo* taxa and populations (e.g., African *Homo erectus*) to assess the evolutionary polarity of this character.

## Conclusion

The human remains associated with Aterian assemblages provide us with a sample of the populations living in the region of the extant Sahara and Magreb where this assemblage is represented, between the Atlantic coast of Morocco, the Mediterranean, and the Nile Valley roughly between 90 and 35 ka. Although today this area is largely unpopulated, this was not always the case. The Sahara, like the Kalahari further south, witnessed episodes of expansion and retraction that directly impacted the landscapes, fauna, and human populations. Isotopic composition of the Sahara-derived dust in marine sediments demonstrates the occurrence of wet periods during which C3 plants (likely trees) developed in the Central Sahara. These wet episodes occurred during the Late Pleistocene and the Holocene, more precisely during MIS 5 (ca. 120–110 ka), MIS 3 (ca. 50–45 ka), as well as during the early Holocene (Castañeda et al. 2009). In contrast, arid episodes are documented during MIS 6 (ca. 170–135 ka) and MIS 4 (ca. 65–55 ka). During the wetter periods, Saharan megalakes developed and considerably extended (Armitage et al. 2007) and river channels extended across what is today a desert to the Mediterranean Coast (Osborne et al. 2008). To what extent these fluctuations could drive groups of early representatives of our species out of Africa and into the Levant and, later, Eurasia is still debated. However, the extant geographical conditions present a biased picture of the situation before and during the last out-of-Africa movement. Much emphasis has been placed on Sub-Saharan populations, who live in areas that are often regarded to be where non-African modern humans originated. In fact, some of the extant sub-Saharan populations might be displaced populations whose hunter-gatherer ancestors lived further north. Circa 50 ka, at a key date for the colonization of the planet by *Homo sapiens*, Africa north of 15° latitude was not empty. It was much greener and more populated than today, mostly occupied by the makers of the Aterian assemblages, whose settlements are found throughout the Sahara and the Magreb.

The dental remains found in the Moroccan Aterian sites provide us with important information on populations that may have played a significant role in the colonization of Eurasia. Our comparative analysis confirms the essentially modern nature of these humans. Until 50 ka, Europe and Africa represented distinct bio-geographical barriers that were peopled by well-separated entities. In terms of metrical as well as non-metrical traits, the dental morphology displayed by the Neandertals can be easily distinguished from that of the Aterians and the UPMH (Bailey et al. 2009). Dental tissue proportions also display a different pattern, with Neandertals characterized by thinner enamel than the Aterians and other modern groups. However, although the Aterian teeth clearly group with modern samples because they lack the traits or trait combinations that are diagnostic of Neandertals, they also display distinctive features. Most spectacular is the size of the Aterian dentitions, especially for the post-canine dentition. Although the comparative sample is very small, this very large size is not observed in the South African MSA teeth, which display much variability. However, to some extent, it is reminiscent of the MPMH of the Near East but also of early *Homo sapiens* in North and East Africa predating the Aterian (Hublin and Tillier 1981; White et al. 2003). Morphologically, this megadontia is expressed in the development of mass-additive traits including extra crests, distal cuspules, Carabelli's cusp, as well as large, often divided, hypocones on the upper teeth. This helps us to set some of the features observed in Neandertals in perspective, highlighting their primitive nature (see discussion in Bailey et al. 2009). The Aterian morphological pattern is also important to consider when interpreting the dental morphology of the first modern humans in Eurasia. Strikingly, a reminiscent pattern is observed on the Peştera cu Oase 1 and 2 specimens, which are the oldest directly dated modern individuals found in Europe to date (Trinkaus et al. 2003; Rougier et al. 2007). These individuals are also missing diagnostic Neandertal features such as the mid-trigonid crest of the lower molars, but display very large post canine dentition with additional cusps such as hypoculids on the lower molars. This observation supports the view that the exceptional dental post-canine robusticity observed on Peştera cu Oase 1 and 2 could be primarily inherited from the immediate African ancestors of the first modern Europeans.

**Acknowledgments**  We are grateful to M. A. El Hajraoui, A. Ben-Ncer, S. Raoui, the "Institut National des Sciences du Patrimoine et de l'Archéologie," and the "Direction du Patrimoine Culturel" (Rabat, Morocco) for allowing and/or facilitating the CT scanning of the fossil material at the "Musée Archéologique de Rabat," and A. Winzer and H. Temming for assistance with scanning, reconstructing computed tomographic data, and illustration.

# References

Armitage, S. J., Drake, N. A., Stokes, S., El-Hawat, A., Salem, M., White, K., et al. (2007). Multiple phases of North African humidity recorded in lacustrine sediments from the Fazzan Basin, Libyan Sahara. *Quaternary Geochronology, 2*(1–4), 181–186.

Bailey, S. E. (2002). *Neandertal dental morphology: Implications for modern human origins*. Ph.D. Dissertation, Arizona State University.

Bailey, S. E. (2006a). Who made the early Aurignacian? Evidence from isolated teeth. *American Journal of Physical Anthropology, 129*(S43), 60.

Bailey, S. (2006b). Beyond shovel-shaped incisors: Neandertal dental morphology in a comparative context. *Periodicum Biologorum, 108*(3), 253–267.

Bailey, S., Weaver, T. D., & Hublin, J.-J. (2009). Who made the Aurignacian and other early Upper Paleolithic industries? *Journal of Human Evolution, 57*, 11–26.

Bouzouggar, A., & Barton, R. N. E. (2012). The identity and timing of the Aterian in Morocco. In J.-J. Hublin & S. McPherron (Eds.), *Modern origins: A North African perspective*. Dordrecht: Springer.

Bräuer, G. (1984). A craniological approach to the origin of anatomically modern *Homo sapiens* in Africa and implications for the appearance of modern Europeans. In F. H. Smith & F. Spencer (Eds.), *The origins of modern humans: A world survey of the fossil evidence* (pp. 327–410). New York: Alan R. Liss.

Bräuer, G. (1988). Osteometrie. In R. Martin & R. Knussmann (Eds.), *Handbuch der vergleichenden Biologie des menschen Band 1: Wesen und Methoden der Anthropologie, 1 Teil: Wissenschaftstheorie, Geschichte, morphologische Methoden. Stuttgart* (pp. 160–232). Jena: Gustav Fisher Verlag.

Burnett, S. E. (1998). *Maxillary Premolar Accessory Ridges (MxPAR) worldwide occurrence and utility in population differentiation*. MA Thesis, Arizona State University.

Cann, R. L., Stoneking, M., & Wilson, A. C. (1987). Mitochondrial DNA and human evolution. *Nature, 325*, 31–36.

Castañeda, I. S., Mulitza, S., Schefuss, E., dos Santos, R. A. L., Damsté, J. S. S., & Schouten, S. (2009). Wet phases in the Sahara/Sahel region and human migration patterns in North Africa. *Proceedings of the National Academy of Sciences of the USA, 106*(48), 20159–20163.

Debénath, A. (1976). Le site de Dar-es-Soltane 2, à Rabat (Maroc). *Bulletins et Mémoires de la Société d'Anthropologie de Paris, 3*, 181–182.

Debénath, A. (1980). Nouveaux restes humains atériens du Maroc. *Comptes Rendus des Séances de l'Académie des Sciences (Paris) Serie D, 290*, 851–852.

Debénath, A., Raynal, J.-P., Roche, J., Texier, J.-P., & Ferembach, D. (1986). Stratigraphie, habitat, typologie et devenir de l'atérien marocain: Données récentes. *L'Anthropologie, 90*(2), 233–246.

d'Errico, F., Vanhaeren, M., Barton, N., Bouzouggar, A., Mienis, H., Richter, D., et al. (2009). Additional evidence on the use of personal ornaments in the Middle Paleolithic of North Africa. *Proceedings of the National Academy of Sciences of the USA, 106*(38), 16051–16056.

Ferembach, D. (1976). Les restes humains de la grotte de Dar-es-Soltane 2 (Maroc), campagne 1975. *Bulletins et Mémoires de la Société d'Anthropologie de Paris, 3*, 183–193.

Gunz, P., Bookstein, F. L., Mitteroecker, P., Stadlmayr, A., Seidler, H., & Weber, G. W. (2009). Early modern human diversity suggests subdivided population structure and a complex out-of-Africa scenario. *Proceedings of the National Academy of Sciences of the USA, 106*(15), 6094–6098.

Hawkins, A. L. (2012). The Aterian of the oases of the Western Desert of Egypt: Adaptation to changing climatic conditions? In J.-J. Hublin & S. McPherron (Eds.), *Modern origins: A North African perspective*. Dordrecht: Springer.

Henshilwood, C., d'Errico, F., Vanhaeren, M., van Niekerk, K., & Jacobs, Z. (2004). Middle Stone Age shell beads from South Africa. *Science, 304*, 404.

Hublin, J.-J., & Tillier, A.-M. (1981). The Mousterian juvenile mandible from Irhoud (Morocco): A phylogenetic interpretation. In C. B. Stringer (Ed.), *Aspects of human evolution* (pp. 167–185). London: Taylor & Francis LTD.

Kono, R. T. (2004). Molar enamel thickness and distribution patterns in extant great apes and humans: New insights based on a 3-dimensional whole crown perspective. *Anthropological Science, 112*(2), 121–146.

Martin, L. (1985). Significance of enamel thickness in hominoid evolution. *Nature, 314*, 260–263.

Martinón-Torres, M., Bermúdez de Castro, J. M., Gómez-Robles, A., Arsuaga, J. L., Carbonell, E., Lordkipanidze, D., et al. (2007). Dental evidence on the hominin dispersals during the Pleistocene. *Proceedings of the National Academy of Sciences of the USA, 104*(33), 13279–13282.

McBrearty, S., & Brooks, A. S. (2000). The revolution that wasn't: A new interpretation of the origin of modern human behavior. *Journal of Human Evolution, 39*, 453–563.

Ménard, J. (1998). Odontologie des dents de la Grotte de Temara. *Bulletin d'Archeologie Marocaine, 18*, 67–97.

Minugh-Purvis, N. (1993). Reexamination of the immature hominid maxilla from Tangier, Morocco. *American Journal of Physical Anthropology, 92*, 449–461.

Olejniczak, A. J. (2006). *Micro-computed tomography of primate molars*. Ph.D. Dissertation, Stony Brook University.

Olejniczak, A. J., Smith, T. M., Feeney, R. N. M., Macchiarelli, R., Mazurier, A., Bondioli, L., et al. (2008a). Dental tissue proportions and enamel thickness in Neandertal and modern human molars. *Journal of Human Evolution, 55*, 12–23.

Olejniczak, A. J., Smith, T. M., Skinner, M. M., Grine, F. E., Feeney, R. N. M., Thackeray, J. F., et al. (2008b). Three-dimensional molar enamel distribution and thickness in *Autralopithecus* and *Paranthropus*. *Biology Letters, 4*, 406–410.

Osborne, A. H., Vance, D., Rohling, E. J., Barton, N., Rogerson, M., & Fello, N. (2008). A humid corridor across the Sahara for the migration of early modern humans out of Africa 120,000 years ago. *Proceedings of the National Academy of Sciences of the USA, 106*(48), 20159–20163.

Piveteau, J. (1957). *Traité de Paléontologie. T. VII: Primates*. Paléontologie Humaine. Paris: Masson.

Raynal, J.-P., & Occhietti, S. (2012). Amino-chronology and an earlier age for the Aterian. In J.-J. Hublin & S. McPherron (Eds.), *Modern origins: A North African perspective*. Dordrecht: Springer.

Richter, D., Moser, J., & Nami, M. (2012). New data from the site of Ifri n'Ammar (Morocco) and some remarks on the chronometric status of the Middle Paleolithic in the Maghreb. In J.-J. Hublin & S. McPherron (Eds.), *Modern origins: A North African perspective*. Dordrecht: Springer.

Roche, J. (1953). La grotte de Taforalt. *L'Anthropologie, 57*, 375–380.

Rougier, H., Milota, S., Rodrigo, R., Gherase, M., Sarcină, L., Molclovan, O., et al. (2007). Peştera cu Oase 2 and the cranial morphology of early modern Europeans. *Proceedings of the National Academy of Sciences of the USA, 104*(4), 1165–1170.

Senyürek, M. S. (1940). Fossil man in Tangier. *Papers of the Peabody Museum of American Archaeology and Ethnology, 16*, 1–27.

Smith, T. M., Olejniczak, A. J., Martin, L. B., & Reid, D. J. (2005). Variation in hominoid molar enamel thickness. *Journal of Human Evolution, 48*, 575–592.

Smith, T. M., Olejniczak, A. J., Reid, D. J., Ferrell, R. J., & Hublin, J.-J. (2006a). Modern human molar enamel thickness and enamel-dentine junction shape. *Archives of Oral Biology, 51*, 974–995.

Smith, T. M., Olejniczak, A. J., Tafforeau, P., Reid, D. J., Grine, F. E., & Hublin, J.-J. (2006b). Molar crown thickness, volume, and development in South African Middle Stone Age humans. *South African Journal of Science, 102*, 513–517.

Smith, T. M., Olejniczak, A. J., Kupczik, K., Lazzari, V., Vos, J., Kullmer, O., et al. (2009). Taxonomic assessment of the Trinil molars using non-destructive 3D structural and developmental analysis. *PaleoAnthropology, 2009*, 117–129.

Stringer, C. B., & Andrews, P. J. (1988). Genetic and fossil evidence for the origin of modern humans. *Science, 239*, 1263–1268.

Texier, J.-P., Porraz, G., Parkington, J., Rigaud, J.-P., Poggenpoel, C., Miller, C., et al. (2010). A Howiesons Poort tradition of engraving ostrich eggshell containers dated to 60,000 years ago at Diepkloof Rock Shelter, South Africa. *Proceedings of the National Academy of Sciences of the USA, 107*(14), 6180–6185.

Trinkaus, E., Moldovan, O., Milota, S., Bilgar, A., Sarcina, L., Athreya, S., et al. (2003). An early modern human from the Peştera cu Oase, Romania. *Proceedings of the National Academy of Sciences of the USA, 100*(20), 11231–11236.

Turner, C., Nichol, C., & Scott, G. (1991). Scoring procedures for key morphological traits of the permanent dentition: The Arizona State University Dental Anthropology System. In M. Kelley & C. Larsen (Eds.), *Advances in dental anthropology* (pp. 13–31). New York: Wiley Liss.

Vallois, H. V., & Roche, J. (1958). La mandibule acheuléenne de Temara. *Comptes-Rendus de l'Académie des Sciences de Paris, 246*, 3113–3116.

Weaver, T. D., Roseman, C. C., & Stringer, C. B. (2008). Close correspondence between quantitative- and molecular-genetic divergence times for Neandertals and modern humans. *Proceedings of the National Academy of Sciences of the USA, 105*(12), 4645–4649.

White, T. D., Asfaw, B., DeGusta, D., Gilbert, H., Richards, G. D., Suwa, G., et al. (2003). Pleistocene *Homo sapiens* from Middle Awash, Ethiopia. *Nature, 423*, 742–747.

# Chapter 14
# The Upper Paleolithic Human Remains of Nazlet Khater 2 (Egypt) and Past Modern Human Diversity

I. Crevecoeur

**Abstract** The Nazlet Khater 2 skeleton was discovered in 1980 during the excavations of the Belgian Middle Egypt Prehistoric Project in the Nile Valley (Egypt). Its association with the early Upper Paleolithic chert mining site of Nazlet Khater 4 (NK 4) (whose exploitation period ranged from 35 to 40 ka) makes it the oldest almost complete Marine Isotope Stage (MIS) 3 modern human skeleton in northern Africa. The Nazlet Khater 2 (NK 2) remains belong to a young adult male. It is well preserved with the exception of the distal part of the legs and the foot. Comparative analyses of the specimen underline the complex morphology of modern humans from this time period. NK 2 exhibits several retained archaic features, notably on the face and the mandible. The inner ear structures display morphological characteristics that stand on the edge of extant human variation. The postcranial remains have strong muscular insertions and are adapted to high biomechanical strength. Furthermore, NK 2 has vertebral and membral lesions. These postcranial characteristics might be related to intensive mining activities. The study of this specimen provides an opportunity to increase our understanding of past modern human diversity during this time period (MIS 3) for which very rare human remains are known.

**Keywords** Egypt • Mining • Modern human • Nazlet Khater • Specialized activities • Upper Paleolithic

## Introduction

The latter part of the Late Pleistocene (<70 ka) is a period crucial to our understanding of the evolution, dispersal, and diversity of anatomically modern humans. Archaeological discoveries show a major increase in technological, social, and cognitive behavioral complexity of some African groups during this period that could have led to major demographical expansion (Mellars 2006; Field et al. 2007). It has also been suggested that these complex behavioral changes were triggered by environmental and climatic shifts in Africa during this period (Wendorf et al. 1993; Partridge et al. 1997; Moreno et al. 2001; Carto et al. 2009). In addition, several genetic studies point out a recent origin of worldwide populations coming from an "African genetic stock," having undergone one or several demographic crises (or bottlenecks) followed by expansion waves anywhere between 65 and 25 ka (Excoffier 2002; Marth et al. 2003; Ramachandran et al. 2005; Garrigan and Hammer 2006; Fagundes et al. 2007; Li et al. 2008; Lohmueller et al. 2008). These results suggest that extant populations might represent only a restricted part of past modern human diversity. Therefore, the study of Late Pleistocene African specimens seems essential to increase our knowledge of this past variation in order to clarify the evolution and biological definition of our species.

However, there is a lack of well-dated and complete modern human remains discovered in Africa—as well as in the rest of the World—during MIS 4 and early MIS 3 (~71–35 ka). The few fossils assigned to this period, such as the Aterian specimens from Northwest Africa or the Taramsa child skeleton, are either fragmentary or not precisely dated (Debénath 1994; Vermeersch et al. 1998). In addition, the few remains from South Africa preceding the MIS 3, like Klasies River Mouth, Blombos, or Border Cave, have delivered incomplete information about past human diversity (e.g., Rightmire and Deacon 1991; Bräuer and Singer 1996). In this region, there is only one well-dated

I. Crevecoeur (✉)
Laboratoire d'Anthropologie des Populations Passées et Présentes, UMR 5199 PACEA/Université Bordeaux 1, Bâtiment B8, Avenue des Facultés, 33405 Talence Cedex, France
and
Laboratory of Anthropology and Prehistory, Royal Belgian Institute of Natural Science (RBINS), 29 rue Vautier, 1000 Brussels, Belgium
e-mail: i.crevecoeur@pacea.u-bordeaux1.fr

J.-J. Hublin and S. P. McPherron (eds.), *Modern Origins: A North African Perspective*, Vertebrate Paleobiology and Paleoanthropology, DOI: 10.1007/978-94-007-2929-2_14, © Springer Science+Business Media B.V. 2012

cranium, the Hofmeyr skull ($36.2 \pm 3.3$ ka; Grine et al. 2007), that yields information about Late Pleistocene modern humans. Its cranial dimensions show stronger affinities with Eurasian Upper Paleolithic specimens rather than with geographically close recent populations (Grine et al. 2007).

In the framework of this problem, the human remains of NK 2 are of particular interest since it represents the only complete skeleton of a modern human from the beginning of the Upper Paleolithic in North Africa. Its study thus offers a real opportunity to assess part of past modern human diversity during the Late Pleistocene (Crevecoeur 2008).

This chapter presents the main morphological and biometrical characteristics of the NK 2 specimen, together with an interpretation of some results in relation to the peculiar archaeological context of this discovery. The previous study of the cranium and mandible of NK 2 has shown the varied phylogenetic affinities of this specimen according to the comparative samples used (Thoma 1984; Bräuer and Rimbach 1990; Pinhasi 2002). Therefore, we have selected comparative groups that are as homogeneous as possible following broad geographical and chronological criteria, and that can be used for all anatomical parts. The purpose was not to establish phylogenetic relationships between NK 2 and the comparative samples, but, rather, to use them to depict something about past human variation in order to characterize the traits of NK 2.

## The Archaeological Context of the Discovery

The primary burial of the NK 2 skeleton was discovered in 1980 near Tahta (Egypt), during the excavation campaign of the Belgian Middle Egypt Prehistoric Project in the Nile Valley (Vermeersch et al. 1984). The individual was laying full length on its back, with its head oriented northward. It was buried in an adjusted desiccation crack filled by sand and large boulders, with a well-preserved bifacial axe laid at the right of the skull. Typologically, this axe is very similar to those found in the industry of NK 4, a nearby Upper Paleolithic chert mining exploitation. Together with sedimentological arguments, this led Vermeersch to associate the burial with the NK 4 extraction site. Radiocarbon dating was not attempted at the time of the discovery since the test for the presence of organic material in a small fragment of cortical bone was negative (Vermeersch 2002). Recently, the skeleton of NK 2 has been directly dated by Electron Spin Resonance (ESR) on enamel tooth fragments to around $38 \pm 6$ ka (Grün 2005, personal communication). This date is associated with a maximized standard error due to the lack of some contextual isotopic data. It is consistent with

the dating of the NK 4 exploitation period, which ranged from 40–35 ka (Vermeersch et al. 2002) and confirms the association between the two sites.

NK 4 is the only Upper Paleolithic site found in the Nile Valley. Sites from MIS 4 and 3 are basically absent in the region because this period is related to a phase of hyper-aridity in Northeast Africa and most of the occupation sites have probably been covered or eroded by later flooding (Vermeersch et al. 1990).

Three different types of digging activities were present at NK 4: ditches (1 m width), vertical shafts (2 m depth) and underground galleries (Vermeersch et al. 1984). Furthermore, axes and gazelle horn cores were used for extraction activities (Vermeersch and Paulissen 1993). The industry of NK 4 is a chronologically and technologically isolated phenomenon in the Nile Valley. Bifacial tools occur in association with a fully-developed blade-production system (Van Peer and Vermeersch 1990). This blade assemblage evokes some Taramsan features and, following Van Peer (2004), it may represent an intrusion of a transformed Nubian Complex into the Nile Valley.

## Biological Identity of Nazlet Khater 2

The NK 2 skeleton is well-preserved, with the exception of the distal part of the legs and the feet, which have been eroded (Fig. 14.1).

The sexual diagnosis is based on both the morphology of the coxal bones following the method of Bruzek (2002), and a discriminant analysis performed on nine pelvic dimensions (Crevecoeur 2008). Both methods gave a consistent diagnosis of masculinity for this individual (posterior probability associated with the squared Mahalanobis distance: $P > 0.99$).

The determination of the age at death relied on bone maturation and the degree of modification of the iliac sacro-pelvic surfaces. All the bony extremities are fused (notably the sternal extremities of the clavicles), but some epiphyseal lines are still visible externally and internally on the femur or the spheno-occipital synchondrosis. The study of the iliac sacro-pelvis surfaces shows no bony modification or remodelling. This observation places NK 2 among young adults between 20 and 29 years of age ($P > 0.8$; according to the method of Schmitt (2005), using the Bayes' theorem).

The stature of this individual was estimated based on the average values of femur and humerus regression equations of Formicola and Franceschi (1996). These equations improve the estimation when dealing with very tall or small individuals. NK 2 has a small stature of 1.61 m according to this method.

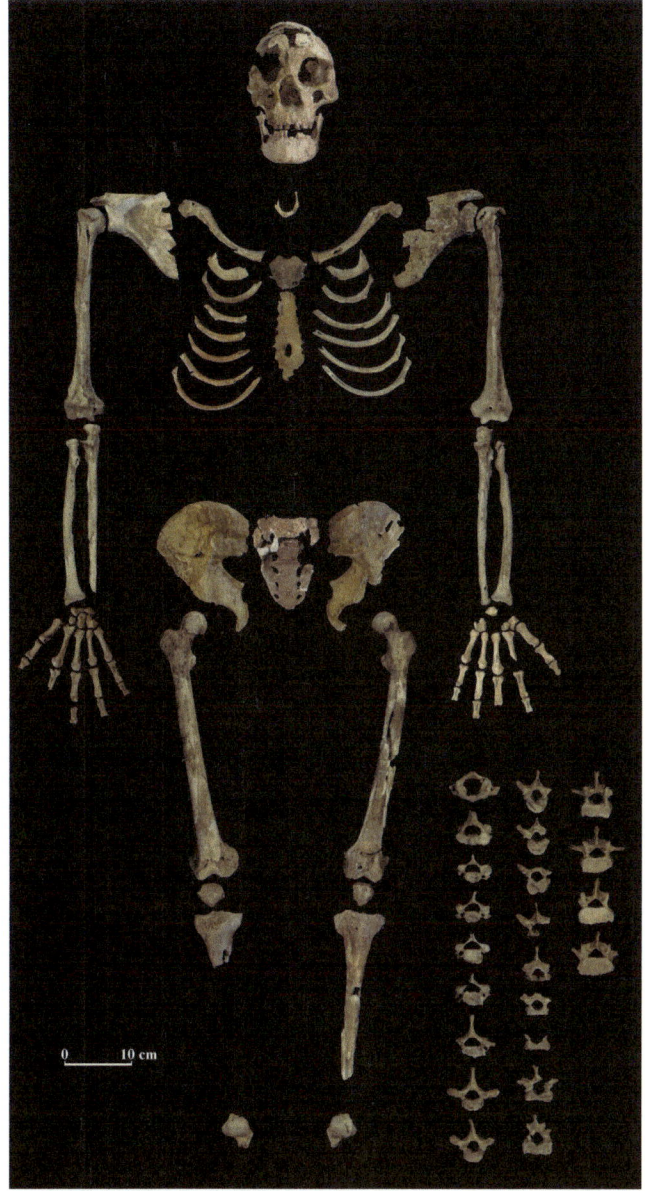

**Fig. 14.1** Preservation of the Nazlet Khater 2 skeleton

## Comparative Samples and Statistical Analyses

The skeleton of NK 2 was compared with various Pleistocene and recent human samples in an attempt to deal with past and present *Homo sapiens* diversity. The first comparative group contains early members of our species from Africa and the Levant dated to the Middle Paleolithic (MPHS). It includes the Qafzeh and Skhul populations, Herto 1, Omo-Kibish 1 & 2, Jebel Irhoud 1 & 2, and Singa. The other comparative groups include European Upper Paleolithic modern humans (EUP); the late Upper Paleolithic populations from North Africa (LUPA, mainly the

Afalou, Taforalt, Jebel Sahaba, and Wadi Halfa series). This latter group is sometimes divided geographically into Northwest and Northeast Africa (LUPAW and LUPAE) when the means and standard deviations were statistically different; 3) the late Upper Paleolithic populations from the Levant (LUPO, mainly Kebaran and Natufian specimens). In addition, we also considered a classic Neandertal sample (NEAND) as a non-modern contemporary group. Finally, we used an extant human sample in different cases to illustrate recent modern variation (EXT).

General morphological and metrical characteristics of the NK 2 skeleton have been assessed by means of uni-, bi-, and multivariate analyses. The measurements employed have been selected from those defined by Martin (Bräuer 1988) and Howells (1973) for the cranium and infra-cranium. The inner ear of NK 2 has been investigated in collaboration with Linda Bouchneb (University of Bordeaux 1). Both right and left bony labyrinths were studied and the linear measurements were taken with Amira 3.1 from planar reformatted images according to the CT-morphometric method of Spoor (1993) and Spoor and Zonneveld (1995, 1998). The cross-sectional properties of the main long bones were examined through scanner slices taken perpendicular to the sagittal and coronal planes, as recommended by Ruff and Leo (1986). The measurements were performed with the ImageJ (2005) software and the add-in *"MomentMacroJv1.2"* (http://www.hopkinsmedecine.org/FAE/mmacro.htm).

The principal component (PCA) and discriminant (DA) analyses were computed with Statistica v.6 (Statsoft France 2002), and each variable was first tested for normality. Both raw and size-adjusted data were used in the multivariate analyses. Shape variables were computed following the Darroch and Mosimann (1985) definition of shape, which best avoids size problems (Jungers et al. 1995).

## Results and Discussion

### *The Cranio-Facial Complex*

Morphologically, the cranium and mandible of NK 2 exhibit various archaic characteristics that may have different etiologies (Fig. 14.2). These are detailed in Table 14.1. The first category could be considered as retained archaic/plesiomorphic features since they are common in *Homo erectus* and Neandertals, without corresponding to an autapomorphy of the latter. The second category of traits is related to the development of the masticatory system. These characteristics are also considered archaic since they are present among earlier members of our genus, but they are suspected to rely on mesologic factors—like mastication

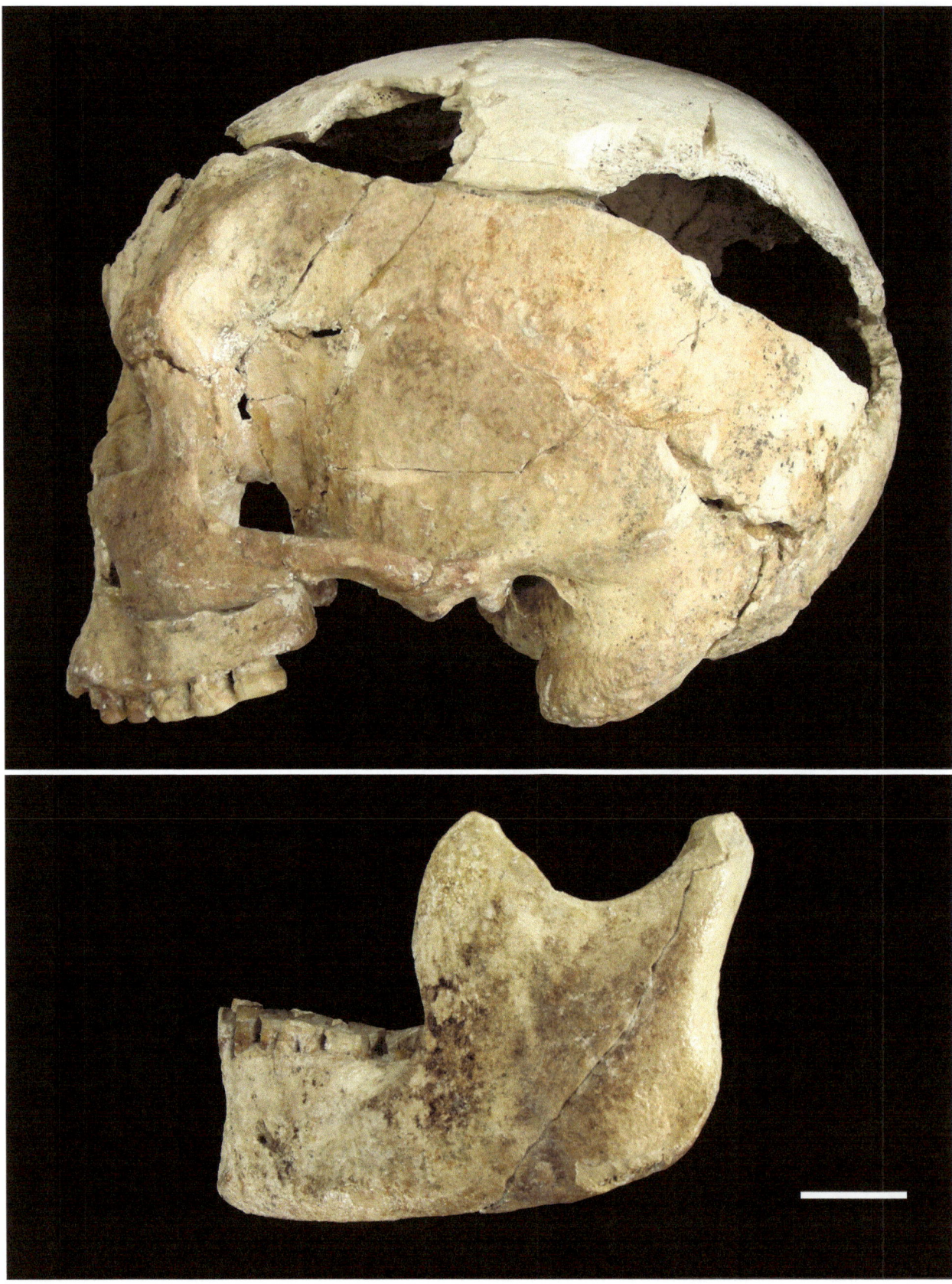

**Fig. 14.2** Lateral (*left*) views of the skull and the mandible of Nazlet Khater 2. *Horizontal line* represents 2 cm

**Table 14.1** List of the archaic features present on NK 2 with literature references to these traits, and their grading

|  | Nazlet Khater 2 | References |
|---|---|---|
| **Plesiomorphic features** | Low position of the maximum cranial breadth |  |
|  | Triangular shape of temporal squama (Form 1) | Elyaqtine (1995) |
|  | Linear parietal border of the temporal squama | ibid. |
|  | Prolongation of the supra-mastoid crests onto the parietal | ibid. |
|  | Anterior bridge on digastric groove (right side) | ibid. |
|  | Medial position of styloid process in relation to the digastrics groove and stylo-mastoid foramen (right side) | ibid. |
|  | Broad ramus, absolutely and relative to mandibular length | Bastir et al. (2004) |
|  | Genioglossal fossa and transverse tori on posterior face of mandibular symphysis | Weidenreich (1936) |
|  | Weak chin (*mentum osseum* rank 4) | Dobson and Trinkaus (2002) |
|  | *Planum triangulare* (left side) | Weidenreich (1936) |
| **Epigenetic (?) archaic traits** (powerful masticatory system) | Cranial vault thickness | Lieberman (1996); Balzeau, (2005) |
|  | Developed maxillar tuberosities | Enlow (1990); Varrela (1992) |
|  | Strong phenozygy (prominent zygomatic arch) | Ibid. |
|  | Developed post-glenoid tubercle | Elyaqtine (1995) |
|  | Zygomatic arch anteriorly rooted (above P4/M1) | Maureille (1994) |

strains—and their expression seems positively correlated with the size of the facial complex (Frieß 1999; Lieberman et al. 2002). The high breadth of the mandibular ramus of NK 2 (Crevecoeur and Trinkaus 2004) is placed in the first category, since it has been reported that the strength of the masticatory system may not be the only factor behind this feature (Bastir et al. 2004).

Among all the traits described in Table 14.1, some are present in very small proportions in extant human populations. For example, the morphology of the mandibular symphysis of NK 2 is encountered in less than 10% of recent humans (n = 473; Ali 2005). Other characteristics are more common in North African late Upper Paleolithic specimens, such as the low position of the maximal cranial breadth, or traits related to a powerful masticatory system. Others, like the bony bridge of the digastric groove, have never been observed in modern humans, although this was found on the early *Homo sapiens* cranium of Ngaloba LH 18 (Elyaqtine 1995).

Metrically, the dimensions of the neurocranium lie within the variation of extant modern humans. However, there are some peculiar characteristics, notably on the frontal bone. The low convexity and the inclination of NK 2's frontal lie at the edge of extant human variation. The NK 2 frontal is oblique, with a bregmatic angle of Schwalbe [M32(2) = 54°] in the lower range of variation of all the recent and Pleistocene modern human samples. The frontal bone is also only slightly curved, as illustrated in Fig. 14.3.

The sagittal frontal index (I.22) of NK 2 is elevated, which characterizes a flat frontal. The value of this index is close to the Neandertal and MPHS means and stands in the upper range of variation for all of the other groups.

NK 2 also has a very short parietal arc (PAA) compared to both the occipital (OCA) and frontal (FRA) ones. The proportion of each arc places NK 2 in the "frontal dominant" category (with FRA = 35.7% of the total sagittal arc). This configuration is actually quite common among our Pleistocene comparative groups (from 43 to 64%). However, what is more unusual is the whole disposition of NK 2, with the OCA longer than the PAA (FRA > OCA > PAA). This feature is common in *Homo erectus* (Hublin 1991), frequent in Neandertals (43%; n = 7), present in one of the MPHS (Skhul IX), but quite rare in the Upper and Late Paleolithic samples (4.3%; n = 94).

The face of NK 2 is particularly robust, with strong height and width dimensions. Compared to the maximum breadth of the neurocranium, the face is extremely broad (M45, bi-zygomatic) so that the transverse cranio-facial index (I.72) of NK 2 is outside of our comparative sample ranges of variation. We performed a PCA analysis on seven facial measurements (cf. Fig. 14.4 caption). The results of the analysis on raw data are plotted in the diagram following the two first principal components that express 72.56% of the total sample variation (Fig. 14.4).

The grey ellipse represents 95% of the variation of an extant Egyptian sample. NK 2 stands outside this ellipse,

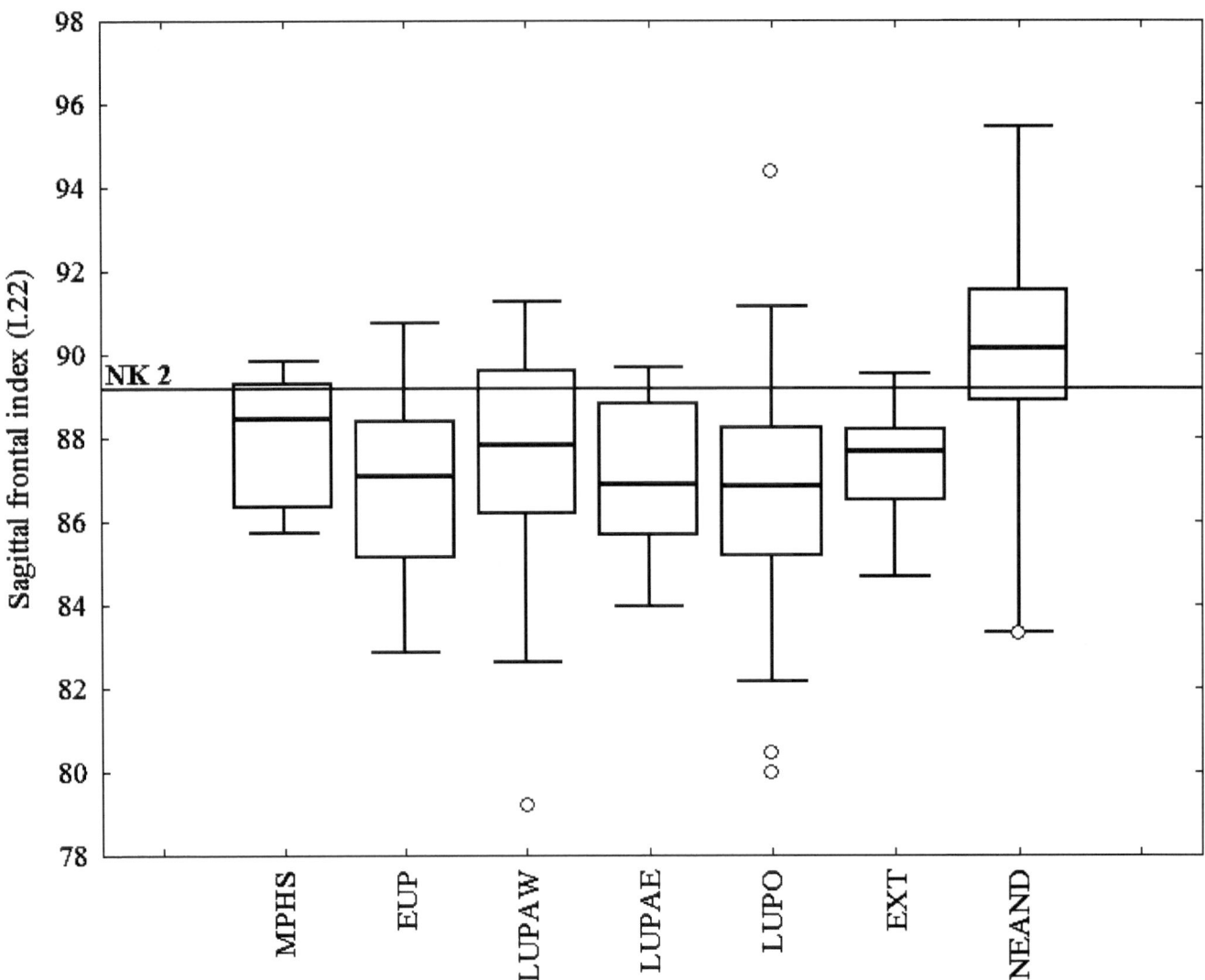

**Fig. 14.3** Box plots of the sagittal frontal index (I.22) for NK 2 and the comparative samples. Boxes display the median, 25th–75th and 5th–95th percentiles. MPHS = Middle Paleolithic *Homo sapiens* (n = 9), EUP = European Upper Paleolithic (n = 38), LUPAW = North-West African Late Upper Paleolithic (n = 54), LUPAE = North-East African Late Upper Paleolithic (n = 34), LUPO = Levant Late Upper Paleolithic (n = 53); EXT = Recent modern human (n = 30), NEAND = Neandertals (n = 7)

close to the MPHS and Neandertal specimens characterized by large faces. The European Upper Paleolithic sample occupies a transitional position between the earlier specimens and the extant population. The dispersion along the first axis illustrates a temporal gradient of reduction of facial dimensions through time. This result is consistent with several studies on diachronic variation of facial dimensions through human evolution, related, for example, to the diminution of masticatory strains (Enlow 1990; Varrela 1992). When performed with size-adjusted variables (shape variables), NK 2 and the MPHS individuals are included within the variation of Upper Paleolithic and recent modern humans samples. This implies that, in terms of shape, NK 2 and the MPHS specimens are more similar to later specimens. It confirms that their position in Fig. 14.4, close to

the Neandertals or Kabwe, is related mainly to the size of their facial skeleton rather than to an archaic conformation (cf. Lieberman 1996; Frieß 1999).

The mandible of NK 2 is extremely robust in all its absolute and relative dimensions. The length of the corpus, and its height and breadth lie in the upper part of the variation of the comparative groups. The symphyseal portion exhibits two pronounced transversal tori (superior and inferior) and a glenioglossal fossa on the posterior side. The anterior part is characterized by a weak chin (Mentum osseum rank 4; Dobson and Trinkaus 2002), as expressed by the value of the mental angle (84°). This configuration is common among early modern humans like the two penecontemporaneous specimens of Oase 1 and Tianyuan 1 (Trinkaus et al. 2003; Shang et al. 2007). The most striking

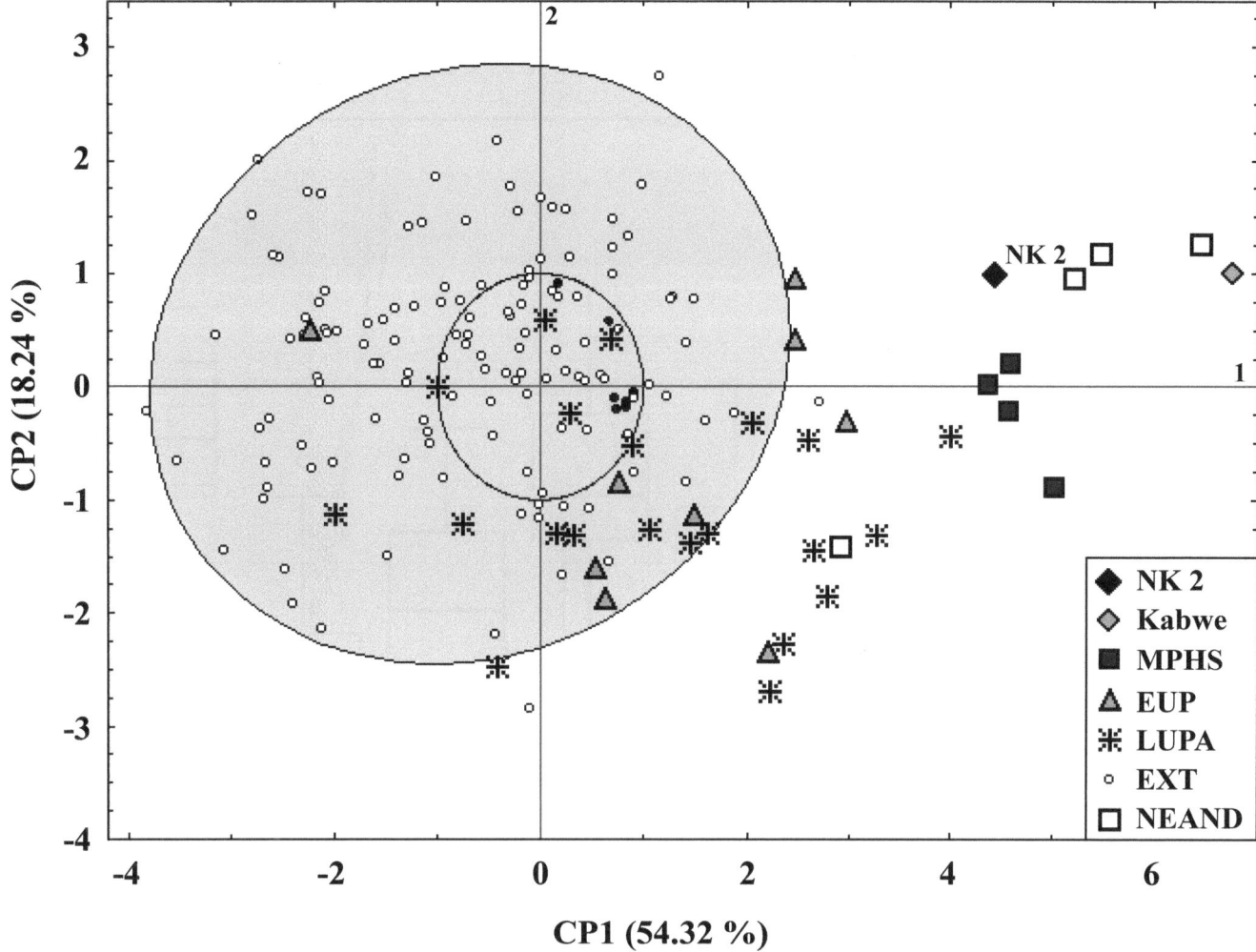

**Fig. 14.4** Bivariate plot of the first two principal components (PC1 & PC2) of the PCA computed from seven raw facial measurements (FMB, ZYB, WMH, NLH, NLB, OBH, MAB; Howells 1973). Comparative sample abbreviations as in Fig. 14.3. LUPA = African Late Upper Paleolithic. Grey ellipse represents 95% of the variability of an extant Egyptian sample (Howells 1996)

trait of the NK 2 mandible is the breadth of its ramus. With a minimal width of 51 mm, NK 2 is separate from all the comparative groups (Fig. 14.5). Even the robust specimen of Oase 1 (46.2 mm) does not display such a broad ramus (Trinkaus et al. 2003).

## The Middle and Inner Ear Structures

The NK 2 cranium includes two well-preserved petrosal bones, enclosing complete right and left bony labyrinths. The interest of this structure lies in its strong genetic component (Spoor 1993; Spoor et al. 1994). The bony labyrinth reaches its adult size between the 17th and 19th week of gestation, and postnatal influences are absent or minimal (Jeffery and Spoor 2004). Therefore, this structure

is important to assess phylogenetic affinities between hominids groups (e.g., Hublin et al. 1996).

Figure 14.6 illustrates the 3D reconstruction of the right labyrinth of NK 2 in lateral view. The shape of the NK 2 labyrinths is characterized by a superiorly-positioned posterior semicircular canal (PSC), and a more horizontally-inclined ampullar line and posterior petrosal surface compared to the extant human sample (for details about the recent human sample, see Bouchneb and Crevecoeur 2009).

With the exception of the width of the anterior semi-circular canal, the absolute dimensions of the NK 2 labyrinth are included in the variation range of the recent human sample. However, they are closer to the MPHS and EUP means than to the recent human and Neandertal ones. On the contrary, the relative dimensions of each semi-circular canal show a particular pattern compared to both the Neandertal and recent human general trends. The proportions of the semi-circular canals of NK 2 differ from the

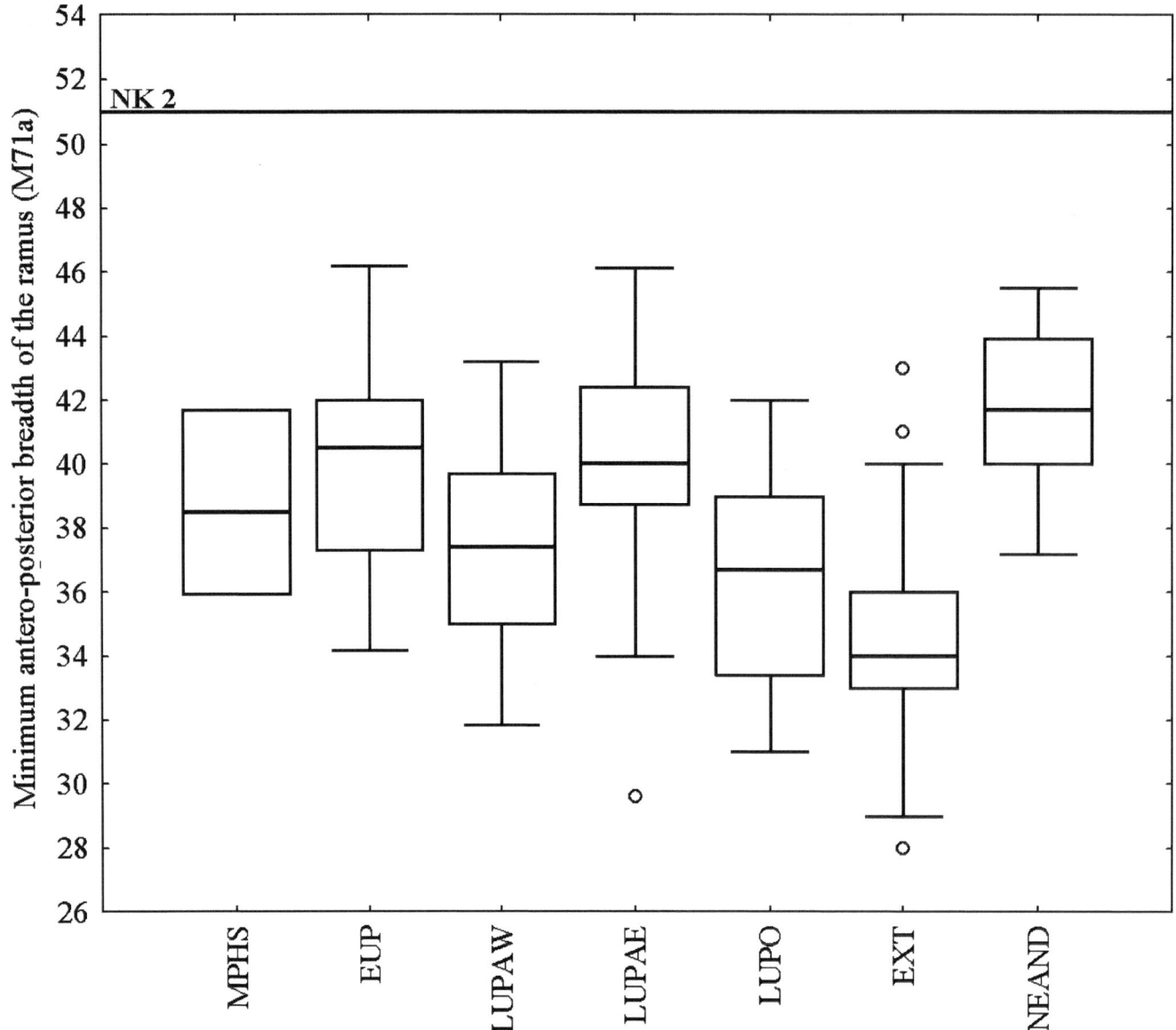

**Fig. 14.5** Box plots of the minimum antero-posterior breadth of the ramus (M71a) for NK 2 and the comparative samples. Boxes display the median, 25th–75th and 5th–95th percentiles. Comparative sample abbreviations as in Fig. 14.3. MPHS (n = 3), EUP (n = 14), LUPAW (n = 31), LUPAE (n = 45), LUPO (n = 77); EXT (n = 49), NEAND (n = 17)

Neandertal pattern, which display an enlarged anterior semi-circular canal, and from extant humans, which display a reduced posterior one. These shape and proportion characteristics are similar to the pattern previously described by Spoor et al. (2003) for European Upper Paleolithic modern humans.

We performed a PCA analysis on six variables of the labyrinth. The two first principal components are plotted in Fig. 14.7 and they express 76% of total variance.

NK 2 stands on the edge of extant human variation like the other Upper Paleolithic specimens and most of the MPHS sample (represented here by the Qafzeh and Skhul individuals). NK 2 and the EUP group are isolated following the second axis, given the previously described morphological and metrical characteristics. Based on a discriminant analysis of 13 variables of the labyrinth, NK 2 has a higher posterior probability of belonging to the EUP group (P $\sim$ 0.8). These results show that NK 2 has a complex labyrinthine set of characteristics that partially distinguish it from recent modern humans and that it shares some of these particularities with European Upper Paleolithic specimens.

Only the right malleus of the middle ear of NK 2 was preserved (Crevecoeur 2007). The morphometry of this

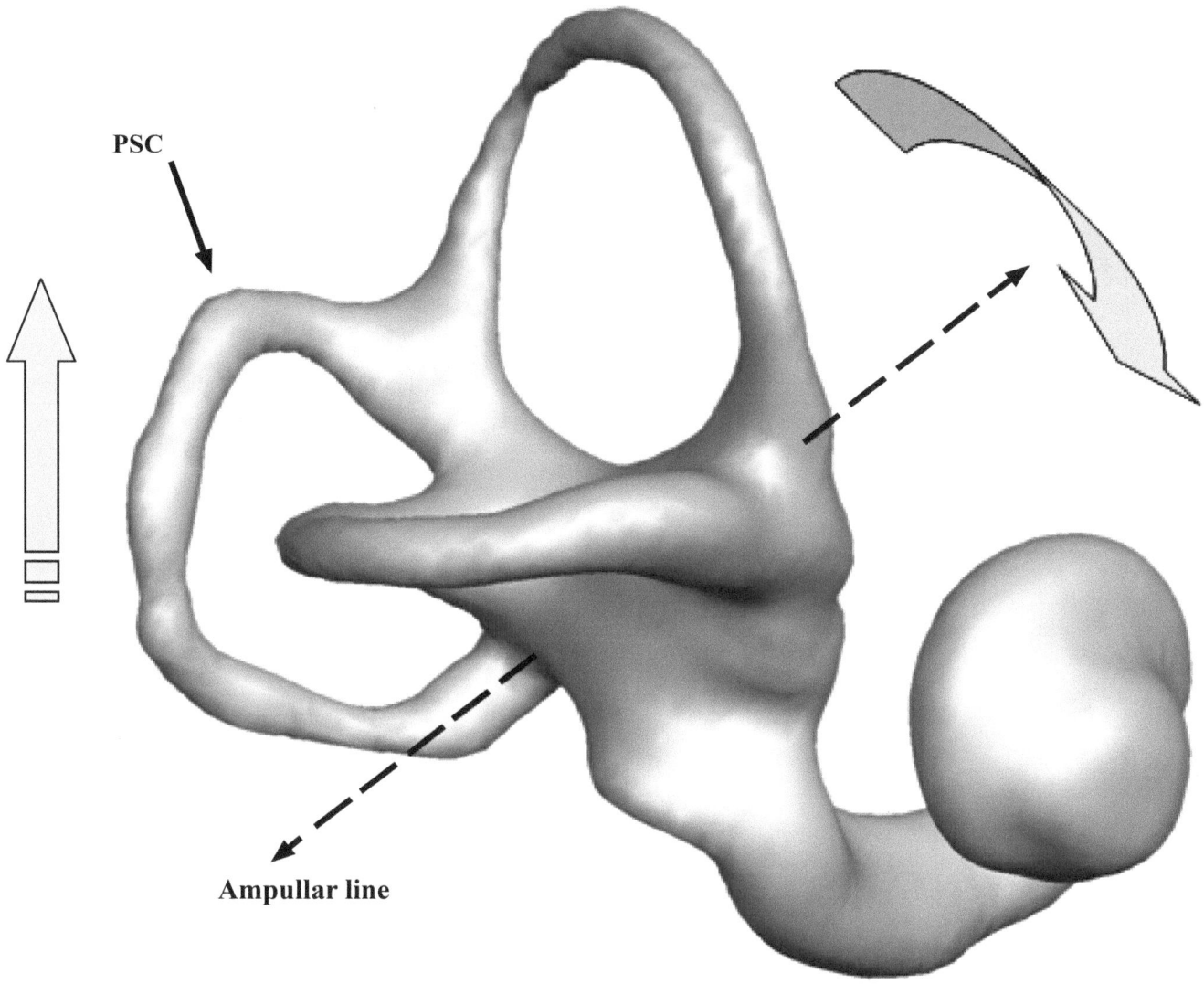

PSC

Ampullar line

**Fig. 14.6** 3D reconstruction of the complete right bony labyrinth of Nazlet Khater 2 in lateral view with the schematic representations of its morphological characteristics. *PSC* posterior semi-circular canal

ossicle lies within the range of extant human variation, but the angle of the manubrium is particularly open. Its dimensions lie at the upper limit of recent human variability, but close to the values of Neandertal specimens as well as Middle to Upper Paleolithic modern humans from Eurasia (Crevecoeur 2007).

These observations about the middle and inner ear structures of NK 2 suggest that the variation expressed by the Late Pleistocene modern humans differs from that of recent humans. The similarities between the NK 2 labyrinth and the European Upper Paleolithic sample, suggested by the discriminant analysis, may support a close relationship between this North African specimen and EUP ones. However, this grouping relies on a very small number of EUP specimens and should be further investigated.

## The Post-Cranium

Characterizing postcranial features is laborious because it is difficult to clearly identify plesiomorphic traits from those influenced by the environment or by an individual's lifestyle (Pearson 2000a, b). The infra-cranial remains of NK 2 are robust, with strong muscular insertions. Some bones display characteristics that can be associated with retained archaic features, like the small, developed styloid process of the ulna (Hambucken 1993), or the marked angle between the neck and shaft of the radius (Pearson and Grine 1997).

In most cases, the features of NK 2 are related to biomechanical advantages. For example, this is the case for the axillary border of the scapula, which possesses two sulci on

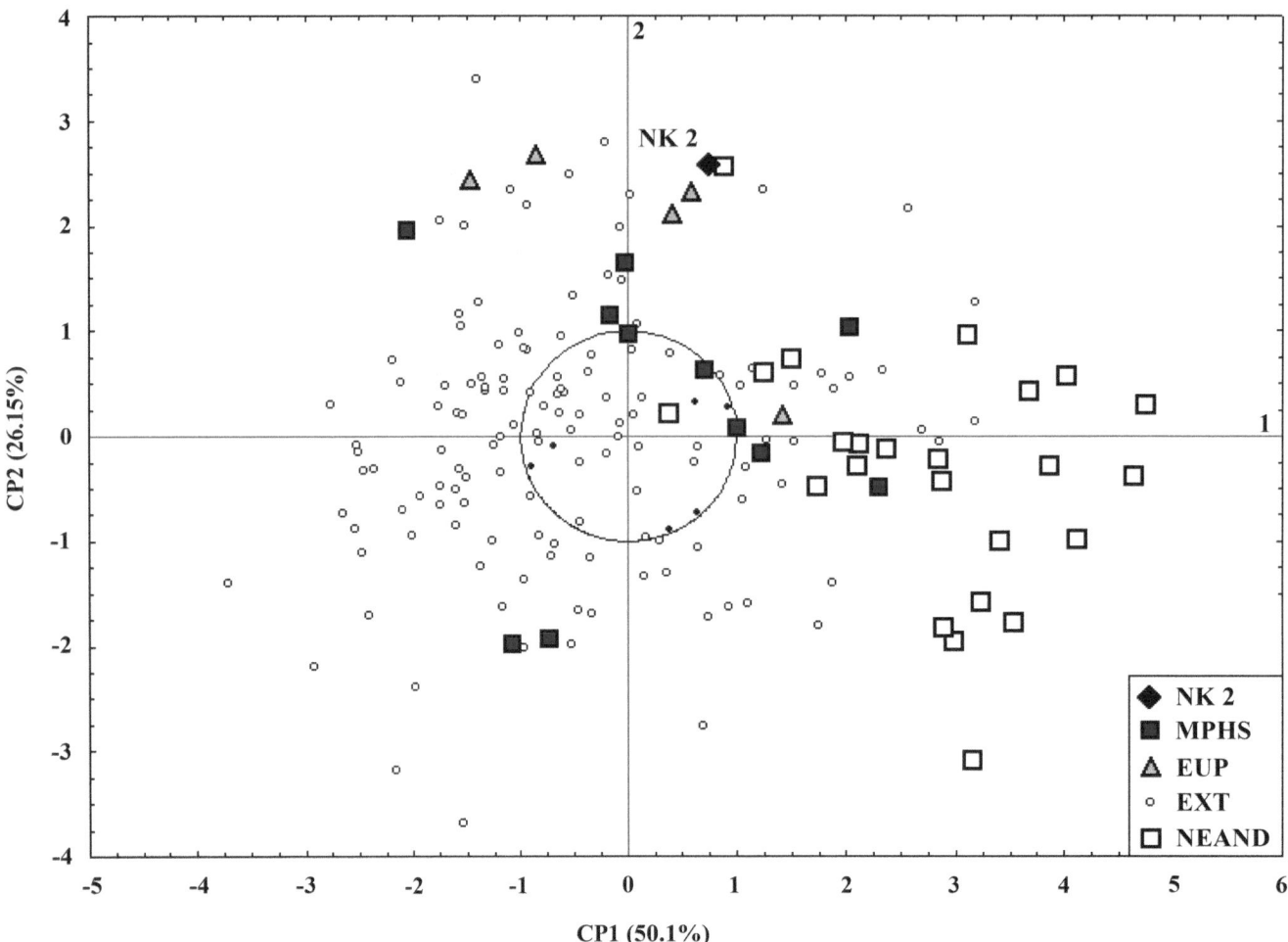

**Fig. 14.7** Bivariate plot of the two first principal components (PC1 & PC2) of the PCA computed for five variables of the bony labyrinth (PSC-%R, PSC-R, LSC-R, LSC-%R, SLI, LSCM < APA; Spoor, 1993). Comparative sample abbreviations as in Fig. 14.3

both side of the medial crest. This morphology is very common among Middle and Upper Paleolithic modern humans, but quite rare currently (Vandermeersch 1981; Trinkaus 2006). This more lateral position of the muscular insertions could allow for more powerful lifting movements of the arm (Vallois 1932).

We have studied the cross-sectional geometric properties of the main long bones (humerus, radius, and femurs). The results for the upper limb show adaptation to high biomechanical strengths. NK 2 possesses a high relative cortical area in the humerus and the radius at 50% of the midshaft length. This trait expresses an adaptation to strong axial compressive and tensile strengths. In addition, the symmetry of the sections accounts for a balanced resistance to all these strains (Ruff and Hayes 1983).

The study of the cross-sectional geometric properties of the femora also suggests a high resistance to all of the different strains. However, the most striking characteristics are related to the shape of the section at both 80 and 50% of the midshaft length. Both ratios of the second moment of

area (following the mediolateral and the anteroposterior axes) are close to 1, which characterizes a prismatic or circular section. In the subtrochanteric area (80%), this configuration has also been described in the Qafzeh-Skhul population, and among Neandertals (Trinkaus and Ruff, 1999). However, according to Trinkaus and Ruff (1999), the similarities in shape between these two groups at that level of the diaphysis (80%) are not due to the same phenomenon. In the case of the Qafzeh and Skhul populations, this may be related to pelvic shape, which could affect medio-lateral bending stresses on the proximal femoral diaphysis (Ruff 1995), rather than to locomotor behavior. It is interesting to note that NK 2 possesses the same pelvic characteristics (narrow pelvis, little iliac flare, high neck-shaft angle of the femur) as these Middle Paleolithic modern humans.

Regarding the mid-diaphyseal section (50%), NK 2 possesses a circular section closer to the mean of Late Paleolithic populations (LUPO and LUPA) rather than earlier specimens (Fig. 14.8). The temporal trend of increasing circularity of the femur midshaft illustrated in

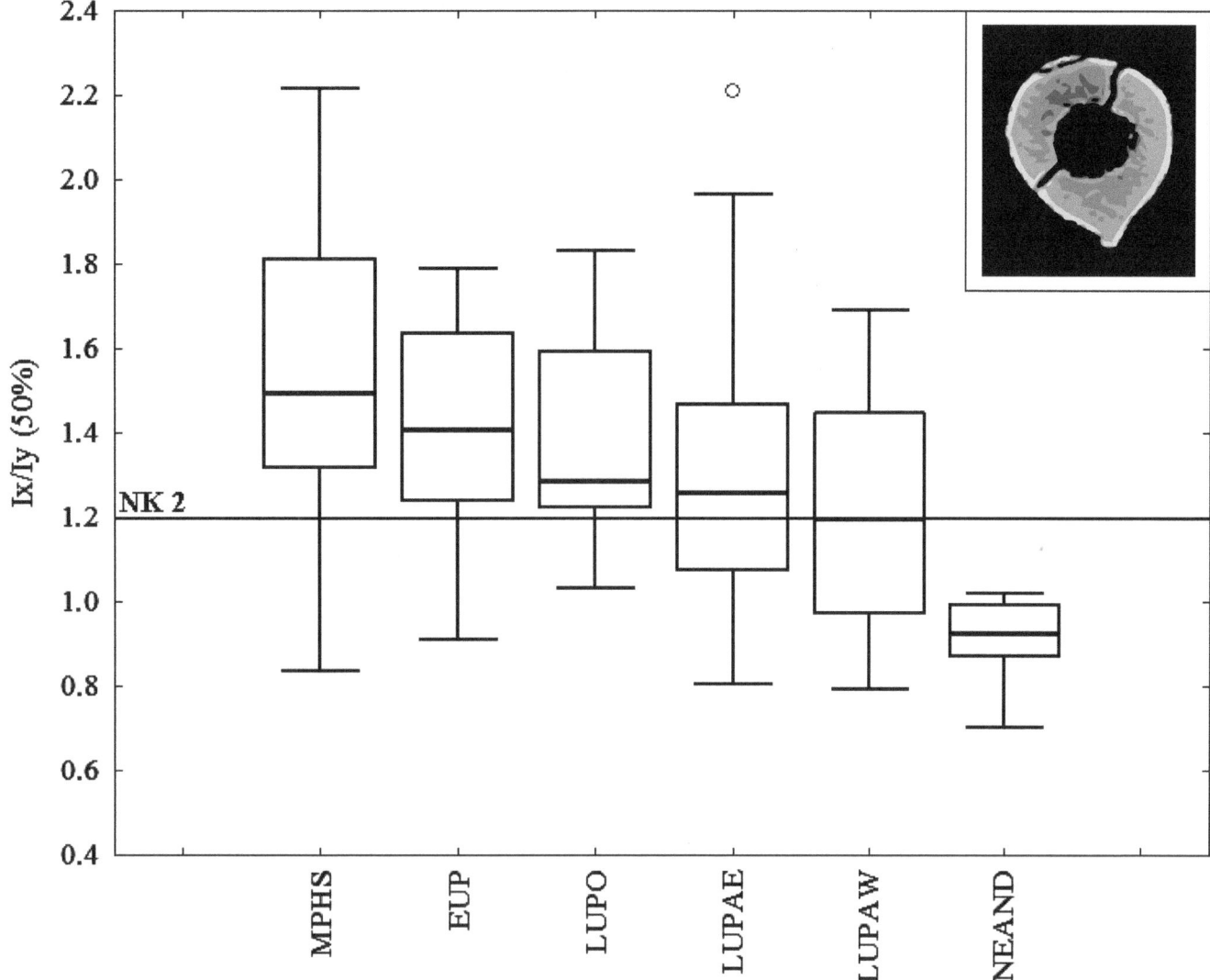

**Fig. 14.8** Box plots of the comparative sample for the ratio of the second moments of area (Ix/Iy) of the femoral midshaft section (50%). Representation of the section of NK 2 is in the upper right corner. Boxes express the median, 75–25 and 5–95 percentiles. Comparative sample abbreviations as in Fig. 14.3. MPHS (n = 8); EUP (n = 23); LUPO (n = 11); LUPAE (n = 32); LUPAW (n = 26); NEAND (n = 8). Data for LUPO and LUPA(E/W) following Shackelford (2005)

Fig. 14.8 has already been described for Late Paleolithic and Mesolithic populations in Europe, Africa, and Asia (Holt 2003; Shackelford 2007). It is generally considered as the expression of reduced mobility compared to the Early Upper Paleolithic morphology (Holt 2003).

## Paleopathology and Specialized Activities

The biomechanical adaptations of NK 2 make sense when they are related to paleopathologies and archaeological context.

The postcranial remains of NK 2 possess a complex set of appendicular and vertebral lesions, unusual for such a young individual. Since NK 2 does not display any symptoms of diffuse idiopathic skeletal hyperostosis (DISH) or spondylarthropathy, it is likely that these lesions have a mechanical origin related to specific activities (Crevecoeur and Villotte 2006).

The vertebral lesions are mainly the result of arthritis. Both articular surfaces and vertebral bodies in the cervical part of the spine are affected with traces of eburnation or with marginal osteophytis on the postero-inferior and superior joint surfaces (Fig. 14.9a).

The membral lesions are more varied. Three categories of activity markers are represented (Dutour 2000):

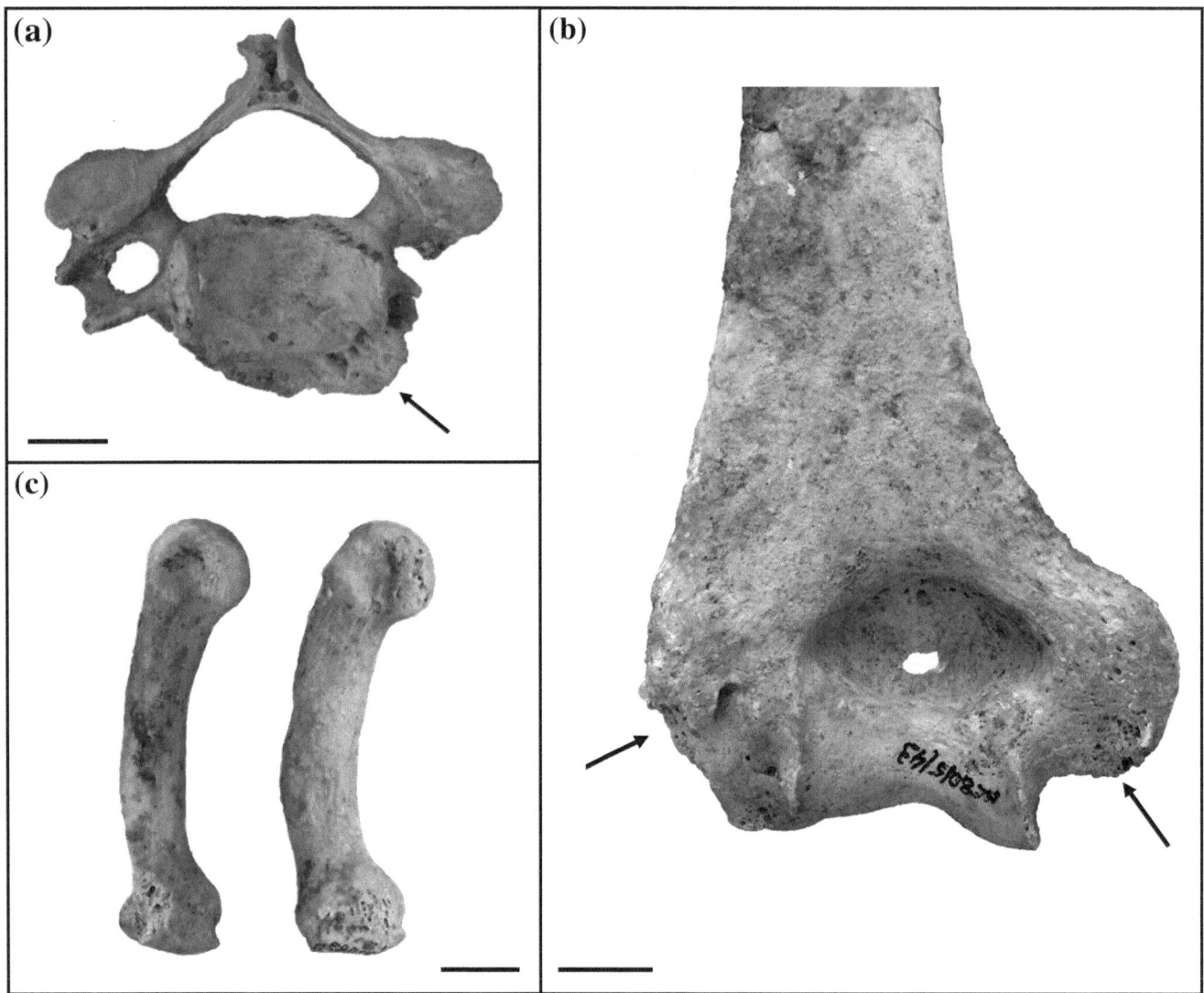

**Fig. 14.9** Illustration of some pathologies of NK 2. Horizontal lines represent 1 cm. **a** Fifth cervical vertebra showing marginal osteophytis on the postero-inferior and superior joint surfaces. **b** Postero-distal view of the *left* humerus distal extremity showing medial and lateral epicondylitis. **c** Midshaft deformation of the *right* fourth metacarpal (*right* on the picture) compared with the *left* one (*left* on the picture)

- the abarticular lesions, like the medial and lateral epicondylitis on the left humerus, which might be related to repeated specific movements (Fig. 14.9b; Commandré 1977; Dutour 1986);
- the articular lesions, such as the marginal osteophytis, on *patella* joint surface;
- and bony lesions *sensu stricto*, like the midshaft deformation of the right fourth metacarpal (Fig. 14.9c).

Some of these lesions are particularly interesting in relation to the archaeological context of a mining site, since many of them have been described among Neolithic, historical, or recent populations that take part in specific activities like mining extractions (Bailly-Maître et al. 1996; Steen and Lane 1998; Pany 2003). The cervical arthritis and rotator cuff lesions of the humerus, also present on NK 2,

might be related to heavy load carrying with forehead straps (Merbs 1983; Bridges 1994; Peyre et al. 1997). The deformation of the metacarpal midshafts has been observed in the Middle-Age mining population of Brandes-en-Oisans and could result from direct impact on the hand or on the rocks while using extracting tools (Bailly-Maître et al. 1996). In addition, we know that in NK 4, horn pits and bifacial handaxes were used for intensive extraction activities and that NK 2 was buried with a similar axe.

Finally, lateral and medial epicondylitis are often associated with intensive and repeated specific movements that induce microtraumatism in the muscular insertion area (Commandré 1977). This kind of lesion is observed significantly more often among mining populations (Steen and Lane 1998).

The extraction activities in the Upper Paleolithic site of NK 4 have been clearly described by Vermeersch and collaborators (Vermeersch et al. 2002). They were intensive and well-organized, using specific extracting structures and tools. The study of NK 2 paleopathologies evokes a coercive lifestyle that subjected this individual to heavy mechanical stress. These observations seem to confirm the archaeological hypothesis of intensive and specialized mining activities in the Nile Valley at the beginning of the Upper Paleolithic.

## Conclusions

The descriptive and comparative analysis of the NK 2 human remains highlight the main characteristics of this unique skeleton. It is an anatomically modern human that displays a complex set of morphometric traits that sometimes put it on the edge of extant modern human variation. The cranium exhibits a combination of plesiomorphic, mesologic, and particular traits (especially regarding its inner ear for the latter). The frequency of these features is quite rare currently, although their occurrence is observed in Middle and Upper Paleolithic specimens. The middle and inner ear structures show closer affinities to the European Upper Paleolithic and the Qafzeh and Skhul individuals than to recent modern humans.

The combination of morphometric features that places NK 2 at the edge of extant variation suggests that past modern human variation may differ from the present one. These results seem consistent with the hypothesis that recent populations could represent only a restricted part of past modern human genetic diversity (e.g., Marth et al. 2003; Fagundes et al. 2007).

Regarding the postcranial remains, the main characteristics express adaptations to high biomechanical strength and specialized activities. Together with the mining archaeological context, these results support the hypotheses of well-organised settlements and planned specialized activities in the Nile Valley at the beginning of the Upper Paleolithic (Van Peer 1998).

Despite gaps in the fossil record from MIS 5 to MIS 3, these observations provide a glimpse of morphological variability that appears to have characterized humanity in the Late Pleistocene, and underline the importance of investigating past modern human diversity and behavior in order to understand the complex evolution of our species.

**Acknowledgments** P. Vermeersch, P. Van Peer, and B. Maureille made the study of Nazlet Khater 2 possible by the entrusting this specimen to I.C. for analysis. We are grateful to the following curators and researchers who have allowed us to examine original fossils and collections in their care: H. de Lumley, D.L. Greene, D. Grimaud-Hervé, M. Judd, R. Kruszynski, S. Louryan, N. Spencer, C. Stringer, D. Van Gerven, and P. Semal. Drs. Hauret and Bar from the St-Augustin Hospital (Bordeaux) facilitated the CT acquisition of the NK 2 temporal remains. This study was supported in part by the "Projet Transition, volet Recherche Région Aquitaine" (convention 20051403003AB). We thank the editors and the anonymous referees for their useful comments and corrections.

## References

Ali, R. (2005). *La variabilité morphologique et métrique de la symphyse et des structures mentonnières dans les populations actuelles et chez les hommes fossiles*. Ph.D. Dissertation, Université Bordeaux 1, France.

Bailly-Maître, M.-C., Simonel, B., Barren, N., & Boulle, E.-L. (1996). Travail et milieu. Incidences sur une population au Moyen-Âge. In Centre de Recherches Archéologiques (Valbonne, Alpes-Maritimes) (Ed.), *L'identité des populations archéologiques* (pp. 211–243). Sophia Antipolis: Editions APDCA.

Balzeau, A. (2005). *Spécificité des caractères morphologiques internes du squelette céphalique chez Homo erectus*. Ph.D. Dissertation, Muséum National d'Histoire Naturelle, France.

Bastir, M., Rosas, A., & Kuroe, K. (2004). Petrosal orientation and mandibular ramus breadth: Evidence for an intergrated petroso-mandibular developmental unit. *American Journal of Physical Anthropology, 123*, 340–350.

Bouchneb, L., & Crevecoeur, I. (2009). The inner ear of Nazlet Khater 2 (Upper Paleolithic, Egypt). *Journal of Human Evolution, 56*, 257–262.

Bräuer, G. (1988). Osteometrie. In R. Martin & R. Knussmann (Eds.), *Handbuch der vergleichenden biologie des menschen (Band I, 1. Teil)* (pp. 160–232). Stuttgart: Gustav Fischer Verlag.

Bräuer, G., & Rimbach, K. W. (1990). Late archaic and modern *Homo sapiens* from Europe, Africa, and southwest Asia: Craniometric comparisons and phylogenetic implications. *Journal of Human Evolution, 19*, 789–807.

Bräuer, G., & Singer, R. (1996). The Klasies zygomatic bone: Archaic or modern? *Journal of Human Evolution, 30*, 161–165.

Bridges, P. S. (1994). Vertebral arthritis and physical activities in the Prehistoric southeastern United States. *American Journal of Physical Anthropology, 93*, 83–93.

Bruzek, J. (2002). A method for visual determination of sex, using the human hip bone. *American Journal of Physical Anthropology, 117*, 157–168.

Carto, S. L., Weaver, A. J., Hetherington, R., Lam, Y., & Wiebe, E. C. (2009). Out of Africa and into an ice age: On the role of global climate change in the late Pleistocene migration of early modern humans out of Africa. *Journal of Human Evolution, 56*, 139–151.

Commandré, F. (1977). *Pathologie abarticulaire*. Laboratoires Cétrane: Levallois-Perret.

Crevecoeur, I. (2007). New discovery of a Upper Palaeolithic ossicle: The right malleus of Nazlet Khater 2. *Journal of Human Evolution, 52*, 341–345.

Crevecoeur, I. (2008). *Etude anthropologique du Squelette du Paléolithique Supérieur de Nazlet Khater 2 (Egypte). Apport à la compréhension de la variabilité passée des hommes modernes*. Egyptian Prehistory Monographs 8. Leuven: Leuven University Press.

Crevecoeur, I., & Trinkaus, E. (2004). From the Nile to the Danube: A comparison of the Nazlet Khater 2 and Oase 1 early modern human mandibles. *Anthropologie (Brno), 42*, 229–239.

Crevecoeur, I., & Villotte, S. (2006). Atteintes pathologiques de Nazlet Khater 2 et activités minières au Paléolithique supérieur.

*Bulletins et Mémoires de la Société d'Anthropologie de Paris, 18,* 165–175.

Darroch, J. N., & Mosimann, J. E. (1985). Canonical and principal components of shape. *Biometrika, 72,* 241–252.

Debénath, A. (1994). L'Atérien du nord de l'Afrique et du Sahara. *Sahara, 6,* 21–30.

Dobson, S., & Trinkaus, E. (2002). Cross-sectional geometry and morphology of the mandibular symphysis in Middle and Late Pleistocene *Homo. Journal of Human Evolution, 43,* 67–87.

Dutour, O. (1986). Enthesopathies (lesions of muscular insertions) as indicators of the activities of Neolithic Saharan populations. *American Journal of Physical Anthropology, 71,* 221–224.

Dutour, O. (2000). Chasse et activités physiques dans la préhistoire : Les marqueurs osseux d'activités chez l'homme fossile. *Anthropologie et Préhistoire, 111,* 156–165.

Elyaqtine, M. (1995). *Variabilité et évolution de l'os temporal chez Homo sapiens. Comparaison avec Homo erectus.* Ph.D. Dissertation, Université Bordeaux 1, France.

Enlow, D. H. (1990). *Facial growth* (3rd ed.). Philadelphia, PA: W. B. Saunders Company.

Excoffier, L. (2002). Human demographic history: Refining the recent African origin model. *Current Opinion in Genetics and Development, 12,* 675–682.

Fagundes, N. J. R., Ray, N., Beaumont, M., Neuenschwander, S., Salzano, F. M., Bonatto, S. L., et al. (2007). Statistical evaluation of alternative models of human evolution. *Proceedings of the National Academy of Sciences of the USA, 104,* 17614–17619.

Field, J. S., Petraglia, M. D., & Lahr, M. M. (2007). The southern dispersal hypothesis and the South Asian archaeological record: Examination of dispersal routes through GIS analysis. *Journal of Anthropological Archaeology, 26,* 88–108.

Formicola, V., & Franceschi, M. (1996). Regression equations for estimating stature from long bones of early Holocene European samples. *American Journal of Physical Anthropology, 100,* 83–88.

Frieβ, M. (1999). Some aspects of the cranial size and shape, and their variation among later Pleistocene hominids. *Anthropologie (Brno), 37,* 239–246.

Garrigan, D., & Hammer, M. F. (2006). Reconstructing human origins in the genomic era. *Nature Reviews Genetics, 7,* 669–680.

Grine, F. E., Bailey, R. M., Harvati, K., Nathan, R. P., Morris, A. G., Henderson, G. M., et al. (2007). Late Pleistocene human skull from Hofmeyr, South Africa, and modern human origins. *Science, 315,* 226–229.

Hambucken, A. (1993). *Variabilité morphologique et métrique de l'humérus, du radius et de l'ulna des Néandertaliens. Comparaison avec l'homme moderne.* Ph.D. Dissertation, Université Bordeaux 1, France.

Holt, B. M. (2003). Mobility in Upper Paleolithic and Mesolithic Europe: Evidence from the lower limb. *American Journal of Physical Anthropology, 122,* 200–215.

Howells, W. W. (1973). *Cranial variation in man: A study by multivariate analysis of patterns of difference among recent humans populations. Papers of the Peabody museum of Archaeology and Ethnology, 67.* Cambridge: Harvard University Press.

Howells, W. W. (1996). Howell's craniometric data on the Internet. *American Journal of Physical Anthropology, 101,* 393–410.

Hublin, J.-J. (1991). *L'émergence des Homo sapiens archaïques : Afrique du nord-ouest et Europe occidentale.* Thèse d'Etat, Université Bordeaux 1, France.

Hublin, J.-J., Spoor, F., Braun, M., Zonneveld, F., & Condemi, S. (1996). A late Neanderthal from Arcy-sur-Cure associated with Upper Palaeolithic artefacts. *Nature, 381,* 224–226.

ImageJ (2005). ImageJ for Windows. Version 1.34.

Jeffery, N., & Spoor, F. (2004). Prenatal growth and development of the modern human labyrinth. *Journal of Anatomy, 204,* 71–92.

Jungers, W. L., Falsetti, A. B., & Wall, C. E. (1995). Shape, relative size, and size-adjustments in morphometrics. *Yearbook of Physical Anthropology, 38,* 137–161.

Li, J. Z., Absher, D. M., Tang, H., Southwick, A. M., Casto, A. M., Ramachandran, S., et al. (2008). Worldwide human relationships inferred from genome-wide patterns of variation. *Science, 319,* 1100–1104.

Lieberman, D. E. (1996). How and why humans grow thin skulls: Experimental evidence for systemic cortical robusticity. *American Journal of Physical Anthropology, 101,* 217–236.

Lieberman, D. E., McBratney, B. M., & Krovitz, G. (2002). The evolution and development of cranial form in *Homo sapiens. Proceedings of the National Academy of Sciences of the USA, 99,* 1134–1139.

Lohmueller, K. E., Indap, A. R., Schmidt, S., Boyko, A. R., Hernandez, R. D., Hubisz, M. J., et al. (2008). Proportionally more deleterious genetic variation in European than in African populations. *Nature, 451,* 994–998.

Marth, G., Schuler, G., Yeh, R., Davenport, R., Agarwala, R., Church, D., et al. (2003). Sequence variations in the public human genome data reflect a bottlenecked popupation history. *Proceedings of the National Academy of Sciences of the USA, 100,* 376–381.

Maureille, B. (1994). *La face chez Homo erectus et Homo sapiens: Recherche sur la variabilité morphologique et métrique.* Ph.D. Dissertation, Université Bordeaux 1, France.

Mellars, P. (2006). Why did modern human populations disperse from Africa ca. 60,000 years ago? A new model. *Proceedings of the National Academy of Sciences of the USA, 103,* 9381–9386.

Merbs, C. F. (1983). Patterns of activities-induced pathology in Canadian Inuit population. *Archeological Survey of Canada, 119,* 1–199.

Moreno, A., Targarona, J., Henderiks, J., Canals, M., Freudenthal, T., & Meggers, H. (2001). Orbital forcing of dust supply to the North Canary Basin over the last 250kyr. *Quaternary Science Reviews, 20,* 1327–1339.

Pany, D. (2003). *An analysis of occupationally-induced stress markers on the skeletal remains from the ancient Hallstatt cemetery.* Diplomarbeit zur Erlangung des akademischen Grades Magistra der Naturwissenschaften an der Universität Wien.

Partridge, T. C., de Menocal, P. B., Lorentz, S. A., Paiker, M. J., & Vogel, J. C. (1997). Orbital forcing climate over South Africa: A 200,000-year rainfall record from the Pretoria Saltpan. *Quaternary Science Reviews, 16,* 1125–1133.

Pearson, O.M. (2000a). Postcranial remains and the origin of modern humans. *Evolutionary Anthropology, 9,* 229–247.

Pearson, O.M. (2000b). Activity, climate, and postcranial robusticity. Implications for modern human origins and scenarios of adaptive change. *Current Anthropology, 41,* 569–607.

Pearson, O. M., & Grine, F. E. (1997). Re-analysis of the hominid radii from Cave of Hearths and Klasies River Mouth, South Africa. *Journal of Human Evolution, 32,* 577–592.

Peyre, M., Middleton, P., Vilet, A., Falda, M., De Lecluse, J., & Rodineau, J. (1997). Lésions de la coiffe des rotateurs. Analyse des gestes sportifs incriminés. Rappel des principes de rééducation. Eléments de prévention. In S. Jacques (Ed.), *Médecine du sport 1997* (pp. 97–101). Paris: Expansion Scientifique Française.

Pinhasi, R. (2002). Biometric study of the affinities of NK–A quantitative analysis of the mandible dimensions. In P. M. Vermeersch (Ed.), *Palaeolithic quarrying sites in Upper and Middle Egypt* (pp. 283–335). Leuven: Egyptian Prehistory Monographs 4.

Ramachandran, S., Deshpande, O., Roseman, C. C., Rosenberg, N. A., Feldman, M. W., & Cavalli-Sforza, L. L. (2005). Support from the relationship of genetic and geographic distance in human populations for a serial founder effect originating in Africa. *Proceedings of the National Academy of Sciences of the USA, 102,* 15942–15947.

Rightmire, G. P., & Deacon, H. J. (1991). Comparative studies of Late Pleistocene human remains from Klasies River Mouth, South Africa. *Journal of Human Evolution, 20*, 131–156.

Ruff, C. B. (1995). Biomechanics of the hip and birth in early *Homo. American Journal of Physical Anthropology, 98*, 527–574.

Ruff, C. B., & Hayes, W. C. (1983). Cross-sectional geometry of the Pecos Pueblo femora and tibiae-a biomechanical investigation: I Methods and general pattern of variation. *American Journal of Physical Anthropology, 60*, 359–381.

Ruff, C. B., & Leo, K. P. (1986). The use of computed tomography in skeletal structure research. *Yearbook of Physical Anthropology, 29*, 181–196.

Schmitt, A. (2005). Une nouvelle méthode pour estimer l'âge au décès des adultes à partir de la surface sacro-pelvienne iliaque. *Bulletins et Mémoires de la Société d'Anthropologie de Paris, 17*, 89–101.

Shackelford, L. (2005). *Regional variation in the postcranial robusticity of Late Upper Paleolithic humans.* Ph.D. Dissertation, Washington University.

Shackelford, L. (2007). Regional variation in the postcranial robusticity of Late Upper Paleolithic humans. *American Journal of Physical Anthropology, 133*, 655–668.

Shang, H., Tong, H., Zhang, S., Chen, F., & Trinkaus, E. (2007). An early modern human from Tianyuan Cave, Zhoukoudian, China. *Proceedings of the National Academy of Sciences of the USA, 104*, 6573–6578.

Spoor, F. (1993). *The comparative morphology and phylogeny of the human bony labyrinth.* Ph.D. Dissertation, Utrecht University.

Spoor, F., & Zonneveld, F. (1995). Morphometry of the primate bony labyrinth: A new method based on high-resolution computed tomography. *Journal of Anatomy, 186*, 271–286.

Spoor, F., & Zonneveld, F. (1998). Comparative review of the human bony labyrinth. *Yearbook of Physical Anthropology, 41*, 211–251.

Spoor, F., Wood, B., & Zonneveld, F. (1994). Implications of early hominid labyrinthine morphology for the evolution of human bipedal locomotion. *Nature, 369*, 645–648.

Spoor, F., Hublin, J.-J., Braun, M., & Zonneveld, F. (2003). The bony labyrinth of Neanderthals. *Journal of Human Evolution, 44*, 141–165.

STATsoft FRANCE (2002). Statistica (logiciel d'analyse de données). Version 6.

Steen, S. L., & Lane, R. W. (1998). Evaluation of habitual activities among two Alaskan Eskimo populations based on musculoskeletal stress markers. *International Journal of Osteoarchaeology, 8*, 341–353.

Thoma, A. (1984). Morphology and affinities of the Nazlet Khater man. *Journal of Human Evolution, 13*, 287–296.

Trinkaus, E. (2006). Modern human versus Neandertal evolutionary distinctiveness. *Current Anthropology, 47*, 597–620.

Trinkaus, E., & Ruff, C. B. (1999). Diaphyseal cross-sectional geometry of Near Eastern Middle Palaeolithic humans: The femur. *Journal of Archaeological Science, 26*, 409–424.

Trinkaus, E., Moldovan, O., Milota, S., Bilgar, A., Sarcina, L., Athreya, S., et al. (2003). An early modern human from the Peştera cu Oase, Romania. *Proceedings of the National Academy of Sciences of the USA, 100*, 11231–11236.

Vallois, H.-V. (1932). L'omoplate humaine. Etude anatomique et anthropologique. *Bulletins et Mémoires de la Société d'Anthropologie de Paris, série, 8*(3), 1–153.

Van Peer, P. (1998). The Nile corridor and the Out-of Africa model. An examination of the archaeological record. *Current Anthropology, 39*, 5115–5140.

Van Peer, P. (2004). Did Middle Stone Age moderns of sub-saharan african descent trigger an Upper Paleolithic revolution in the lower Nile Valley? *Anthropologie (Brno), 42*, 215–225.

Van Peer, P., & Vermeersch, P. M. (1990). Middle to Upper Palaeolithic transition: The evidence for the Nile Valley. In P. G. Mellars (Ed.), *The emergence of modern humans: An archaeological perspective* (pp. 139–159). Edinburgh: Edinburgh University Press.

Vandermeersch, B. (1981). *Les hommes fossiles de Qafzeh (Israël).* Paris: Editions du CNRS.

Varrela, J. (1992). Dimensional variation of craniofacial structures in relation to changing masticatory-functional demands. *European Journal of Orthodontics, 14*, 31–36.

Vermeersch, P. M. (2002). Two Upper Palaeolithic burials at Nazlet Khater. In P. M. Vermeersch (Ed.), *Paleolithic quarrying sites in Upper and Middle Egypt* (pp. 273–282). Leuven: Egyptian Prehistoy Monograph 4.

Vermeersch, P. M., & Paulissen, E. (1993). Palaeolithic chert quarrying and minig in Egypt. In L. Krzyńaniak, M. Kobusiewicz, & J. Alexander (Eds.), *Environmental change and human culture in the Nile Basin and Northern Africa until the second millennium B.C* (pp. 337–349). Poznan: Poznań Archaeological Museum.

Vermeersch, P. M., Paulissen, E., Gijselings, G., Otte, M., Thoma, A. B., Van Peer, P., et al. (1984). 33,000-yr old chert mining site and related Homo in the Egyptian Nile Valley. *Nature, 309*, 342–344.

Vermeersch, P. M., Paulissen, E., & Van Peer, P. (1990). Palaeolithic chert exploitation in the limestone stretch of Egyptian Nile Valley. *The African Archaeological Review, 8*, 77–102.

Vermeersch, P. M., Paulissen, E., Stokes, S., Charlier, C., Van Peer, P., Stringer, C., et al. (1998). A Middle Paleolithic burial of a modern human at Taramsa Hill, Egypt. *Antiquity, 72*, 475–484.

Vermeersch, P. M., Paulissen, E., & Vanderbeken, T. (2002). Nazlet Khater 4. An Upper Palaeolithic underground chert mine. In P. M. Vermeersch (Ed.), *Paleolithic quarrying sites in Upper and Middle Egypt* (pp. 211–271). Leuven: Egyptian Prehistoy Monograph 4.

Weidenreich, F. (1936). *The mandible of Sinanthropus pekinensis: A comparative study.* Peking: The geological survey of China.

Wendorf, F., Schild, R., & Close, A. E. (1993). Summary and conclusions. In F. Wendorf & R. Schild (Eds.), *Egypt during the Last Interglacial: The Middle Paleolithic of Bir Tarfawi and Bir Sahara East* (pp. 552–573). New York: Plenum Press.

# Chapter 15
# Middle Pleistocene Diversity in Africa and the Origin of Modern Humans

G. Bräuer

**Abstract** Different views exist on the pattern of Middle Pleistocene evolution in Africa. Some favor a splitting into two or more species, for example, *Homo heidelbergensis*, *Homo helmei*, and *Homo sapiens*, whereas others see evidence for a continuously evolving lineage of only one chronospecies, *Homo sapiens sensu lato*. This latter view then considers the one chronospecies to be separated further into several subspecies, grades, steps, paleo-demes or other entities. The question is, which of these diverse perspectives is best supported by the current evidence? There is also some disagreement about the geographic pattern of the anatomical modernization process. Although there is clear evidence that northern, eastern, and southern Africa were involved, it appears difficult to assess the distinct roles of the different regions within this long-term process. Interregional migration, for example, during periods of a "green Sahara" might have led to complex patterns.

**Keywords** Archaic *Homo sapiens* • Grade • *Homo heidelbergensis* • *Homo helmei* • *Homo sapiens* • Modernization process • Species

## Introduction

Most current phylogenies suggest a speciation event in Africa at around 0.8 Ma, when *Homo erectus sensu lato* gave rise to a new species, for which different names like *Homo sapiens*, *Homo heidelbergensis*, and *Homo rhodesiensis* are used (e.g., Rightmire 1998; Tattersall and Schwartz 2000; Bräuer 2001, 2007; Hublin 2001; Stringer 2002; Foley and Lahr 2003). This species might have expanded into Europe at around 0.6 Ma, evolving there

towards the Neandertals, whereas the evolution in Africa led to the origin of anatomically modern humans. Such a split of the Neandertal and modern human lineages is also in agreement with the current evidence from Neandertal mitochondrial and nuclear DNA (Krings et al. 2000; Green et al. 2006, 2010; Noonan et al. 2006).

Regarding Europe, there is broad consensus that "the development of the Neandertal morphology results from an *accretion* phenomenon beginning in the middle of the Middle Pleistocene, around 450,000 BP or a bit before" (Hublin 1998, p. 301; see also Dean et al. 1998; Stringer 2002). New dating evidence for the pre-Neandertals from Atapuerca/Sima de los Huesos suggests that the process of Neandertal evolution might have already started at around 0.5 Ma (Bischoff et al. 2007), and thus only shortly after the speciation event in Africa and the expansion of the new species into Europe. The accretion process can be characterized as a "progressive accumulation of Neandertal features" (Klein 2000, p. 24), and as Hublin (1998, p. 302) pointed out, "considering the mosaic nature of the accretion phenomenon, tracing clear divisions along the pre-Neandertals/Neandertals lineage is quite artificial." Manzi (2004, pp. 21–22) also regards the Neandertal lineage as an anagenetic sequence, "which could be more reasonably considered a sequence of chrono-subspecies." In view of the well documented continuous process in Europe without any clear divisions, Hublin (1998, p. 302) suggested the inclusion of all the specimens involved in this Neandertalization process in one taxon, *Homo neanderthalensis* or *Homo sapiens neanderthalensis* (see also Harvati 2007). In fact, there is scant evidence for a speciation event along the pre-Neandertal/Neandertal lineage and, thus, the splitting into two species, *Homo heidelbergensis* and *Homo neanderthalensis* (Fig. 15.1).

It is beyond the scope of this chapter to deal with the European evidence in more detail (see Bräuer 2006, 2008a, b). Yet it is obvious that the different views on the taxonomic diversity of the Middle Pleistocene Europeans are influenced by the possible scenarios of the evolution of

G. Bräuer (✉)
Abteilung für Humanbiologie, Universität Hamburg,
Allende-Platz 2, 20146 Hamburg, Germany
e-mail: guenter.braeuer@uni-hamburg.de

J.-J. Hublin and S. P. McPherron (eds.), *Modern Origins: A North African Perspective*,
Vertebrate Paleobiology and Paleoanthropology, DOI: 10.1007/978-94-007-2929-2_15,
© Springer Science+Business Media B.V. 2012

**Fig. 15.1** Phylogenetic scheme
after Rightmire (2002, p.126)

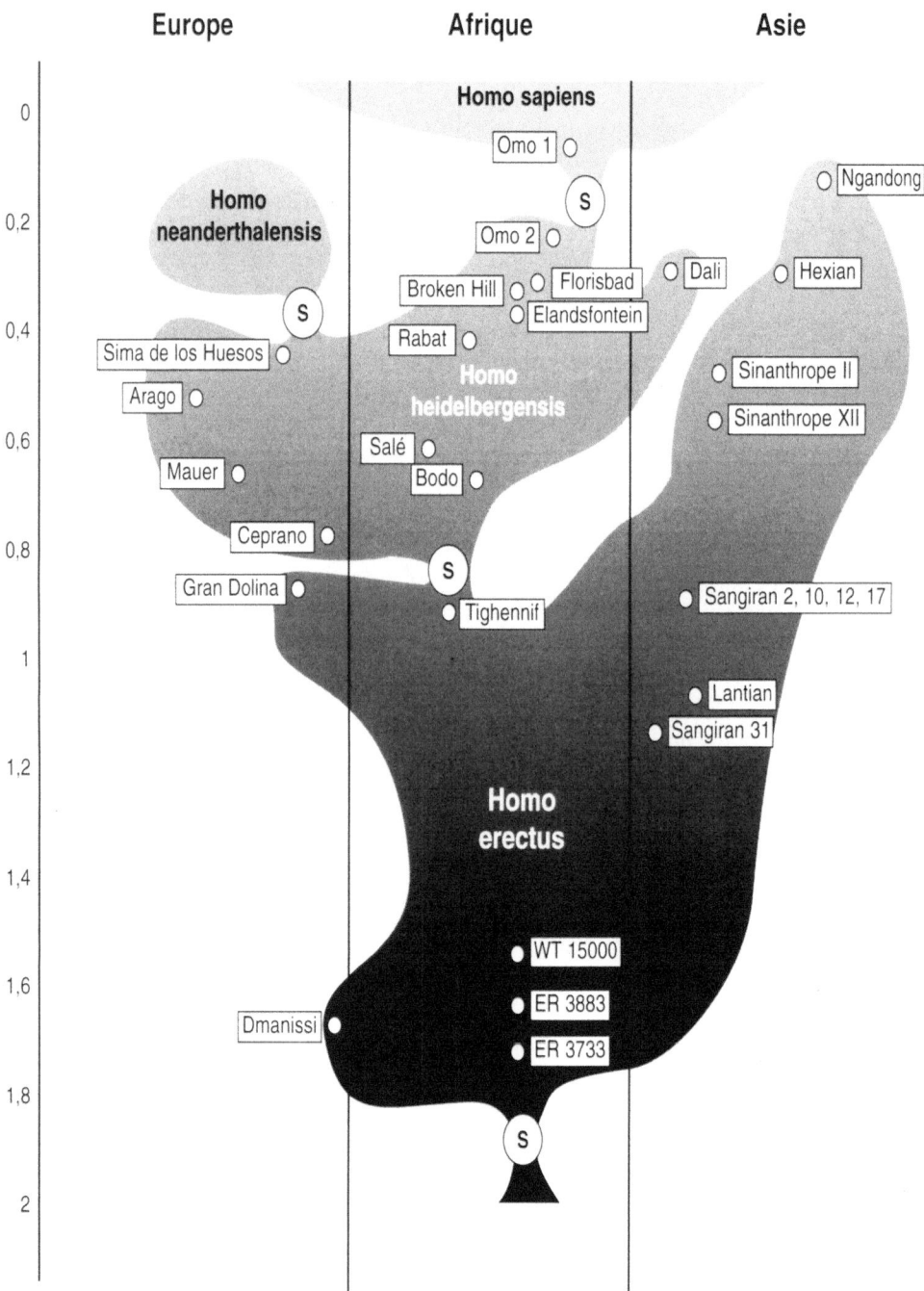

post-*erectus* hominins in Africa. Thus, the present chapter
focuses on the mode and pattern of anatomical modernization in Africa. It also discusses the question of whether
the current fossil evidence from Africa is in agreement with
the assumption of an anagenetic Middle Pleistocene lineage
of one evolving species, *Homo sapiens*—largely comparable to the process in Europe—or whether a distinction of the
African lineage into two or more different species would be
more appropriate.

## Diversity and Continuity in Africa

During most of the 1980s and 1990s, the predominant view
suggested only one post-*erectus* species, *Homo sapiens
sensu lato*, which included modern humans as well as post-
*erectus* archaic humans from Africa and Eurasia (e.g.,
Stringer et al. 1979; Rightmire 1983; Bräuer 1984, 1992;
Klein 1989; Clarke 1990; Bilsborough 1992; Dean 1993;

Hublin 1996; Dean et al. 1998). In the 1990s, a tendency arose among some researchers (e.g., Tattersall 1992; Groves and Lahr 1994)—mainly based on the cladistic approach—to split the polymorphic species *Homo sapiens* into several distinct species. Rightmire (1996) also felt that both the concept of intraspecific Middle Pleistocene evolution in Africa and parts of Eurasia and the distinction of subspecies have limited utility, as it is not certain how many subspecies should be distinguished. In addition, there is disagreement on the criteria by which such groups can be identified. Thus, he suggested sorting fossils by their morphological similarity using metrical and non-metrical characteristics, "even if it may never be established that these phena represent biological species as defined for living groups" (p. 33). Based on phenetic similarities between Middle Pleistocene specimens from Africa (Bodo, Kabwe, Saldanha, Ndutu) and Europe (Petralona, Arago), Rightmire (1996, 1998) suggested that this assemblage, together with the Mauer jaw, be assigned to *Homo heidelbergensis*. He also included more derived near-modern specimens like Florisbad and Omo 2 within this species (Rightmire 2002). *Homo heidelbergensis* is consequently regarded as ancestral to two further species, *Homo sapiens* (restricted to modern humans) in Africa and *Homo neanderthalensis* (or Neandertals) in Europe (Fig. 15.1).

However, as outlined above, a splitting of the pre-Neandertal/Neandertal lineage into two species is difficult to demonstrate. Also, the problem raised by Rightmire regarding the uncertainties in the number of possible subspecies and their distinctions cannot be solved by a split into the two species. *Homo heidelbergensis* would still include a morphologically variable spectrum of African fossils like Bodo, Florisbad, and Omo 2, as well as a number of European pre-Neandertals and archaic Chinese specimens. *Homo sapiens*, on the other hand, would cover anatomically modern humans including Omo 1, the roughly contemporaneous and closely related Omo specimen from Ethiopa (Rightmire 2002) (Fig. 15.1). Even if we would consider the African specimens alone, few experts appear to support such a distinction. Stringer (2002), for example, suggests a split that includes the more derived archaic specimens like Florisbad, Omo 2, and Jebel Irhoud among the *Homo sapiens*—as "archaic" *Homo sapiens*. Foley and Lahr (2003), in contrast, favor the splitting of the African lineage into three species: *Homo heidelbergensis*, a more derived *Homo helmei* (including Florisbad), and *Homo sapiens*. Yet the suggestion that *Homo helmei* is ancestral to *Homo sapiens* and *Homo neanderthalensis* is problematic, not only because it also includes certain European pre-Neandertals, but also because Neandertal features were already present prior to the hypothesized appearance of the species (McBrearty and Brooks 2000; Stringer 2002; Bräuer 2007, 2008a). Although Tattersall and Schwartz (2000) use the

name *Homo heidelbergensis* for African and non-African fossils, they are not really convinced that the specimens within this taxon represent one species: "Closer scrutiny will probably show that the material allocated to *H. heidelbergensis* actually embraces more than just one species. For the moment, however, *H. heidelbergensis* serves as a useful umbrella for specimens whose characteristics include a sizeable brain of around 1,200 ml or so" (p. 165). Under the heading, "Other members of the '*Archaic Homo sapiens*' Group from the Levant and Africa," Schwartz and Tattersall (2003, pp. 600–601) deal with a diversity of specimens like Florisbad, Ndutu, Laetoli H.18, and Salé, which they do not include in *Homo heidelbergensis* although they emphasize their differences from *Homo sapiens*.

However, Schwartz and Tattersall (2003) do not consider *Homo sapiens* to mean anatomically modern humans in the same way that other authors have. Instead, they recently suggested a new definition of *Homo sapiens* based on nine cranial features, which they regard among hominoids to be autapomorphic for *Homo sapiens*. This definition led them to the conclusion that "a relatively large contingent of forms that have in the past been identified as 'modern *Homo sapiens*' (more or less)... fall short of this narrower definition" (Schwartz and Tattersall 2003, p. 599). Among others, these humans include most, but not all, of the Klasies River Mouth specimens, Omo 1, Singa, and some of the Qafzeh specimens. Concerning Qafzeh, Schwartz and Tattersall (2003, p. 600) concluded that "Qafzeh 1, 2, 9, and 11 are clearly *Homo sapiens*, but the rest are equally clearly not." Few experts might be ready to support such a splitting into different "modern species." Therefore, the question arises whether the features can be supported or rejected by tests. For example, one of the suggested traits is the inverted-T-shaped chin. Yet recent skeletal material indicates that there is not only great variability in chin morphology, but also many specimens lack an inverted-T-shaped chin. Other features are rather variable as well (e.g., the lateral placement of the styloid process, or the narrow and high occipital plane), making a simple assessment of presence/absence difficult (Bräuer 2008a; Pearson 2008). Also, Foley (2002) emphasized that variability and polymorphisms in both ancestral and descendent taxa often make autapomorphies difficult to demonstrate. Moreover, he stated that "it is unlikely that the species concept itself will be the most useful tool for unravelling what is in effect,... a very small-scale event, especially in its later stages (the last half million years)" (p. 33).

In fact, this short review of some of the current attempts to split the African post-*erectus* lineage into different species reveals that the definitions of the suggested entities or divisions are largely subjective. There is obviously no more agreement on the criteria by which these "species" have been identified than with regard to any subspecies, or other intraspecific divisions. Foley (2001, pp. 9–10) highlights this problem when he concedes that continuity rather than discontinuity "is the reason for the persistent problem of delimiting the taxonomic units in later stages of human evolution and gives rise to the question of whether the species concept,... is sufficiently fine-tuned to cope with evolution at this scale. The lineages of later human evolution seem to show simultaneously *continuously evolving lineages* and very distinctive derived endpoints..." From all this, the question arises whether an intraspecific interpretation of the African lineage would indeed be more reasonable and more in agreement with the facts than a fashionable splitting into a number of arbitrarily defined entities claimed to be species.

## Pattern of the African Lineage

When we consider the process of African Middle Pleistocene evolution in general, it is evident that, by 0.7 or 0.6 Ma, hominins existed that differed from *Homo erectus*, as is clearly indicated by the Bodo cranium from Ethiopia. This specimen exhibits a number of derived traits found in later near-modern or modern humans, or as Rightmire (1996, p. 32) put it, "it is clear that the Bodo cranium cannot be excluded from a population that is advanced anatomically in comparison to *H. erectus*. This finding is strengthened when brain size is considered. It is further supported by the observation that Bodo shares with Broken Hill (Kabwe) a number of facial traits found in more modern humans." From about 0.7–0.6 Ma to about 0.2–0.15 Ma, when anatomically modern humans appeared, the fossil record documents the morphological changes leading to the fully modern pattern. Yet how can this process be described adequately?

Many experts recognize three different entities or groups. Foley and Lahr (2003) distinguish an early morph represented by specimens like Bodo, Kabwe, and others, a later morph including, for example, Florisbad and Laetoli H.18, and a third including anatomically modern humans. Stringer's (2002) three groups include about the same African specimens as those considered by Foley and Lahr, but he is convinced that the later morph belongs to *Homo sapiens* (as "archaic" *Homo sapiens*). This entity, or "Group 2" according to McBrearty and Brooks (2000), is considered by these authors to be "intermediate" in morphology between "Group 1," to which they count Bodo, Kabwe, Salé, and others, and "Group 3," *Homo sapiens sensu stricto* or anatomically modern humans. Smith (2002) also regards the second or later group as a "transitional group" between earlier hominins and anatomically modern humans.

A division of the African post-*erectus* lineage into three groups is not a recent idea. Such a pattern was already suggested in the early 1980s on the basis of a comprehensive study of the fossil material (Bräuer 1982, 1984, 1989). This and subsequent research provided evidence for evolutionary continuity between these groups and an increase of derived modern *Homo sapiens* or transitional modern conditions. Such derived features already occurred in the early post-*erectus* group, suggesting that the whole lineage of modernization should be classified as *Homo sapiens* (e.g., Bilsborough 1992; Hublin 1996; Bräuer et al. 1997; Dean et al. 1998; Klein 1999; Jurmain et al. 2000; Bräuer 2001, 2007, 2008a, b; Grimaud-Hervé and Vialet 2002; Smith 2002; Turbón 2006). The three distinguishable groups of this lineage were often regarded as grades of an evolving chronospecies *Homo sapiens*: early archaic *Homo sapiens*, late archaic *Homo sapiens*, and anatomically modern *Homo sapiens*. Each of the grades includes specimens of a similar evolutionary level, with both chronological and morphological overlap between the grades (Fig. 15.2). The mode of a continuous anagenetic evolution without speciation events has been supported by a new study of the Middle Pleistocene hominins, showing clear temporal trends from early archaic up to modern *Homo sapiens* for a large number of metrical and non-metrical features (Mbua and Bräuer 2012). These include, for example, major dimensions of the frontal, parietal, and occipital bones as well as non-metrical traits of the supraorbitals, parietal expansion, temporal squama and mastoid region, occipital shape, and facial features. Many of these changes also correlate with an increase in the rounding of the vault or a retraction of the face. Lieberman et al. (2002) also found these aspects to be significant in the process of anatomical modernization. The changes in cranial shape can be observed in a varying mosaic-like pattern over some hundreds of thousands of years (Bräuer 2008a; Mbua and Bräuer 2012) and are thus hardly the result of a few anatomical adjustments during a short time period (see comments in Balter 2002). In another recent approach to the evolution of modern morphology, Pearson (2008) also arrived at the conclusion that the derived modern features did not appear as a set in the fossil record but gradually in a mosaic-like fashion. He identified several steps based on specific morphological features or changes by which the long process of evolution of modern anatomy can be described. Initial changes are already seen in an increase in brain size among late African *H. erectus*. Subsequent post-*erectus* features include, among others, vertically shorter faces, smaller brow ridges, more vertical frontal bone, rounded occipital, and progressive chin development. Pearson (2008) emphasized that these steps can be regarded as belonging to one evolving lineage and are steps in the process of our species' evolution. In the following sections,

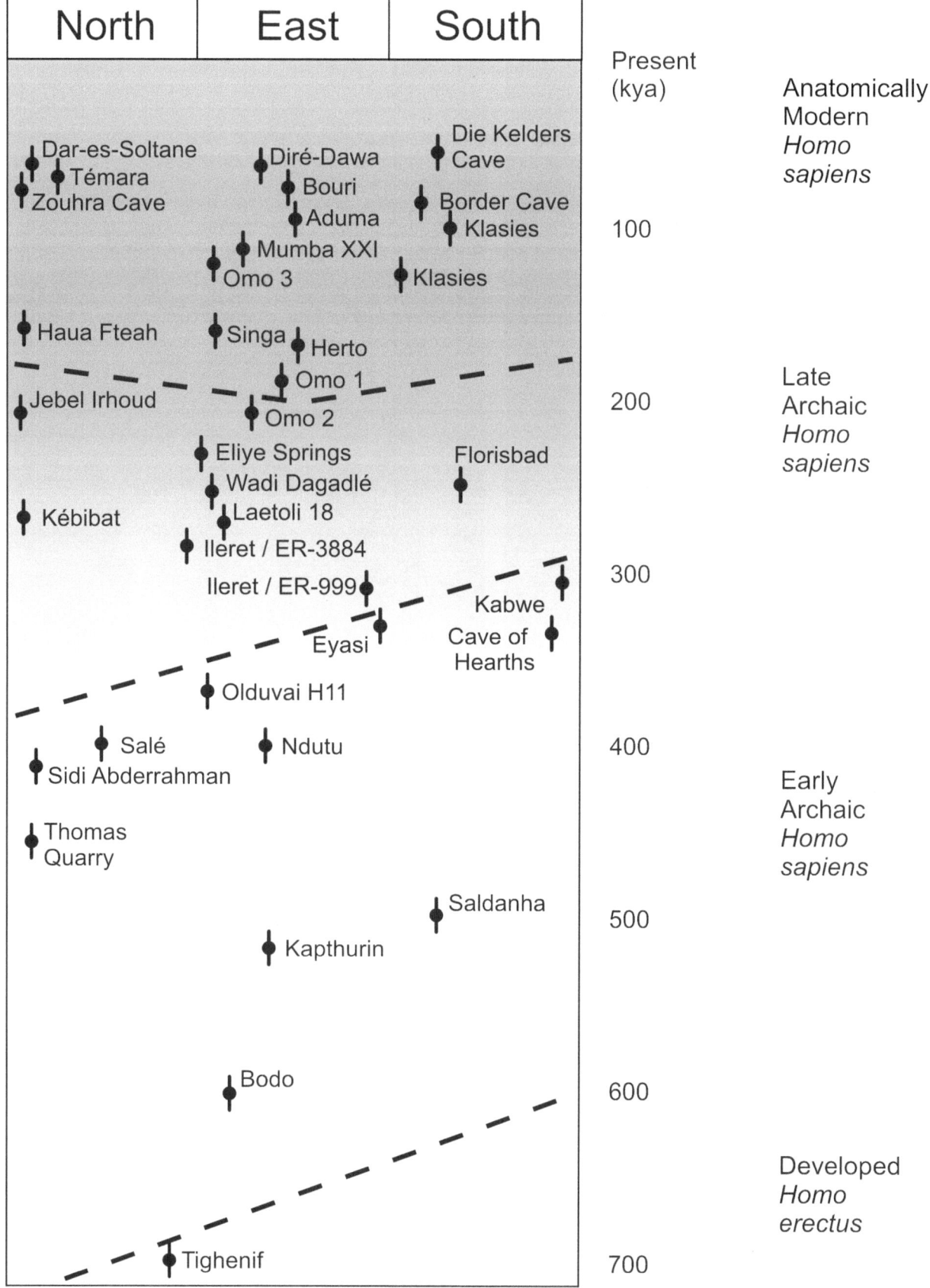

**Fig. 15.2** The African lineage of *Homo sapiens* evolution. Three grades or groups can be distinguished that overlap both chronologically and morphologically

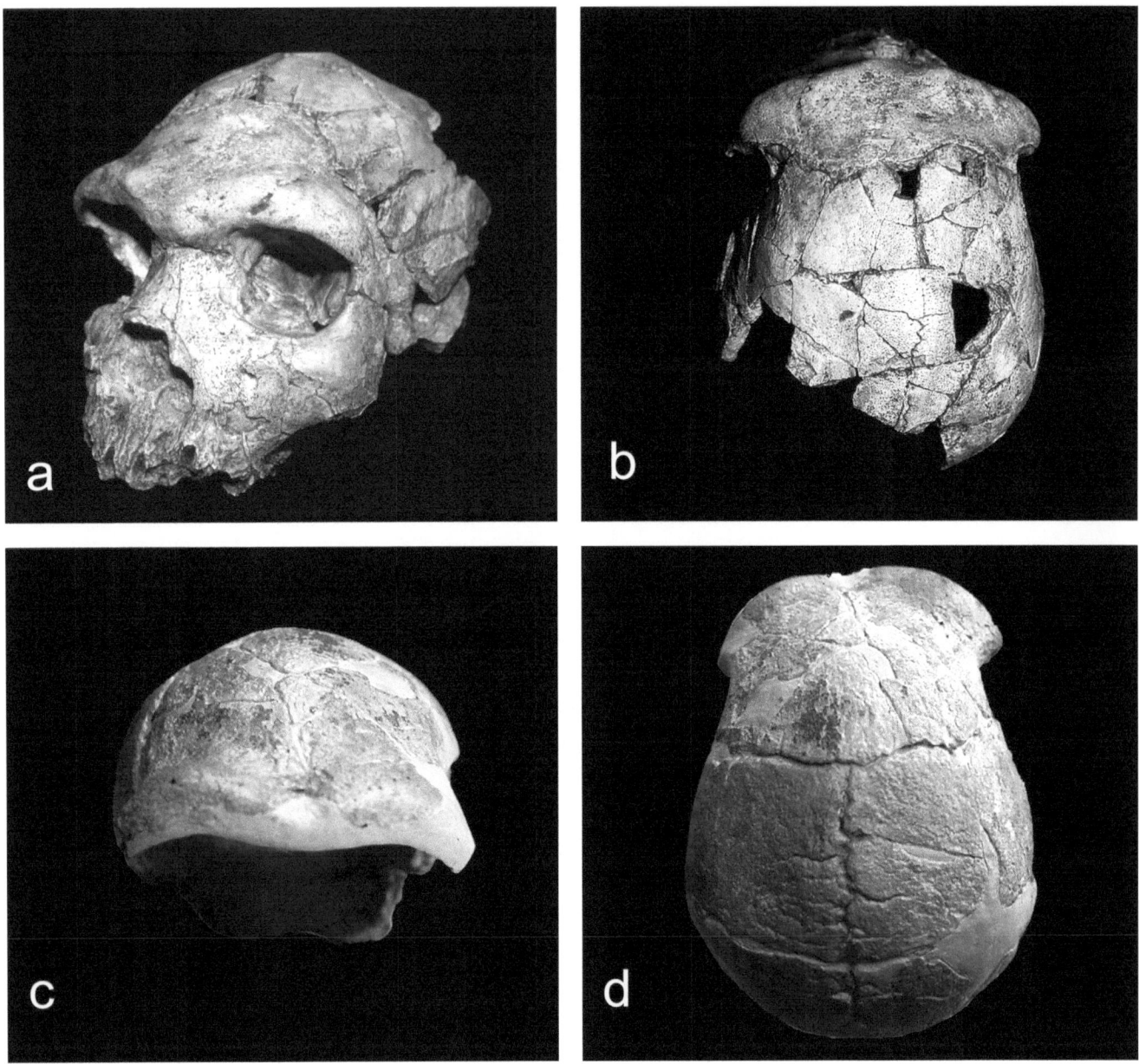

**Fig. 15.3** Early archaic *Homo sapiens*. **a** and **b** Bodo, Ethiopia; **c** and **d** Saldanha, South Africa

the three widely recognized groups or grades of this process are described, with emphasis on apomorphic modern or transitional modern features. This description is not restricted to our own observations but includes assessments by other authors as well.

## Early Specimens

The early archaic *Homo sapiens* specimens currently available roughly date to between 0.6 and 0.3 Ma

(Fig. 15.2). A closer look at some of these specimens reveals derived character states that unite them with later near-modern or modern humans (Bräuer, 2008a). The Bodo cranium from Ethiopia (Fig. 15.3) represents the oldest well-preserved specimen of this group, with an age of about 0.65 to 0.6 Ma (Clark et al. 1994). It has a large cranial capacity of about 1,300 cc, associated with an "expansion of the parietal walls relative to the bitemporal width [which] is characteristic of *H. sapiens*" (Rightmire 1996, p. 30). The coronally expanded frontal bone (Mbua and Bräuer 2012) exhibits proportions of the squama similar to those of the much younger near-modern specimen Omo 2, and "the squamosal suture describes a high arch as in more modern

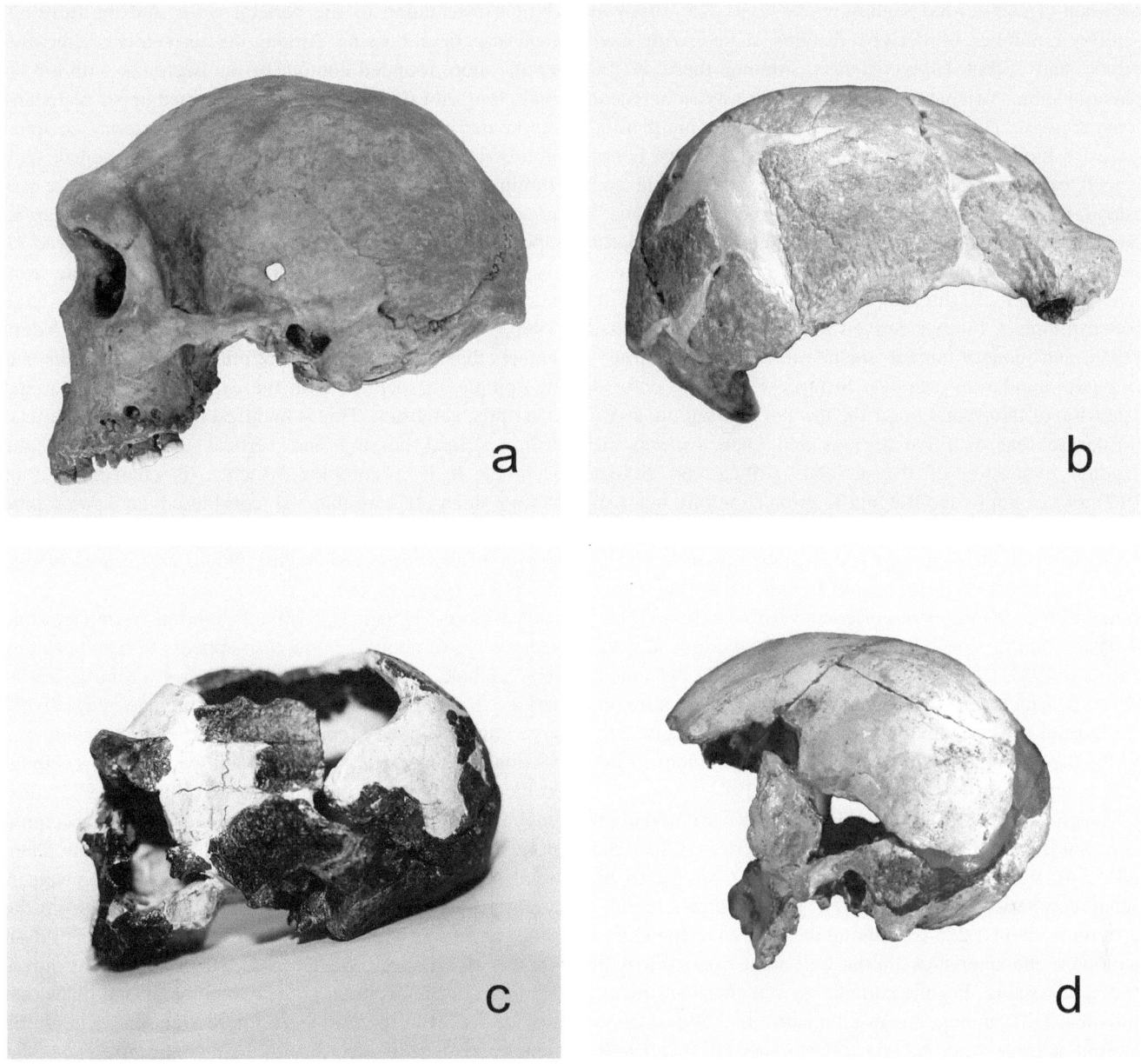

**Fig. 15.4** Early archaic *Homo sapiens*. **a** Kabwe, Zambia; **b** Saldanha, South Africa; **c** Ndutu, Tanzania; **d** Salé, Morocco (photo: J.-J. Hublin)

humans" (Rightmire 1996, p. 30). The supraorbital torus shows a mid-orbital division into medial and lateral segments, representing a transitional modern feature also present in a similar pattern in later archaics (Laetoli H.18), and the torus also attenuates laterally (Mbua and Bräuer 2012). Rightmire (2001a) even regards Bodo's supraorbital torus as a derived condition present in recent *Homo*. Although the face is heavily built, Rightmire (1996, 1998) identified other characteristics that both the Kabwe cranium and modern humans share. These are the vertical margin of the nasal aperture and the incisive canal opening into the front of the hard palate. Interestingly, Harvati and Hublin (2012) also found close affinities between Bodo and modern

humans in a recent analysis of the face using geometric morphometrics (in which, however, the area of the canine fossa was not included). According to Rightmire (2000, p. 68), the "frontal bone proportions, the squamous temporal, the cranial base and the face [of Bodo] show resemblances to recent *Homo*" (see also Rightmire 2007). Thus, it is not surprising that after her detailed study of the Bodo specimen, Adefris (1992) classified it as "archaic *Homo sapiens*."

The large similarities between Bodo and the considerably younger Kabwe cranium (presumably around 0.3 Ma or even somewhat younger) from Zambia (Figs. 15.4 and 15.5) have been widely emphasized. This large-brained specimen (1,280 cc) led Rightmire (1990, p. 225, 1996) to identify a number of derived features shared with later archaic and recent *Homo sapiens*. Among these is the sphenoid spine "oriented in about the same way as in recent *Homo sapiens*. The inferior border of the tympanic plate is thin as in modern humans... The juxtamastoid ridge is not as well expressed as in European Neandertal crania and instead resembles that seen in many modern humans." Kabwe and Bodo share a similar vertical margin of the nasal aperture and positioning of the incisive canal, both of which were regarded by Rightmire (1996) as synapomorphies with modern humans. Further derived features shared with later archaic and modern humans include the very low petrous-tympanic angulation (see also Stringer 1984), the coronal expansion of the frontal bone, the low parietal sagittal angle (similar to that of Eliye Springs and Omo 1), and the diagonal expansion of the parietals (Mbua and Bräuer 2012). For example, the Bregma-asterion chord/arc index of Kabwe is identical to that of the late archaic Kenyan specimen from Eliye Springs and smaller than in Saldanha and even Laetoli H.18 (Mbua and Bräuer 2012). The upper scale of the occipital bone is oriented vertically. In addition, its basicranial flexion is similar to that of modern humans (Laitman 1985). There is also a second cranial specimen, Kabwe 2, which could be contemporaneous to the cranium. This partial maxilla obviously appears more modern than that of the Kabwe 1 cranium due to its clearly defined but weak canine fossa (see Fig. 14 in Bräuer 1984).

Closely related to Kabwe and Bodo is the partial cranium from Saldanha (Elandsfontein), South Africa (Figs. 15.3 and 15.4), which dates to about 0.6 to 0.4 Ma based on faunal comparisons (Klein 1999). The estimated cranial capacity is about 1,225 cc. Among the derived near-modern or modern characteristics are the well arched parietals with moderate bossing. For the parietal sagittal chord/arc index, Saldanha falls among modern humans, like the Afalou-Taforalt samples from Algeria and Morocco (Bräuer 1984). More recently, Haile-Selassie et al. (2004) found that Saldanha is quite similar to the early modern humans from Qafzeh, as well as to sub-Saharan modern humans with regard to the parietal arch shape index (Bregma-lambda chord/biasterionic breadth). Saldanha's frontal squama is coronally enlarged and the supraorbital torus attenuates laterally. The occipital is less angulated than generally seen in *Homo erectus* and the transverse torus is reduced and restricted centrally (Bräuer 2008a; Mbua and Bräuer 2012).

Another specimen generally assigned to this group or evolutionary grade is the Ndutu cranium from Tanzania (Figs. 15.4, 15.5), which dates to around 0.4 Ma or slightly earlier (Klein 1999; McBrearty and Brooks 2000). The braincase of this presumably female specimen, which has a capacity of approximately 1,100 cc, exhibits obvious similarities to the other early archaics. These include the rather

vertical orientation of the parietal walls and the development of parietal bossing. Among the derived characteristics are the more rounded contour of the occipital, with a relatively long and slightly posteriorly inclined upper scale, and a more derived torus morphology, with an external occipital protuberance (Rightmire 1990). Lahr (1996) agrees with Hublin (1978) that the external occipital protuberance is a modern human apomorphy. The temporal squama is curved superiorly. The root of the styloid process is preserved on one side and the styloid sheath also does not deviate from modern anatomy (Rightmire 1990). According to Clarke (1990, pp. 726–727), "the most diagnostic of the Ndutu features that would suggest an affinity with *Homo sapiens* is the complex associated with the expansion of the parietal area of the cerebrum. This is manifest in the vertical parietal walls, parietal bossing and vertical upper scale of the occipital. It is a complex which... is characteristic of modern man. It can thus be considered an apomorphic character complex of *Homo sapiens* and it follows that it is taxonomically correct to classify Ndutu as a representative of archaic *Homo sapiens*..."

The more fragmentary Eyasi 1 cranium from Tanzania, with an age of about 0.3 Ma, also appears to belong to this early archaic group (Fig. 15.5). It shows a similar set of derived *sapiens*-like features. These include a relatively large cranial capacity (ca. 1,280 cc), parietal expansion and bossing, rather vertical lateral walls, a rounded occipital profile, and a weak occipital torus. Hublin (1978) also emphasized the presence of a derived external occipital protuberance in this specimen. From the lateral view, there are also general similarities in shape to the slightly younger, late archaic Laetoli H.18 specimen, which comes from the same region in Tanzania (see Fig. 5 in Bräuer 1984). Another Eyasi specimen, Eyasi 2, which consists of only a large part of the occipital bone, exhibits a nuchal plane that provides an even more modern impression, underlining the complex pattern of early-late archaic continuity.

Northern Africa has also provided relevant hominin remains from this time period (Fig. 15.2). The most diagnostic specimen is the relatively small-brained cranium from Salé, Morocco (Figs. 15.4, 15.5), which dates to about 0.4 Ma. Morphological analyses by Hublin (1985, 2001, p. 109) showed that, in addition to a number of plesiomorphic retentions, "Salé displays clear synapomorphies with *Homo sapiens*." Among these are the convexity and orientation of the frontal bone (inferred from the frontal crest), the rounded shape of the temporal squama, parietal bossing, the vertical orientation of the lateral walls of the vault, a well marked tuberculum pharyngeum, the presence of a sphenoid spine, and the orientation of the preglenoid plane. The more rounded occipital bone and the poorly developed transverse torus might have been influenced by pathological alterations during ontogenetic development

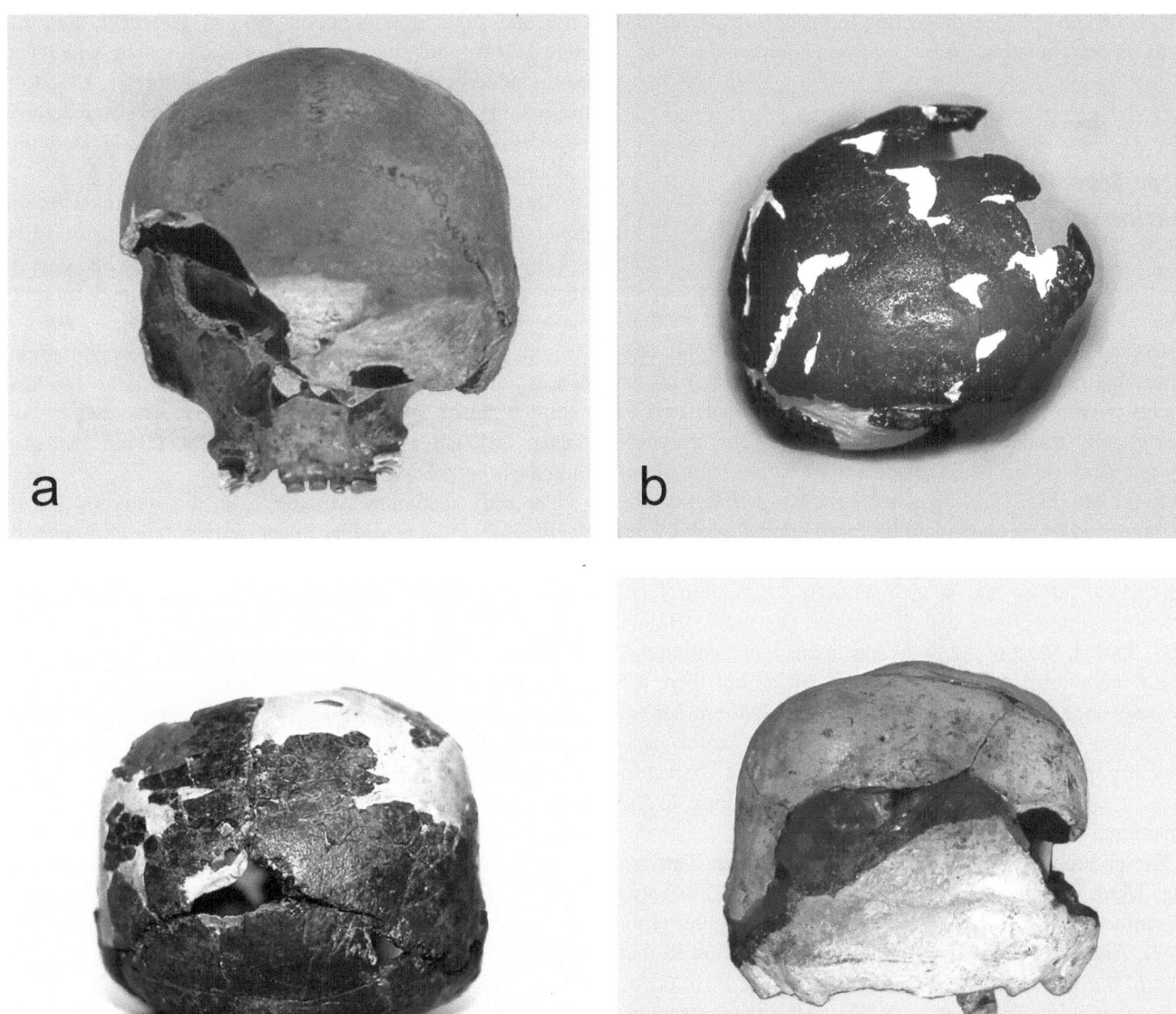

**Fig. 15.5** Early archaic *Homo sapiens*. **a** Kabwe, Zambia; **b** Eyasi, Tanzania; **c** Ndutu, Tanzania; **d** Salé, Morocco (photo: J.-J. Hublin)

and cannot be regarded as normal conditions (Hublin 1985). In addition, Rightmire (1990) pointed out that the Salé cranium displays some of the features of more modern humans. These include the short basioccipital, the glenoid cavity, and the tympanic bone. He further emphasized the substantial bossing of the parietal walls and the modern appearance of the prominent plate-like sphenoid spine. Moreover, Salé exhibits obvious similarities to the Ndutu cranium (Clarke 1990) (Figs. 15.4, 15.5).

Taken together, the evidence from these relevant early specimens leaves little doubt that, based on the existing derived modern or transitional modern features, the evolution of modern anatomy had begun at this early post-*erectus*

time. According to Hublin (2001), the mosaic of ancestral and derived features as observed on these African specimens has been traditionally assigned to "archaic *Homo sapiens*." If these early specimens can be clearly separated from later representatives of early *Homo sapiens*, it would seem appropriate to Hublin to use the term *Homo rhodesiensis*. Yet he emphasized that the question remains as to whether the distinction between *Homo rhodesiensis* and *Homo sapiens* results from a speciation event or if we are dealing with grades. Since there is good evidence for a continuous mosaic-like process of modernization in the late archaic group as well, it appears reasonable to include the early archaics with the species *Homo sapiens sensu lato* (see also Klein 1999), representing a chrono-subspecies that could be designated as *Homo sapiens rhodesiensis*.

## Later Transitional and Early Modern Specimens

The fossil remains of the subsequent late archaic group or grade clearly show an increase in near-modern or modern *sapiens* apomorphies and a decrease in plesiomorphic features. This group should therefore also be included in our species, *Homo sapiens*, as an archaic transitional form to anatomically modern humans. The near-modern morphology is evident in all of these specimens dating from about 0.3 to 0.2 Ma (Fig. 15.2). In general, the late archaics can be characterized by an even more enlarged cranial capacity, an essentially modern vault morphology, a more reduced supraorbital torus, and a near-modern or modern face, including canine fossa and inframalar curvature (Bräuer 2007, 2008a). Most of the fossil specimens representing this evolutionary level come from eastern Africa but there are also relevant specimens from northern and southern Africa. In fact, the morphology of these specimens is largely anatomically modern and only a few plesiomorphic characteristics or metric affinities indicate that they belong to the transitional group or grade.

The cranium KNM-ER 3884 from Ileret, East Turkana (Fig. 15.6), for which a direct U/Th age of ca. 0.27 Ma was determined, also belongs to the late archaics (Bräuer et al. 1997). Analysis of this large-brained cranium showed that most of the vault cannot be distinguished from a Holocene African sample (Bräuer et al. 1992a). However, the moderately developed but continuous supraorbital torus (see also Schwartz and Tattersall, 2003) deviates from the largely modern morphology of the specimen. In spite of the fragmentary face, there can be little doubt that the cranium is close to modern anatomy.

The same holds true for the well-preserved Tanzanian cranium Laetoli H.18 (Fig. 15.6), which might have a similar age of around 0.25 Ma (Manega 1995; Bräuer 2008a). According to Rightmire (2001b, p. 233), it "is less archaic than earlier Middle Pleistocene specimens, and it exhibits several features that foreshadow the modern condition." The cranium shows a modern face with a canine fossa and a well-arched inframalar curvature, a near-modern braincase with a capacity of ca. 1,350 cc, parietal bosses yet with a relatively large Bregma-asterion chord/arc index (Mbua and Bräuer 2012), and a rounded occipital bone. It is mainly the frontal bone that retains a number of plesiomorphic characteristics. Among these are the rather flat and narrow squama and a supraorbital torus that is relatively

thick and rounded in its lateral segment. However, the torus shows clear indications of a mid-orbital division, which can be regarded as a transitional modern apomorphy. Based on the extant morphological pattern of ancestral and derived features, Magori and Musiba (2006) also designated this specimen as a late archaic *Homo sapiens*.

Another well preserved East African cranium, which is generally associated with this group, comes from Eliye Springs, West Turkana (Fig. 15.6). Although the date is unclear, the specimen also exhibits a near-modern morphology (Bräuer and Leakey 1986; Bräuer et al. 2004a). Among the derived features are the well developed parietal bosses, the coronally expanded frontal, and the rounded occipital lacking a transverse torus (see also Mbua and Bräuer 2012). Preliminary reconstruction of the face also indicates the presence of a canine fossa.

The only specimen available from the southern part of the continent that is similar in age and evolutionary level of anatomical modernization is the partial cranium from Florisbad, South Africa (Fig. 15.6). Based on dental remains, it was directly dated by ESR to about 0.26 Ma (Grün et al. 1996). Its derived morphology is evident in the coronally much expanded frontal bone and modern postorbital condition (Mbua and Bräuer 2012), as well as the basically modern face with well-developed canine fossa. A supraorbital torus is present, but it is only weakly developed and slightly projecting. However, the cranial wall is rather thick, similar to Laetoli H.18.

A near-modern morphology is also present in the two crania and the child's mandible from Jebel Irhoud, Morocco, which appear to date to around 0.19 or 0.17 Ma (Grün and Stringer 1991; Smith et al. 2007a). The derived morphology of these specimens led Dean et al. (1998) to refer to them as "archaic" moderns. Alternatively, the Jebel Irhoud specimens were designated as "archaic" *Homo sapiens* (Stringer 2002) or as late archaic *Homo sapiens* (Bräuer 2001). No matter which assignment one prefers, it is obvious that these specimens belong to a near-modern transitional evolutionary level or group. In view of the known geographic and individual variation among the late archaic specimens, the Jebel Irhoud crania can be regarded as northwestern African representatives of this group. The nearly complete Jebel Irhoud 1 cranium (Fig. 15.7) exhibits a basically modern or near-modern morphology in most aspects (Hublin 2001). The low, wide face shows weak canine fossae and strongly arched inframalar curvatures. PCA analysis of facial morphology revealed deviations from modern conditions as seen in the Afalou/Taforalt samples (Bräuer 1984). Another obvious archaic feature is the well-developed continuous supraorbital torus. Other aspects of the vault (e.g., the moderately curved parietal sagittal profile and the absence of an external occipital protuberance; see Lahr 1996), as well as multivariate

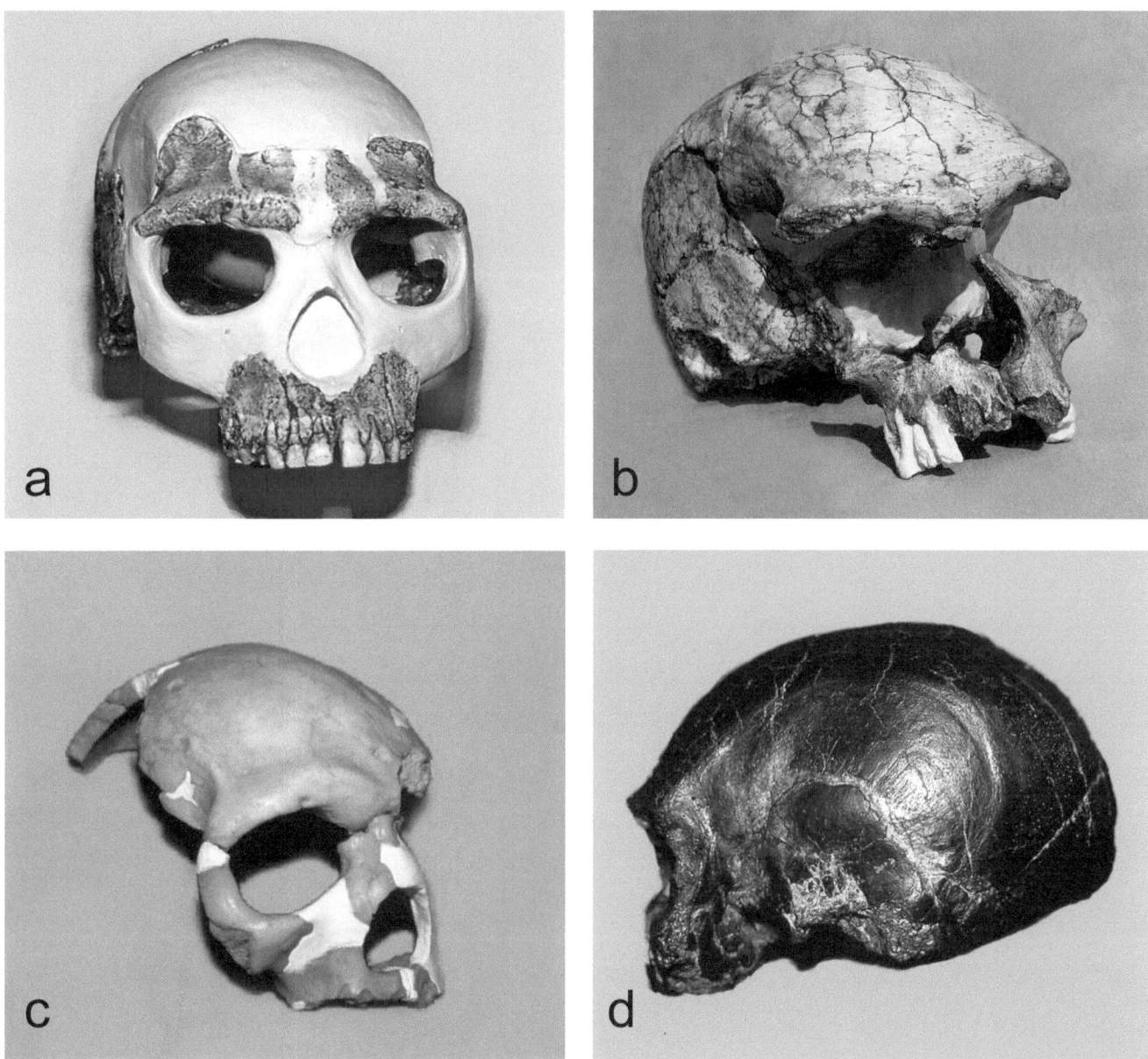

**Fig. 15.6** Late archaic *Homo sapiens*. **a** Ileret/ER 3884, Kenya; **b** Laetoli H.18, Tanzania; **c** Florisbad, South Africa; **d** Eliye Springs, Kenya

affinities, support a classification of Irhoud 1 as an archaic *Homo sapiens* specimen (Mbua and Bräuer 2012). Although lacking the face, Jebel Irhoud 2 (Fig. 15.7) exhibits great similarities in vault morphology to Irhoud 1. The frontal is wider and more flexed sagittally than that of Irhoud 1. Irhoud 2 also exhibits a supraorbital torus that, in contrast to Irhoud 1, shows a mid-orbital depression, but lacks a modern-like supraorbital trigone (Hublin 2001; Mbua and Bräuer 2012). The differences between the two specimens are likely due to intrapopulational variation.

Possible evidence for an early-late archaic continuity can be seen in the specimen from Kébibat (Rabat), Morocco, which dates to around 0.25 Ma or somewhat earlier (Klein

1999; Hublin 2001). The cranial fragments and partial mandible belong to a 13–15 year old juvenile. The occipital has a rounded profile, lacking a transverse torus, and the cranial walls are moderately thick. In view of some more ancestral features, especially regarding the dentition and mandibular body, the option still exists that the remarkably derived features of the occipital could simply be due to the young age of the specimen (Hublin 2001).

Regarding the transition from late archaic to early anatomically modern *Homo sapiens*, the available specimens from Ethiopia appear to provide further indications of this mosaic-like process. Although the Omo Kibish remains were already discovered in the late 1960s, recent field work and Ar/Ar

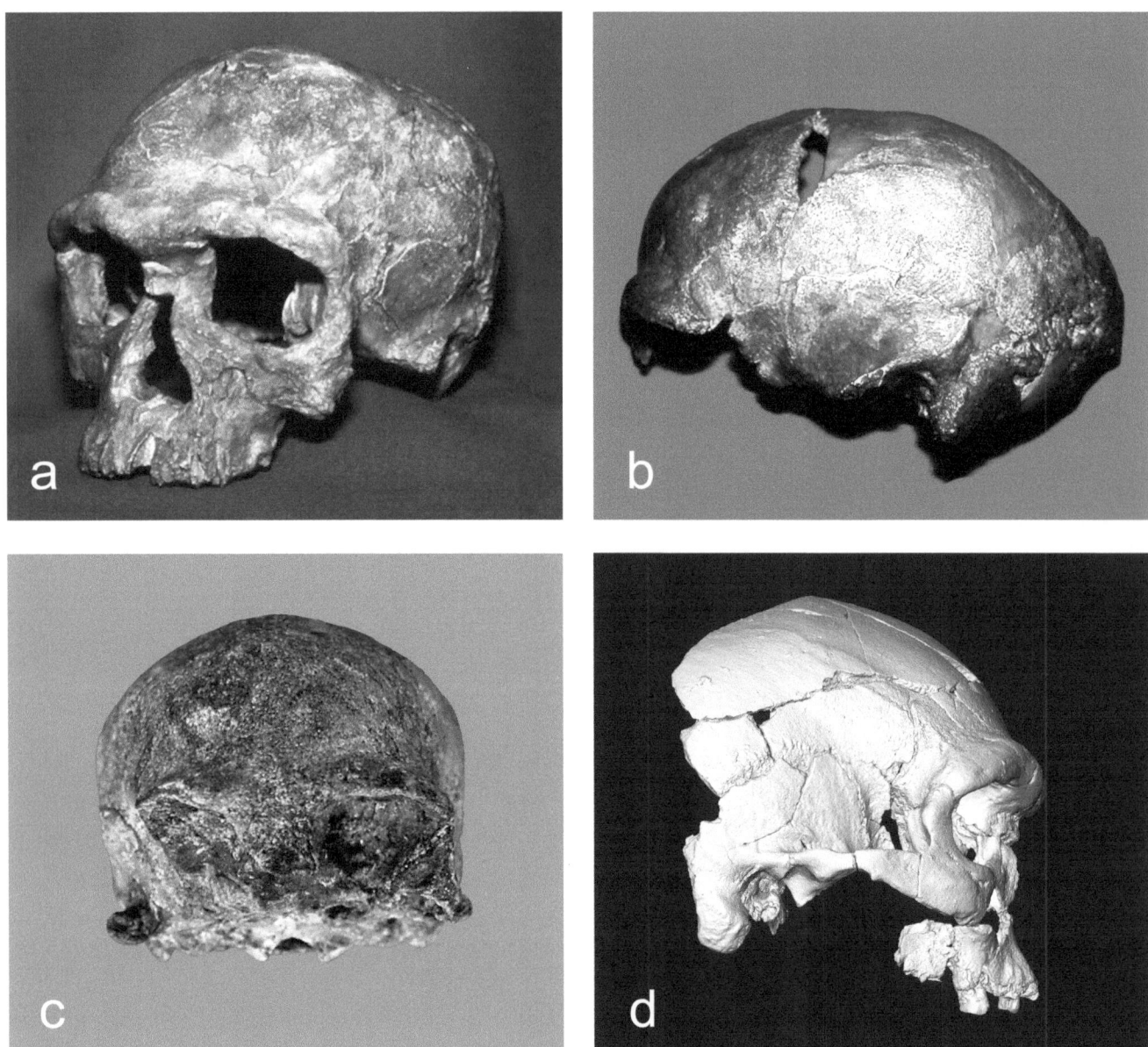

**Fig. 15.7** Late archaic and early modern *Homo sapiens*. **a** Jebel Irhoud 1, Morocco; **b** Jebel Irhoud 2, Morocco; **c** Singa, Sudan; **d** Dar es-Soltan 5, Morocco (photo: J.-J. Hublin)

dating has suggested that both the Omo 1 skeleton and the Omo 2 cranium date to about 0.2 Ma (McDougall et al. 2005, 2008). Whereas Omo 1's skull is anatomically modern (Fig. 15.8) and, according to the recent dating evidence, the oldest known modern human, Omo 2 (Fig. 15.8) shows predominantly modern features along with some archaic traits. It exhibits a robust, but basically modern supraorbital morphology. It is clearly visible that the superciliary arch protrudes relative to the more flattened lateral segment (Fig. 15.8). The frontal bone shows extensive coronal expansion and lacks a postorbital constriction. Modern aspects are also present in major dimensions of the parietal and occipital bones (Gunz and Harvati 2007; Mbua and

Bräuer 2012). The angulation of the occipital could be an archaic feature, but this feature must be considered in the light of the Herto specimens (see below). Included among the archaic traits in Omo 2 is the marked mid-sagittal keeling. In spite of some morphological differences between the two Omo crania, they are obviously closely related. According to Trinkaus (2005), the differences could be due to considerable variation within a population. It hardly appears justifiable to classify the two large-brained specimens as two different species, as indicated in Fig. 15.1.

Most important for understanding the transition to modern humans is the set of more recently discovered cranial remains from Herto, Ethiopia, which were dated by

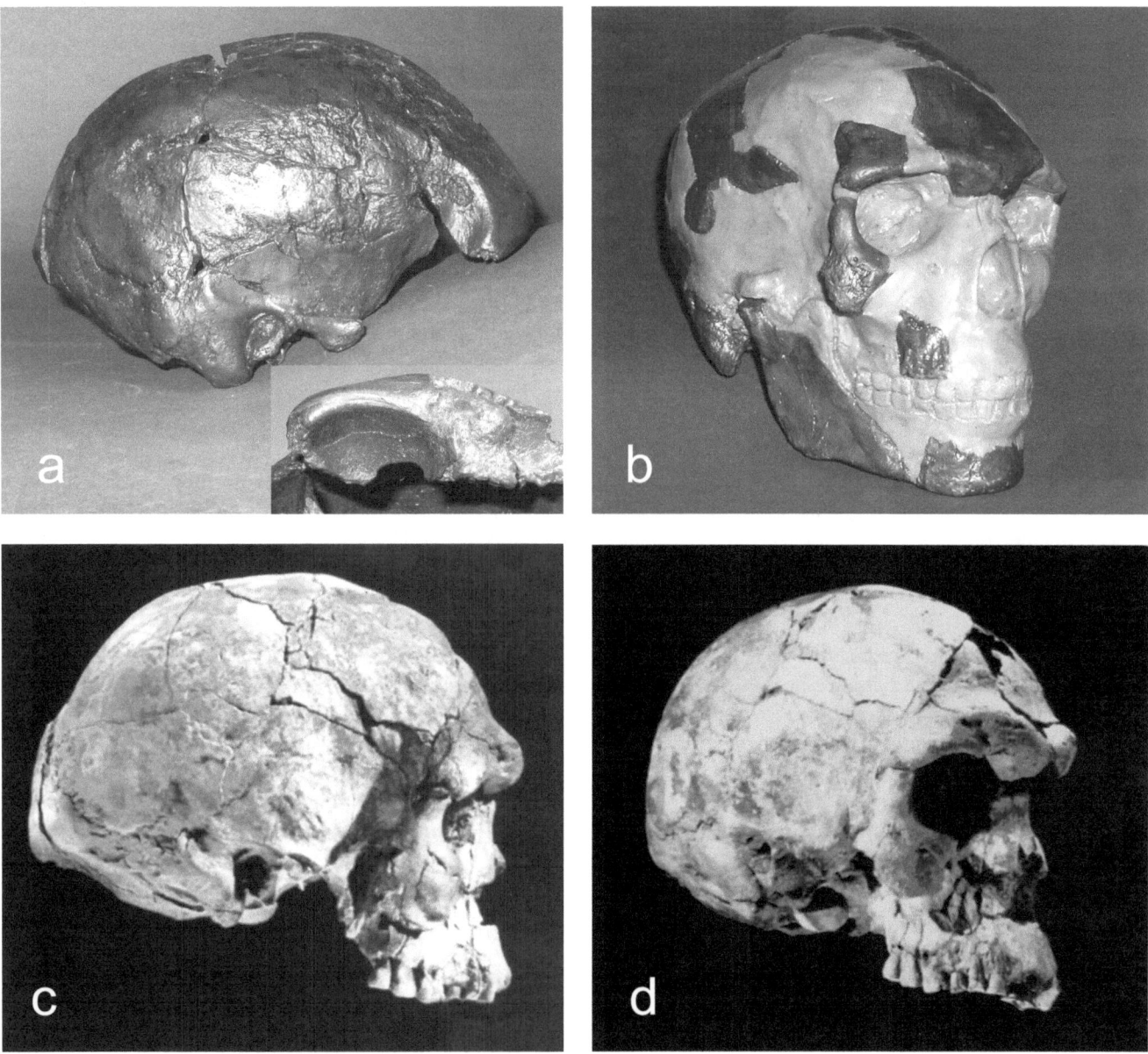

**Fig. 15.8**  Late archaic and early modern *Homo sapiens*. **a** Omo 2, Ethiopia (section shows the basically modern supraorbital pattern); **b** Omo 1, Ethiopia; **c** and **d** Herto BOU-VP-16/1, Ethiopia (photos: T.D. White)

Ar/Ar to about 0.16 Ma (White et al. 2003). The large robust cranium BOU-VP-16/1 (Fig. 15.8) exhibits a basically modern cranial shape, a robust but modern supraorbital morphology, and a modern face. The somewhat angulated occipital bone might not necessarily reflect an archaic feature because the occipital of the second adult specimen from Herto, BOU-VP-16/2, is not so angled in sagittal profile (White et al. 2003), pointing to a large intrapopulational variation for this complex. Finally, the Singa cranium from Sudan (Fig. 15.7), with an age of about 0.15 Ma, also belongs to this early modern human spectrum in eastern Africa (Bräuer 2008a). The strongly pronounced

parietal bosses, however, are due to pathological diploic widening (Spoor et al. 1998).

The earliest evidence of anatomically modern humans in southern Africa dates to about 0.12 Ma and, thus, is later than in eastern Africa. These oldest known modern South Africans derived from the Klasies River Mouth Caves on the southern coast. There is a rather large collection of fragmentary, but mostly diagnostic, cranial and postcranial remains from a number of individuals that were found at various levels of the deposits (Deacon and Shuurman 1992). The oldest member of the Klasies deposits (LBS Member) dates to about 0.12 Ma and provides maxillary and dental remains. These fall both metrically and morphologically within the range of variation of Holocene Africans (Bräuer et al. 1992b). A nearly complete mandible from somewhat higher in the deposits dates to about 0.1 Ma and is anatomically modern, as are the other cranial fragments from the site (Bräuer and Singer 1996; Bräuer 2001, 2008a). Another relevant early modern specimen from South Africa is the partial cranium from Border Cave, which probably dates to about 0.09 Ma or somewhat earlier, depending on which layer it is ultimately derived from (Grün and Beaumont 2001). The diagnostic elements of this fragmentary cranium are basically anatomically modern. Although the supraorbitals are not well divided into medial and lateral segments, they are very much reduced and only weakly projecting (Bräuer 1984; Schwartz and Tattersall 2003), clearly differing from the tori or torus-like structures of the late archaics. Since the fossil record of the preceding late archaics is very scarce in southern Africa, it remains unclear whether regional evolution occurred from populations currently represented by the Florisbad specimen to anatomically modern humans, or whether migration and gene flow from eastern Africa also contributed to the change towards fully modern humans in the South. In view of the gap in time, it could also be possible that the presently known remains from Klasies and Border Cave do not represent the first modern humans in that region.

Regarding the northwestern part of the continent, there is evidence for the presence of late archaic or near-modern humans associated with a Mousterian industry at around 0.2 Ma, as documented by the Irhoud remains. There are also early, anatomically modern specimens from Morocco, including the partial cranium and associated mandible from Dar es-Soltan 2, the cranial and mandibular remains from Témara and some more fragmentary specimens (El Harhoura/Zouhra Cave, Mughuret el 'Aliya). All these moderns derived from deposits associated with Aterian lithic technology. The major problem regarding these Aterian sites is the paucity of secure dating evidence, as Bouzouggar and Barton (2012) have pointed out. New dating approaches of Moroccan sites indicate that the Aterian might date to between 0.14 and 0.04 Ma (Bouzouggar and Barton 2012; Richter et al. 2012). Based on the current dating evidence, Harvati and Hublin (2012) assume a possible age of about 0.07 to 0.05 Ma for all the human specimens that derived from rather late Aterian deposits (Hublin 2000). For this part of the continent, the question also arises as to whether regional continuity from archaic *Homo sapiens* to anatomically modern humans is likely to have occurred. Since early modern humans from northern and eastern Africa exhibit close morphological similarities and share general robusticity, reliable dating evidence is necessary for developing realistic scenarios. It would thus make a difference if the cranium from Dar

es-Soltan 2 were 0.04 or 0.14 Ma. Until the Aterian specimens are more reliably dated, it is hardly possible to decide which of these modern humans are, in fact, early representatives or much later ones. Morphology alone is not a sufficient indicator here, since the diagnostic remains of the Aterians fall largely within the range of variation of the terminal late Pleistocene, but, nevertheless, the samples from Afalou/Taforalt are rather robust (Ferembach 1976a, b; Bräuer 1984).

The adult male partial cranium Dar es-Soltan 5 (Fig. 15.7) represents a robust, but modern, human. The superciliary arches are well defined and protruding, clearly showing a modern condition (see also Schwartz and Tattersall 2003). "None of the mandibular or cranial features observed in Dar es-Soltan 5 exclude it from modern variation" (Hublin 1992, p. 186). PCA analyses based on fronto-facial measurements show the closest affinities to Upper Paleolithic Europeans as well as to final late Pleistocene and Holocene Africans, and, specifically for the face, also to specimens from Skhul and Qafzeh (Bräuer and Rimbach 1990). The Upper Aterian remains from Témara include a mandible as well as occipital, parietal, and frontal parts. Ferembach (1976a) described the specimens as clearly anatomically modern. The morphology and metrical features of the occipital cannot be distinguished from the variation seen within the Afalou/Taforalt samples (Bräuer 1984; Hublin 1992). The same holds true for the supraorbital fragment that exhibits a modern flattened supraorbital trigone (Hublin 1992). Despite the fact that the morphology of Dar es-Soltan 5 and Témara is difficult to distinguish from the range of variation of later northern Africans and that these specimens could be around either 0.05 or 0.1 Ma, it is nevertheless possible that regional continuity existed from late archaic *Homo sapiens*, represented by the Irhoud specimens, to these early modern humans, as indicated by the large teeth, for example (Ferembach 1976b; Hublin 2000). Regarding a possible regional trend of megadonty in northern Africa, it has to be considered that the Middle Pleistocene dental records from other parts of the continent are even sparser. Nevertheless, there is overlap in molar crown size between, for example, Rabat, Témara, Mughuret el 'Aliya, and eastern and southern specimens like Kabwe, Eyasi, Herto, Cave of Hearths, and Klasies. In addition, the premolars and third molars of the Baringo/Kapthurin mandibles (Fig. 15.2) are among the largest in African specimens from this time period.

Based on the currently available fossil and dating evidence from Africa as a whole, it appears that the process of anatomical modernization in eastern Africa is rather well documented morphologically and chronologically and this evidence points to an origin of modern anatomy as early as 0.2 Ma. So far, this is earlier than can be documented for any other region. On the other hand, more complex and perhaps more realistic scenarios assuming migration and gene flow during periods of green Sahara between western and eastern regions and into the Near East are also possible (Larrasoaña 2012; Smith 2012). In order to demonstrate such complex patterns, the fossil evidence from those parts of the continent need to be considered as a whole.

## Conclusions and Discussion

The present chapter has addressed several aspects of the current debate on the pattern of Middle Pleistocene evolution in Africa. One major point is the question of whether the fossil record favors a continuous, mosaic-like Middle Pleistocene evolutionary lineage without any speciation events. The presence of derived modern and transitional modern features among the early post-*erectus* specimens would support this view. As demonstrated in detail, various researchers have identified a large number of such apomorphic *sapiens* traits in all relevant early specimens, including Bodo, Kabwe, Saldanha, Ndutu, Eyasi, and Salé. The presence of many derived *sapiens*-like features in these specimens has been further supported by a recent study of the Middle Pleistocene hominins from Africa (Mbua and Bräuer 2012). It thus appears plausible to include these early specimens within the species *Homo sapiens sensu lato* (see also Klein 1999), as is generally done for the subsequent archaic group of specimens dated at about 0.3 to 0.2 Ma. This view would also be in agreement with Howell's (1999) suggestion of paleo-demes: spatially and temporally bound entities below the species level. Howell (1999) recognized an early p-deme represented by Kabwe and related specimens like Saldanha, Ndutu, Bodo, Baringo/Kapthurin, and Eyasi.

The more derived morphological pattern of the late archaics is also found in northern, eastern, and southern Africa with specimens like Irhoud, Omo 2, Ileret/ER 3884, Eliye Springs, Laetoli H.18, and Florisbad. These specimens have often been described as representing an intermediate or transitional group between the earlier specimens and modern humans (e.g., McBrearty and Brooks 2000; Grimaud-Hervé and Vialet 2002; Smith 2002; Bräuer 2007). Howell (1999) also distinguished a later p-deme including most of the same specimens. Thus, in my view (Fig. 15.9), the current evidence from Africa appears in good agreement with the assumption of a last speciation event at about 0.8 or 0.7 Ma when *Homo erectus* gave rise to a new species that should be designated *Homo sapiens* (*sensu lato*).

As demonstrated by the presence of modern apomorphies among the early specimens, the process of *Homo sapiens* evolution in Africa is connected to the process of anatomical modernization and can be shown to have begun in the early Middle Pleistocene. Since we are dealing with one continuous

lineage or chronospecies, we favor the term "grade" for the three distinguishable but overlapping entities. We do not see any convincing evidence (Bräuer 2008a; Mbua and Bräuer 2012) for a split into three species, *Homo heidelbergensis*, *Homo helmei*, and *Homo sapiens*, as it can hardly be supported for several reasons (see above). A distinction into two species, *Homo heidelbergensis sensu* (Rightmire 1998, 2002) and *Homo sapiens*, also appears problematic, since this would suggest a speciation event between near-modern humans like Omo 2 and modern humans like Omo 1. Moreover, Rightmire's perspective would suggest a distinction between anatomically modern humans or *Homo sapiens* on the one hand, and a rather variable *Homo heidelbergensis* species, including such diverse specimens as Omo 2 and Bodo, pre-Neandertals, and Chinese archaics like Dali and Maba, on the other. This can hardly be regarded as a more conclusive approach than a splitting of one chronospecies into several subspecies, grades, or paleo-demes. Instead, there can be little doubt that the late archaics should be included within *Homo sapiens* (e.g., Smith 2002; Stringer 2002; Turbón 2006). However, we see little evidence in view of the numerous derived features shared with later as well as modern humans for separating only the early archaics on the species level. Rather, Hublin (2001) leaves unresolved the issue of whether specimens like Bodo, Kabwe, Ndutu, Saldanha, and Salé should be distinguished as a separate species (*Homo rhodesiensis*) from later representatives of early *Homo sapiens* such as Irhoud, or should instead represent a grade within one species, *Homo sapiens sensu lato*. He assumes that "during the second half of the Middle Pleistocene, a phenomenon of accretion occurred in Africa, leading to the emergence of 'primitive modern humans'" (Hublin 2000, p. 159), which include the Irhoud specimens. Stringer (2002) also favors an "accretional mode" of *Homo sapiens* evolution in Africa.

The process of anatomical modernization and that of Neandertalization obviously occurred largely in parallel in a mosaic-like or accretional mode (Fig. 15.9). Neandertal apomorphies accumulated over a long time period—beginning at about 0.5 Ma—without any indication of a clear speciation event (Hublin 1998). Any subdivisions, such as early pre-Neandertals, late pre-Neandertals, proto-Neandertals, and classic Neandertals (e.g., Condemi 2003), serve only to describe this intraspecific evolutionary process. We see a comparable situation for Africa. Here as well, there is no clear evidence for speciation events and thus, the use of grades or other useful intraspecific entities appears appropriate to describe the pattern within this continuously evolving lineage (Bräuer 2008a). Current alternative suggestions on how to split the African anagenetic sequence into several species appear problematic and contradictory and can hardly be regarded as solutions to the subspecies *versus* species problem. As outlined elsewhere (Bräuer 2008a), evidence from living primates also adds

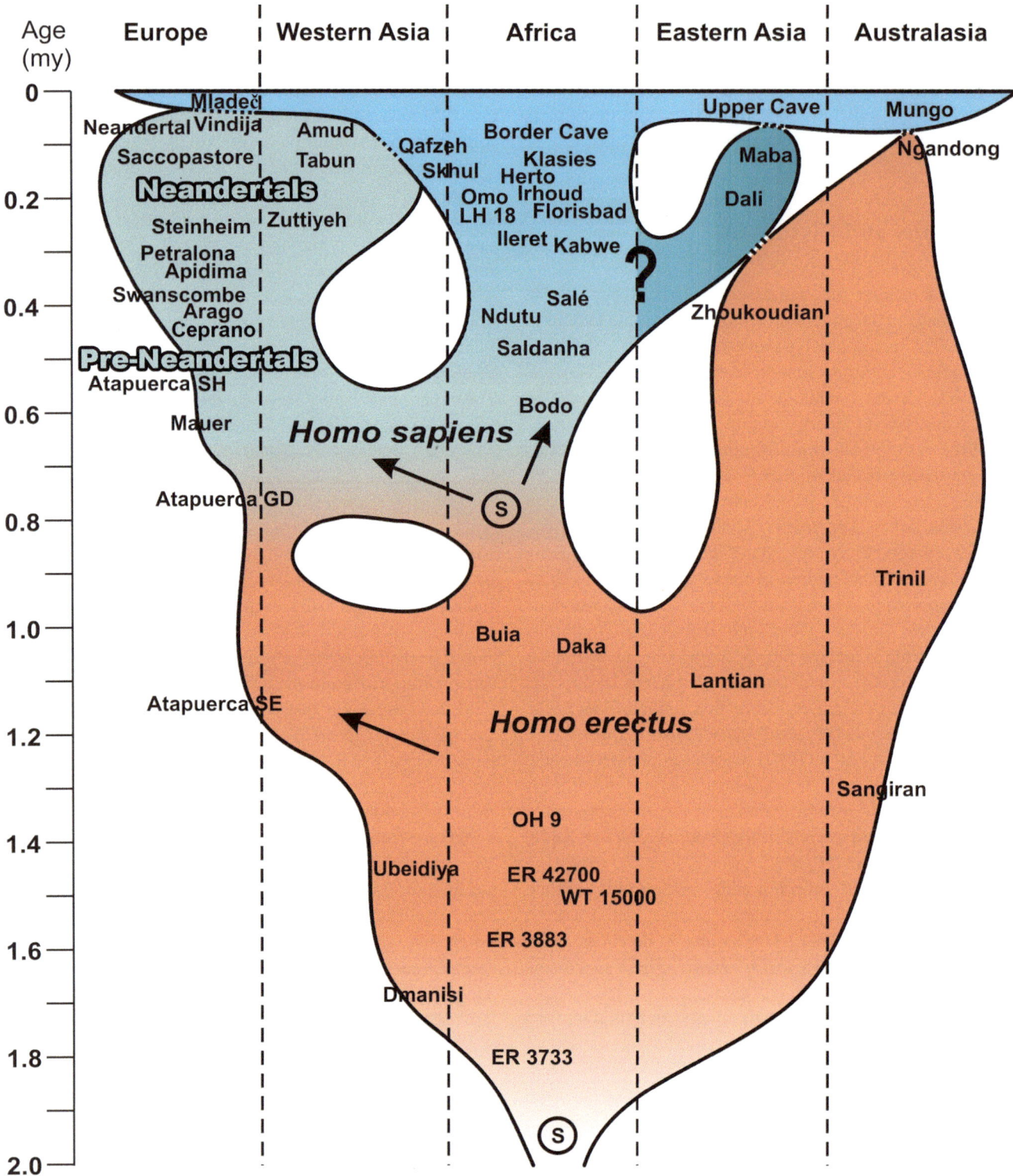

**Fig. 15.9** Phylogenetic scheme (Bräuer)

plausibility to European pre-Neandertals/Neandertals and African pre-modern/modern lineage as belonging to one polytypic *Homo sapiens* species (Jolly 2001, 2003; see also Cartmill and Smith 2009).

Recent results of the draft sequence of the Neandertal genome, which is based on more than 4 billion nucleotides, suggest that "between 1 and 4% of the genomes of people in Eurasia are derived from Neandertals" (Green et al.

2010, p. 721). This small but significant amount of gene flow is regarded by these authors as "a challenge to the simplest version of an Out of Africa model for modern human origins" (p. 721) (i.e., complete replacement/no gene flow). The new results obviously support Out of Africa scenarios, assuming some degree of gene flow, as has been advocated by a number of researchers for a long time (e.g., Bräuer 1984; Bräuer and Stringer 1997; Bräuer et al. 2004b; Smith et al. 2005; Trinkaus 2005; Bräuer 2006; Cartmill and Smith 2009). In fact, molecular research started with a very strong bias against mixture (see Gibbons 2010). Thus, the current molecular evidence for Neandertal—modern interbreeding might also have implications for the species controversy. One could indeed now follow Lieberman's (2003, p. 665) suggestion: "If one can find good evidence that humans and Neandertals interbred and that modern humans have some Neandertal autapomorphies, then, by all means, let's include Neandertals in our species." Regarding the species question, recent research has also focused on possible Neandertal—modern differences in life history and enamel growth processes (Ramirez Rozzi and Bermudez de Castro 2004; Guatelli-Steinberg et al. 2005; Smith et al. 2007b; Guatelli-Steinberg and Reid 2008). However, current indications point to large variation in modern human dental development as well as great overlap with the few Neandertal data available. Moreover, extrapolations to some "overall pace" of Neandertal life histories based on rates of dental development appear to not be justified because of dissociations among dental development, somatic growth, and life history variables occurring in closely related species or different populations of a single species (Guatelli-Steinberg 2009).

The other relevant aspect considered in this chapter is the question of geographic patterning of the anatomical modernization process. As has been shown, based on the current evidence, it is very difficult, if not impossible, to reconstruct the roles of the major geographic regions within this long-term process. Although it is clearly evident that northern, eastern, and southern Africa were involved in the process, the eastern region currently appears to best document a mosaic-like continuous evolutionary pattern, especially the transitional period to early modern *Homo sapiens*, who appeared here very early at about 0.2 Ma. Regarding northwestern Africa, there is also reliable evidence for the presence of early and late archaic humans similar in grade to what we know from eastern and southern Africa. In addition, the Aterian people could indeed be regional descendants. Yet a better dating framework for the Mousterian and Aterian of northern Africa and more reliable dates for the associated human remains are needed to answer the question of whether such regional continuity to fully modern humans occurred in this area as well or whether there was a more complex pattern of interregional migration and gene flow between northwestern, northeastern, and eastern regions, especially during periods of a green Sahara. Also, further discoveries of fossil hominins could help to better understand the role northern Africa played in the process of *Homo sapiens* evolution and the origin and expansion of modern humans.

**Acknowledgments** I would like to thank Jean-Jacques Hublin and Shannon McPherron for inviting me to this interesting conference, as well as the other members of the organizing team for their kind support. I also thank my friends and colleagues during the conference for their stimulating conversation some of which I also tried to address in this contribution. Many thanks go to Frederik Jessen, Angelika Kroll and Eszter Schoell for their great support with the final version of the manuscript. I am also grateful to Jean-Jacques Hublin and Tim White for permission to use several photographs. Last but not least, I thank the numerous colleagues and friends for their kind cooperation and generous support during my studies of the African hominin material over the last three decades. The chapter benefited from the constructive comments of the three reviewers.

# References

Adefris, T. (1992). *A description of the Bodo cranium: An archaic Homo sapiens* cranium from Ethiopia. Ph.D. Dissertation, New York University.

Balter, M. (2002). What made humans modern? *Science, 295*, 1219–1225.

Bilsborough, A. (1992). *Human evolution*. London: Blackie Academic & Professional.

Bischoff, J. L., Williams, R. W., Rosenbauer, R. J., Aramburu, A., Arsuaga, J. L., García, N., et al. (2007). High-resolution U-series dates from the Sima de los Huesos hominids yields $600 \pm {}^{8}_{66}$ kyrs: Implications for the evolution of the early Neanderthal lineage. *Journal of Archaeological Science, 34*, 763–770.

Bouzouggar, A., & Barton, R. N. E. (2012). The identity and timing of the Aterian in Morocco. In J.-J. Hublin & S. McPherron (Eds.), *Modern origins: A North African perspective*. Dordrecht: Springer.

Bräuer, G. (1982). Early anatomically modern man in Africa and the replacement of the Mediterranean and European Neanderthals. In H. de Lumley (Ed.), *I. Congrès International de Paléontologie Humaine* (Resumés: 112). Nice: Centre National de la Recherche Scientifique/Louis-Jean Scientific and Literary Publications.

Bräuer, G. (1984). A craniological approach to the origin of anatomically modern *Homo sapiens* in Africa and implications for the appearance of modern Europeans. In F. H. Smith & F. Spencer (Eds.), *The origins of modern humans: A world survey of the fossil evidence* (pp. 327–410). New York: Alan R. Liss.

Bräuer, G. (1989). The evolution of modern humans: A comparison of the African and non-African evidence. In P. Mellars & C. B. Stringer (Eds.), *The human revolution* (pp. 123–154). Edinburgh: Edinburgh University Press.

Bräuer, G. (1992). Africa's place in the evolution of Homo sapiens. In G. Bräuer & F. H. Smith (Eds.), *Continuity or replacement—controversies in Homo sapiens evolution* (pp. 83–98). Rotterdam: Balkema.

Bräuer, G. (2001). The KNM-ER 3884 hominid and the emergence of modern anatomy in Africa. In P. V. Tobias, M. A. Raath, J. Moggi-Cecchi, & G. A. Doyle (Eds.), *Humanity from African naissance to coming millennia* (pp. 191–197). Firenze: Firenze University Press.

Bräuer, G. (2006). The African origin of modern humans and the replacement of the Neanderthals. In R. W. Schmitz (Ed.),

*Neanderthal 1856–2006* (pp. 337–372). Mainz: Philipp von Zabern.

Bräuer, G. (2007). Origin of modern humans. In W. Henke & I. Tattersall (Eds.), *Handbook of paleoanthropology. Volume III: Phylogeny of hominids* (pp. 1749–1779). Berlin: Springer.

Bräuer, G. (2008a). The origin of modern anatomy: By speciation or intraspecific evolution? *Evolutionary Anthropology, 17*, 22–37.

Bräuer, G. (2008b). "*Homo heidelbergensis*" und die Entwicklung des Menschen im Mittelpleistozän. *Museo (Heilbronn), 24*, 8–21.

Bräuer, G., & Leakey, R. E. (1986). The ES-11693 cranium from Eliye Springs, West Turkana, Kenya. *Journal of Human Evolution, 15*, 289–312.

Bräuer, G., & Rimbach, W. K. (1990). Late archaic and modern *Homo sapiens* from Europe, Africa and Southwest Asia: Craniometric comparisons and phylogenetic implications. *Journal of Human Evolution, 19*, 789–807.

Bräuer, G., & Singer, R. (1996). The Klasies zygomatic bone: Archaic or modern? *Journal of Human Evolution, 30*, 161–165.

Bräuer, G., & Stringer, C. (1997). Models, polarization and perspectives on modern human origins. In G. A. Clark & C. M. Willermet (Eds.), *Conceptual issues in modern human origins research* (pp. 191–201). New York: Aldine de Gruyter.

Bräuer, G., Leakey, R.E., & Mbua, E. (1992a). A first report on the ER-3884 cranial remains from Ileret/East Turkana. In G. Bräuer & F. H. Smith (Eds.), *Continuity or replacement—controversies in Homo sapiens evolution* (pp. 111–119). Rotterdam: Balkema.

Bräuer, G., Deacon, H.J., & Zipfel, F. (1992b). Comment on the new maxillary finds from Klasies River, South Africa. *Journal of Human Evolution, 23*, 419–422.

Bräuer, G., Yokoyama, Y., Falguères, C., & Mbua, E. (1997). Modern human origins backdated. *Nature, 386*, 337–338.

Bräuer, G., Groden, C., Gröning, F., Kroll, A., Kupczik, K., Mbua, E., et al. (2004a). Virtual study of the endocranial morphology of the matrix-filled cranium from Eliye Springs, Kenya. *The Anatomical Record, 276A*, 113–133.

Bräuer, G., Collard, M., & Stringer, C. (2004b). On the reliability of recent tests of the Out of Africa hypothesis for modern human origins. *The Anatomical Record, 279A*, 701–707.

Cartmill, M., & Smith, F. H. (2009). *The human lineage.* New York: Wiley and Sons.

Clark, J. D., de Heinzelin, J., Schick, K. D., Hart, W. K., White, T. D., WoldeGabriel, G., et al. (1994). African *Homo erectus*: Old radiometric ages and young Oldowan assemblages in the Middle Awash Valley, Ethiopia. *Science, 264*, 1907–1910.

Clarke, R. J. (1990). The Ndutu cranium and the origin of *Homo sapiens*. *Journal of Human Evolution, 19*, 699–736.

Condemi, S. (2003). Les Néandertaliens. In C. Susanne, E. Rebato, & B. Chiarelli (Eds.), *Anthropologie Biologique. Evolution et Biologie Humaine* (pp. 271–279). Bruxelles: De Boeck.

Deacon, H., & Shuurman, R. (1992). The origins of modern people: The evidence from Klasies River. In G. Bräuer & F. H. Smith (Eds.), *Continuity or replacement—controversies in Homo sapiens evolution* (pp. 121–130). Rotterdam: Balkema.

Dean, D. (1993). *The Middle Pleistocene Homo erectus/Homo sapiens transition: New evidence from Space Curve Statistics.* Ph.D. Dissertation, City University of New York.

Dean, D., Hublin, J.-J., Holloway, R., & Ziegler, R. (1998). On the phylogenetic position of the pre-Neandertal specimen from Reilingen, Germany. *Journal of Human Evolution, 34*, 485–508.

Ferembach, D. (1976a). Les restes humains atériens de Témara (Campagne 1975). *Bulletins et Mémoires de la Société d'Anthropologie de Paris 3, série XIII*, 175–180.

Ferembach, D. (1976b). Les restes humains de la grotte de Dar-es-Soltane 2 (Maroc), Campagne 1975. *Bulletins et Mémoires de la Société d'Anthropologie de Paris 3, série XIII*, 183–193.

Foley, R. (2001). In the shadow of the modern synthesis? Alternative perspectives on the last fifty years of paleoanthropology. *Evolutionary Anthropology, 10*, 5–14.

Foley, R. (2002). Adaptive radiations and dispersals in hominin evolutionary ecology. *Evolutionary Anthropology, 11*(Supplement 1), 32–37.

Foley, R., & Lahr, M. M. (2003). On stony ground: Lithic technology, human evolution, and the emergence of culture. *Evolutionary Anthropology, 12*, 109–122.

Gibbons, A. (2010). Close encounters of the prehistoric kind. *Science, 328*, 680–684.

Green, R. E., Krause, J., Ptak, S. E., Briggs, A. W., Ronan, M. T., Simons, J. F., et al. (2006). Analysis of one million base pairs of Neanderthal DNA. *Nature, 444*, 330–336.

Green, R. E., Krause, J., Briggs, A. W., Maricic, T., Stenzel, U., Kircher, M., et al. (2010). A draft sequence of the Neandertal genome. *Science, 328*, 710–722.

Grimaud-Hervé, D., & Vialet, A. (2002). Les *Homo sapiens* archaïc africains. In D. Grimaud-Hervé, F. Marchal, A. Vialet, & F. Détroit (Eds.), *Le deuxième homme en Afrique. Homo ergaster. Homo erectus* (pp. 127–132). Paris: Èditions Artcom.

Groves, P. C., & Lahr, M. M. (1994). A bush not a ladder. *Speciation and replacement in human evolution. Perspectives in Human Biology, 4*, 1–11.

Grün, R., & Beaumont, P. (2001). Border Cave revisited: A revised ESR chronology. *Journal of Human Evolution, 40*, 467–482.

Grün, R., & Stringer, C. B. (1991). Electron spin resonance dating and the evolution of modern humans. *Archaeometry, 33*, 153–199.

Grün, R., Brink, J. S., Spooner, N. A., Taylor, L., Stringer, C. B., Franciscus, R. G., et al. (1996). Direct dating of Florisbad hominid. *Nature, 382*, 500–501.

Guatelli-Steinberg, D. (2009). Recent studies of dental development in Neandertals: Implications for Neandertal life histories. *Evolutionary Anthropology, 18*, 9–20.

Guatelli-Steinberg, D., & Reid, D. J. (2008). What molars contribute to an emerging understanding of lateral enamel formation in Neandertals vs modern humans. *Journal of Human Evolution, 54*, 236–250.

Guatelli-Steinberg, D., Reid, D. J., Bishop, T. A., & Larsen, C. S. (2005). Anterior tooth growth periods in Neandertals were comparable to those of modern humans. *Proceedings of the National Academy of Sciences USA, 102*, 14197–14202.

Gunz, P., & Harvati, K. (2007). The Neanderthal "chignon": Variation, integration, and homology. *Journal of Human Evolution, 52*, 262–274.

Haile-Selassie, Y., Asfaw, B., & White, T. D. (2004). Hominid cranial remains from Upper Pleistocene deposits at Aduma, Middle Awash, Ethiopia. *American Journal of Physical Anthropology, 123*, 1–10.

Harvati, K. (2007). Neanderthals and their contemporaries. In W. Henke & I. Tattersall (Eds.), *Handbook of paleoanthropology. Volume III: Phylogeny of hominids* (pp. 1717–1748). Berlin: Springer.

Harvati, K., & Hublin, J.-J. (2012). Morphological continuity of the face in the late Middle and Upper Pleistocene hominins from northwestern Africa—a 3-D geometric morphometric analysis. In J.-J. Hublin & S. McPherron (Eds.), *Modern origins: A North African perspective.* Dordrecht: Springer.

Howell, F. C. (1999). Paleo-demes, species clades, and extinctions in the Pleistocene hominin record. *Journal of Anthropological Research, 55*, 191–243.

Hublin, J.-J. (1978) *Le torus occipital transverse et les structures associées: Evolution dans le genre Homo.* Ph.D. Dissertation, Université Paris VI.

Hublin, J.-J. (1985). Human fossils from the North African Middle Pleistocene and the origin of *Homo sapiens*. In E. Delson (Ed.), *Ancestors: The hard evidence* (pp. 283–288). New York: Alan R. Liss.

Hublin, J.-J. (1992). Recent human evolution in northwestern Africa. *Philosophical Transactions of the Royal Society of London B, 337*, 185–191.

Hublin, J.-J. (1996). The first Europeans. *Archaeology, 1*(2), 36–44.

Hublin, J.-J. (1998). Climatic changes, paleogeography, and the evolution of the Neandertals. In T. Akazawa, K. Aoki, & O. Bar-Yosef (Eds.), *Neandertals and modern humans in Western Asia* (pp. 295–310). New York: Plenum Press.

Hublin, J.-J. (2000). Modern-non modern hominid interactions: A Mediterranean perspective. In O. Bar-Yosef & D. Pilbeam (Eds.), *The geography of Neandertals and modern humans in Europe and the greater Mediterranean* (pp. 157–182). Cambridge: Harvard University Press.

Hublin, J.-J. (2001). Northwestern African Middle Pleistocene hominids and their bearing on the emergence of *Homo sapiens*. In L. Barham & K. Robson-Brown (Eds.), *Human roots. Africa and Asia in the Middle Pleistocene* (pp. 99–121). Bristol: Western Academic & Specialist Press.

Jolly, C. J. (2001). A proper study for mankind: Analogies from the Papionin monkeys and their implications for human evolution. *Yearbook of Physical Anthropology, 44*, 177–204.

Jolly, C. J. (2003). Comment to: Species concepts, reticulation, and human evolution by T. W. Holliday. *Current Anthropology, 44*, 662–663.

Jurmain, R., Nelson, H., Kilgore, L., & Trevathan, W. (2000). *Introduction to physical anthropology* (8th ed.). Belmont, CA: Wadsworth/Thomson.

Klein, R. G. (1989). *The human career*. Chicago: The University of Chicago Press.

Klein, R. G. (1999). *The human career. Human biological and cultural origins* (2nd ed.). Chicago: The University of Chicago Press.

Klein, R. G. (2000). Archeology and the evolution of human behaviour. *Evolutionary Anthropology, 9*, 17–36.

Krings, M., Capelli, C., Tschentscher, F., Geisert, H., Meyer, S., von Haeseler, A., et al. (2000). A view of Neanderthal genetic diversity. *Nature Genetics, 26*, 144–146.

Lahr, M. M. (1996). *The evolution of modern human diversity. A study of cranial variation*. Cambridge: Cambridge University Press.

Laitman, J. T. (1985). Later Middle Pleistocene hominids. In E. Delson (Ed.), *Ancestors: The hard evidence* (pp. 265–267). New York: Alan R. Liss.

Larrasoaña, J.C. (2012). A Northeast Saharan perspective on environmental variability in North Africa and its implications for modern human origins. In J.-J. Hublin & S.P. McPherron (Eds.), *Modern origins: A North African perspective*. Dordrecht: Springer.

Lieberman, D. E. (2003). Comment to Holliday, T.W. : Species concepts, reticulation, and human evolution. *Current Anthropology, 44*, 664–665.

Lieberman, D. E., McBratney, B. M., & Krovitz, G. (2002). The evolution and development of cranial form in *Homo sapiens*. *Proceedings of the National Academy of Sciences of the United States of America, 99*, 1134–1139.

Magori, C. C., & Musiba, C. M. (2006). *Further morphological considerations of the Ngaloba skull (LH 18) from Laetoli, northern Tanzania*. Johannesburg: Paper at the African Genesis Symposium.

Manega, P. C. (1995). *New geochronological results from the Ndutu, Naisiusiu and Ngaloba Beds at Olduvai and Laetoli in Northern Tanzania: Their significance for evolution of modern humans*. Italy: Bellagio Conference.

Manzi, G. (2004). Human evolution at the Matuyama-Brunhes boundary. *Evolutionary Anthropology, 13*, 11–24.

Mbua, E., & Bräuer, G. (2012). Patterns of Middle Pleistocene hominin evolution in Africa and the emergence of modern humans. In S. C. Reynolds & A. Gallagher (Eds.), *African Genesis: Perspectives on Hominin Evolution*. (pp. 394–422). Cambridge: Cambridge University Press.

McBrearty, S., & Brooks, A. S. (2000). The revolution that wasn't: A new interpretation of the origin of modern human behavior. *Journal of Human Evolution, 39*, 453–563.

McDougall, I., Brown, F., & Fleagle, J. G. (2005). Stratigraphic placement and age of modern humans from Kibish, Ethiopia. *Nature, 433*, 733–736.

McDougall, I., Brown, F. H., & Fleagle, J. G. (2008). Sapropels and the age of hominins Omo I and Omo II, Kibish, Ethiopia. *Journal of Human Evolution, 55*, 409–420.

Noonan, J. P., Coop, G., Kudaravalli, S., Smith, D., Krause, J., Alessi, J., et al. (2006). Sequencing and analysis of Neanderthal genomic DNA. *Science, 314*, 1113–1118.

Pearson, O. M. (2008). Statistical and biological definitions of "anatomically modern" humans: Suggestions for a unified approach to modern morphology. *Evolutionary Anthropology, 17*, 38–48.

Ramirez Rozzi, F. V., & Bermudez de Castro, J. M. (2004). Surprising rapid growth in Neanderthals. *Nature, 428*, 936–939.

Richter, D., Moser, J., & Nami, M. (2012). New data from the site of Ifri n'Ammar (Morocco) and some remarks on the chronometric status of the Middle Paleolithic in the Maghreb. In J.-J. Hublin & S. McPherron (Eds.), *Modern origins: A North African perspective*. Dordrecht: Springer.

Rightmire, G. P. (1983). The Lake Ndutu cranium and early *Homo sapiens* in Africa. *American Journal of Physical Anthropology, 61*, 245–254.

Rightmire, G. P. (1990). *The evolution of Homo erectus. Comparative anatomical studies of an extinct human species*. Cambridge: Cambridge University Press.

Rightmire, G. P. (1996). The human cranium from Bodo, Ethiopia: Evidence for speciation in the Middle Pleistocene? *Journal of Human Evolution, 31*, 21–39.

Rightmire, G. P. (1998). Human evolution in the Middle Pleistocene: The role of *Homo heidelbergensis*. *Evolutionary Anthropology, 6*, 218–227.

Rightmire, G. P. (2000). Middle Pleistocene humans from Africa. *Human Evolution, 15*, 63–74.

Rightmire, G.P. (2001a). Comparison of Middle Pleistocene hominids from Africa and Asia. In L. Barham & K. Robson-Brown (Eds.), *Human roots. Africa and Asia in the Middle Pleistocene* (pp. 123–133). Bristol: Western Academy and Specialist Press.

Rightmire, G.P. (2001b). Diversity in the earliest "modern" populations from South Africa, Northern Africa and Southwest Asia. In P.V. Tobias, M.A. Raath, J. Moggi-Cecchi & G. A. Doyle (Eds.), *Humanity from African naissance to coming millennia* (pp. 231–236). Witwatersrand: Witwatersrand University Press.

Rightmire, G. P. (2002). Les plus anciens *Homo erectus* d'Afrique et leur rôle dans l'évolution humaine. In D. Grimaud-Hervé, F. Marchal, A. Vialet, & F. Détroit (Eds.), *Le deuxième homme en Afrique: Homo ergaster, Homo erectus* (pp. 123–126). Paris: Editions Artcom.

Rightmire, G. P. (2007). Later Middle Pleistocene *Homo*. In W. Henke & I. Tattersall (Eds.), *Handbook of paleoanthropology. Volume III: Phylogeny of hominids* (pp. 1695–1715). Berlin: Springer.

Schwartz, J. H., & Tattersall, I. (2003). *The human fossil record. Vol. 2: Craniodental morphology of genus Homo (Africa and Asia)*. Hoboken, NJ: Wiley-Liss.

Smith, F. H. (2002). Migrations, radiations and continuity: Patterns in the evolution of Middle and Late Pleistocene humans. In W. C. Hartwig (Ed.), *The primate fossil record* (pp. 437–456). Cambridge: Cambridge University Press.

Smith, J. R. (2012). Spatial and temporal variation in the nature of Pleistocene pluvial phase environments across North Africa. In J.-J. Hublin & S. P. McPherron (Eds.), *Modern origins: A North African perspective*. Dordrecht: Springer.

Smith, F. H., Jankovic, I., & Karavanic, I. (2005). The assimilation model, modern human origins in Europe, and the extinction of the Neandertals. *Quaternary International, 137*, 7–19.

Smith, T. M., Tafforeau, P., Reid, P. J., Grün, R., Eggins, S., Boutakiout, M., et al. (2007a). Earliest evidence of modern human life history in North African early *Homo sapiens*. *Proceedings of the National Academy of Sciences of the United States of America, 104*, 6128–6133.

Smith, T. M., Toussaint, M., Reid, D. J., Olejniczak, A. J., & Hublin, J.-J. (2007b). Rapid dental development in a Middle Paleolithic Belgian Neanderthal. *Proceedings of the National Academy of Sciences USA, 104*, 20220–20225.

Spoor, F., Stringer, C., & Zonneveld, F. (1998). Rare temporal bone pathology of the Singa calvaria from Sudan. *American Journal of Physical Anthropology, 107*, 41–50.

Stringer, C. B. (1984). The definition of *Homo erectus* and the existence of the species in Africa and Europe. *Courier Forschungs-Institut Senckenberg, 69*, 131–144.

Stringer, C. B. (2002). Modern human origins: Progress and prospects. *Philosophical Transactions Royal Society London B, 357*, 563–579.

Stringer, C. B., Howell, F. C., & Melentis, J. K. (1979). The significance of the fossil hominid skull from Petralona, Greece. *Journal of Archaeological Science, 6*, 235–253.

Tattersall, I. (1992). Species concepts and species identification in human evolution. *Journal of Human Evolution, 22*, 341–349.

Tattersall, I., & Schwartz, J. H. (2000). *Extinct humans*. Boulder, CO: Westview Press.

Trinkaus, E. (2005). Early modern humans. *Annual Review of Anthropology, 34*, 207–230.

Turbón, D. (2006). *La evolución humana*. Barcelona: Editorial Ariel.

White, T. D., Asfaw, B., DeGusta, D., Gilbert, H., Richards, G. D., Suwa, G., et al. (2003). Pleistocene *Homo sapiens* from Middle Awash, Ethiopia. *Nature, 423*, 742–747.

# Index

J.-J. Hublin and S. P. McPherron (eds.), *Modern Origins: A North African Perspective,*
Vertebrate Paleobiology and Paleoanthropology, DOI: 10.1007/978-94-007-2929-2,
© Springer Science+Business Media B.V. 2012